World Sustainability Series

Series Editor

Walter Leal Filho, European School of Sustainability Science
and Research, Research and Transfer Centre "Sustainable Development
and Climate Change Management", Hamburg University of Applied Sciences,
Hamburg, Germany

Due to its scope and nature, sustainable development is a matter which is very interdisciplinary, and draws from knowledge and inputs from the social sciences and environmental sciences on the one hand, but also from physical sciences and arts on the other. As such, there is a perceived need to foster integrative approaches, whereby the combination of inputs from various fields may contribute to a better understanding of what sustainability is, and means to people. But despite the need for and the relevance of integrative approaches towards sustainable development, there is a paucity of literature which address matters related to sustainability in an integrated way.

Notes on the quality assurance and peer review of this publication

Prior to publication, the quality of the works published in this series is double blind reviewed by external referees appointed by the editor. The referees are not aware of the author's name when performing the review; the referees' names are not disclosed.

More information about this series at http://www.springer.com/series/13384

Walter Leal Filho · Eugene V. Krasnov ·
Dara V. Gaeva
Editors

Innovations and Traditions for Sustainable Development

Springer

Editors
Walter Leal Filho ⓘ
HAW Hamburg
Hamburg, Germany

Manchester Metropolitan University
Manchester, UK

Dara V. Gaeva
Centre for Scientific and Technical
Information
Immanuel Kant Baltic Federal University
Kaliningrad, Russia

Eugene V. Krasnov
Immanuel Kant Baltic Federal University
Kaliningrad, Russia

Eryong Xue and Jian Li share the co-first authorship and contribute equally in this book.

ISSN 2199-7373   ISSN 2199-7381 (electronic)
World Sustainability Series
ISBN 978-3-030-78827-8   ISBN 978-3-030-78825-4 (eBook)
https://doi.org/10.1007/978-3-030-78825-4

© The Editor(s) (if applicable) and The Author(s), under exclusive license to Springer Nature Switzerland AG 2021
This work is subject to copyright. All rights are solely and exclusively licensed by the Publisher, whether the whole or part of the material is concerned, specifically the rights of translation, reprinting, reuse of illustrations, recitation, broadcasting, reproduction on microfilms or in any other physical way, and transmission or information storage and retrieval, electronic adaptation, computer software, or by similar or dissimilar methodology now known or hereafter developed.
The use of general descriptive names, registered names, trademarks, service marks, etc. in this publication does not imply, even in the absence of a specific statement, that such names are exempt from the relevant protective laws and regulations and therefore free for general use.
The publisher, the authors and the editors are safe to assume that the advice and information in this book are believed to be true and accurate at the date of publication. Neither the publisher nor the authors or the editors give a warranty, expressed or implied, with respect to the material contained herein or for any errors or omissions that may have been made. The publisher remains neutral with regard to jurisdictional claims in published maps and institutional affiliations.

This Springer imprint is published by the registered company Springer Nature Switzerland AG
The registered company address is: Gewerbestrasse 11, 6330 Cham, Switzerland

Preface

The growth of digital technologies in all areas of educational, scientific and industrial activities should be accompanied by appropriate innovations, which take into account the rapidly changing situations in both the economy and the financial sector on a global scale.

The sustainability of development models largely depends on the balance between social, economic and environmental processes. This is especially so in sectors where innovation can play a key role in IT, energy or environmental protection, to name a few examples. However, many countries and regions still adhere to obsolete procedures in the production of goods and services.

This is in stark contrast with what is needed today, where businesses need to rapidly adapt their procedures to meet ever-increasing challenges and demands, partly as a result of the COVID-19 pandemic, so as to increase their competitiveness.

This book highlights the vital necessity for combining sustainable development processes from different areas, with applications in areas such as science, education and production sectors. These sectors have previously been separated by linguistic and technological barriers. Breaking down these barriers will allow an interdisciplinary and transdisciplinary flow of information, leading to greater efficiency, and toward a more real resilient and sustainable economy development.

This book fills in the gap in respect of publications addressing aspects of innovation and sustainable development and focuses on a range of areas, such as:

I. Gradual transition to innovative development
II. Continuity of technology in education, science and industry
III. Convergency directions, interdisciplinary relations in scientific research
IV. Digital technologies for sustainable development
V. Global trends and regional aspects of innovation and traditions in environmental management
VI. International legal regulations and environmental and economic relations among business communities.

We thank the authors and reviewers for their contribution. We hope that the contributions on this volume will provide a timely support in combining innovations and traditions for sustainable development and will foster the global efforts toward taking better advantage of the many opportunities which innovation in specific areas may offer.

Enjoy the reading.

Hamburg, Germany/Manchester, UK	Walter Leal Filho
Kaliningrad, Russia	Eugene V. Krasnov
Kaliningrad, Russia	Dara V. Gaeva
Winter 2021–2022	

Contents

Gradual Transition to Innovative Development

The Hybrid Innovation System Principles for Resilient Innovative Growth 3
Anna Alekseevna Mikhaylova and Dmitry Vitalievich Hvaley

Energy Development Alternatives—From Global to Local Scale 23
Eugene V. Krasnov and Izumrud R. Ragulina

Urban Green Innovation Ecosystem to Improve the Environmental Sustainability 37
José G. Vargas-Hernández, Karina Pallagst, and Jessica Davalos

Social Innovation for Sustainable Development 47
Skaidrė Žičkienė and Zita Tamasauskiene

Transformational Transition of Sustainable Development Based on Circular Green Economy. An Analysis Based on the Theory of Resources and Capabilities 69
José G. Vargas-Hernández, Jorge Armando López-Lemus, and Marlene de Jesús Morales Medrano

The Ecosystem Approach to Assessing the Quality of the Urban Environment and Managing Urban Development 87
Nataliya V. Yakovenko and Igor V. Komov

Continuity of Technologies in Education, Science and Industry

The Next Steps for the Baltic Universities' Cooperation in Accordance with the Development of Education and the Upbringing of Students 109
Elena Kropinova and Eugene Krasnov

Continuing Education at Faculty of Geography (Lomonosov Moscow State University): Trend for Sustainable Development 127
Valentina Toporina and Alexandra Goretskaya

The Culture of Learning in Organisations: What is the Current Perspective for Sustainable Development? 143
Orlando Petiz Pereira, Maria João Raposo, Miloš Krstić, and Oleksii Goncharenko

Interplay of Traditions and Innovations in Teaching Sustainability Issues: National and Global Discourses 161
Dzintra Iliško

Towards a More Sustainable Transport Future—The Cases of Ferry Shipping Electrification in Denmark, Netherland, Norway and Sweden 177
Maciej Tarkowski

Digitalisation and Communication for Sustainable Development

Overcoming Digital Inequality as a Condition for Sustainable Development 195
Vladimir Krivosheev

Design Thinking and Collaborative Digital Platforms: Innovative Tools for Co-creating Sustainability Solutions 207
Diane Pruneau, Viktor Freiman, Michel T. Léger, Liliane Dionne, Vincent Richard, and Anne-Marie Laroche

Hybridization of Time: Towards Temporal Sustainability of the Digital Economy 227
Bohdan Jung and Tadeusz Kowalski

Impact of Uncompetitive Coexistence of Innovation and Tradition on Sustainable Development in McLuhan's Media Theory 245
Saulius Keturakis

Sustainable Event Management: New Perspectives for the Meeting Industry Through Innovation and Digitalisation? 259
Dirk Hagen

Global Trends and Regional Aspects of Environmental Management

Towards Sustainable and Responsible Regional Innovation Policy—the Case of Tampere Region 279
Anna Martikainen, Mika Kautonen, and Mika Raunio

Sustainable Development of Russia's North-Western Border Areas and Their Neighbors: A Study of Landscape Effects on the Settlement Patterns of Villages and Towns 295
Elena A. Romanova

Innovations for Sustainable Production of Traditional and Artisan Unrefined Non-centrifugal Cane Sugar in Mexico 313
Noé Aguilar-Rivera and Luis Alberto Olvera-Vargas

Traditions and Innovations in the North Caucasus Nature Management 331
Khava Zaburaeva and Evgeny Krasnov

Revitalization of Local Traditional Culture for Sustainable Development of National Character Building in Indonesia 347
Cahyono Agus, Sri Ratna Saktimulya, Priyo Dwiarso, Bambang Widodo, Siti Rochmiyati, and Mulyanto Darmowiyono

Modern Technologies in Tourism as a Tool to Increase International Tourism Attractiveness and Sustainable Development of the Kaliningrad Region 371
Anna V. Belova, Nikolay Belov, and Ivan Gumeniuk

Cross-Border Cooperation Programmes as a Sustainable Tool for Tourism Development: The Case of the Kaliningrad Region 383
Anna V. Belova and Irina V. Fedina-Zhurbina

New Approaches to Sustainable Management of Wetland and Forest Ecosystems as a Response to Changing Socio-Economic Development Contexts 395
M. G. Napreenko, O. A. Antsiferova, A. V. Aldushin, A. K. Samerkhanova, Y. K. Aldushina, P. N. Baranovskiy, T. V. Napreenko-Dorokhova, V. V. Panov, and E. V. Konshu

SkyrosIsland in the Front Line of Sustainable Development Promotion 417
Valentina Plaka, Chrysoula Sardi, Iliana–Dimitra Psomadaki, Olga Kouleri, and Constantina Skanavis

Transformations of Trolleybus Transport in Belarus, Russia and Ukraine in 1990–2020 429
Marcin Połom

Legal Instruments for Business Communities

Business Communities as a Tool for Sustainable Development 447
Rafael Gustavo de Lima and Samara da Silva Neiva

Sustainability Practices Among Russian Business Communities: Drivers and Barriers Towards Change (The Cases of Moscow and Kazan) 467
Polina Ermolaeva and Ksenia Agapeeva

Short-termism—The Causes and Consequences for the Sustainable Development of the Financial Markets 485
Małgorzata Janicka, Artur Sajnóg, and Tomasz Sosnowski

Gradual Transition to Innovative Development

The Hybrid Innovation System Principles for Resilient Innovative Growth

Anna Alekseevna Mikhaylova and **Dmitry Vitalievich Hvaley**

Abstract Around the globe, cities act as nodes of innovation growth accumulating significant human, scientific, technical, entrepreneurial, investment, infrastructure, institutional, and other potentials. Generally, cities are attractors of resources, alienating them from nearby territories. The effects of this process are especially acute in the countryside, facing the problem of depopulation and decline. This is also true for small and medium-sized towns (SMSTs) located at a short distance to a large city—the center of attraction. On the other hand, urban sprawl provides an impetus for the innovative development of neighboring territories. The flow of innovation and knowledge from the center to the periphery takes place within the framework of the diffusion-absorption model. In this situation, it is very important to study the issue of sustainable innovative development of the urban–rural system based on the balanced use of rural resources and the effects of the innovative development of a city. In the framework of this study, based on the case of the Kaliningrad region, the processes of the innovative development of a large city—the regional center, and the adjacent territories are evaluated. We study the effects of the mutual influence of the urban–rural settlements in the transition to an innovative economy. Estimations are made over the development dynamics of the transport and ICT infrastructure in the intercity context of the Kaliningrad region—the most important systems for the transfer of knowledge and innovation. The conclusion is made about the heterogeneity of the innovation space of the Kaliningrad region, as well as differences in the functional roles of urban and rural settlements in it. The idea of considering the city and neighboring rural settlements as a holistic innovation system is put forward. The research concludes with policy implications to implement the hybrid innovation system principles for resilient innovation growth.

Keywords Urban agglomeration · Rural settlements · Sustainable development · Innovation diffusion · Regional resilience

---

A. A. Mikhaylova (✉) · D. V. Hvaley
Immanuel Kant Baltic Federal University, 14 A. Nevskogo Str., Kaliningrad 236041, Russia

© The Author(s), under exclusive license to Springer Nature Switzerland AG 2021
W. Leal Filho et al. (eds.), *Innovations and Traditions for Sustainable Development*,
World Sustainability Series, https://doi.org/10.1007/978-3-030-78825-4_1

1 Introduction

Cities are increasingly perceived as centers of innovation activity, the hubs of intersection between communication channels and knowledge flows, and the places where resources necessary for generating innovations are accumulated (Appio et al. 2019; Carvalho and Winden 2017; Wong et al. 2018). A tendency is emerging towards the formation of urban agglomerations and megacities—the networks of cities of different size and administrative status, which are closely interconnected by economic, social, and cultural ties (Glaeser et al. 2016; Li et al. 2018; Zhao et al. 2017). Active urbanization and urban growth are affecting the countryside (Berdegue and Soloaga 2018; He et al. 2019; Mikhaylov et al. 2019). On the one hand, rural areas face the problem of depopulation as a result of the outflow of the working-age population to cities, and, consequently, a decrease in domestic resources for sustainable development. On the other hand, as the agglomeration grows, the reverse process of its decentralization begins with the overflow of part of the accumulated intellectual, human, investment capital to the periphery. On a regional scale, a large urban agglomeration acts as a driver of the socio-economic development of the rest of its territory, including by redistributing tax revenues from urban to rural settlements. The latter are closely related to the city, therefore, ensuring their sustainable development cannot be considered separately, but only in the context of development dynamics of a single urban–rural territorial system.

This study is aimed at identifying the effects of the urban agglomeration on the sustainable innovative development of the rest of the region. The focus is given to transport accessibility and information and communication connectivity of municipal areas as the most important factors for the formation of knowledge flows and ensuring mobility of human capital within the region. We believe that the efficiency of innovative development of a region is affected not only by the ability to generate knowledge and innovations inherent in a small number of its cities but also by the existence of a formed system of intra-regional communication channels that ensure diffusion of knowledge and innovations between a city and a countryside. Thus, our research perspective on innovation and infrastructure falls beyond the SDG 9 on to the intersection with SDGs 8 and 11—ensure a decent quality of life within local communities against increasing urbanization and economic clustering within large cities.

The main goal of the study is to test the research hypothesis stating that more active development will be characteristic of municipalities in two cases: either with better transport accessibility to the core city of the metropolitan area or with a higher level of development of information and communication infrastructure. We assume that the factor of information and communication connectivity of a municipality can to some extent compensate for its transport remoteness or isolation. Therefore, the objectives are to differentiate the agglomeration belts within the urban agglomeration and to measure the agglomeration externalities depending on physical and virtual proximity to its core.

A few recent studies have elaborated on a similar issue in a slightly different context. For instance, a study of Ning et al. (2016) has relied on patents and FDI data to evaluate intra-regional knowledge spillover across 181 cities spread around China. Boussauw et al. (2018) suggest using spatial visioning tools for combating negative consequences of urban growth and optimizing agglomeration externalities in the case of the Flanders region. An important observation is made by Cottineau et al. (2019) with regards to the scale of observation applied during the delineation of territorial units. By measuring the regression of wages versus population and jobs scholars have estimated urban agglomeration externalities between large and small cities France. Our study is performed on the materials of the Kaliningrad region of Russia. This region is of considerable interest in the context of testing the hypothesis set, since it is an exclave. The organization of transport and information and communication links in the intra-regional space is more significant for an exclave territory than for in-country regions with greater opportunities for communication outside their borders.

2 Kaliningrad Urban Agglomeration

The Kaliningrad region is the westernmost region of Russia, located on the shores of the Baltic Sea, surrounded by EU countries. On land, the Kaliningrad region borders Poland and Lithuania (Fig. 1). The exclave position of the region results is its reduced transport accessibility and the presence of institutional and transport barriers to the

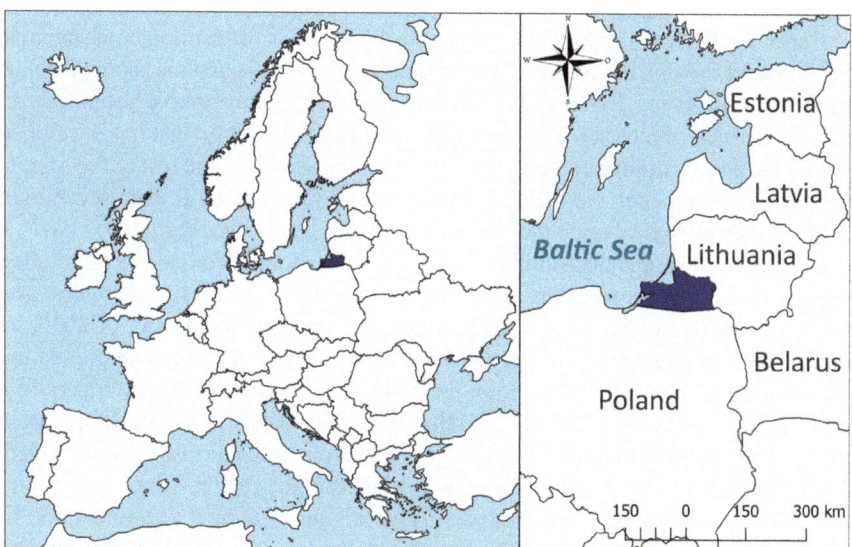

Fig. 1 The Kaliningrad region on a map of Europe

movement of goods and people (Anokhin and Fedorov 2019; Domaniewski and Studzińska 2016; Gumenyuk et al. 2019; Karpenko et al. 2014). Given the limited labor market and capital, as well as energy resources in the region, no conditions have been formed for the development of large labor-intensive and energy-intensive manufacturing industries (Ataev 2018; Kazakova et al. 2017; Mikhaylova 2018). The regime of the Special Economic Zone in the Kaliningrad region led to the formation of predominantly assembly plants without deep processing, and, therefore, with low added value (Fyodorov 2018; Gareev 2013; Gareev and Voloshenko 2015; Nilov 2018). The region also experiences some geopolitical pressure due to the increased presence of NATO forces in the Baltic (Maass 2020; Nieto 2011; Zohn 2019).

The current specifics of the Kaliningrad region are further affected by national and global trends in socio-economic development, including those related to the impact of the COVID-19 pandemic and difficulties in the political dialogue between Russia and the European Union (EU) (Diener and Hagen 2011; Gänzle and Müntel 2011; Joenniemi and Prawitz 1998; Richard et al. 2015). The innovative way is seen as the preferred development model for the Kaliningrad region in the long term (Mikhaylova 2019; Mikhaylova et al. 2019; Zemtsov and Smelov 2018). Sustainability of the regional economy can be achieved through the clustering of knowledge-intensive and innovative industries, as well as the active development of the innovative services sector, primarily information and communication and tourism (Aleynikova 2020; Roos et al. 2020; Zemtsov et al. 2018). Also, the region already has a certain accumulated scientific and technological potential in both traditional and new fields: Earth sciences, chemical and biological sciences, engineering, medicine and health sciences (Druzhinin et al. 2020; Mikhaylov et al. 2019, 2020). Creating conditions for its subsequent commercialization will strengthen the economic stability of the region.

Currently, the Kaliningrad region faces the difficult task of ensuring its sustainable development through the mobilization of domestic resources and the search for new mechanisms of socio-economic growth. One of such mechanisms may be building an effective system of interaction between the urban agglomeration that has developed around the administrative center—the city of Kaliningrad, and the rest of the region, which will make better use of the human, natural, infrastructural, entrepreneurial, scientific, technological and other resources accumulated in the region.

For the last 10 years, the Kaliningrad region has been characterized by population growth, including due to external migration. In 2019, in terms of population, the region passed the threshold of 1 million people, and at the beginning of 2020, its total population was 1,012,512 people. On average in 2011–2020 annual population growth was 0.8%. At the same time, the Kaliningrad region is characterized by a strong polarization around the city of Kaliningrad, where 49% of the total population and 69% of economic entities (without individual entrepreneurs) are concentrated. The Kaliningrad urban agglomeration was formed in the region, which along with Kaliningrad, which is its core, includes 13 urban and 610 rural settlements of 12 neighboring municipalities (Fig. 2). Three urban districts are adjacent to Kaliningrad and form its first agglomeration belt: Guryevsky, Zelenogradsky, Svetlovsky. The second agglomeration belt to the west of Kaliningrad in the coastal zone is the

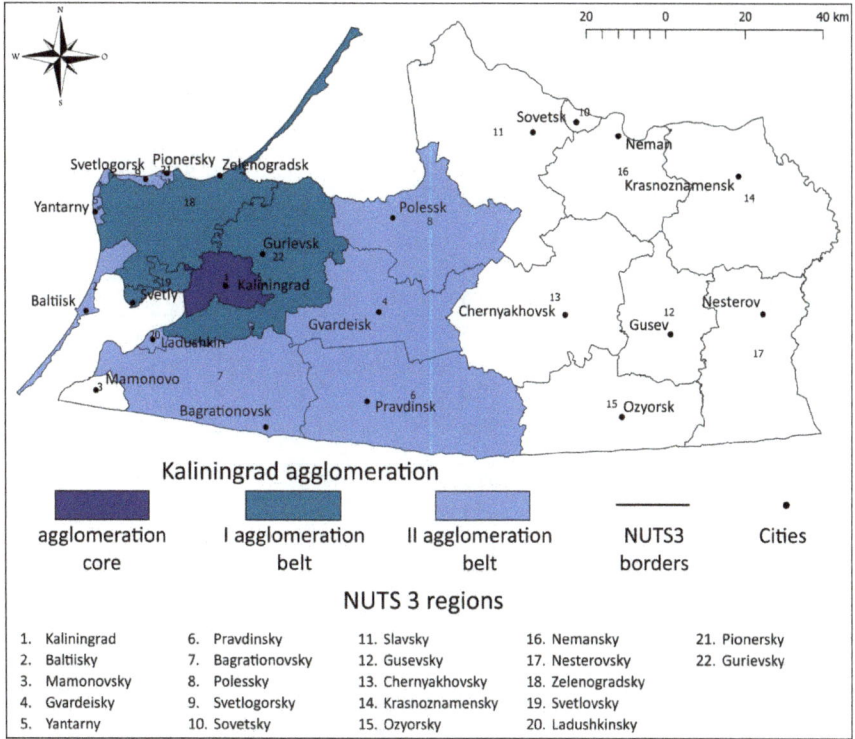

Fig. 2 Urban agglomeration of the Kaliningrad region

Pionersky, Baltiysk, Svetlogorsky, Ladushkinsky, Yantarny city districts, and in the east deep into the Kaliningrad region—Bagrationovsky, Gvardeysky, Polessky, and Pravdinsky city districts.

The total population of the Kaliningrad urban agglomeration at the beginning of 2020 amounted to 804,669 people or 79.5% of the total population of the region, of which 80.8% is the urban population. In general, the agglomeration around Kaliningrad concentrates 82.6% of the urban and 68.4% of the rural population of the Kaliningrad region. The highest population density is in the core and in the first agglomeration belt: 77.9% of the total agglomeration population is concentrated here. Table 1 presents data on the uneven development of the territory of the Kaliningrad region in the context of the agglomeration process.

The most active development in 2012–2019 was achieved by 3 municipalities of the 1st belt, which increased the volume of housing construction and the number of employees of large and medium enterprises. It is these municipalities that account for the second-largest volume of investment in fixed assets and the volume of goods produced by manufacturing companies after Kaliningrad. With the distance from the agglomeration core, the economic attractiveness of municipalities decreases. In this regard, the objective of more actively transferring knowledge, competencies,

Table 1 The growth of the Kaliningrad urban agglomeration of the Kaliningrad region

Indicator	Year	Municipalities				Other
		Agglomeration	Including			
			Core	1 belt	2 belt	
Share in the total land area, %	2012	51.5	1.5	22.9	27.1	48.5
Share of organizations, without individual entrepreneurs, %		89.8	68.9	11.3	9.6	10.2
	2020	91.4	69.9	12.5	9.0	8.6
Share of goods produced by large and medium-sized manufacturing enterprises, %	2012	93.5	77.8	9.3	6.5	6.5
	2019	94.3	65.5	24.7	4.2	5.7
Share of fixed capital investments by organizations of the municipality, %	2012	79.0	45.8	28.7	4.4	21.0
	2018	95.5	67.1	22.0	6.4	4.5
Share on commissioning of residential buildings, %	2012	97.2	68.5	19.3	9.4	2.8
	2018	97.1	55.6	33.1	8.4	2.9
Share of the average number of employees of large and medium-sized organizations (without external part-time workers), %	2012	85.7	65.9	8.3	11.5	14.3
	2019	85.9	61.5	10.9	13.5	14.1

Source based on data from the Territorial authority of the Federal State Statistics Service for the Kaliningrad Region—Kaliningradstat URL: https://kaliningrad.gks.ru/main_indicators

technologies, and innovations accumulated in the core of the agglomeration to the periphery, as well as strengthening the digital proximity of the periphery and the core through the improvement of information and communication technologies is becoming crucial in maintaining sustainable development.

3 Methodology

In order to evaluate the agglomeration externalities and the mutual influence of the agglomeration-forming city and adjacent municipalities in the aspect of the regional innovation system development, we examine the existing channels for the transfer of knowledge, information, and diffusion of innovations within the region. Based on the review of relevant literature, we have outlined two general ways of communication and networking: (a) by transport and (b) by using information and communication technologies (ICT). The debate on the role of physical proximity is becoming even more prominent with increased affordability of ICT and digitalization of business, public services and daily routines (He et al. 2020; Ievoli et al. 2019; Martinus et al. 2020; Martinus and Sigler 2018; Yu et al. 2018). With that, the geographical location and the distance to 'hotspots' remains critical in gaining numerous positive externalities. In our study, we focus on the analysis of the physical or virtual involvement of urban and rural settlements in innovative processes within the region, as well as on the definition of the functions that they perform.

Two key aspects for evaluating a communication channel are identified: The infrastructural capabilities of its use and the interest of the population in its use. The study is divided into 2 stages. At the first stage, we investigate the development of the transport infrastructure of the region and the transport accessibility of its municipalities. In the second stage, we evaluate the level of development of information and communication infrastructure and the representation of the population and economic entities of municipalities in the Internet space. Table 2 presents the system of indicators used in the analysis.

Features of the collection and aggregation of data on bus and rail links between Kaliningrad and other settlements of the Kaliningrad region implies an assessment of the possibility of pendulum labor migration by the population. The transport accessibility of the municipality relative to the core of the urban agglomeration is estimated for the standard working week from Monday to Friday. When forming the sample, the existing differences in the movement of public transport on weekdays and weekends are taken into account. The key factors of transport accessibility are the number of round trips and the temporary remoteness of the municipality from the administrative center of the region. The number of round trips is calculated as the maximum number of possible round trips from the municipality to Kaliningrad and back during the day. Travel time from Kaliningrad to the municipality is equal to travel time from the same town to Kaliningrad. Its duration is calculated as the average travel time for all available routes: For bus communication—regular intercity buses, for railway communication—commuter trains.

We have analyzed 189 official bus routes, covering 100 urban and rural settlements of the region. According to the railway communication, the data covers 10 routes of the suburban trains in the direction of Kaliningrad, including 87 departure stations in more than 80 settlements of the region. Note that rail transport has high passenger capacity, therefore, in terms of total passenger traffic, it successfully competes with bus transport in the main directions. Moreover, railway communication is weakly or not represented at all in the peripheral districts of the region.

Data on the Internet coverage of the Kaliningrad region is presented by aggregating the total data on 3G and 4G coverage of the territory of the municipalities of the Kaliningrad region, available on the official websites of the largest telecommunication operators in Russia: Beeline, Megafon, MTS, and Tele2. By combining coverage maps of all telecom operators in the QGIS program, a single Internet coverage of the Kaliningrad region is created using the distribution method. Further, for each municipality, the proportion of the territory that has Internet coverage is calculated using the specialized software.

Data on the representation of the population and companies of the Kaliningrad region on the Internet are uploaded from the official website of the largest social network Vkontakte using internal service search algorithms. This includes depersonalized data on the sex and age structure of the population and the intensity of

Table 2 Assessment of innovative development of cities of the Kaliningrad region and its impact on the countryside

Factor	Indicator	Period/Data source
Transport system		
Transport infrastructure development	1. The total length of local public roads	2012–2019/Kaliningradstat, URL: https://kaliningrad.gks.ru/main_indicators
	2. Public bus service availability (number of round-trip bus services and average travel time)	As of 9.https://doi.org/10.2019/ official website of the Kaliningrad bus station, URL: https://avl39.ru/routes/registry
	3. Railway connectivity (the number of direct trains scheduled per day and average travel time)	As of 20.12.2019/official site of the Kaliningrad suburban passenger company, URL: http://kppk39.ru/index.php/raspisanie
Population coverage	1. Passengers transported by regular buses and commuter trains	2017–2018/Information on the activities of passenger road transport in the Kaliningrad region. Statistical Bulletin. Kaliningradstat 2012–2019/Official website of the Ministry of Infrastructure Development of the Kaliningrad region URL: http://infrastruktura.gov39.ru
Information and communication system		
Development of information and communication infrastructure	1. The density of 3G and 4G Internet coverage, % of the territory of the municipality	As of 22.https://doi.org/10. 2019/official websites of telecommunication companies Beeline, Megafon, MTS, and Tele2
Population coverage	1. Representation of the population in the Internet	As of 7.07.2020/Vkontakte social network, URL: https://vk.com (search by people)
	2. Representation of companies in the Internet	As of 7.07.2020/Vkontakte social network, URL: https://vk.com (search by communities)

Internet commerce in 30 urban settlements of the Kaliningrad region, including 23 cities. Data on the "Internet population" of the region are uploaded by gender and age (by categories: Under 18 years old, 18–65 years old, and older than 65 years old). To assess the intensity of Internet commerce, we searched for communities by categories: "Restaurants", "Shops", "Tourism and Recreation", "Professional Services", "Household Services", and "Culture".

4 Results and Analysis

4.1 Development of the Transport System

The total length of local public roads in the Kaliningrad region as of the beginning of 2020 was over 4.4 thousand km, increasing over 2012–2019 by 1.8 times. The most active increase (more than 2 times) in the length of roads was carried out in the municipalities peripheral to the Kaliningrad agglomeration: Mamonovsky, Ozersky, Nemansky, Gusevsky, and Chernyakhovsky. In second place by increasing the length of local public roads were municipalities of the 2nd agglomeration belt featuring the growth of 30–40%. The smallest increase in the length of roads was registered at the territory of Kaliningrad and the municipalities of the 1st agglomeration belt: By 1–3% in 2013–2019, with the exception of the Zelenogradsk urban district, which showed an increase of 25%.

While in 2012 the Kaliningrad urban agglomeration accounted for 64.4% of the total length of the region's roads, by 2019 it decreased to 59.5%. The redistribution in the structure occurred due to a reduction in the share of the core (from 23.4 to 13%) and municipalities of the 1st agglomeration belt (from 22.2 to 18.5%). At the same time, the share of municipalities in the 2nd agglomeration belt (from 18.7 to 28.1%) and the peripheral areas (from 35.6 to 40.5%) increased. In general, the presented dynamics testifies to the development of the transport sector of the region's municipalities that are remote from Kaliningrad, which contributed to the increase in their transport accessibility. At the same time, the leaders in the density of paved roads are still the municipalities of the Kaliningrad urban agglomeration with a high share of the urban population: Pionersky, Kaliningrad, Svetlogorsk, Yantarny, Svetlovsky, as well as the city of Sovetsk—the second-largest city in the region located on the border with Lithuania.

The study of transport accessibility of the municipalities of the Kaliningrad region from the perspective of pendulum labor migration was carried out on the basis of 2 types of public transport: Bus and trail. Kaliningrad has direct bus access from 100 settlements of the Kaliningrad region (Fig. 3). About half of these settlements are located 30 km from Kaliningrad, including 70% in the 1st agglomeration belt. Another 31% of the settlements are located in the zone of 30–50 km from Kaliningrad, including 80% in the 2nd agglomeration belt. Thus, 78% of all settlements covered by direct bus connections with Kaliningrad are located no further than 50 km from it.

For settlements located in a 30 km zone from the agglomeration core the average travel time to Kaliningrad is about 1 h, and for those located in a 30–50 km zone, it is 1.5 h. The most attractive for the implementation of pendulum labor migration seems to be the Guryevsky urban district: 24 of its settlements in the 30 km zone are connected with Kaliningrad by at least 2 direct buses per day. The total number of roundtrips between Kaliningrad and the Guryevsky urban district is 489. The bus service is most active with the city of Guryevsk, the villages of Kosmodemyanskoye and Konstantinovka. The high transport accessibility of the Guryevsky urban district

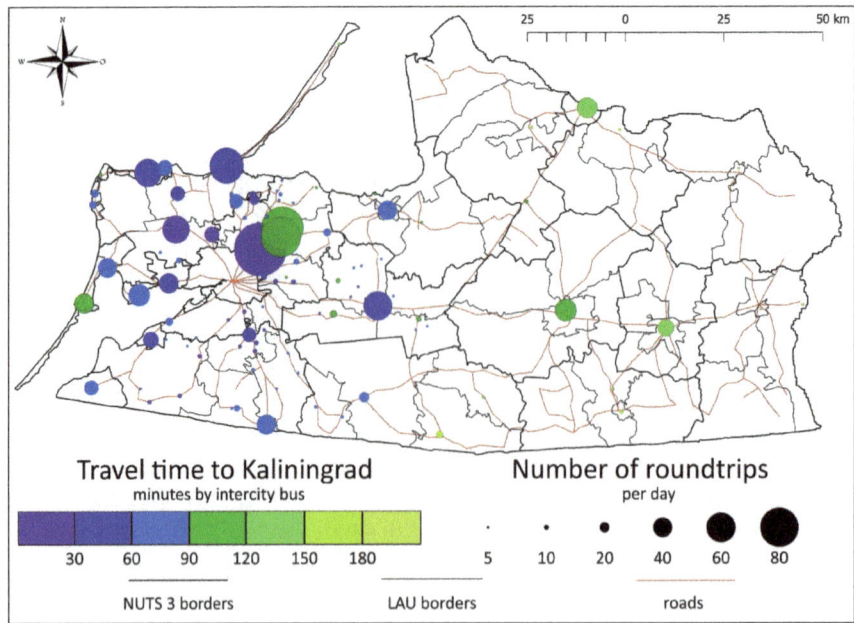

Fig. 3 Bus service between Kaliningrad and other settlements of the region. *Source* The official website of the Kaliningrad bus station, as of 9.10.2019. URL: https://avl39.ru/routes/registry

to the core of the agglomeration acted as a driver of active migration growth in 2012–2019: Over 1.7 thousand people a year. This ranks Guryevsk 2nd by migration attractiveness after Kaliningrad among the municipalities of the region.

Also, good conditions for labor pendulum migration are characterized by individual settlements of the Svetlovsky urban district, located 30 km from Kaliningrad. This is the city of Svetly and the village of Vzmorye, connected with Kaliningrad with more than 40 circular bus services, and the village of Lublino with 11 roundtrips per day. However, with regard to migration attractiveness, the settlements of Svetlovsky urban district significantly lose to Guryevsky urban district, and in some years (2017 and 2018) it even had an outflow of population. A similar situation is observed in Ladushkino. Despite good transport connectivity with the core of the agglomeration, Svetly did not form favorable conditions for attracting the population.

Among the settlements of Zelenogradsky urban district, which, as well as the Guryevsky and Svetlovsky municipalities, is included in the 1st agglomeration belt, the largest number of roundtrip bus connections with Kaliningrad are of Zelenogradsk (73), the villages of Pereslavskoye (57) and Romanovo (30). Good transport accessibility and proximity to Kaliningrad in combination with a developed social infrastructure and resort status allowed Zelenogradsk urban district to rank 3rd in terms of migration attractiveness for the population among the municipalities of the region.

In 2019, the population growth in the municipality amounted to more than 1 thousand people. Between Kaliningrad and Zelenogradsk a stream of pendulum labor migration has been established.

A similar population growth trend is characteristic of the Svetlogorsk urban district, whose administrative center is the city of Svetlogorsk—a resort that is closely connected with Kaliningrad by 57 circular bus routes. However, due to the greater actual distance from Kaliningrad as compared to Zelenogradsk, the city is less suitable for labor pendulum migration. Also, in the case of all resort towns of the region, an increased load on the public transportation system is observed during the tourist season. The movement of a significant number of holidaymakers from Kaliningrad to resort cities and back during the day determine a large number of trips between the core of the agglomeration and settlements on the coast.

In the 50–100 km zone, there are individual cities and villages of the 1st and 2nd agglomeration belts located farthest from the core of the agglomeration: Krasnoflotskoe and Morskoye rural areas in Zelenogradsk urban district; Krasnoye village in the Guryevsky urban district; the city of Znamensk and the village of Gordoye in Gvargeysk urban district; the city of Pravdinsk, Zheleznodorozhny, villages Ermakovo and Mozyr in Pravdinsky urban district; Saranskoye village in the Polessky urban district. Settlements that are not part of the Kaliningrad agglomeration are more than 100 km away from its core—the city of Kaliningrad. Direct public transport connections outside the Kaliningrad agglomeration is mainly available with the administrative centers of the municipalities: Sovetsk, Gusev, Slavsk, Ozersk, Neman, Nesterov, Krasnoznamensk. The largest number of possible daily roundtrips between Kaliningrad is available for Chernyakhovsk (45), Sovetsk (42), and Gusev (34).

Rail transport covers a smaller number of settlements in the region than the bus (Fig. 4). Coverage area—17 out of 22 municipalities of the Kaliningrad region (except for Krasnoznamensky, Ozersky, Pravdinsky, Nemansky, and Yantarny urban districts). The advantages of railway transport are that it has a much larger capacity, less travel time, including due to the absence of delays due to traffic jams during rush hour and the tourist season in the resort areas. The most developed railway transport connection between Kaliningrad and settlements in the north-west of the region, primarily located on the coast.

The best directions for the implementation of pendulum labor migration using rail transport are settlements in the suburbs of Kaliningrad (10 km), as well as in the Guryevsk and Zelenogradsk municipalities. Between Kaliningrad and Guryevsk there are up to 8 round trips per day, and between Kaliningrad and Zelenogradsk up to 7 round trips per day. On average in 2012–2020, according to the Ministry of Infrastructure Development of the Kaliningrad region, the total number of passengers transported by rail in suburban traffic annually amounted to more than 3 million people, during peak years—up to 5 million people. The average annual demand for suburban bus services is slightly higher—about 5.1 million passengers a year, of which a little less than 2 million passengers were transported to rural areas.

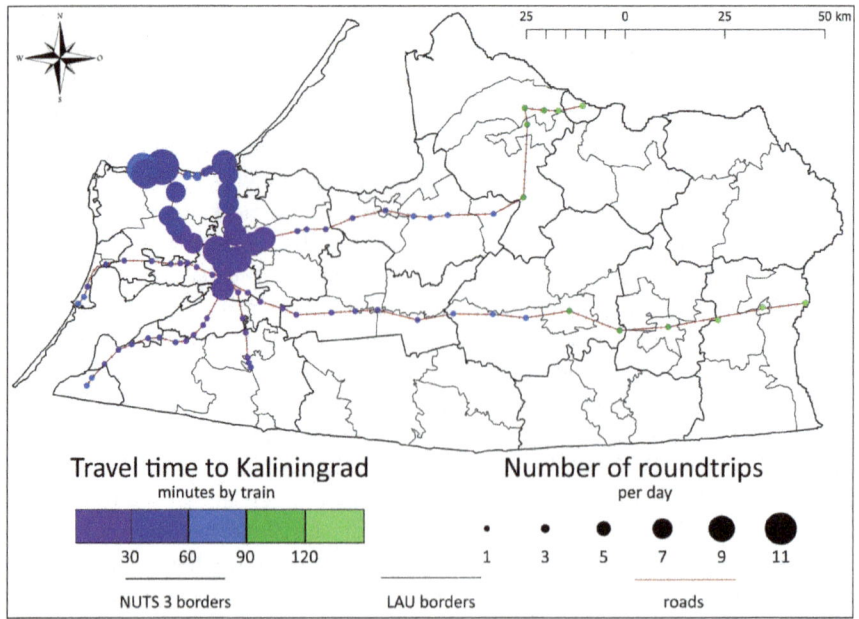

Fig. 4 Railway communication between Kaliningrad and other settlements of the region. *Source* The official site of the Kaliningrad suburban passenger company, as of 20.12.2019. URL: http://kppk39.ru/index.php/raspisanie

4.2 Development of an Information and Communication System

The coverage of the territory of the Kaliningrad region with 3G and 4G Internet communications by 2020 was not fully implemented (Fig. 5). While 3G technology has become more widespread, the 4G technology is limited around the largest cities in the region. The best Internet connection coverage is implemented inside the Kaliningrad metropolitan area, where both 3G and 4G communication standards are presented. The density of Internet coverage in the eastern peripheral regions of the region is lower, often only a 3G communication standard is presented, and there are also significant blind spots.

In the context of individual municipalities, the entire territory is completely covered by Internet connection (3G and/or 4G in total) in only 2 cities of the Kaliningrad region: Pionersky and Sovetsk, with an area of up to 50 km^2 each. In Ladushkin, Svetlogorsk, and Neman municipalities, this indicator is equal to 93–98% of the territory. These are also small municipalities of up to 50 km^2 with the exception of the Neman urban district. Another 5 municipalities (Kaliningrad, Bagrationovsky, Yantarny, Gusevsky, and Mamonovsky), in which more than half of the entire population of the region live, the density of coverage varies from 80 to 89%. The lowest density of the total Internet coverage of 35% is found in Zelenogradsky

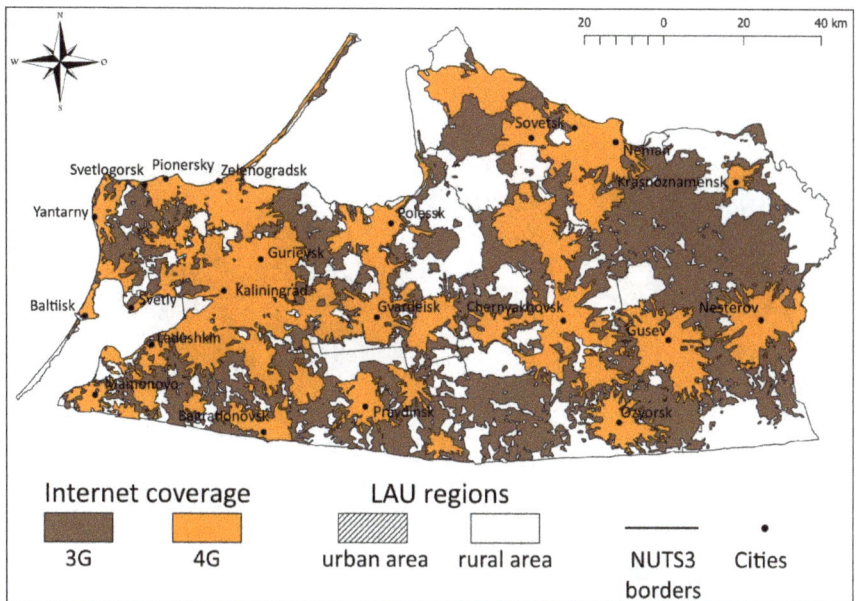

Fig. 5 Density of Internet coverage on the territory of the Kaliningrad region, 2020. *Source* based on the official websites of telecommunication companies Beeline, Megafon, MTS, and Tele2, as of 22.10.2019

district, which is due to its large size (over 2.0 thousand km^2) and the location of the Curonian Spit National Park on its territory.

To assess the demand for information and communication services, we analyzed the level of digitalization of the population and companies of the region based on the data of the largest social network Vkontakte (Fig. 6). In total, data were downloaded on 949,839 unique profiles of people who indicated their place of residence in a particular settlement of the Kaliningrad region, and 10,502 companies in the field of trade, catering, culture, and services.

We have identified a disproportion in the distribution of the "virtual" population of the Kaliningrad region. The correlation coefficient between the indicators of "distance from Kaliningrad to the settlement" and "the level of digitalization of the population, calculated as the ratio between the number of virtual profiles of people and the actual number of the population" is 0.674. The closer to the core of the agglomeration, the smaller the ratio between the number of the real population and the number of profiles in the social network that indicated this settlement as a place of residence. At the same time, Kaliningrad accounts for 1.7 times more virtual profiles of people than real residents. This suggests that the closer the inhabitants of the region live to Kaliningrad, the more they feel like Kaliningraders and indicate Kaliningrad as their place of residence. The ratio between virtual and real residents of more than 50% was observed only in cities remote from the Kaliningrad agglomeration: Nesterov, Chernyakhovsk, Ozersk, Slavsk, Neman, Gusev, Sovetsk,

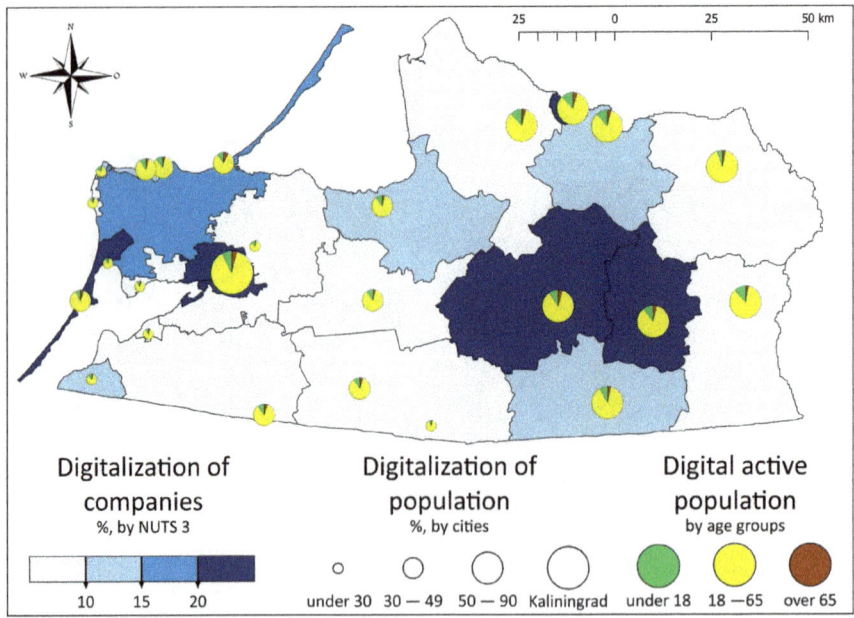

Fig. 6 Distribution of the "virtual" population and economic entities of the Kaliningrad region, 2020. *Source* based on Vkontakte social network, as of 7.07.2020. URL: https://vk.com

Krasnoznamensk. The level of representation of companies of the region's municipalities in the Internet space is low, primarily online stores. For tourist cities of the Kaliningrad region, it is also relevant to post information about travel services and catering places on social networks. At the same time, no correlation was found between Internet coverage and the number of digital profiles of municipal companies.

5 Conclusion

The agglomeration effect has a significant impact on shaping economic and settlement systems within a region, determining its development. The region as an integral territorial entity consists of a system of urban and rural settlements of different sizes and functions. The stability and efficiency of the functioning of this system largely depend on the level of connectivity of its individual elements. Connectivity can be formed both as a result of good transport accessibility and other proximity factors: institutional, economic, social, cultural and tourism, etc. The best result is generally achieved when several factors are combined. Our research focuses on the study of the information and communication technology (ICT) factor in combination with an assessment of the development of transport infrastructure.

Studies on agglomerations and urban development show that the main nodes of regional growth are large administrative centers (Fang and Yu 2017; Wang et al. 2020; Zeng et al. 2016). They act as the centers of attraction for various types of resources necessary for sustainable development, pulling them from other less developed settlements. Thus, the underdeveloped periphery of the region is being formed. At the same time, in settlements that have close ties with the core of the agglomeration, significant positive agglomeration externalities can appear. First of all, this applies to settlements that have good transport connectivity with the city-core of the agglomeration. Our hypothesis was to check whether digitalization and the development of information and communication technology (ICT) infrastructure compensate for the imbalance in the territorial development of the region, whether it can prevent the formation of vast periphery around the agglomeration.

In this study, we analyzed the impact of urban agglomeration on other municipalities in the region using data from the Kaliningrad region of Russia. This is an exclave region with difficult transport accessibility to other regions of Russia, therefore, sustainable development by increasing the efficiency of the use of internal resources is of great importance. In our research, we relied on the concepts of physical and virtual proximity. We assessed the degree of transport and digital connectivity of the city of Kaliningrad—the core of the agglomeration and other settlements in the region, included and not included in the Kaliningrad agglomeration.

Our study showed that the closest connectivity between settlements was achieved within the Kaliningrad urban agglomeration. This was manifested both from the standpoint of the development of suburban bus and rail services, and the virtual unity of the residents of Kaliningrad and neighboring settlements. The hypothesis regarding the fact that more active development will be characteristic of municipalities with better transport accessibility to the core city of the agglomeration has been confirmed. It is for these settlements that the largest migration influx of the population was observed, accompanied by more active volumes of housing construction.

The strong influence of information and communication technologies (ICT) in the development of municipalities is currently not traced and requires additional research. Perhaps this is due to the low innovation level of the economy of the region's municipalities. At the same time, our results showed that the further the settlements are located from the core of the agglomeration, the stronger the independent identity of the population is traced in the virtual space relative to their place of residence. For instance, only at distant cities of Chernyakhovsk, Gusev, Sovetsk, and other people indicate their actual place of residence, while the first and even second agglomeration belts are inalienable from the agglomeration core. Currently, the level of development of information and communication infrastructure in the region remains insufficient, especially in the east of the region and in rural areas. For more active innovative development of the territory, it is required to improve communication channels between the Kaliningrad agglomeration and the peripheral municipalities of the region.

The results obtained are of great practical importance in the development of policy documents in the field of sustainable development of municipalities in the Kaliningrad region, as well as in drafting the socio-economic, innovation, and digital policies in the region. The study has identified municipal areas with positive development dynamics and those prone to peripheralization due to low connectivity with Kaliningrad. The sustainable development of the latter requires more active government support.

This study also makes a significant contribution to the development of the concept of urban agglomeration in the aspect of studying the transport and digital connectivity of the agglomeration core with other settlements, depending on the territorial remoteness. We have found not only strong economic ties but also a single virtual identity of the population between the agglomeration core and the neighboring settlements with good transport accessibility. Further research in this area should be aimed both at conducting similar studies based on different types of agglomerations and at deepening the conclusions obtained by identifying information and communication links within the region, assessing their strength and impact on the regional economy.

Acknowledgements The reported study was funded by RFBR according to the research project № 18-310-20016 "Coastal cities in innovation spaces of the European part of Russia".

References

Aleynikova Y (2020) Global new economy: structure and perspectives in Kaliningrad Region. In: Leal Filho W, Borges de Brito P, Frankenberger F (eds) International business, trade and institutional sustainability. World Sustainability Series. Springer, Cham, pp 831–838. https://doi.org/10.1007/978-3-030-26759-9_49

Anokhin AA, Fedorov GM (2019) Regions as international development corridors in the Western borderland of Russia. Vestnik of Saint Petersburg University. Earth Sci 64(4): 545–558 (in Russian). https://doi.org/10.21638/spbu07.2019.403

Appio FP, Lima M, Paroutis S (2019) Understanding smart cities: Innovation ecosystems, technological advancements, and societal challenges. Technol Forecast Soc Chang 142:1–14. https://doi.org/10.1016/j.techfore.2018.12.018

Ataev ZA (2018) Isolated energy system of Kaliningrad oblast. Izvestiya Rossiiskaya Akademii Nauk, Seriya Geograficheskaya 1:101–110. https://doi.org/10.7868/S2587556618010095

Berdegué JA, Soloaga I (2018) Small and medium cities and development of Mexican rural areas. World Dev 107:277–288. https://doi.org/10.1016/j.worlddev.2018.02.007

Boussauw K, Van Meeteren M, Sansen J, Meijers E, Storme T, Louw E, Derudder B, Witlox F (2018) Planning for agglomeration economies in a polycentric region: envisioning an efficient metropolitan core area in Flanders. Eur J Spat Dev 69:1–26. https://doi.org/10.30689/EJSD2018:69.1650-9544

Carvalho L, Van Winden W (2017) Planned knowledge locations in cities: studying emergence and change. Int J Knowl Based Dev 8(1):47–67. https://doi.org/10.1504/IJKBD.2017.082429

Cottineau C, Finance O, Hatna E, Arcaute E, Batty M (2019) Defining urban clusters to detect agglomeration economies. Environ Plan B Urban Anal City Sci 46(9):1611–1626. https://doi.org/10.1177/2399808318755146

Diener A, Hagen J (2011) Geopolitics of the Kaliningrad exclave and enclave: Russian and EU perspectives. Eurasian Geogr Econ 52(4):567–592. https://doi.org/10.2747/1539-7216.52.4.567

Domaniewski S, Studzińska D (2016) The small border traffic zone between Poland and Kaliningrad region (Russia): the impact of a local visa-free border regime. Geopolitics 21(3):538–555. https://doi.org/10.1080/14650045.2016.1176916

Druzhinin AG, Kuznetsova TY, Mikhaylov AS (2020) Coastal zones of modern Russia: delimitation, parametrization, identification of determinants and vectors of Eurasian dynamics. Geogr Environ Sustain 13(1):37–45. https://doi.org/10.24057/2071-9388-2019-81

Fang C, Yu D (2017) Urban agglomeration: an evolving concept of an emerging phenomenon. Landsc Urban Plan 162:126–136. https://doi.org/10.1016/j.landurbplan.2017.02.014

Fyodorov GM (2018) The social and economic development of Kaliningrad. In: Joenniemi P, Prawitz J (eds) Kaliningrad: The European amber region, London, Routledge, p 32–56. https://doi.org/10.4324/9780429447624

Gänzle S, Müntel G (2011) Europeanization beyond Europe? Eu impact on domestic policies in the Russian enclave of Kaliningrad. J Balt Stud 42(1):57–79. https://doi.org/10.1080/01629778.2011.538513

Gareev T (2013) The special economic zone in the Kaliningrad region: development tool or institutional trap? Balt J Econ 13(2):113–129. https://doi.org/10.1080/1406099X.2013.10840535

Gareev TR, Voloshenko KYu (2015) Features of balance model development of exclave region. Econ Reg 2:113–124. https://doi.org/10.17059/2015-2-9

Glaeser EL, Ponzetto GAM, Zou Y (2016) Urban networks: connecting markets, people, and ideas. Pap Reg Sci 95(1):17–59. https://doi.org/10.1111/pirs.12216

Gumenyuk IS, Voloshenko KY, Novikova AA (2019) Scenarios of increasing the economic efficiency of the Kaliningrad regional transport system. Balt Reg 11(2):51–72. https://doi.org/10.5922/2079-8555-2019-2-4

He SY, Wu D, Chen H, Hou Y, Ng MK (2020) New towns and the local agglomeration economy. Habitat Int 98:102–153. https://doi.org/10.1016/j.habitatint.2020.102153

He Y, Zhou G, Tang C, Fan S, Guo X (2019) The spatial organization pattern of urban-rural integration in urban agglomerations in China: an agglomeration-diffusion analysis of the population and firms. Habitat Int 87:54–65. https://doi.org/10.1016/j.habitatint.2019.04.003

Ievoli C, Belliggiano A, Marandola D, Milone P, Ventura F (2019) Information and communication infrastructures and new business models in rural areas: the case of Molise region in Italy. Euro Countrys 11(4):475–496. https://doi.org/10.2478/euco-2019-0027

Joenniemi P, Prawitz J (1998) Kaliningrad: a double periphery? In: Joenniemi P, Prawitz J (eds) Kaliningrad: The European Amber Region, p 226–260. https://doi.org/10.4324/9780429447624-11

Karpenko A, Krasnov E, Simons G (2014) Crisis management challenges in Kaliningrad. Routledge, London, p 216. https://doi.org/10.4324/9781315574899

Kazakova NA, Bolvacheva AI, Gendon AL, Golubeva GF (2017) Value added analysis and trend forecasting in the manufacturing industry in Kaliningrad oblast. Stud Russ Econ Dev 28(2):160–168. https://doi.org/10.1134/S107570071702006X

Li D, Ma J, Cheng T, van Genderen JL, Shao Z (2018) Challenges and opportunities for the development of megacities. Int J Digit Earth 12(12):1382–1395. https://doi.org/10.1080/17538947.2018.1512662

Maass A-S (2020) Kaliningrad: a dual shift in cooperation and conflict. In: East European politics. Online first. p 14. https://doi.org/10.1080/21599165.2020.1763313

Martinus K, Sigler TJ (2018) Global city clusters: theorizing spatial and non-spatial proximity in inter-urban firm networks. Reg Stud 52(8):1041–1052. https://doi.org/10.1080/00343404.2017.1314457

Martinus K, Suzuki J, Bossaghzadeh S (2020) Agglomeration economies, interregional commuting and innovation in the peripheries. Reg Stud 54(6):776–788. https://doi.org/10.1080/00343404.2019.1641592

Mikhaylov AS, Mikhaylova AA, Lachininskii SS, Hvaley DV (2019) Coastal countryside innovation dynamics in North-Western Russia. Eur Countrys 11(4):541–562. https://doi.org/10.2478/euco-2019-0030

Mikhaylov A, Kuznetsova TYu, Peker IYu (2019) Knowledge geography: human geography approach to measuring regional divergence of knowledge capital. Proc Eur Conf Knowl Manage (ECKM) 2:738–745. https://doi.org/10.34190/KM.19.239

Mikhaylov AS, Wendt JA, Peker IY, Mikhaylova AA (2020) Spatio-temporal patterns of knowledge transfer in the borderland. Balt Reg 12(1):132–155. https://doi.org/10.5922/2079-8555-2020-1-8

Mikhaylova AA (2018) The dimension of innovation in the economic security of Russian regions. Eur J Geogr 9(4):88–104

Mikhaylova AA (2019) In pursuit of an innovation development trajectory of the Kaliningrad region. Balt Reg 11(3):92–106. https://doi.org/10.5922/2079-8555-2019-3-5

Mikhaylova AA, Mikhaylov AS, Savchina OV, Plotnikova AP (2019) Innovation landscape of the Baltic region. Adm Si Manag Public 33:165–180. https://doi.org/10.24818/amp/2019.33-10

Nieto WAS (2011) Assessing Kaliningrad's geostrategic role: the Russian periphery and a Baltic concern. J Balt Stud 42(4):465–489. https://doi.org/10.1080/01629778.2011.621737

Nilov KN (2018) The special economic zone in the Kaliningrad region: towards a more effective legal regime. Balt Reg 10(4):74–87. https://doi.org/10.5922/2079-8555-2018-4-5

Ning L, Wang F, Li J (2016) Urban innovation, regional externalities of foreign direct investment and industrial agglomeration: evidence from Chinese cities. Res Policy 45(4):830–843. https://doi.org/10.1016/j.respol.2016.01.014

Richard Y, Sebentsov A, Zotova, M (2015) The Russian exclave of Kaliningrad. Challenges and limits of its integration in the Baltic region. CyberGeo. 719. https://doi.org/10.4000/cybergeo.26945

Roos NGA, Voloshenko KY, Drok TE, Farafonova YY (2020) An economic complexity analysis of the kaliningrad region: identifying sectoral priorities in the emerging value creation paradigm. Balt Reg 12(1):156–180. https://doi.org/10.5922/2079-8555-2020-1-9

Wang H, Zhang B, Liu Y, Liu Y, Xu S, Zhao Y, Hong S (2020) Urban expansion patterns and their driving forces based on the center of gravity-GTWR model: a case study of the Beijing-Tianjin-Hebei urban agglomeration. J Geog Sci 30(2):297–318. https://doi.org/10.1007/s11442-020-1729-4

Wong C-Y, Ng B-K, Azizan SA, Hasbullah M (2018) Knowledge structures of city innovation systems: Singapore and Hong Kong. J Urban Technol 25(1):47–73. https://doi.org/10.1080/10630732.2017.1348882

Yu H, Jiao J, Houston E, Peng Z-R (2018) Evaluating the relationship between rail transit and industrial agglomeration: an observation from the Dallas-fort worth region, TX. J Transp Geogr 67:33–52. https://doi.org/10.1016/j.jtrangeo.2018.01.008

Zemtsov SP, Baburin VL, Kidyaeva VM (2018) Innovation clusters and prospects for environmental management in Russia. Geogr Nat Resour 39(1):10–15. https://doi.org/10.1134/S1875372818010002X

Zemtsov SP, Smelov YA (2018) Factors of regional development in Russia: geography, human capital and regional policies. Zhournal Novoi Ekonomicheskoi Assotsiacii 4(40):84–108. https://doi.org/10.31737/2221-2264-2018-40-4-4

Zeng C, Zhang A, Xu S (2016) Urbanization and administrative restructuring: a case study on the Wuhan urban agglomeration. Habitat Int 55:46–57. https://doi.org/10.1016/j.habitatint.2016.02.006

Zhao SX, Guo NS, Li CLK, Smith C (2017) Megacities, the world's largest cities unleashed: major trends and dynamics in contemporary global urban development. World Dev 98:257–289. https://doi.org/10.1016/j.worlddev.2017.04.038

Zohn Y (2019) Unthawed: Post-Cold War economic ties between Kaliningrad and Europe. J Balt Stud 50(3):327–349. https://doi.org/10.1080/01629778.2019.1590434

Anna Mikhaylova holds a doctoral degree in Human geography from Immanuel Kant Baltic Federal University (2017). She is currently a senior research fellow at the Center for Baltic region research at the Regional development institute of the Immanuel Kant Baltic Federal University,

Russia. Over the past 10 years she has published over 100 papers on regional growth and resilience in journals of Springer-Nature, Elsevier, Inderscience, Walter de Gruyter, Pleiades Publishing, etc. Her current research interests are focused on place-adaptive policies of socio-economic and innovation development, giving a particular attention to the issues of security and sustainability.

Dmitry Hvaley is a master's degree student in Regional policy and territorial management at the Immanuel Kant Baltic Federal University. He holds the position of research assistant at the Laboratory of innovation geography at the Regional development institute of the Immanuel Kant Baltic Federal University, Russia. Being an early-stage scholar, Mr. Hvaley has already co-authored numerous publications in high-ranked journals. His research focus is on spatial patterns of human activity—factors of borderland, coastalization, metropolizalion, etc.

Energy Development Alternatives—From Global to Local Scale

Eugene V. Krasnov and Izumrud R. Ragulina

Abstract The article discusses the prospects of using different energy sources as well as the problems of balancing generation, consumption and restoring energy resources. It also reveals innovative methods for assessing bioenergy potential by increasing the energy equivalent of photosynthesis and energy production as well as the use of biomass energy: the introduction of the most promising and scientifically based forest-forming species into forest-deficient regions, the cultivation of high-calorie plants, the use of short-cycle willow plantations as biofuel. The authors present constructive proposals for maintaining the geo-energy balance based on the correct determination of the structure and parameters of the regional energy generating industry. The proposals take into account the universal human requirements for sustainable development, the right choice and implementation of the optimal management methodology for the regional energy complex, including its bioenergy potential It is noted that together with the growth in the volume of consumed resources and the advancement of new technologies, the efficiency of energy management decreases. It results in the pollution of air, water, soil, etc. Therefore, it is necessary to introduced measures of sustainable development. The authors introduced the concept of the energy potential of society.

Keywords Traditional and modern sources · Multi-scale analysis optimisation model · Bioenergy perspectives · Generation and consumption balance · International cooperation

1 Introduction

Solar energy has been and remains the main traditional source of humanity life for many centuries. Initially, people heated themselves by fires and burnt plant biomass

---

E. V. Krasnov
Immanuel Kant Baltic Federal University, Kaliningrad, Russia

I. R. Ragulina (✉)
Geographic Sciences, Baltic State Fishing Fleet Academy of Kaliningrad State Technical University, Kaliningrad, Russia

© The Author(s), under exclusive license to Springer Nature Switzerland AG 2021
W. Leal Filho et al. (eds.), *Innovations and Traditions for Sustainable Development*, World Sustainability Series, https://doi.org/10.1007/978-3-030-78825-4_2

(mainly wood). Later they began to use the energy of water and wind. Dependence on renewable sources consistently decreased with the use of fossil fuels—coal, oil and natural gas. Consumers got access to a large amount of energy, despite the costs of its production and transportation. However, at the end of the twentieth century and the beginning of the twenty-first century, the trend changed again in developed countries: the consumption of coal, oil and even nuclear energy began to decrease and the interest in the alternative sources of energy started to grow, including personalized solar vortex power plants (Solovyev 2018).

2 Methodology

A comparative analysis of the production and consumption of energy in the world and in Russia, in particular, revealed a continuing trend towards the use of traditional energy sources (Sadovnichy et al. 2012). However, in Russia, unlike other countries, natural gas dominates as a source of energy, rather than coal. In addition, the role of atomic energy is significant (Alexanrovich et al. 2011). However, at the regional level, renewable energy sources are becoming increasingly important. Special attention is paid to the problems of generation and consumption of energy in the Kaliningrad region—the Russian exclave on the South-Eastern coast of the Baltic Sea, optimisation of the energy complex and energy conservation (Beley 2018).

The management of the regional energy complex should be carried out on the system level. At the same time, it is possible to save financial resources during the next few years solely through the implementation of the methodology aimed at the optimal management of power consumption without significant capital investments. Gnatyuk (2010) proposed a sustainable energy three-level of generation energy system clustering several hydrotechnical, thermal and wind sources). The proposed approach is based on the cumulative effect of the system of distribution of energy sources by capacity (H-distribution) (Fig. 1).

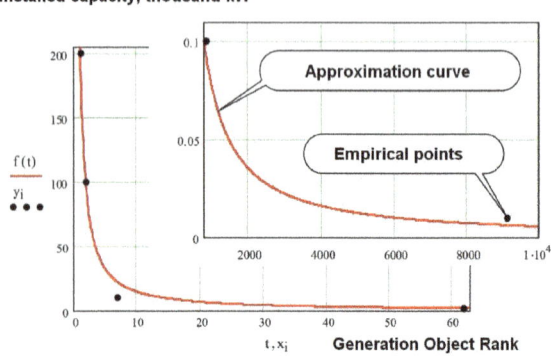

Fig. 1 The principal model of optimisation parameter-based distribution energy generation facilities in the regional power system energy, Gnatyuk (2010)

He developed a methodology for a systematic analysis of the rank distribution of energy consumption in the Kaliningrad region for the next 5–7 years in order to achieve a more balanced generation (Gnatyuk 2010). In a long-term investment strategy, the correct determination of the structure and parameters of the regional generating complex is of key importance, while the correct choice and implementation of the methodology for optimal management, including its bioenergy potential is of vital importance for the universal requirements of sustainable development.

Uncontrolled growth of power consumption is one of the main destabilizing factors in the development of the regional generating complex (including the Kaliningrad one). The efficiency of capital investments in the development of generation decreases over time, and new needs for electric power in the absence of energy saving factors continue to grow steadily almost linearly.

The algorithm optimal management of the energy complex should be carried out by the regional government at the system level within the framework of a related methodology in several stages (Fig. 2).

Traditional and innovative methods for assessing bioenergy potential by increasing the energy equivalent of photosynthesis and energy production are presented as well as the use of biomass energy: the introduction of the most promising and scientifically based technologies for forest management, the cultivation of high-calorie plants, as well as the use of short-cycle willow plantations as biofuel.

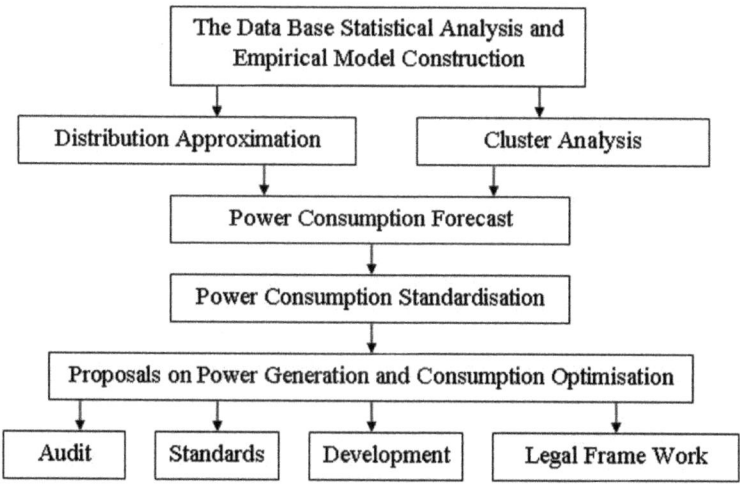

Fig. 2 The algorithm of regional energy consumption management in Kaliningrad District (by Gnatyuk 2010 with author modification)

3 Values of Hydrocarbon and Alternative Energy

Electricity production at a global scale is significantly more dependent on traditional coal—36.4% and natural gas—23.3% respectively. In Russia, natural gas still dominates—46.6% with nuclear sources ranking second—18.7% (Fig. 3).

Russia is planning to increase the construction of hydropower and renewable energy generation facilities. Yet, about 70% of primary energy is generated at CHPs operating on traditional fuels (Fig. 4), and the share of renewables is only 0.02% of the consumed energy resources.

From the comparison of the presented data, it is possible to see large differences between the consumption of energy sources in Russia and in the world. For example, the cost of natural gas in Russia exceeds 53.7%, while on a world scale it is only 24.2%.

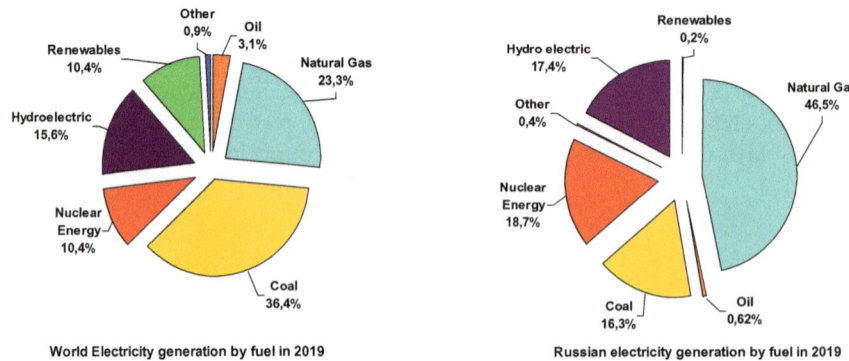

Fig. 3 Electricity generation by fuel in the world and in Russia, 2019. *Source* Authors, according to Statistical Review of World Energy (2020)

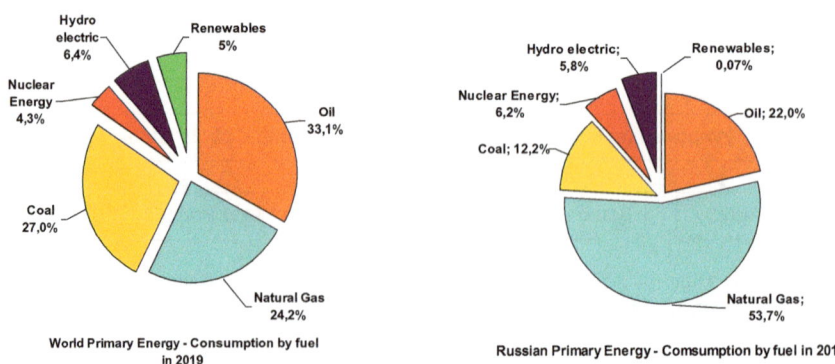

Fig. 4 Primary energy consumption by fuel in the world and in Russia, 2019. *Source* Authors, according to Statistical Review of World Energy (2020)

4 Traditional Assessment Approaches

According to Ragulina (2008) the total potential of biomass resources in the Kaliningrad region reaches 4591 TJ (T = 10^{12}) per year, which is equivalent to 156.7 thousand tons of standard fuel (fuel equivalent). A more complete use may lead to an increase in its share in the region's heat power industry from 0.4 to 19.2% (Table 1).

Table 1 demonstrates the ratio of the share of different bioenergy resources in the bioenergy potential of the Kaliningradregion: forest residues—20.0%, industrial wood residues—16.0%, willow plantations—17.6%, peat—12.7%, biogas—8.8% and municipal solid waste—24.9%.

Based on the integrated assessment of the bioenergy potential and taking into account all types of biofuel in the Kaliningrad region, four types of territories have been identified depending on the bioenergy potential: high bioenergy potential—250–350 TJ, medium—150–250 TJ, low—100–150 TJ and the lowest—less than 100 TJ (Fig. 5).

The highest bioenergy potential was calculated for the regional centre—the city of Kaliningrad and its environs. A relatively high potential is characteristic of the northern and central parts of the region (Neman, Nesterov, Polessk, Pravdinsk, Slavsk and Chernyakhovsk municipalities). This is 43.7% of the entire territory of the region. The western and north-eastern parts of the region have medium bioenergy potential: Bagrationovsk, Guryevsk, Zelenogradsk, Krasnoznamensk and Ozersk Municipalities, comprising 43.9% of the region's territory. Low bioenergy potential is characteristic of Gvandejsk and Gusev municipalities (9.6%), and the lowest indicator is for the Baltijsk, Svetliy and Svetlogorsk urban districts (1.3% of the region's territory).

The total regional potential of biomass resources is 4591.4 TJ per year. The use of biomass in the region's thermal power industry will improve the ecological situation in the region, reduce CO_2 emissions by 431.6 thousand tons, and SO_2—by 2167 tons

Table 1 Bioenergy potential of the Kaliningrad region

Type of biofuel	Potential, TJ, (T = 10^{12})	% of consumed thermal energy	Equivalent, thousand tons			Emission reduction (coal equivalent), t/year	
			Coal	Fuel oil	Reference fuel	CO_2	SO_2
Forestresidues	917.5	3.85	36.6	21.9	31.3	86.249	433.08
Industrial wood residues	732.9	3.07	29.2	17.5	25.0	68.896	345.95
Willowchips	810.0	3.39	32.3	19.3	27.6	76.140	382.32
Peat	585.8	2.46	23.3	14	20.0	55.061	276.47
Biogas	404.1	1.69	16.1	9.6	13.8	37.982	190.72
MSW	1141.1	4.78	45.5	27.2	38.9	107.265	538.61
Total	4591.4	19.24	182.9	109.6	156.7	431.593	2167.1

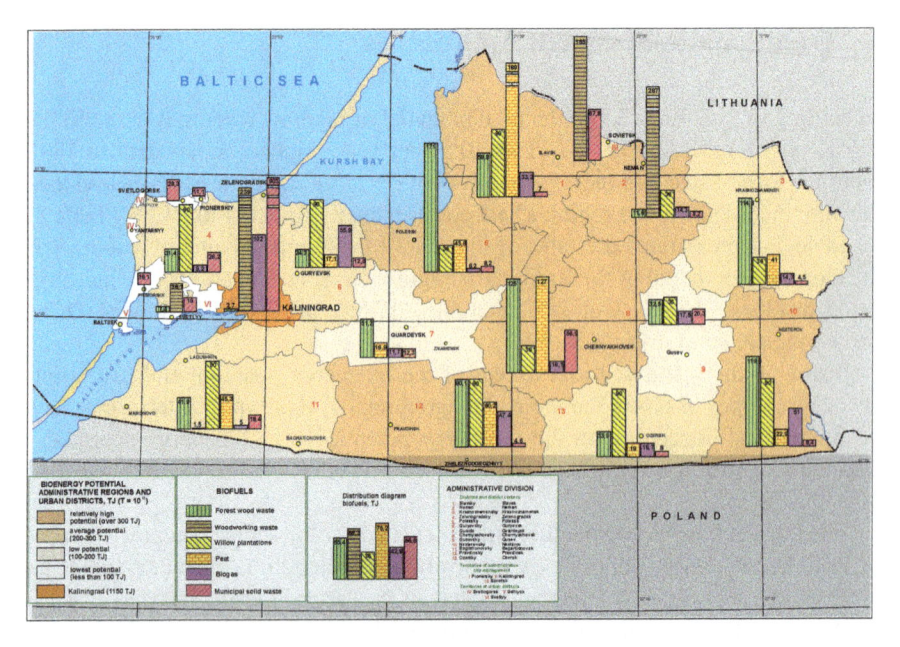

Fig. 5 Integral assessment of the Kaliningrad region bioenergy indicators. *Source* Author

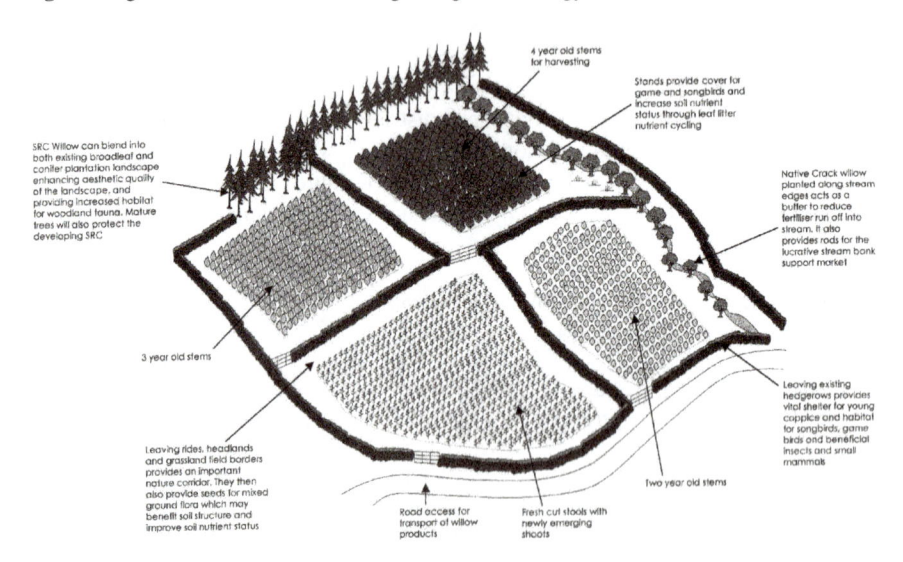

Fig. 6 An example of short rotation coppice plantation design in Great Britain. *Source* www.salix.org.uk

per year, as well as reduce the use of coal by 183 thousand tons and fuel oil—by 110 thousand tons (Krasnov 2003; Ragulina 2008).

The review article by Teterina and Shakun (2014) provides only some estimates of the potential of alternative energy sources in the Kaliningrad region. The authors rightly believe that the greatest potential is presented by the biomass of agricultural waste and municipal solid waste.

5 Innovative Ways of Using Bioenergy

Modern renewable energy generation includes high-tech methods of using biomass, hydropower with a capacity of up to 10–30 MW (hydroelectric power station); geothermal (Geo-TPP), wind (WPP), solar (SPP), tidal (TPP), and other types of generation facilities that are quickly developing in Western Europe, the USA, China, Japan, Australia and New Zealand.

The energy equivalent of photosynthesis and energy production do not remain constant as the calorific value of the main plant species changes from season to season (Dyakonov et al. 2019). The Legumes (astragalus, lupine) stand out among the «productive» forces. The generative parts of plants (fruits, seeds, nuts) turned out to be the highest-calorie ones.

Kaliningrad is a forest-deficient region. Forests cover only 19% (300, 8 thousand ha) of its territory. This is significantly lower than in Lithuania- 35%. Kaliningrad-forests belong to the zone of mixed coniferous-deciduous forests where the oak and the linden prevail. There are other forest-forming species—the beech, hornbeam, ash tree, and elms. The Weymouthpine, red oak and giant Thuja are considered to be the most promising species to be introduced to Kaliningradforest (Drobiz 2019).

Maintaining the positive dynamics of the forest cover and increasing the share of valuable species in the environmental and forestry activities of the region is the basis for the balanced development of the region. It is important to combine traditional and innovative techniques in forest management. It is proposed to make a wider use of phytomelioration techniques, planting willow polysadas around agricultural fields, taking into account the importance of restoration of the regional ecosystem as a whole, land reclamation, and the restoration of the microclimate of forest landscapes.

Short rotation willow plantations are an important source of renewable fuel. The willow crop harvested annually on 1 ha of arable land is equivalent to 11 tons of coal. A rod is mowed (4–7 m high) on average once every 3–4 years. Then the stems are processed into wood chips, which are burnt to generate heat and electricity. The total lifetime of a willow plantation is 24–25 years, then the soil has to be restored and prepared for a new plantation of willow or another crop. The willow plantation area should be sufficient so that about 1/4 of its total area can be felled annually, and 3/4 of the area (Fig. 4) would be occupied by young plantings and plantings of tree biomass to be felled next year (Randerson et al. 2011).

In many North European countries, fast-growing species of willow (*Salix viminalis, S. Dasyclados*, hybrids *"Tora"*(*S. viminalis x S. schwerinii*) are the basis for short-cycle reproduction of energy biomass on lands irrigated with under-treated sewage (Wishowski, 2003). Willows absorb more than 30% of nitrogen and 40% of phosphorus, providing a decrease in the concentration of nitrogen in the groundwater to 3–7 mg / l, phosphorus—0.1–0.4 mg/l and BOD_5 mg O_2/l up to the standards of treated water discharged into natural water bodies (Ragulina 2010).

The dense and very deep willow root system is suitable for absorbing not only nutrients, but also metals contained in sewage sludge. Under ideal combustion conditions, most of the nitrogen is released as NO_x, while heavy metals will remain in the ash. This is an important condition for the use of sewage sludge when growing willows has an environmentally beneficial effect (Ragulina 2010). When burnt, wood emits the same amount of CO_2 as the plant used in the growth process (CO_2-neutral fuel).

6 New Opportunities: International Cooperation

Challenges and opportunities of biofuels development were discussed at several Danish-Polish workshops on biomass for energy, organized by Gdansk University of Technology (organizers—Piotr Kowalik (Poland), Bjarne Rasmussen (Denmark), Lutz Mez (Germany et al.). These workshops were a good example of cooperation between scientists and practitioners who are striving to combine their efforts across borders to solve regional environmental problems. Special importance was given to landscape and energy planning, wetland conservation, energy efficiency and energy saving (Krasnov 2007a).

These workshops were held following the official agreement of cooperation between the Vojevodship of Gdansk (Poland) and Storstroem County (Denmark). Another important document was an agreement of cooperation on environmental studies signed by Roskilde University Centre, Free University of Berlin and Politechnical University of Gdansk.

The Baltic Sea region benefitted from the extension of cooperation agreement between Poland, Denmark and Germany. More partners from Sweden, Estonia, Russia, Great Britain and other countries joined their efforts in tackling environmental problems of the region (1991–2003).

The most efficient and environmentally friendly biofuels are produced from carbon-containing materials of plant and animal origin, including marine plankton and benthic algae. Smokeless furnaces are increasingly used to generate heat and electricity at modular automated power plants. Currently, the production capacities of modern wind power plants exceeds hundreds of megawatts. Over the past 20 years, their production has grown more than 20 times and annually increases by 15–25%. The power of photovoltaic converters grows annually by 20–30% (Krasnov et al. 2016).

It is generally advisable to motivate small and medium-sized businesses to use alternative energy thus reducing greenhouse gas emissions (CO_2, CO, CH_4, NO_x,

SO_2, etc.), minimizing the negative effects of climate change and even smoothing out the inevitable conflict of interest between energy markets and energy resources.

Analyzing the cycles of matter and energy as showed by Timofeev-Resovsky (1972) and many other authors mankind can increase the absorption of solar energy by vegetation, by increasing the density of green cover on all suitable areas of the earth's surface and in water bodies. It is possible to increase the biological productivity of the entire biosphere by using more 'productive' plant species for these purposes. With a general increase in the density of green cover, the density of the animal population of the Earth will also increase. Balancing energy consumption and restoration of resources will result in deriving more benefits from the biosphere material-energy cycle, change and improve the state of the biosphere and its communities (Hagett et al. 2005, Krugman 1999).

7 Geoenergy Balance: From Ideas to Constructive Proposals

In the arid zone of Transnistrian Moldova, the imbalance of the land fund is especially pronounced. Although forests perform nature conservation functions, biodiversity is declining everywhere in the region. The basis of a new method for assessing the effectiveness of the functioning of forest ecosystems Marunich (2017) laid down a number of sequential procedures: determination of the capacity of a forest ecosystem using thermodynamic indicators (solar energy, precipitation potential, phytomass, etc.), energy spent on reforestation (in joules) using nature-friendly technology. The best option is reforestation by self-sowing and seedlings of oak petiolate with a significant reduction in the proportion of anthropogenic energy by reducing the use of heavy equipment and using a special software information system.

In areas of excessive moisture, sapropel is used—a valuable organic substance, consisting mainly of carbohydrates, proteins and fats. In Japan, it is used to produce food, cattle feed and fertilizers. In the confectionery industry, it is used as a substitute for gelatin and agar. In the near future, biotechnological engineers should take up posts at the "exits" from the large biosphere cycle, whose main task will be to capture biogens in the form of energetically more useful molecules of proteins, fats and carbohydrates (Timofeev-Resovsky 1972).

In developed countries, the system of national accounts (SNA)—a set of interrelated indicators of the most common results and proportions of economic development was developed following new macroeconomic policies and the development of measures to regulate the market economy. However, in many developing countries, there is a discrepancy between the SNA with the real level of public welfare. It can be explained by a number of reasons. GDP growth comes from external resources. Over the past hundred years, the consumption of zinc, chromium copper, magnesium, molybdenum has grown 2–8 thousand times. Their extraction required enormous energy consumption, accompanied by the destruction and movement of giant rock masses, which led to an increase in anthropogenic load and an imbalance in the functioning of natural systems (Krasnov 2007b).

Simultaneously with the growth in the volume of consumed resources and the advancement of new technologies, the efficiency of management decreased. If in the pre-industrial era one calorie was spent on manufacturing a product containing 100 cal of energy, then now 10–40 cal are spent on rice production, from 50 to 1000 cal—on fruit growing, 3.5 thousand on the production of beef, and greenhouses 'consume' 50 thousand calories. At the same time, any type of production is accompanied by air, water and pollution, soil, etc. A significant share of GDP does not reflect the production of goods and services, but the compensation costs for combating losses and problems in the system of environmental and economic relations.

8 Energy Potential of Society

The amount of energy available and involved in the turnover can be characterized by the concept of energy potential—the totality of available means and capabilities. The energy potential of society means not only energy of any kind but also the possibility of saving energy resources in order to achieve sustainable development (Sustainable Development 2005).

The systemic importance of energy integrated into socio-economic activity was marked during the crisis in the late 1970s. Odum noted that although the role of the ecosystem approach for global development cannot be overestimated, the success of economic decisions and actions depends not only on social or economic factors but also on the capacity of natural ecosystems (Odum 1978). Energy entering the system is converted into a product stream with a 'feedback loop', as a result of which the production of high-quality energy continues.

Ecosystems that receive additional energy from natural sources have the highest productivity. The accumulated biomass characterizes the efficiency of the solar energy conversion. The potential energy flows necessary for the functioning of ecosystems are almost unlimited. The energy produced by the ecosystem on output autocatalytically replenishes its consumption in the input stream and inside the system without restrictions, which leads to an increase in the total energy. At this stage, they are still not taken into account.

Self-regulation can be carried out through energy balance management: if energy input exceeds the consumption, then the system works better. If consumption exceeds energy input, then the system degrades. In some ecological-economic models, no special governing body was provided. However, with their help, only quantitative characteristics are determined—areal, height, volumetric, and dimensional. Increasingly, environmental performance is expressed in terms of an "energy currency" (in joules) that is not subject to fluctuations and, therefore, is universal (Sozinov et al. 1985).

The geoenergy approach was proposed by Tomsk School of Professor Pozdnyakov (2008), who assessed the functioning of the agroecosystem of the steppe zone in Central Kazakhstan and Siberia. On the basis of the intersectoral data, energy equivalents of energy carriers (fossil fuels) were calculated; energy consumption in terms of labour resources and the cost efficiency of production were determined. In the future, the contribution of natural energy to agricultural production should be taken into account.

Academician Vernadsky (1932) was among the first researchers to study the problem of the energy expression of natural productive forces based on the quantitative accounting of energy costs. He considered it necessary and possible to fully quantify the potential energy of the country, especially so big as Russia.

Harnessing the growth of energy consumption and striving for sustainable development are more urgent tasks than tackling the problem of the lack of biological resources (Marfenin 2006). The general understanding of the 'limits of growth' poses a no less global task for the United Nations—the development of a new market methodology (Weizsaecker et al. 1997), taking into account the ratio of energy revenues and costs and the efficiency of environmental management. On this basis, it will be possible to develop technologies of the future and to ensure not only the achievement of a reasonable level of business profitability and the well-being of society, but also its biogeoenergy stability and security.

9 Conclusion

A comparative multi-scale analysis of traditional and alternative energy sources has revealed numerous imbalances in their use at the global, national and even municipal levels. Recent progress in the energy sector of developed countries is based on the growing consumerism and the dependence of the population on it. In developing countries, cheap labour of the population is widely used. In the market economy, cash or digital money is a measure of the value of goods and services. However, acting as a measure of the value of a commodity, money itself became a commodity whose value is determined not only by the ratio of supply and demand, but also by financial speculation. Developed countries receive huge profit from issuing their own paper currency, although besides money, natural energy resources, products, intellectual and physical labour, and other systemically important parameters should be taken into account (Kotlyakov 2011; Zaburaeva 2014).

References

Alexandrovich IM, An NV, Dolgikh EV, etc (2011) Global energy development/under the editorship of O.L. Kuznetsova and others. Russian Academy of Natural Sciences and others. Economics, Moscow 124p

Beley VF (2018) Analysis of the energy sector of the Kaliningrad region, taking into account the trends in the development of energy in the world. VI Baltic Maritime Forum: materials of the International Baltic Maritime Forum: in 6 volumes. BFFSA Publishing House, Kaliningrad, pp 838–846

Drobiz MV (2019) Mapping the spatial and temporal dynamics of natural-economic systems of the Kaliningrad region. Geography and cartography 80(1, 2010):136–145

Dyakonov KN, Reteyum AYu (2019) The role of geography in the transition to sustainable development. Ecological and geographical problems of the transition to a green economy. Materials of the International Scientific Seminar and the 23rd session of the Joint Scientific Council on Fundamental Geographical Problems at the IAAS and the Scientific Council on Fundamental Geographic Problems of the Russian Academy of Sciences, 2019. pp 21–28

Environmental Management and Sustainable Development (2006) World Ecosystems and Problems of Russia. KMK Scientific Press Ltd., Moscow, p 448

Gnatyuk VI (2010) On the development strategy of the regional electric power complex of the Kaliningrad region. Baltic Region 1(3):78–91

Haggett P, Skopin A (2005) General geography: global synthesis. Dorset Press, London, pp 2005–2352

Kotlyakov VM (2011) Innovative and integrated processes in the regions and countries of the CIS. Media-Press, Moscow, 216p

Krasnov EV (2007a) Environmental protection and ecology, training in quality certification systems. Kaliningrad region: on the way to regional MBA/MPA programs: Materials of training courses of the Integrated cross-modular curriculum of postgraduate education in the field of business administration and public administration. Kaliningrad: Publishing House of the RSU named after I. Kant, 2:272–281 (2007)

Krasnov EV, Borodin AI (2007b) The role of environmental factors in the system of national accounts: energy approach. Problems of managing the socio-economic processes of the regions: Materials III Int. scientific prakt. conference -Kaliningrad: Publishing House of the Higher School of Economics, pp 153–164

Krasnov EV, Ragulina IR (2003) Possibilities of using short rotation willow plantations for bioenergy development in the Kaliningrad region. Proceedings of the 8th Polish-Danish Workshops on Biomass for energy. Gdansk, pp 67–71

Krasnov EV, Ragulina IR (2016) Renewable energy development: history, modernity and forecasts. IV Baltic Maritime Forum: Proceedings of the International Maritime Forum, May 22–28, 2016, Kaliningrad: BFFSA Publishing House, pp 946–953

Krugman P (1999) The role of geography in development. Int Reg Sci Rev 22(2):142–161

Marfenin NN (2006) Sustainable development of humanity: a textbook. Publishing House of Moscow State University, M624p

Marunich NA, Kochurov BI, Khaziakhmetova YuA, Krasnov EV (2017) Environmentally balanced land structure and energy efficiency of forestry in Transnistria. Geog Nat Res 4:197–202

Odum H, Odum E (1978) Energy basis for man and nature: trans. from English, Moscow, Progress, 379p

Pozdnyakov AV, Shurkina KA (2008) A new methodological approach to the analysis of the functioning of agroecosystems. Bulletin of the Tomsk State University, Tomsk, vol 316. pp 206–212

Proceedings of the VII International Scientific Conference «Innovations in Science and Education-2009». Kaliningrad: Kaliningrad State Technical University, October 20–22, 2009, in two volumes, part 1, (2009) 425p

Russia and its regions: integrated potential, risks, ways of transition to sustainable development (2012) KMK Scientific Press Ltd., Moscow, 448p

Rational nature management: international programs, Russian and foreign experience (2010) KMK Scientific Press Ltd., Moscow, p 412.

Rio + 20 and energy-ecological development of the world in the XXI century (2012) Russian Academy of Natural Sciences, Moscow 332p

Ragulina IR (2010) Prospects for the creation of willow plantations to improve the condition of soil and water resources (on the example of the Kaliningrad region). Materials of the III Int. scientific prakt. conf. «Ideas V.V. Dokuchaev and modern approaches to the study of the natural environment, the development of regional socio-ecological problems», May 27–28, Smolensk: Universum, pp 84–91

Ragulina IR, Krasnov EV (2008) Bioenergy potential of sustainable development of the Kaliningrad region and its geoinformation support. Sustainable development of territories: GIS theory and practical experience: Materials int. conf. «InterCarto-InterGIS 14», T. 2, - Saratov (Russia), Urumqi, Lhasa (China), pp 38–46

Randerson PF, Moran C, Bialowiec A (2011) Oxygen transfer capacity of willow (*Salix viminalis L.*). Biomass Bioenergy 35(5):2306–2309. https://doi.org/10.1016/j.biombioe.2011.02.018

Sadovnichy VA, Akayev AA, Korotaev AV, Malkov SYu (2012) Modeling and forecasting world dynamics. Fundamental Scientific Council. researched Presidium of the Russian Academy of Sciences «Economics and Sociology of Knowledge». ISPI RAS, Moscow 360p

Solovyev AA, Solovyev DA, Shilova LA (2018) Solar-vortex power plants: principles of effective work and technical requirements on the preparation of initial data for design. MATEC Web Conf 196:04075-1-04075–6

Sozinov AA, Novikov YuF (1985) The energy price of the industrialization of the agricultural sector. Nature 5:11–19

Statistical Review of World Energy (2020) https://www.bp.com/en/global/corporate/energy-eco nomics/statistical-review-of-world-energy.html

Sustainable Development of Agriculture and Rural Areas (2005) Foreign experience and problems of Russia. KMK Scientific Press Ltd., Moscow, pp 2005–2615

Teterina N, Shakun V (2014) The use of renewable energy sources in the Kaliningrad region: current state and prospects. Vestnik of Baltic Federal University 1:167–174

Timofeev-Resovskiy NV (1972) Biosphere and humanity. J UNESCO Courier 1:29–32

Vernadsky VI (1932) The problem of time in modern science. News of the USSR Academy of Sciences, 7 series. Department of Mathematical Natural Sciences. Geology Series 4:511–541

Weizsaecker EU, von Lovins LH (1997) Factor four: doubling wealth—halving resource use. The New Report to the Club of Rome. L.

Wishowski R (2003) Biomass for energy in Sweden. In: Proceedings of the 8th polish-danish workshops on biomass for energy. Gdansk, pp 201–210

Yafasov AY (2010) Modernization: innovative solutions in economics and management: Sat. Art. / North-West Academy of Public Administration, St. Petersburg, p 312

Zaburaeva HSh (2014) Geoecological potential of sustainable development: the evolution of concept and its structure. Bulletin of the Baltic Federal University. I. Kant, No 1:175–181

Eugene V. Krasnov Dr. Dr. habil. Professor at Immanuel Kant Baltic Federal University, Kaliningrad, Russia.

Izumrud R. Ragulina PhD in Geographic sciences, Professor at the Baltic State Fishing Fleet Academy of Kaliningrad State Technical University, Kaliningrad, Russia.

Urban Green Innovation Ecosystem to Improve the Environmental Sustainability

José G. Vargas-Hernández, Karina Pallagst, and Jessica Davalos

Abstract This chapter is aimed to analyze the relationships between environmental sustainability, urban ecosystems and green innovation. The purpose of this paper focuses on the assumption to reviewing the work done on green areas innovation in urban settings and identifying the social, niche, incremental and radical types of innovation, as well as the dimensions of sustainable urban ecosystems associated with the concept, encompassing the risks inherent to administrative, technological and operational innovations. The method employed is the critical analytical review of literature and further discussion on the issues focusing the city´s experience on managing the formulation, generation, development, implementation and evaluation of new behaviors and ideas in green innovation. It is concluded that the green innovation is directly related with the environmental sustainability and urban ecosystems. The interest of this analysis lies in providing support to urban settlements in managing the risks inherent in green area innovation, incremental or radical as a community's management would experience in relation to the environmental sustainability in urban ecosystems.

Keywords Environmental sustainability · Urban ecosystems · Green innovation

1 Introduction

Cities and urban ecosystems are facing important global challenges at local scale for a sustainable development. Academic research on green area innovation is a new shift in the environmental debate in public and private city spaces towards social

---

J. G. Vargas-Hernández (✉) · J. Davalos (✉)
University Center for Economic and Managerial Sciences, University of Guadalajara, Periférico Norte 799 Edif. G-201-7, Núcleo Universitario Los Belenes, Zapopan, Jalisco 45100, México

K. Pallagst
IPS Department International Planning Systems, Faculty of Spatial and Environmental Planning, Technische Universität Kaiserslautern, Pfaffenbergstr. 95, 67663 Kaiserslautern, Germany
e-mail: karina.pallagst@ru.uni-kl.de

good and environmental value fostered by innovation and new technologies (Berger ET AL. 2007). However, empirical literature of innovative urban green spaces is very scant.

The unit of analysis in the present study is green areas innovation projects and its relation and applications for environmental sustainability, since theory and practice of green innovation has become a strategic priority for urban settlements and local communities. Moreover, green innovation is one of the key factors of community development, government management and city governance to achieve economic growth, social development, environmental sustainability, and a better quality of life. Community initiatives are independent of government but can influence government actions to benefit low-income groups under a framework of governance. Local governments supporting community and non-governmental organizations in a wide range of urban community programs contribute to improve basic services, housing, environmental sustainability management, infrastructure, and micro-finance for enterprise development (Boonyabancha 2004; Stein 2001).

Urban innovation ecosystem is a comprehensive conceptual and theoretical framework to a structural urban living laboratory enabled to identify as the next generations of a smart city acting as agent of urban change (Hann and Hawken 2018). A territorial urban innovation ecosystem can be spatially specific for local and city governments working close with other partners such as universities, business, civil society and organizations, communities, etc. (Selada 2017). Local governments may propose actions in urban planning aligned under a policy process linking local intervention policies on funding, resources and tools aimed to improve its environmental sustainability, capitalizing on the mitigation and adaptation policies to leverage climate change efforts (Lombardi et al. 2018).

The article is organized in the following way. Firstly, the relation between green innovation and environmental sustainability is presented, followed by a description of the different innovation types. After that, the role of local governments in the context of green innovation application is shown, to finally present the conclusions from the analysis.

2 Green Innovation and Environmental Sustainability

To address urban development challenges, innovative urban planning and design must have a green innovation orientation interconnected and innovative development stages and processes (Adams 2006; Blewitt 2012). The process of urban innovation is inherent to the practice of sustainable development. The conceptual framework for green urban area innovation is based on a multi-faceted process where the key types of environmental development are in interaction, like natural resources, materials, energy, and pollution, and others, which have an impact on the environment at different stages: analysis, formulation, implementation, disposal and evaluation (Roy et al. 1996).

Fig. 1 Innovation diagram. *Source* Prepared by the authors

Environmental sustainability is an integral part of green area innovation requiring greater demands of ecological performance out of eco-efficient green technologies with reduction and mitigation of green negative environmental impacts.

Innovation in environmental sustainability modifies the competitive dynamics of urban spaces supported that innovation modifies the environmental sustainability impacts through the interaction with some features of the external environment (Moyano Fuentes et al. 2017).

Urban spaces driven by the different types of innovation that constitute enablers leading to higher levels of environmental sustainability and have potentially diverse possibilities to improve the ecosystem. The incentive systems for the different types of environmental sustainability innovation are affected by regulations (Kemp et al. 2000).

This analysis places more emphasis on the social, niche, incremental and radical types of innovation that are more connected to environmental sustainability. However, the various other types of innovation are still linked to further natural resource consumption and economic efficiency and growth at the cost of environmental sustainability (Hansen and Grosse-Dunker 2012).

In Fig. 1 it can be appreciated the different types of innovation that can be applied to accomplish environmental sustainability.

2.1 Social Innovation

Urban social innovation identifies urban resources for sustainable development, economic growth and social development (Cozens 2008; Bélissent 2010).

Social innovation of urban green areas improves the quality of life and attractiveness of the city and develops the ability to sustain human-dominated local ecosystem services.

Community greening innovation is a community-based tool to promote social learning, adaptive management and resilience of the green ecosystems (Tidball and Krasny 2007).

Green innovations in urban farming development as well exhibit the characteristics of social innovations contributing to a sustainable urban environment and urban food movements. Urban farming is using diverse technologies such as hydroponics or aquaponics to promote green and environmental innovation programs. Urban farming production serves to offer processing, cooking and selling fresh food produce to the local market and enhancing the food business with high quality, sustainable and innovative approaches. Urban farming as an innovation is an incubator that uses special forms for promoting new concepts of organizing urban life and consumption around sustainable food production. These special forms can be diverse integrated designs to reproduce built environments such as modular containers and components, greenhouses, and others.

Integration of community gardens through an innovative approach facilitates opportunities for residents to participate more actively in urban green space planning processes to provide ecosystems services, environmental and sustainable education, alternative and accessible forms for physical activities, bridges interactions between different social groups, enhance local ecological and environmental outcomes.

2.2 Niche Innovation

Environmental sustainable urban planning and development as a theoretical framework supported by transitional theory can be adapted to specific issues of urban land use and infrastructure. Theoretical analytical transitions are the result of interactions and learning processes between the levels of niche-innovations, socio-technical regimes and socio-technical landscapes aimed to improve performance and support from groups. The socio-technical regimes and landscapes can become destabilized due to inner pressures and tensions originated from outside that may result in niche-innovations. Niche innovations redefine the limits to environmental and ecological development goals in terms that decoupling economic growth from environmental degradation can be achieved (Smith et al. 2010).

2.3 Incremental Innovation

Incremental green area innovations are more related to the increasing use of existing key green dimensions such as eco-efficiency, the use of materials and processes with a lower environmental impact (Hellström 2007). Incremental green area innovations are characterized by small and incremental improvements of previous green area versions and their reliance on small changes on existing technologies, process, and others. Green areas research projects originated from business sources expected to respond to the market forces maybe more oriented by incremental innovation.

2.4 Radical Innovation

Radical green innovations include the use of new technologies, components, processes, and others, aimed to reduce and mitigate the overall environmental impact. Green areas innovations are radical if they are new and offer new features based on radical new technologies to the population of communities and neighborhoods. Local communities and governments can find and deal with local demands of citizens by responding with innovative projects in green areas in some specific local urban spaces, managing risks attached and providing greater benefits. More innovative projects are found to be associated with longer waiting times before having some results (Roberts and Hauptman 1987). In this sense, the radicalness of innovation in urban green areas needs to be conceptualized and operationalized in the varying degrees of dimensional urban innovation spaces (Roberts and Berry 1985).

Innovative undertakings require working where the organization, community or local government has more experience to be regarded as radical innovation. Green area projects are more or less radical innovative undertakings on some differentiated dimensions such as the development degree of technology employed, technology costs, technological uncertainty, technical inexperience, and business inexperience. Radical innovations in green areas are taking less time to be terminated and less tied to market needs represent risks and therefore are less supported by communities and local governments (Souder 1987).

Consistent theory on radical green areas innovation with empirical research experienced by practitioners is relevant for urban societies, economies and firms. Research on green innovation projects originated exclusively within R&D is more likely to be radical innovation. As well, green area projects originated from R&D in areas where there is less experience and practice and greater uncertainty are considered more radical innovations.

Radical innovation has implications for community, organizations, local governments and business development strategies pursued by firms opening new markets (Knight 1967; Roberts and Berry 1985). Also, radical innovation in green areas creates more change and has more impact on the community and neighborhoods. The amount of change attached to the green innovation projects where there is not extensive technological knowledge and experience make it very costly. However, radical innovation in green areas brings significant changes such as emerging, transforming and disappearing the old and conventional forms and processes (Kaplan 1999; Van de Ven Polley et al. 1999).

2.5 Urban Innovation Ecosystem

Urban innovation ecosystems are characterized according their social orientation towards environmental and sustainable development. This innovative green urban planning and design policies is the result of an interrelated concepts and issues of urban growth based on innovation ecosystems and environmental social sustainability built on green city. The urban innovation ecosystems aim towards design and implementation of innovative environmental and sustainable development process in cities and manage the portfolio of innovation assets, resources, facilities, information flows, knowledge, and others, through mechanisms such as the partnerships among economic agents and actors governing the access to users and developers.

Urban innovation ecosystems should promote participation of citizens and organizations, firms, and local governments in the planning and development of urban economic activities and utilities. Urban innovation can be harnessed in a highly politicized urban environment making urban governance more sensitive to urban green ecosystem dynamics facing social, ecological, environmental and sustainable uncertainties.

Technology applications in urban innovation ecosystems in federated platforms, but dependent on contexts and user locations can provide support for new urban e-services.

Urban innovation ICT programs can stimulate citizens, communities, social organizations, business and other societal applications, scaling-up real-life deployment projects to large-scale levels. Innovative ICT-based services with user-driven innovation link smart cities with experimental infrastructure and facilities the design of new applications and green services. The co-creation of green services in different areas requires the employment of innovative devices and customized sensors used by citizens.

ICT-based applications in urban innovation deploying broadband infrastructure plays an important role to enhancing citizens' quality of life. However, ICT has a limited role at processing and integrating real-time information where processes are not based to any great extent on handling transactions in such areas as local public e- government communication between authorities and citizens, innovation, entrepreneurship and social inclusion, education, and others (González and Rossi 2011).

Emphasis on the assessment of urban ecosystem services does not motivate and support the innovations of specific ecosystem services in urban and infrastructure development activities (Ahern et al. 2014). In this urban ecosystem of innovation, innovative companies coexist with research institutes, training and tech transfer centers, urban green areas, infrastructure and facilities.

Innovation ecosystems can be fostered and developed through collaboration frameworks by future internet test-beds integrated with elements of Living Lab environments. Cooperation frameworks include some elements of sharing access to experimentation facilities and technological knowledge resources. The method of Living Lab concept is used as a model for organizing and conducting innovation projects, programs and experiments. Living labs are organized around experimental research projects driven by internet integrated in smart city programs with common resources for research in green innovation ecosystems and environments.

Living Lab integrates services in urban contexts where users and citizens define and prioritize elements of urban cultural heritage and explore security and private issues for the safety of urban environments. The Living Lab-convergent service platforms are developed in a discovery-driven arena setting. Projects requiring radical technological innovation are associated with longer waiting times to complete and before there is a desired product (Roberts and Hauptman 1987; Schoonhoven et al. 1990).

Living Labs Interfaces projects engage users in co-creation innovation processes whereas Future Internet Experimentation (FIRE) projects involve end users and communities in assessing the socio-economic impacts of technological changes and controlled innovation technologies. FIRE and Living Lab projects use some models for resources sharing in experimentation and innovation opportunities in user's communities. Living labs provide product and services of real social innovation and improvements and lower the risks for future use based on public data of the city.

3 The Importance of Innovation at Local Government

The process of urban planning is aimed to create and develop an inclusive economy centered on innovative improvements at local level (Smart City Edinburgh 2011). The green urban planning has three priorities: the economic growth and development, social equity and the environmental sustainability. Local governments at the same time that other levels of federal and state authorities and agencies can take different approaches based on environmental economics and technologies innovations and political commitment to incentive and motivate involvement of actors and stakeholders to find specific innovative solutions to urban planning. Transforming green infrastructure models into action require new approaches of innovative planning and policy making of programs.

A sustainable city is supported by an urban planning and designing paradigm. Urban design and planning supported by local governments are urged to spur technical innovations, innovative practices and contribute to develop new business models and marketing strategies aimed to find green resource-efficient solutions for climate change adaptation and mitigation while restructuring and reinventing local economies.

Innovative forms of funding urban green areas innovation are the financial sources for support coming from the business sector, private donors, agencies promoting development, social organizations, non-governmental organizations, etc. To implement these innovative urban green space strategies, it is necessary to develop management strategies for financing maintenance and developing multi-functional urban green spaces, including technological and cultural innovations.

4 Conclusions

Urban areas are considered the object of innovation ecosystems able to empower citizens with capabilities to design, co-create and develop best urban working and living spaces. Innovation in urban green areas can provide solutions by defining quality of service and operational standards and results. In this chapter, it has been proposed and analyzed a model of urban innovation ecosystem based on four types of innovation: The social, niche, incremental and radical innovation types which can contribute to develop the structural fundaments to improve the environmental sustainability.

An innovation approach based on environmental integration of urban areas combining housing, urban green areas and business activities in a kind of eco-neighborhood could be supported by a public–private partnership.

One type of urban innovation ecosystems may constitute Public–Private-People-Partnership ecosystems aimed to co-create, experiment and validate scenarios supported technology platforms user-driven involving and providing opportunities citizens, small, medium and large businesses, corporations, local governments, and any other stakeholder.

This present analysis sheds light on environmental sustainability and urban ecosystems as dimensions of green innovation, which includes the management and use of natural resources, energy minimization, materials reduction, and pollution prevention.

Innovation strategies of cooperation supported by a network and strategic alliances ensure long-term viability of urban projects (Bélissent 2010). Urban farming can bring some ecological and environmental benefits besides the revenues. The innovative farming start-ups are supported by interdisciplinary academic knowledge, business experiences, urban developers, city and local government agencies and financial investors. Large scale high-tech commercial farming initiated by start-ups is facing specific urban land-use challenges regarding urban permitting, zoning, designing and constructing. Tacit knowledge is necessary for projects and contributes to the innovation (Nonaka and Takeuchi 2007) and creativity to negotiate transition towards integrated risk management where the participants are less likely to negotiate from entrenched positions (Pahl-Wostl et al. 2007). Farming as a green innovation requires interdisciplinary cooperative exchanges among networks of actors.

Green areas innovation offers other relevant opportunities for environmental sustainability, improvement of living conditions, waste reduction, new business creation, and others. In order to achieve these goals, it is required to set strategies, policies and targets to move forward and implement the projects. They should be supported by tools to measure environmental impact such as life cycle analysis at each stage of development.

Innovation is fostered through the collaborative innovation processes between the interaction of clusters and networks, therefore it is important to develop collaboration in green areas innovation with actors and agents of local social and ecological systems. Collaboration within the innovation is an ongoing interaction process between technology, research and applications development, validation and practical utilization. Cooperation relationships frameworks and synergy linkages should be developed between urban innovation policies, local government ICT, future internet research and open users.

Future internet technologies integrate augmented reality services in cultural heritage, safety and security with networks of video-cameras to monitor urban spaces. They can give support to a platform for monitoring and governance processes of social interactions, development of mobility behaviors, participatory civic decisions, learning natural and cultural heritage and delivery of e-government services. Future internet research is a competitive offer proving its added value to citizens/users. It enables co-creation of innovative scenarios by users and citizens who may contribute to build new applications and public data to the open city.

Local communities and governments should initiate selectively some green area projects with different degrees of innovation dimensions following a profile according to a strategic urban planning. Development of sustainable communities in smart cities can be supported by design, transfer and implementation through collaborative urban planning of innovative urban policies.

Future priorities for research in urban innovation should identify and develop principles of sustainable urban green planning and development to provide support to policy makers in local governments to design and implement mechanisms for more resilient communities and neighborhoods in cities. Green resilience innovation of urban life in contemporary cities can be supported in digital cities with the implementation of cyberspace to cities.

References

Adams WM (2006) The future of sustainability: re-thinking environment and development in the 21st century. Report of the Iucn renowned thinkers meeting, 29–31 January 2006. *IUCN*. Online at: http://www.iucn.org/members/future_sustainability/docs/iucn_future_of_sustanability.pdf

Ahern J, Cilliers S, Niemelä J (2014) The concept of ecosystem services in adaptive urban planning and design: a framework for supporting innovation. Landsc Urban Plan 125:254–259

Bélissent J (2010) Getting clever about smart cities: new opportunities require new business models. Forrester Research. Inc., Cambridge

Berger IE, Cunningham PH, Drumwright ME (2007) Mainstreaming corporate social responsibility: developing markets for virtue. Calif Manage Rev 49(4):132–157

Blewitt J (2012) Understanding sustainable development. Routledge

Boonyabancha S (2004) A decade of change: from the urban community development office to the community organization development institutes in Thailand. Empowering Squatter Citizen, 25–49

Cozens PM (2008) New urbanism, crime and the suburbs: a review of the evidence. Urban Policy and Research 26(4):429–444

González JA, Rossi A (2011) New trends for smart cities, open innovation mechanism in smart cities. European commission within the ICT policy support programmes

Hann H, Hawken S (2018) Introduction: innovation and identity in next-generation smart cities. City Cult Soc 12:1–4

Hansen EG, Grosse-Dunker F (2012) Sustainability-oriented innovation. In: Idowu SO, Capaldi N, Zu L, Das Gupta A (eds) Encyclopedia of corporate social responsibility: Heidelberg, Germany. Springer, New York. Available at SSRN: https://ssrn.com/abstract=2191679

Hellström T (2007) Dimensions of environmentally sustainable innovation: the structure of eco-innovation concepts. Sustain Dev 15(3):148–159

Kaplan SM (1999) Discontinuous innovation and the growth paradox. Strat Lead 27(2):16–21

Kemp R, Smith K, Becher G (2000) How should we study the relationship between environmental regulation and innovation? In: Hemmelskamp J, Rennings K, Leone F (eds) Innovation-oriented environmental regulation. ZEW economic studies, vol 10. Physica, Heidelberg

Knight KE (1967) A descriptive model of the intra-firm innovation process. J Bus 40(4):478–496

Lombardi M, Laiola E, Tricase C, Rana R (2018) Toward urban environmental sustainability: the carbon footprint of Foggia's municipality. J Clean Prod 186:534–543

Moyano Fuentes J, Maqueira-Marín M, Bruque-Cámara S (2017) Process innovation and environmental sustainability engagement: an application on technological firms. J Clean Prod 171:844–856. https://doi.org/10.1016/j.jclepro.2017.10.067

Nonaka I, Takeuchi H (2007) The knowledge-creating company. Harv Bus Rev 85(7/8):162

Pahl-Wostl C, Craps M, Dewulf A, Mostert E, Tabara D, Taillieu T (2007) Social learning and water resources management. Ecol Soc 12(2)

Roberts EB, Berry CA (1985) Entering new businesses selecting strategies for success. Sloan Manage Rev, 3–17

Roberts EB, Hauptman O (1987) The financing threshold effect on success and failure of biomedical and pharmaceutical start-ups. Manage Sci 33(3):381–394

Roy R, Wield D, Gardiner JP, Potter S (1996) Innovative product development. The Open University

Selada C (2017) Smart cities and the quadruple helix innovation systems conceptual framework: the case of Portugal. In: De Oliveira Monteiro S, Carayannis E (eds) The quadruple innovation helix nexus. Palgrave studies in democracy, innovation, and entrepreneurship for growth. Palgrave Macmillan, New York

Smith A, Voß JP, Grin J (2010) Innovation studies and sustainability transitions: the allure of the multi-level perspective and its challenges. Res Policy 39(4):435–448

Schoonhoven CB, Eisenhardt KM, Lyman K (1990) Speeding products to market: waiting time to first product introduction in new firms. Admin Sci Q 35(1)

Smart City Edinburgh (2011) Smart City Edinburgh. Recovered from http://www.edinburgh.gov.uk

Souder W (1987) Managing new product innovation. Lexington, MA Lex

Stein A (2001) Participation and sustainability in social projects: the experience of the Local Development Programme (PRODEL) in Nicaragua. Environ Urban 13(1):11–35

Tidball KG, Krasny ME (2007) From risk to resilience: what role for community greening and civic ecology in cities. Soc Learn Towards More Sustain World, 149–164

Van de Ven AH, Polley DE, Garud R, Venkataraman S (1999) The innovation journey. Oxford University Press. New York

Social Innovation for Sustainable Development

Skaidrė Žičkienė and Zita Tamasauskiene

Abstract The aim of this paper is to discuss the essence of social innovation, to disclose the importance of social innovation in the realization of Sustainable Development Goals (SDG), and to examine the relationship between social innovation and Sustainable Development in a panel of 11 European countries over the period 2015–2019. Social innovation is an extremely important factor dealing with societal challenges, from the national to global scales, as society is aging, people suffer from chronic diseases, alcohol, drugs, they are confronted with poverty and social exclusion, families face violence, and environmental problems also affect modern society. In this context, social innovation can become an effective way of addressing existing problems and preventing their deepening in the future as social innovation works across the globe and supports all 17 SDGs helping to create new ways and encouraging systems for sustainable development. The results of the regression analysis revealed that there is a positive and statistically significant relationship between sustainable development and social innovation. This finding is robust to the inclusion of different control variables into regression analysis.

Keywords Social innovation · Sustainable development · Sustainable development goals · Sustainable development goals index · Social progress index

1 Introduction

Great attention to social problems and the promotion of social innovation is given at the institutional EU level. The establishment of a quantitative target concerning poverty and social exclusion has been one of the major novelties introduced by the Europe 2020 Strategy: social innovation was presented as a key area for facilitating

S. Žičkienė (✉) · Z. Tamasauskiene
Institute of Regional Development, Vilnius University Šiauliai Academy, P. Višinskio str. 25, 76352 Šiauliai, Lithuania
e-mail: skaidre.zickiene@sa.vu.lt

Z. Tamasauskiene
e-mail: zita.tamasauskiene@sa.vu.lt

its achievement (Sabato et al. 2015). Social innovation gets reflection in a wide range of policy initiatives of the EC: European Platform against Poverty and Social Exclusion, flagship Initiative Innovation Union, Social Innovation Europe Initiative, Social Business Initiative, Employment Package, and Social Investment Package which frame and fund a new approach to social policies.

Despite the fact, that social innovation is gaining speed, there is no definite consensus about the term "social innovation", and a wide spectrum of definitions and interpretations can be found, this indicates the apparent novelty of social innovation as a construct (McGowan and Westley 2015). The diversity of social innovation encompasses all spheres of life and has very wide boundaries—from gay partnerships to new ways of using mobile phone texting, and from new lifestyles to new products and services (Mulgan et al. 2006), they can appear in education, health care, finance sector, take a form of movement or become a novel form of organization of work.

Social innovation is considered to be suitable for solving many of the most challenging problems specified in the 2030 Agenda for Sustainable Development, as social innovation initiatives are not limited to addressing the welfare and social inclusion challenges, but also concern issues of environmental protection and sustainable development (BEPA 2011). The role of forward-looking social innovators is key in seizing the SDGs' potential by understanding the integral and interdependent nature of the Goals, as innovators gain a whole-systems perspective and become enablers of a vitality by which society, ecologies, and economies can co-evolve and thrive (East 2018). Nonetheless, there is little knowledge of which SDGs social innovation already addresses (Eichler and Schwarz 2019). Challenges remain in measuring the longer-term outcomes of social innovation, creating social innovation impact metrics that can demonstrate to policymakers its effectiveness and sustainability in delivering services, meeting social needs, and addressing societal challenges (BEPA 2014; Caulier-Grice et al. 2012).

The paper examines the relationship between social innovation and sustainable development, using panel ordinary least square (OLS) effects regression for 11 new EU countries. Sustainable Development Goals Index (SDGI) was used as a dependent variable. Measuring social innovation (independent variable) is challenging and any measure is inevitably controversial. Only some attempts are found in the development of measurement of social innovation and social innovation index (Bund et al. 2013; Schmitz et al. 2013; Hoelscher et al. 2015; Benedek et al. 2016) Social innovation index was created by the Economist Intelligence Unit but computed only for 2016, that is why the Social Progress Index (PPI) was used as a proxy for social innovation. Results suggest that social innovation has a positive impact on sustainable development.

The rest of the paper is organized as follows: Sect. 2 describes the concept of social innovation and provides classification systems; Sect. 3 establishes the relationship between social innovation and SDGs; Sect. 4 presents measurements of social innovation; Sect. 5 presents the methodology; Sect. 6 discusses the results and finally Sect. 7 gives conclusions.

2 The Concept of Social Innovation and Its Classification

Social innovation, like any other innovation, starts with the idea driven by urgent changes. Innovation can be stimulated by the need for new products or services or new/different ways to fulfil these needs. Social innovation as innovation, in general, strives to implement novel ideas that create value for society and directly lead to societal benefit. The focus on social innovation stems from the apparent inability of existing institutions and traditional forms of the private market to deliver a higher quality of life, social justice, and well-being towards the vulnerable groups in society. In most cases, large, established non-profits, businesses, and even governments are producing social innovation (Phills et al. 2008), but innovation can originate from people not involved in social business, social innovation are often born with the initiative from committed citizens with social visions, will and drive, by those, who have the social problems or unsatisfied needs (Ellis 2010).

Social innovation has gained increasing attention, nevertheless, social innovation being described as ill-defined and as a buzzword (Pol and Ville 2009), as heterogeneous whose boundaries are unclear (Bonifacio 2014), and constitutes a broad and inconsistent range of meanings (Neumeier 2012). Social Innovation can concern conceptual, process or product change, organisational change and changes in financing, and it can deal with new relationships with stakeholders and territories in order to solve social problems (Forum on Social Innovations); it can get a shape of a principle, an idea, a piece of legislation, a social movement, an intervention, or some combination of them (Phills et al. 2008); it can be framed as a set of interesting citizen-driven experiments and practices or as a potential solution to some of the grand societal challenges of the twenty-first century (Slee 2019).

Bureau of European Policy Advisers (BEPA 2011) defines social innovation as new ideas that simultaneously meet social needs (more effectively than alternatives) and create new social relationships or collaborations, and are not only good for society but also enhance society's capacity to act. Phills et al. (2008) point out, that social innovation is a novel solution to a social problem that is more effective, efficient, sustainable, or just than existing solutions, and creates value to society as a whole rather than to private individuals.

Innovation usually takes the form of an outcome or process. Scientists stress the importance of the process dimension of social innovation, underlining that it is not just about pursuing an innovative social outcome, but also about an innovative endeavour, a novel way to shape patterns of social interaction to achieve a social goal (Bonifacio 2014); the collaboration that crosses traditional roles and boundaries between citizens, civil society, the state and the private sectors, the creation of new relationships and building people's capacity to do things differently in future (Reynolds et al. 2017); the possibility to realize better inclusion of excluded groups and individuals in various spheres of society at various spatial scales (Moulaert et al. 2005).

The academic attention is paid to specific forms of social innovation resulting from cooperation between business, consumers, communities, movements, NGOs, state institutions, etc. Such innovations are the following: user innovation, consumer innovation, consumer-driven innovation, as they stress the role of consumers who want and can innovate; grassroots innovation with vision, that innovation processes must be more inclusive towards local communities in terms of knowledge, processes and outcomes; open innovation which combines social cooperation, social cohesion, social tolerance, as well as multi-product/multi-innovation, co-creation, co-design/participatory design, participatory innovation, practice-based innovation, collaborative innovation (Bhalla 2011; Friedman and Angelus 2011; Melkas and Harmaakorpi 2012; Swink 2015).

Social innovation is of different nature, it happens in many different ways, and has diverse impacts; therefore, various classifications/typologies are available in the scientific literature. The typology proposed by Brandon and Lombardi (2010) refers to the context within which the innovation originates: (1) socio-cultural (innovation of non-formal institutions, etc.); (2) socio-political (innovation of governance, policies, etc.); (3) socio-ideological (innovation of ideological frameworks, mindsets, paradigms, etc.); (4) socio-ethical (innovation of ethical/normative frameworks, etc.); (5) socio-economic (innovation of economic models, business models, etc.); (6) socio-organizational (innovation of organisational arrangements, etc.); (7) socio-technical (innovation of human-technology interaction, etc.); (8) socio-ecological (innovation of human–environment interaction, etc.); (9) socio-analytical (innovation of analytical and sense-making frameworks, etc.); (10) socio-juridical (innovation of legal frameworks and laws, etc.).

Less detailed social innovation typology is given by Brooks (1982). He distinguished innovation that are almost purely technical, socio-technical and social, and classifies social innovation as the market, managerial, political, and institutional innovations. While TEPSIE (2014) offers the typology of social innovation which covers the following categories: (1) new services and products; (2) new practices; (3) new processes; (4) new rules and regulations.

Slee (2019) identified nine main types of social innovation based on the research made in Scotland: (1) new institutional/organisational forms; (2) established forms in new places; (3) new constituencies of interest for established social institutions; (4) revitalisation, renewal, and change of established social entities; (5) new practices and processes in social institutions; (6) known social practices in new places; 7) new social products and services; (8) known products or services in new places; (9) new enabling/support policies. In this classification, attention must be drawn to the reference of "new place", as the author links the novelty of social innovation with novel institutional structures/networks/governance, the effective innovative social practices, or the place where the innovation occurs.

Most important social innovation (all over the world) can be classified into 7 wide groups (Jiang and Thagard 2014):

1. Education (Kindergarten, University of the Third Age, Distance learning).
2. Health care (Nursing, Helpline, Hospice care).
3. Law and regulation (Patent law, International Labour Standards, Emissions trading).

4. Technology (Telephone, Internet, Facebook).
5. Social movements (Fair trade, Environmentalism, Feminism).
6. Finances (Pensions, Insurance, Microcredit).
7. Organizations and methods of organizing (Census, Think tank, Scientific management).

Bureau of European Policy Advisers (BEPA) classified social innovation into three broad categories: (1) grassroots social innovation that respond to pressing social demands not addressed by the market and are directed towards vulnerable groups in society; (2) a broader level of social innovation that addresses societal challenges in which the boundary between "social" and "economic" blurs and which are directed towards society as a whole; (3) the systemic type of social innovation that relates to fundamental changes in attitudes and values, strategies and policies, organizational structures and processes, delivery systems and services (BEPA 2011, p. 12). Bonifacio (2014) classifies social innovation distinguishing three hierarchical levels: micro level, meso level, and macro level. At micro level, the role of generating social innovation is delegated to the individual entrepreneur, through the meso level to public/private partnerships, and at the macro level governments and institutions innovate patterns of social interaction to generate social value through policies, laws, and institutional reforms. These two classifications are close to each other as in each category/hierarchical level, the same actors are specified.

3 Social Innovation and Sustainable Development Goals

The 2030 Agenda for Sustainable Development adopted by the General Assembly on 25 September 2015 is a plan of action for people, planet and prosperity, based on the realization of SDGs in building a sustainable society and improving people's quality of life through solving global, social and environmental problems (The 2030 Agenda for Sustainable Development 2015). The SDGs are indivisible and balance the economic, social, and environmental dimensions of sustainable development, they represent a global consensus on objectives for addressing the world's most pressing social and environmental issues.

The Goals are certainly very ambitious, as their implementation will create a world without poverty, hunger and disease, a world free of fear and violence; a world with universal literacy, health care, and social protection; a world with the access to safe drinking water and improved sanitation, where food is sufficient, safe, affordable and nutritious; a world where human habitats are safe and have access to sustainable energy; a world of universal respect for human rights, the rule of law, justice and equal opportunities; a world where sustainable economic growth and decent work is ensured; a world where everybody uses natural resources sustainably and lives in harmony with nature; a world in which democracy, good governance and the rule of law are essential for sustainable development (UN General Assembly 2015, pp. 3–4).

Despite being a global agenda, the realization of the SDGs implies different approaches, models and tools available to each country following national peculiarities and priorities in achieving sustainable development, and enables everyone to contribute in fulfilling SDGs, giving excellent chance for social entrepreneurs and social enterprises, non-profit and non-governmental organizations, foundations to realize their potential, for businesses—to become responsible, for governments—to act in the interests of the whole society, for universities—to reveal science-based approaches to make cost-effective solutions.

Although the term "innovation" is mentioned only in Goal 9 (Build resilient infrastructure, promote inclusive and sustainable industrialization and foster innovation), different forms of innovation are relevant carrying out all 17 SDGs. Social innovation is increasingly accepted as an integral part of the innovation system to achieve SDGs. Like in developed countries, social innovation is starting and becoming recognised in many developing countries as it helps to meet social needs in a new way that involves collaboration and empowerment. Social innovation works with ordinary people rather than just doing something to them as passive recipients, also developing their own capabilities around and ownership of the service, and thereby transforming their social relations and improving their access to power and resources (Millard 2018).

There is no doubt that anyone can get involved in the realization of SDGs as Goals are often named as people's goals by providing innovative initiatives. In this case, however, it is very important how well the public is informed about the SDGs, as governments in order to achieve the Goals need to engage with citizens, listen to their needs and mobilise them into action.

The survey concerning the level of SDGs awareness among the global public was conducted in 2019 by The World Economic Forum. About 20,000 people between the ages of 16 and 74 from 28 countries responded. The findings revealed that 26% of respondents were "somewhat or very familiar" with the UN SDGs, and higher levels of SDGs awareness were fixed in developing countries. The three countries with the highest levels of awareness were: (1) Turkey (92%); (2) Mainland China (90%); (3) India (89%), and only around 1-in-10 of people surveyed in Japan, France, Italy, Canada, and the UK said they were familiar with the Goals (Fleming 2019; Tedeneke 2019). The survey disclosed that the public tends to support SDGs that are associated with direct human needs: the provision of food, water, health, or energy. These needs are most obvious and easily understood, when goals such as gender equality, reduced inequality and industry, innovation and infrastructure, were among the lowest-ranked (see Table 1).

There are only some research works examining the links between social innovation and SDGs. After analysing the definition of social innovation, Eichler and Schwarz (2019) conducted the classification of social problem/need according to one or more SDGs. They analysed 115 scientific articles and found out, that the SDGs assigned most frequently, independently of the economic development level of the country, was "good health and well-being" (SDG3), followed by "partnerships for the goals" (SDG17) and "decent work and economic growth" (SDG8). The research outcomes only partly are in line with the results of the public survey, as in both cases only

Table 1 Globally, the highest and lowest-ranked SDGs in perceived importance, 2019

The top-ranked SDGs	The lowest-ranked SDGs
Zero hunger (SDG2)	Gender equality (SDG5)
Clean water and sanitation (SDG6)	Reduced inequality (SDG10)
Good health and well-being (SDG3)	Industry, innovation and infrastructure (SDG9)
Affordable and clean energy (SDG7)	Responsible consumption and production (SDG12)
Life below water (SDG14)	Peace, justice and strong institutions (SDG16)

Source Tedeneke (2019)

"good health and well-being" was mentioned as one of the most important goals. The most widespread SDGs addressed in social innovation case studies in developed countries were "partnerships for the goals" (SDG17), "sustainable cities, communities" (SDG11), and "good health and well-being" (SDG3). In developing countries, "decent work and economic growth" (SDG8), "good health and well-being" (SDG3), and "no poverty" (SDG1) were the most frequently assigned SDGs. In developing countries, no case study was assigned to "affordable and clean energy" (SDG7), while in developed countries no case study dealt with "life below water" (SDG14). About the same percentage of case studies in developed and in developing countries targeted the SDGs "responsible consumption and production" (SDG12) and "reduced inequalities" (SDG10) (Eichler and Schwarz 2019, pp. 9–10).

Social innovation comes in different forms and is a source of ideas, basis for work, an opportunity for partnership, and cooperation realizing SDGs. During the first UN summit on the SDGs (24–25 September 2019, New York) since the adoption of the 2030 Agenda, Political Declaration "Gearing up for a decade of action and delivery for sustainable development" was adopted and world leaders called for a decade of action to deliver the SDGs by 2030 and announced actions they are taking to advance the agenda. SDGs "Acceleration Actions" are initiatives voluntarily undertaken to accelerate the SDGs implementation by national governments and any other non-state actors working individually or in partnership (see examples in Table 2). The role of social innovation in this process is extremely important.

The social innovation listed in Table 2 are not necessarily entirely new solutions, some of them are new solutions in a certain area/locally, some of them cover a wider/global spectrum. Most innovations involve not one but several SDGs and this is obvious as all SDGs are interconnected.

Table 2 Examples of social innovation to accelerate fulfilment of SDGs

Goals	Project title	Project aim
Goal 4, 12	World's largest lesson Nigeria	To engage all children in learning about the SDGs and using this as a stimulus for SDG advocacy and local community action
Goal 17	Waves	To create an easy-to-use robot-advisor which allows generating a personal investment portfolio (creating sustainable investments)
Goal 5, 16	Dissemination of the criminal justice system	To contribute to the dissemination of the Criminal Justice System so that it becomes part of the Mexican culture (women, girls and boys)
Goal 16	Rule of law matters project in Jordan	To improve youth and community knowledge of the rule of law, build their civic skills, empower them with confidence to engage in strengthening a rule of law culture
Goal 1–17	Promote cycling	To promote cycling for everyone regardless of age, gender, religion, social status, or economic standing
Goal 1–17	Youth co: lab (Philippines)	To enable young people to develop their own solutions towards the SDGs
Goal 11	Lamphope	To make alternative lodging available for those interested in tourism rural, social, ecological medical or missionary purpose (platform)
Goal 1, 5, 8, 10, 12, 16	A scaling-up of Sweden's Feminist Foreign Policy	To strengthen women's and girls' rights, representation and resources
Goal 4, 5, 9, 10, 17	Pipol Konek (Free Wi-Fi for All)	To enable women and men from disadvantaged communities to improve their educational attainment, deepen engagement in governance processes, etc., through increased access to public Wi-Fi hotspots

(continued)

Table 2 (continued)

Goals	Project title	Project aim
Goal 1, 5, 3, 10	Inclusive and differentiated social protection for women in Mexico	To increase access to social protection for domestic workers, women in rural communities through the consolidation of a national care strategy based on humanitarian assistance for historically forgotten populations
Goal 6, 16, 17	The SDG Impact accelerator	To bring together start-ups that work on SDG-relevant solutions with their users and potential partners, creating a unique, action-oriented partnership platform
Goal 10, 11, 16	Participation model and participatory budgeting	To help city employees consider how the operations and services could be planned in better co-operation with the residents (Helsinki)
Goal 1, 5, 8, 17	ITC she trades initiative	To economically empower women through trade, increase women entrepreneurs' access to financial resources
Goal 12, 17	The green deal—commitments	To realize a voluntary agreement between the state and the business sector: decrease the usage of plastic bags, decrease CO_2 emissions, etc.
Goal 1, 2, 5, 8, 12–17	Fish forward	To raise awareness of the social and environmental impacts of fish consumption
Goal 3	Youth mental health	To ensure positive psychology to become a household term
Goal 5, 7	Women in energy expert platform	To become a networking platform for women working in sustainable energy, and a key driver to enhance gender balance and diversity in the sector
Goal 3, 11, 13, 15, 17	Trees in cities challenge	To plant more trees in urban areas

Source SDG Acceleration Actions https://sustainabledevelopment.un.org/sdgactions

4 The Measurements of Social Innovation

The social innovation process is considered as rather complex and the research on the measurement of social innovation is still in its early stages (Hoelscher et al. 2015). The importance of measuring social innovation is related to the possibility to determine their effectiveness and disclose their potential in meeting social needs.

Researchers working in TEPSIE project singled out three interrelated levels to measure social innovation and proposed indicators (Bund et al. 2013; Schmitz et al. 2013):

- *Entrepreneurial Activities*: investment activities, start-ups and death rates, collaboration and networks. Examples: expenditure in innovation by the social economy, start-ups of firms dedicated to social purposes, environment to start a company.
- *Field Specific Outputs and Outcomes*: education, health/care, employment, housing, societal capital and networks, political participation, environment. Examples: equal opportunities, access to quality of health facilities, earnings, social cohesion, preservation of natural capital.
- *Framework Conditions*: resources framework, institutional framework, political framework, societal framework. Examples: public social expenditure as a percentage of GDP, the number of volunteers, e-readiness, solidarity, legislative background for starting a social organisation, human rights, national innovation strategies, corruption perception, requests to the EU Parliament, membership in humanitarian organisations, etc.

Benedek et al. (2016) proposed a methodology for measuring social innovation potential (index) and suggested three types of indicators: input, output and impact. Input indicators are organized across 4 factors (Institutional System, Location Factors, Human Conditions and Activity) and a set of 10 indicators is constructed. Output indicators are organized across 4 factors (Economic, Social, Cultural and Health) and a set of 13 indicators is proposed. Impact indicators are organized across 6 factors (Social conditions, Family relationship, Sense of security, Social infrastructure, Living conditions, Environmental conditions) and a set of 11 indicators is prepared. Scientists calculated the innovation index for the 19 NUTS3 counties of Hungary (2007–2013). In terms of social and economic innovation potential NUTS3 territories were grouped into four clusters. The research revealed that there was weak social innovation in the territories with low economic innovation.

The Social Innovation Index was constructed by the Economist Intelligence Unit Research Team and calculated only for 2016. The Index includes seven quantitative data points and 10 qualitative scores grouped into four pillars. Each pillar is given a different weight in the overall score: Policy and Institutional Framework (44.44%), Financing (22.22%), Entrepreneurship (15%), Society (18.33%). Data points within each pillar are normalised (from 0–100, where 0 = worst and 100 = best). The Social Innovation Index was calculated for 2016 and measured the capacity for social innovation across 45 countries-G20 and OECD nations, together with some other countries. The United States appeared on the top of the Social Innovation Index 2016, scoring 79 out of 100. The Philippines took the last position scoring 27 out of 100 (Old problems, new solutions, 2013).

One more attempt to measure social innovation is the creation of Regional Social Innovation Index (RESINDEX). The aim of RESINDEX (2013) pilot project was to develop an exploratory model of indicators of Social Innovation and test it within the context of the Basque Autonomous Community. RESINDEX model is designed around three indices: Potential Capacity for Innovation Index, Social Orientation Index, and Social Innovation Index (SII). SII is a synthetic unit of measure made up of four factors in the implantation of innovative social projects: knowledge acquisition, development of innovative social projects, impact of innovative social projects, and governance of innovative social projects. The index was constructed from the statistical management of a predefined questionnaire where scores ranged from 0 to 100. The capacity of knowledge and social innovation in four types of key regional agents: businesses, non-profit organizations, universities, and technological centres and whole regions were evaluated. The results revealed that only for technological centres SII was medium (31–70 points) except governance (low), while SII of all other agents was low (below 30 points).

5 Methodology

To conduct the analysis, we need measures of social innovation, sustainable development also econometric methods for identifying the relationship between social innovation and sustainable development. This section describes the variables, discusses the econometric methods, provides summary statistics, regression analysis, and results.

5.1 The Data Set and Variables

We gathered a panel dataset that includes 11 new EU member countries: Bulgaria, the Czech Republic, Croatia, Estonia, Hungary, Latvia, Lithuania, Poland, Romania, Slovakia, and Slovenia for the period from 2015 to 2019 on an annual basis. According to Baltagi (2008, p. 7), "Panel data give more informative data, more variability, less collinearity among the variables, more degrees of freedom and more efficiency".

The reasons for focusing on New EU countries are threefold. Firstly, these postsocialist countries are most comparable because similar political and economic changes were required to become members of the EU. Second, in these countries, the impact of social innovation on sustainable development may be different compared to higher developed EU countries, due to structural differences in fundamental economic and political institutions. Finally, by focusing on a certain group of countries, we may reduce sample heterogeneity. Because of data limitations, our panel is unbalanced, and numbers of observations vary over time for different countries.

Dependent variable A key challenge for any study investigating the impact of social innovation on sustainable development is the question of how to measure these variables. To quantify sustainable development, we use the SDGI (see Table 3).

Table 3 Data description and sources

Indicator name	Short name of the variable	Description	Data source
Sustainable development goals index	SDGI	Assesses where each country stands in regard to achieving the SDGs	Kroll and Annan (2015) SDG index and dashboards (2016) The sustainable development goals report (2017) SDG index and dashboards (2018) Sustainable development report (2019)
Social progress index	SPI	Measures the extent to which countries provide for the social and environmental needs of their citizens	Porter et al.(2015) Porter et al. (2016) Porter et al. (2017) Social progress index. Executive summary (2018) Social progress index. Executive summary (2019)
Index of economic freedom	EFI	The index scores nations on 12 broad factors of economic freedom	2020 Index of economic freedom https://www.heritage.org/index/explore?view=by-region-country-year
GDP per capita based on PPP	GDP	GDP converted to international dollars using purchasing power parity rates and divided by the total population	IMF: World Economic Outlook (WEO), October 2019
Population share with tertiary education	EDUC	The share from 15 to 64 years of age with finished tertiary education (levels 5–8)	Eurostat
Unemployment rate	UNEMPLOY	The number of unemployed persons as a percentage of the labour force	IMF: World Economic Outlook (WEO), October 2019

(continued)

Table 3 (continued)

Indicator name	Short name of the variable	Description	Data source
CO_2 emissions intensity	CO_2	Carbon dioxide emissions intensity is calculated as total emissions (kg) per 1000-dollar GDP	Global GHG and CO_2 Emissions https://knoema.com/EDGARED2019/global-ghg-and-co2-emissions

Since the SDGI is not completely comparable over time, obtained results should be interpreted with care. SDGI ranges from 0 to 100, 0 indicating the lowest and 100 the greatest achievement fulfilling SDGs. The analysis relates sustainable development, measured by the SDGI, to social innovation and a few control variables.

Explanatory Variables Social innovation can be measured, using quantitative and qualitative methods or a combination of them, constructing sets of indicators or creating indices, but so far there is no one generally accepted method, measurements are performed on a local scale as pilot projects. To measure social innovation, we would ideally like to have a generalized indicator such as the Social Innovation Index. But this index was calculated only for 2016. This means that there is no long-term database that can be used to assess social innovation in different countries. Due to the lack of data within countries, a comprehensive Social Progress Index (SPI) represents "what is broadly meant by social innovation". Therefore, we assume that SDGI measures sustainable development and the SPI measures social innovation.

Control Variables To assess the strength of the linkage between sustainable development and social innovation in regression analysis, we also include several control variables. In the choice of control variables, we aim to choose variables as proxies for economic, social and environmental sustainability. The analysis includes such control variables: economic sustainability measured by an index of economic freedom and GDP per capita based on PPP; social sustainability measured by population share with tertiary education and unemployment rate; environmental sustainability measured by CO_2 emissions intensity. Variables used in the analysis, together with their description and the data sources, are presented in Table 3.

The summary statistics of dependent, independent and control variables are presented in Table 4.

The mean value for key independent variable SPI is 78.9 with a maximum value of 85.8. This provides the insights on the level of SPI in 11 new EU member countries. The average SDGI during the analysed period is 73.8 and the maximum value is 81.9.

We check for multicollinearity among regressors. Table 5 shows pairwise correlations between various indicators included in regression analysis. The correlation matrix presented confirms the strong correlation (0.8685) between GDP and SPI. This suggests that these indicators may contain common information, which may lead to multi-collinearity and overparameterization problems. Correlation among independent variables can lead to large standard errors for the OLS estimates

Table 4 Descriptive statistics of variables, 2015–2019

Variable	Descriptive statistics				
	Mean	Median	S.D.	Min	Max
SDGI	73.8	74.8	5.37	55.5	81.9
SPI	78.9	79.3	4.03	68.4	85.8
EFI	68.8	68.3	5.25	59.1	79.1
GDP	29.7	29.9	4.98	19.3	38.8
EDUC	25.3	24.4	6.08	15.0	37.4
UNEMPLO	7.17	6.70	2.98	2.20	17.1
CO_2	0.271	0.225	0.121	0.150	0.610

Source Own compilation

Table 5 Cross country correlation coefficients between selected variables

Variables	SDGI	SPI	EFI	GDP	EDUC	UNEMPLOY	CO_2
SDGI	1						
SPI	0.5453	1					
EFI	0.0296	0.1943	1				
GDP	0.4766	0.8685		1			
EDUC	0.2458	0.4126	0.5658	0.3557	1		
UNEMPLOY	−0.2443	−0.3324	−0.4366	−0.4550	−0.0387	1	
CO_2	0.0772	0.3076	0.3076	0.0811	0.3668	−0.2338	1

Source Own compilation

(Wooldridge 2013). Because of this multi-collinearity problem, we exclude GDP variable from our regression analysis. Table 5 confirms that other variables are also strongly correlated. All correlation coefficients (except between unemployment rate and other variables) are positive and changes between 0.03 and 0.7.

5.2 Empirical Methodology

In order to decide whether to apply the least-squares method (OLS), fixed effects (FE), or random effect (RE), the panel diagnostics tests were performed. A joint significance test shows that the pooled OLS model is adequate, in favour of the FE alternative and the Breusch-Pagan test shows that the pooled OLS model is adequate, in favour of the RE alternative. Therefore, pooled OLS is the most appropriate econometric method for the estimation of the association between sustainable development and social innovation. This method is also used because of short time-series data.

The effect of social innovation on sustainable development is described by the following equation:

$$SDGI_{it} = \alpha + \delta_2 td2_t + \cdots + \delta_T tdT_t + \beta SPI_{it} + \gamma' Z_{i,t} + \varepsilon_{it}$$

where $SDGI_{it}$—the dependent variable of interest, which is SDGI in country i and year t; $t = 2,3,...,T$; $i = 1,2....N$; α—constant; td—time variables which absorb the impact of time on research results; SPI_{it}—the social progress index. β is the independent variable's impact on the dependent variable, and ε_{it} is an error term. $Z_{i,t}$ denotes a set of control variables: EFI, EDUC, UNEMPLOY, CO_2. To facilitate the interpretation of coefficients and to transform a model into linear, we use all dependent and independent variables in logs. Different control variables are included in various model specifications. We analyse the data using the GRETL program.

To obtain a deeper insight into the results, we split the 11 New EU Member countries into two subsamples according to the level of social innovation which is proxied by SPI and re-estimate the model. Thus, we split the 11 New EU Member countries into 2 different sub-groups. The "higher" social innovation group includes 5 countries having an average SPI during the analysed period of 83 or higher. The "higher" social innovation countries include the Czech Republic, Estonia, Slovenia, Slovakia, and Poland. The 6 countries in our dataset, below this threshold, are classified as "lower" social innovation countries. These countries are Croatia, Latvia, Hungary, Lithuania, Bulgaria, and Romania.

To investigate whether the impact of social innovation on sustainable development varies across countries with different levels of social innovation, we construct one additional variable: $(SPI_{it} D_i)$. This variable is an interaction term between the social innovation measure with a dummy variable, which is a binary variable and coded to either zero or one. The OLS model using dummy variables is called a least square dummy variable (LSDV) model.

It is described by the following equation:

$$SDGI_{it} = \alpha + \delta_2 td2_t + \cdots + \delta_T tdT_t + \beta_1 SPI_{it} + \beta_{1D} SPI_{it} D_i + \gamma' Z_{i,t} + \varepsilon_{it}$$

where D_i is a dummy variable, which is equal to 1 for a "higher" social innovation country group and 0 for a "lower" social innovation country group. β_1 shows the impact of social innovation on the dependent variable (e.g. sustainable development) in countries with a "lower" social innovation and β_{1D} shows the difference of the x impact on y in the "higher" social innovation countries compared with "lower" social innovation countries. The impact in the "higher" social innovation country group is calculated as $\beta_1 + \beta_{1D}$.

In assessing the impact of social innovation on sustainable development and aiming to ensure the validity of research results, we verified the following issues:

- Whether there is no serial correlation in the idiosyncratic errors. Wooldridge test for autocorrelation in panel data shows that there is no autocorrelation of errors.

- Whether errors are homoskedastic. We use the White test and heteroscedasticity robust (HAC) standard errors in this research.
- Whether there is no multicollinearity among regressors. To detect multicollinearity, we apply the variance inflation factors (VIF).

6 Results and Analysis

6.1 Descriptive Analysis

In Fig. 1 we initially visualize the relationship between sustainable development which is measured by SDGI and social innovation measured by SPI.

The first descriptive look at the potential link between these variables suggests the existence of a positive association between both indicators. However, when interpreting the data provided by Fig. 1, it should be stressed that it is very likely that sustainable development depends not solely on the level of social innovation. It means that the empirical information shown in Fig. 1 should be interpreted with caution because other variables also affect sustainable development. Concerning these potential issues, in the next section, we submit a multivariate regression analysis on the relationship between sustainable development and social innovation.

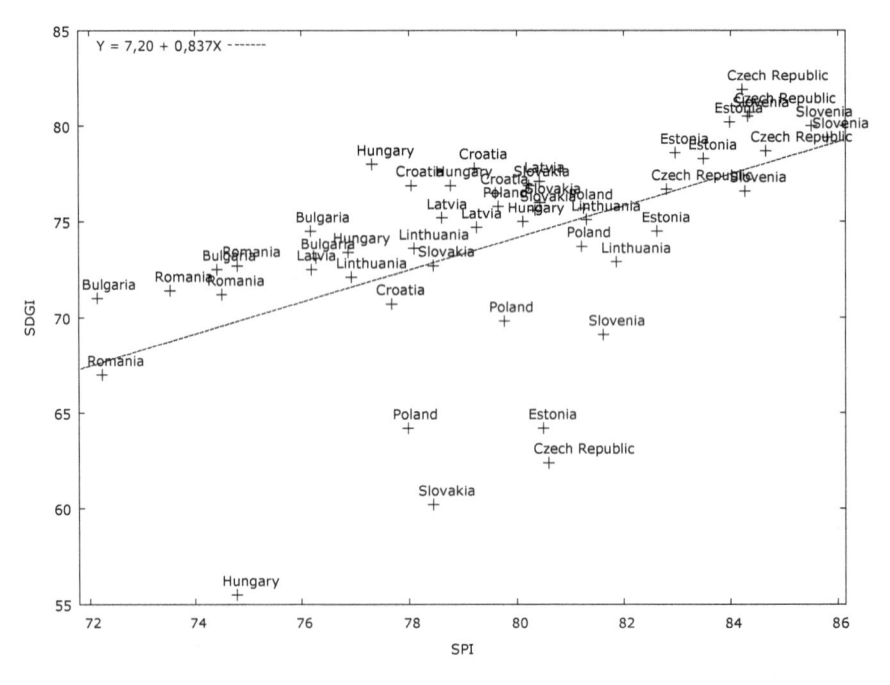

Fig. 1 SDGI versus SPI (with least squares fit). *Source* Own compilation

6.2 Results

Using the OLS model, we estimate five model specifications by testing the impact of different control variable(s) in each forward step. Model specifications 1–5 in Table 6, with heteroskedasticity consistent standard errors, estimate the impact of social innovation on sustainable development, using the SPI as a measure of social innovation. Specification 1 shows that the SPI has a significant positive effect on the SDGI. An increase in SPI by 1% increases the SDGI approximately by 0.72–0.8% in various model specifications at a 1% level of significance. This depends on control variables included in different model specifications (see 1, 2, 3, 4, 5 specifications in Table 6).

Model specification 2 adds an EFI along with the SPI. The effect of the SPI remains positive and statistically significant but contrary to the commonly hypothesized positive effect EFI has a negative effect on the SDGI, but this effect is not statistically significant.

In model specification 3 we add EDUC and in model specification 4 we add UNEMPLOY. Results are consistent with previous and show statistically significant impact of SPI on SDGI in both model specifications. The results of these model specifications also show that the coefficient for population share with tertiary education

Table 6 Sustainable development and social innovation: OLS regression analysis

Dependent variable	SDGI				
	(1)	(2)	(3)	(4)	(5)
SPI	0.779*** (0.085)	0.800*** (0.087)	0.794*** (0.127)	0.748*** (0.121)	0.720*** (0.096)
EFI		−0.076* (0.038)	−0.079* (0.044)	−0.143** (0.050)	−0.153** (0.055)
EDUC			0.003 (0.023)	0.028 (0.021)	0.023 (0.019)
UNEMPLOY				-0.017 (0.012)	−0.013 (0.012)
CO_2					0.014 (0.010)
Const	0.731* (0.377)	0.959* (0.448)	0.990 (0.631)	1.416** (0.627)	1.610** (0.525)
n	50	50	50	39	39
p value of testing H0: errors are not serially correlated	0.6524	0.8873	0.9075	0.2961	0.2999
Adjusted R-squared	0.879	0.882	0.880	0.875	0.875

Source Own compilation

Note Standard errors in parentheses. * Significant at the 10 percent level, ** significant at the 5 percent level, *** significant at the 1 percent level. Dependent and all explanatory variables are in logs. All key assumptions of multiple regression analysis are met. In order to save space, the coefficients of the time dummies are not reported in the table.

is positive but not significant, whereas the coefficient for the unemployment rate is negative and insignificant. Specification 5 shows that the impact of the SPI remains highly significant and positive when additional control variable CO_2 is added as a regressor. Regression analysis reveals that a highly significant and positive relationship between the SDGI and SPI is consistent with the preliminary demonstration provided by Fig. 1. It should be mentioned that the time dummies are not reported in Table 6, but they are statistically significant in all model specifications.

To test whether the impact of SPI on SDGI is different in two country groups, we use LSDV model. Table 7 provides the results of our estimate of the impact of social innovation on sustainable development, where we enter the variable of the interaction of SPI and the country's level of innovations, i.e. it verifies whether the impact of social innovation on sustainable development differs between countries with different level of social innovation.

Parameter estimates of individual regressors using various specifications of the LSDV are only slightly different from those in the pooled OLS. Regression results show that the impact of social innovation on sustainable development does not vary

Table 7 Sustainable development and social innovation: LSDV regression analysis

Dependent variable	Sustainable development goals Index				
	(1)	(2)	(3)	(4)	(5)
Const	0.655 (0.733)	0.892 (0.730)	0.927 (0.999)	1.504 (1.086)	1.285 (0.950)
SPI	0.797*** (0.171)	0.816*** (0.149)	0.809*** (0.207)	0.728*** (0.214)	0.800*** (0.193)
SPI * Higher	−0.000 (0.004)	−0.000 (0.003)	−0.000 (0.004)	0.000 (0.004)	−0.002 (0.004)
EFI		−0.076* (0.039)	−0.078 (0.049)	−0.144** (0.058)	−0.152** (0.055)
EDUC			0.002 (0.024)	0.029 (0.025)	0.017 (0.026)
UNEMPLOY				−0.017 (0.012)	−0.013 (0.011)
CO_2					0.018 (0.011)
Observations	50	50	50	39	39
Adj. R^2	0.876	0.880	0.877	0.871	0.871
p value of testing H0: errors are not serially correlated	0.7778	0.9887	0.1049	0.3434	0.1894

Source Own compilation

Note Standard errors in parentheses. * significant at the 10 percent level, ** significant at the 5 percent level, *** significant at the 1 percent level. Dependent and all explanatory variables are in logs. All key assumptions of multiple regression analysis are met. In order to save space, the coefficients of the time dummies are not reported in the table with the level of social innovation. Countries with higher levels of social innovation have not experienced a faster achievement of sustainable development than countries with lower levels of social innovation over the period 2015–2019. This can be explained by the fact that the difference in levels of social innovation in both country groups is not big: the average index in the "lower" SPI group during the analysed period was 76.3 and in the "higher" SPI group—81.98.

7 Conclusions

Social innovation necessarily takes on a number of forms and operates in a range of fields. Different types of social innovation are close to each other as one leads to another or some request for previous.

Social innovation steams from the cooperation between business and state institutions, business and consumers, communities and NGOs, they can be carefully planned as well as entirely spontaneous and locally led and delivered, created by ordinary people who know what they really need.

The innovation lies at the heart of the SDGs, and ambitious SDGs can be achieved by bringing together a variety of stakeholders from developed and developing counties to solve global, social and environmental problems through cooperation, partnership, and mutual assistance.

The measurement of social innovation takes into account different stages of the innovation process, identification of dimensions, and choosing of the indicators. Measurement attempts, using a set of indicators or constructing a social innovation index, are not sufficiently developed, as there is no consensus on social innovation and a lack of data to perform the measurement.

We analysed the impact of social innovation on sustainable development. The analysis was based on panel data for 11 New EU countries over the period 2015–2019. The results using the OLS regression model demonstrate the positive relationship between social innovation and sustainable development when we measure social innovation by SPI. The positive relationship between the SDGI and SPI is reported in various model specifications. The analysis demonstrates that an increase in SPI by 1% increases the SDGI approximately by 0.8% at a 1% level of significance. The investigation results suggest that social innovation in New EU countries are associated with higher achievement of sustainable development. The influence of SPI on SDGI is in line with the theoretical expectation and consistently positive across all estimations.

Using the LSDV model, we also examined whether the impact of social innovation on sustainable development depends on the level of social innovation. The results show that the impact of SPI on SDGI does not depend on the country's level of social innovation.

References

Baltagi BH (2008) Econometric analysis of panel data. Wiley, John & Sons, p 335

Benedek J, Kocziszky G, Somosi MV, Balaton K (2016) Generating and measuring regional social innovation. J Econ Lit 12:14–25

Bhalla G (2011) Collaboration and co-creation: new platforms for marketing and innovation. Springer, New York, p 205

Bonifacio M (2014) Social innovation: a novel policy stream or a policy compromise? An EU perspective. Eur Rev 22(01):145–169

Brandon P, Lombardi P (2010) Evaluating sustainable development in the built environment (2nd ed). Wiley-Blackwell, Chichester, UK, p 280

Brooks H (1982) Social and technological innovation. In: Lunstedt SB, Colglazier WE (eds) Managing innovation. Pergamon Press, Elmsford, New York, p 30

Bund E, Hubrich DH, Schmitz B, Mildenberger G, Krlev G (2013). Blueprint of social innovation metrics—contributions to an understanding of opportunities and challenges of social innovation measurement. A deliverable of the project TEPSIE. European Commission, Brussels. https://archiv.ub.uni-heidelberg.de/volltextserver/18700/1/D2.4_final.pdf (Last accessed 11 Feb 2020)

Bureau of European Policy Advisers (BEPA) (2011) Empowering people, driving change. Social innovation in the European Union. Luxembourg: Publications Office of the European Union

Bureau of European Policy Advisers (BEPA) (2014) Social innovation. A decade of changes. Publications Office of the European Union, Luxembourg

Caulier-Grice J, Davies A, Patrick R, Norman W (2012) social innovation overview: a deliverable of the project: "The theoretical, empirical and policy foundations for building social innovation in Europe" (TEPSIE), European commission—7th framework programme. European Commission, DG Research, Brussels. (Last accessed 15 Feb 2020)

East M (2018) 5 ways for social innovators to use the power of the SDGs. https://medium.com/@gaiaeducation/5-ways-for-social-innovators-to-use-the-power-of-the-sdgs-d531180d57f2 (Last accessed 20 Apr 2020)

Eichler GM, Schwarz EJ (2019) What sustainable development goals do social innovations address? A systematic review and content analysis of social innovation literature. Sustainability 11(522):1–18

Ellis T (2010) The new pioneers: Sustainable business success through social innovation and social entrepreneurship. Wiley, p 247

Fleming S (2019) What does the world really think about the UN sustainable development goals? World Economic Forum. https://www.weforum.org/agenda/2019/09/un-sustainable-development-goals/ (Last accessed 10 Mar 2020)

Forum on social innovations. Organisation for economic co-operation and development. https://www.oecd.org/fr/cfe/leed/forum-social-innovations.htm (Last accessed 5 Mar 2020)

Friedman M, Angelus H (2011) Best practices in collaborative innovation. How manufacturers and retailers can profit from collaborative innovation. Kalypso White Paper. https://viewpoints.io/uploads/files/Best_Practices_in_Collaborative_Innovation_1.pdf (Last accessed 25 Mar 2020)

Hoelscher M, Mildenberger G, Bund E, Gerhard U (2015) A methodological framework for measuring social innovation. Hist Soc Res 40(3):48–78

Jiang M, Thagard P (2014) Creative cognition in social innovation. Creat Res J 26(4):375–388

Kroll C, Annan K (2015) Sustainable development goals: are the rich countries ready? Bertelsmann stiftung and sustainable development solutions network. https://www.bertelsmann-stiftung.de/en/publications/publication/did/sustainable-development-goals-are-the-rich-countries-ready/ (Last accessed 25 May 2020)

McGowan K, Westley F (2015) At the root of change: the history of social innovation. In: Nicholls A et al (eds) New Frontiers in Social Innovation Research. Macmillan Publishers Limited, UK, p 299

Melkas H, Harmaakorpi V (2012) Practice-based innovation: insights, applications and policy implications. Springer, Berlin, p 452

Millard J (2018) How social innovation underpins sustainable development. In: Howaldt J, Kaletka C, Schröder A, Zirngiebl M (eds) Atlas of Social Innovation. TU Dortmund University, Dortmund, pp 40–43

Moulaert F, Martinelli F, Swyngedouw E, Gonzalez S (2005) Towards alternative model(s) of local innovation. Urban Stud 42(11):1969–1990

Mulgan G, Wilkie N, Tucker S, Ali R, Davis F, Liptrot T (2006) Social silicon valleys a manifesto for social innovation: what it is, why it matters and how it can be accelerated. The Basingstoke Press. https://youngfoundation.org/wp-content/uploads/2013/04/Social-Silicon-Valleys-March-2006.pdf (Last accessed 17 Apr 2020)

Neumeier S (2012) Why do social innovations in rural development matter and should they be considered more seriously in rural development research? Sociol Rural 52:148–169

Old problems, new solutions: measuring the capacity for social innovation across the world (2013) https://eiuperspectives.economist.com/sites/default/files/Social_Innovation_Index.pdf (Last accessed 20 May 2020)

Phills JAJ, Deiglmeier K, Miller DT (2008) Rediscovering social innovation. Stanf Soc Innov Rev 6(4):34–43

Pol E, Ville S (2009) Social innovation: buzz word or enduring term? J Socio-Econ 38(6):878–885

Porter ME, Stern S, Green M (2015) Social progress index 2015. Social Progress Imperative, Washington

Porter ME, Stern S, Green M (2016) Social progress index 2016. Social Progress Imperative, Washington

Porter ME, Stern S, Green M (2017) Social progress index 2017. Social Progress Imperative, Washington

RESINDEX: Regional Social Innovation Index (2013) Innobasque, basque innovation agency, parque tecnológico de Bizkaia, p 84. http://www.simpact-project.eu/publications/indicators/2014_RESINDEX_eng.pdf (Last accessed 17 Apr 2020)

Resolution adopted by the general assembly on 25 September 2015 transforming our world: the 2030 agenda for sustainable development. United Nations. https://www.un.org/en/development/desa/population/migration/generalassembly/docs/globalcompact/A_RES_70_1_E.pdf (Last accessed 19 Feb 2020)

Reynolds S, Gabriel M, Heales C (2017) Social innovation policy in Europe: where next? Social Innovation Community. https://www.siceurope.eu/sites/default/files/field/attachment/social_innovation_policy_in_europe_-_where_next.pdf (Last accessed 10 May 2020)

Sabato S, Vanhercke B, Verschraegen G (2015) The EU framework for social innovation—between entrepreneurship and policy experimentation. Improve Working Paper No. 15/21. Herman Deleeck Centre for Social Policy—University of Antwerp, Antwerp. https://www.researchgate.net/publication/287994531_The_EU_framework_for_social_innovation_Between_entrepreneurship_and_policy_experimentation (Last accessed 12 Mar 2020)

Schmitz B, Krlev G, Mildenberger G, Bund E, Hubrich D (2013) Paving the way to measurement—a blueprint for social innovation metrics. A short guide to the research for policy makers. A deliverable of the project: "The theoretical, empirical and policy foundations for building social innovation in Europe" (TEPSIE), European commission—7th framework programme, Brussels: European commission, DG research. https://www.siceurope.eu/sites/default/files/field/attachment/TEPSIE%20Policy%20Paper%20Measurement%20Blueprint%20(WP2).pdf (Last accessed 21 Feb 2020)

SDG index and dashboards (2016) A global report. Bertelsmann stiftung and sustainable development solutions network

SDG index and dashboards (2018) Global responsibilities. Implementing the goals. Bertelsmann stiftung and sustainable development solutions network

Slee B (2019) An inductive classification of types of social innovation. Scottish Aff 28(2):152–176

Social innovation theory and research: a summary of the findings from TEPSIE (2014) A deliverable of the project: the theoretical, empirical and policy foundations for building social innovation in Europe' (TEPSIE), European commission—7th framework programme, brussels:

european commission, DG research. https://iupe.files.wordpress.com/2015/11/tepsie-research_report_final_web.pdf (Last accessed 18 Feb 2020)

Social progress index. Executive summary (2018) Social progress imperative, Washington

Social progress index. Executive summary (2019) Social progress imperative, Washington

Sustainable development report (2019) Transformations to achieve the sustainable development goals. Bertelsmann stiftung and sustainable development solutions network

Swink M (2015) Building collaborative innovation capability. Res Technol Manage 49(2):37–47

Tedeneke A (2019) Global survey shows 74% are aware of the sustainable development goals. World Economic Forum. https://www.weforum.org/press/2019/09/global-survey-shows-74-are-aware-of-the-sustainable-development-goals/ (Last accessed 12 May 2020)

The sustainable development goals report (2017) UN, New York

Wooldridge JM (2013) Introductory econometrics: a modern approach. Cengage Learning, Boston, p 910

Professor Skaidrė Žičkienė received her Master degree in Economics from Vilnius University, Lithuania, and Ph.D. degree in Social Sciences (Economics) from Vilnius Gediminas Technical University, Lithuania. She is currently Professor at Vilnius University Šiauliai Academy, Regional Development Institute working in the Master Program in Management, also in doctoral studies. Her current research interests include sustainable development, corporate social responsibility, social innovation management and environmental economic.

Professor Zita Tamasauskiene received her Master degree in Economics from Vilnius University, Lithuania, and Ph.D. degree in Social Sciences (Economics) also from Vilnius University. She is currently Professor at Vilnius University Šiauliai Academy, Regional Development Institute, working in the Master Program in Economics and with doctoral students. Her current research interests include social policy, income inequality, human capital.

Transformational Transition of Sustainable Development Based on Circular Green Economy. An Analysis Based on the Theory of Resources and Capabilities

José G. Vargas-Hernández, Jorge Armando López-Lemus, and Marlene de Jesús Morales Medrano

Abstract The main objective of this chapter is to analyze the transformational transition of the model of the circular and green economy (CGE) from the point of view of the resources and capacities of the organization and its impact on sustainable development. Regarding the design and methodology: used in this research was descriptive and correlational through the VRIO framework (Value, rarity, inimitability and organization). The results obtained show that CGE represents an essential factor within an operational level of transformational transition with a sustainable development impact. Also, when used as an internal resource of the company, it becomes a competitive advantage through resources and capabilities in transformational transitions towards sustainable development. The findings are relevant and of great value because there are currently not enough investigations that are focused on the variables analyzed on the business sector.

Keywords Circular economy · Resources and capacities · Strategy · Sustainable development · Transformational transition · VRIO analysis · JEL D24 · D2

J. G. Vargas-Hernández (✉)
Department of Administration University Center for Economic and Managerial Sciences, University of Guadalajara, Periférico Norte 799 Edificio G-201-7, Núcleo Universitario Los Belenes CUCEA, Zapopan, Jalisco C.P. 45100, México
e-mail: josevargas@cucea.udg.mx

J. A. López-Lemus
Department of Multidisciplinary Studies, University of Guanajuato, Av. Universidad s/n, Col Yacatitas, Yuriria, Guanajuato 38944, México

M. de Jesús Morales Medrano
Centro Universitario de Ciencias Económicas Administrativas, Universidad de Guadalajara, Periférico Norte 799, Universitario Los Belenes, Zapopan, Jalisco 45100, México

1 Introduction

In recent decades, the care of the planet has begun to appear on international political agendas as a matter of urgent concern, since we have begun to notice the consequences of the decisions taken by past generations to obtain economic benefits without worrying about the damage to the environment they caused (Brundtland 1987; Cezarino et al. 2019). As such, it has been decided that it is time to worry and take measures to survive in a planet of limited resources with a population that does not stop growing.

Beyond the individual responsibility, those who can make a noticeable change and chain reaction are the companies that, regardless of size or classification, are important actors in the global scope since they have an active role in the degradation or preservation of its environment close to social, economic and environmental level. Then, it can be considered that it is of vital importance that companies begin to have the main goal of achieving sustainability.

That is why the circular and green economy (CGE) represents an essential factor that arises from the reengineering of existing products or the creation of new products that are generated (Ünal et al. 2018). According to the Ellen MacArthur Foundation, an organization dedicated to the study and dissemination of the circular economy, the beginning of the concept as such has not been registered, but is the result of an evolution of several schools of thought such as Regenerative Design, economics of performance, industrial ecology, biomimetics, blue economy and natural capitalism.

In this sense, regenerative design it is a school of thought that focuses on the theory of systems oriented to design processes. That is, the theory emphasizes the fact that existing processes are modified to improve, eliminate or adhere new sources of energy and/or materials (Morlet et al. 2016). Therefore, the regenerative design has a base derived from the ecology of the systems that is responsible for providing a biokinetics in the ecosystems (Ballie and Woods 2018) with the aim of achieving an ecological economy system (Gleason Espíndola et al. 2018) that is viable and closed for any industry (Liakos et al. 2019). It also seeks to ensure that the resulting system does not generate waste, which is completely effective, to achieve this it is necessary to redesign the culture of human habitats (Heaven Grown s.f.).

According to the European Commission, the regenerative design will impact the processes at the social level through the generation of jobs, in economic competitiveness (Ecointeligencia 2017), in the new distribution in the use of resources and waste. The CE promotes a performance economy in companies (Kumar et al. 2019) through four main objectives to extend the useful life of existing products, generate new products that are considered durable from their design, think of campaigns or activities for the prevention of waste and reuse of these (Bocken et al. 2016; Ellen MacArthur Foundation 2019). Therefore, the green circle economy is one of the essential sustainability factors in SMEs.

2 Green Economy, Green Growth and Sustainable Development

The green economy concept is called the next oxymoron after sustainable development because they overlap each other (Green and McCann 2011). Conference on Sustainable Development (Rio+20) in 2012 agreed that in the context of sustainable development, the concept of green economy should be promoted. The transition to a green economy has economic and social justifications for public and private actors to contribute offering opportunities for investments and green procurement by providing new market-based incentives and mechanisms (Shimova 2019; Popkova et al. 2018).

It is because the green economy prioritizes well-being for the present and for future generations and the efficiency of improved technology is not necessarily sufficient. Likewise, a green economy is essential for sustainable development (Popkova et al. 2018; Shimova 2019), improving social equity, human well-being, reducing ecological scarcity and environmental risks. Green economy supports sustainable development aimed to replace the social and environmental costs of the current economic model that is reaching limits in terms of greenhouse gas emissions, use of natural resources, water, land, forests (Barry 2010).

Green economy (GE) transformation and inclusive greener growth strategy initiatives are needed to pursue the economic and social benefits of sustainable development (Zsolnai 2002) while reducing negative environmental, inequality and poverty impacts, sustainable management of natural resources, reduce greenhouse gas emissions, climate change, resilience to natural disasters, improve public services (Barry 2010; Jones and Wynn 2019). Green economy and sustainable development strategies strengthen the resilience of communities and regions.

In this sense, the GE supports sustainable development aimed at replacing the social and environmental costs (Gliedt and Parker 2007) of the current economic model that is reaching limits in terms of greenhouse gas emissions, use of natural resources, water, land, forests, etc. Initiatives are needed to pursue the economic and social benefits of sustainable development while reducing negative impacts on the environment, inequality and poverty, sustainable management of natural resources, greenhouse gas emissions, change climate, resistance to natural disasters, improvement of public services (Popkova et al. 2018). GE and sustainable development strategies strengthen the resilience of communities and regions such as SMEs (Bıçakcıoğlu et al. 2019; Gliedt and Parker 2007).

Green growth responds to critical emerging issues by facing the global challenge of environmental sustainability. However, the elements of the green economy concept are already integrated in strategic documents focused on achieving sustainable green growth, rather than merely achieving a green economy. Green growth is based upon the sustainable development strategies used to support the transition to green economy defined as the process of improving the economic, social and cultural and environmental well-being of future generations.

Likewise, the GE promotes economic growth and development, ensures the natural assets that provide environmental resources and services for the benefit of humanity's well-being, focusing on synergies and compensation between sustainable environmental and economic development. The interactions between society and the environment drive change and the transformation of the green economy as an opportunity to achieve sustainable development and human well-being. It is worth seeking the green economy as an opportunity to prioritize well-being and sustainable development for the present and for future generations.

According to Klingenberg and Kochanowski (2015) GE is one of the tools used to achieve the sustainable (Shimova 2019) development goals (SDGs) of eradicating poverty, hunger and food security, good health and wellness, education, gender equality and empowerment of women, water and sanitation, energy, economic growth, infrastructure, industrialization, inequality, cities, sustainable consumption and production, climate change, oceans, biodiversity, associations (Popkova et al. 2018).The socio-ecological and SDG indicators of the green economy described go beyond GDP as a transformative concept to measure wellness (Ferguson 2014; Fioramonti 2014; Stiglitz et al. 2009). The visions of the green economy are relevant to the legitimacy and the global green economy under the SDGs.

It should be noted that ecological sectors and industries (Shurrab et al. 2019) have the potential to become engines of ecological growth by reducing the use of fossil fuels as climate-resistant development. Some of the most relevant sectors for the green economy are agriculture (Sbicca 2019), energy, water management, tourism, waste (Haldar 2019). In this sense, green growth policies promote economic growth and environmental development by ensuring that natural resources provide environmental services for human wellness.

Green economy is a critical component of sustainable development which implies a change in the social construct of all the economy sectors. Details about the scale of the specific sectors of the green economy development and implementation are required to be explicitly for absolute decoupling. Rural and urban development programs are a vehicle for enabling the transition from traditional rural and urban economy to a rural and urban green economy. Locally led development programs and group actions can often respond more effectively to local needs in the transition to the green economy. Urban green brands are already developing focusing more on green growth and low carbon economy which may be stronger and attract greater interest.

Green economy approaches offer economic opportunities of low carbon transitions under the framework of an agreement of a collective carbon finance goal (United Nations 2015). Resource and energy efficiency supports green products, services and low-carbon green economy as part of the transition to deliver economic, social and environmental benefits. Low carbon actions are part of a transition towards green economies can make the most of the resources available. The transition towards a green economy by investing and preserving the natural capital to generate growth, create jobs and eradicate poverty. A green job is defined as one that works with information, data, technologies, and materials, and requires specialized knowledge, training, skills, and experience for activities that minimize environmental impact.

The green economy concept is mutually complementary between different dimensions of sustainable development and poverty eradication to enhance convergence through different approaches, among which are the internalization of externalities, systemic economic structure, reconciling social goals, policies and objectives and the macroeconomic framework of development strategy (UNDESA 2010).

The design of green economy to contribute to sustainable development is away from the dysfunctionalities of traditional mainstream economy and results in human well-being and equitable access to opportunities safeguarding economic and environmental economic integrity. Green economy contributes to sustainable development with different forms of implementation for different countries. Resource efficiency is a green economy process is supported by environmental awareness and technological green innovation, although the increase in consumption may occur when efficiency gains are lost leading to the so called Jevons paradox, which may be addressed by specific policies on fiscal mechanisms and education.

3 Green Economy Principles

Guiding principles of a Green Economy helping practitioners in the application of the green economy concept are sustaining that it is a means for sustainable development, is equitable, just and fair, creating green jobs, protects biodiversity and ecosystem services, provide green resources and efficient green energy within the ecological limits, delivers well-being, access to essential services, poverty reduction, livelihoods, social protection. The principles of a green economy that according to GEC deliver a sustainable, inclusive and participative green economy are: sustainability, justice, dignity, healthy planet, inclusion, good governance and accountability, resilience, efficiency and sufficiency and generations.

The sustainable development and green economy strategies and policies is a model based on the core principles of economic efficiency, equity, social inclusion, environmental sufficiency and accountability which requires dialogue among all the involved stakeholders and participative policy design.

Also, these principles consider the measurement of green economy using appropriate metrics and indicators, internalizes externalities, improves governance and the rule of law being democratic, participatory, inclusive, transparent and accountable and more other principles. A good indicator of the relevance of green growth and economy for a specific society is to identify the number or share of population involved in any form, as employee, consumer, etc.

Equitable green economy is linked to sustainable development supported by principles and informed by policy and market decisions. Green economy principles must be integrated in sustainable development programs and initiatives such as in pollution prevention and sustainable production and consumption. Green economy principles aimed to develop a fair and inclusive economy to provide a better quality of life are the sustainability, justice, dignity, healthy planet, inclusion, good governance and accountability, resilience, efficiency and sufficiency and generations principles (Green Economy Coalition 2012).

Green economy principles can be applied to urban sustainable development by coordinating some deliverables with other organizations and donors relating to methodologies, platforms, best practices and tools that contribute to poverty eradication (UNCSD 2012).

Green economy principles must be developed after engaging in discussion with the different stakeholders to meet the vision, priorities and needs of each sector. A set of green economy principles emerging from dialogue can serve as guidelines for making decisions which can be applied across sectors and institutions to operationalize a green economy. Social equity, ecological limits and community ownership are core principles for green economy.

4 Transformational Transition of Green Economy

The transition towards green economy can solve the interconnected economic, social, and environmental crises. Green economy and transformation as enabling approaches are related to potentially create dynamic change (Pelling and Navarrete 2011; Pelling et al. 2014). Green economies integrate economic, social and environmental activities. The concept of green economy represents a transition for more environmentally friendly and resource-efficiency technologies to tackle environmental degradation by reducing carbon emissions and mitigating the effects of climate change (Jänicke 2012).

Transition to green economies require that green activities and investments in the interlinkages between rural–urban areas can contribute to green economic growth. Public and private funding of green economy has to be scaled up at all institutional levels and sectors supporting sustainable and responsible green investments in green business and companies, clean technology, green investments, etc. A green company is defined as a company that produces goods and services designed to reduce their environmental impact.

Green technologies should be developed with government financial support and subject to wider dissemination. Private investment flows in green technology should ensure that achieve full potential in spin-off benefits. Any percentage of global GDP invested to green economy sectors increase the growth, employment and reduce water, energy, etc. Public and private investments in green economy, promote revenue growth and employment from a rational use of natural and financial resources and energy efficiency that reduce carbon emissions and pollution and prevent loss of biodiversity and environmental services.

Development programs structured according priorities supports transition to the green economy in practice although the term may not be explicitly used, planning authorities may use different approaches and measures. Design and implementation of rural and urban and urban development programs support the transition of business activities to the green economy and the environmental performance. Development programs can provide financial support to support the transition towards a green economy with impact in long-term business opportunities. Business can foster practices that contribute to the green economy transition.

Equitable green economy is a transformation process in constant dynamic progression, although it has been questioned if green economy is equitable. One of the four green economy typologies is green transformation of economic growth through political interventions (Death 2015, 2216). Building on Ferguson's typologies based on weak/strong green economy, the UNEP's concept is more transformational providing enabling conditions for green economy transitions (UNEP 2011). Transformational green economy renders strong green economy and growth concepts deployed as organizing principles for climate change (Pelling et al. 2014).

The term green economy comprises the application of some economic instruments which requires social, institutional and political contexts to harness economic activities in support of sustainable development goals. Design of development plans can contribute to the transition to the green economy activities such as mitigation of climate warming, sustainable water and waste management, sustainable infrastructure, ecosystems services and buildings, investment in natural resources and capital, renewable energy feedstocks and energy efficiency, green research, green tourism and eco-innovation, agricultural and forest land management, forestry and fisheries (bio economics) green manufacturing and supply chain green public procurement, etc.

The transition to the green economy makes sure that agriculture and forestry are both economically and environmentally sustainable activities for the long-term. For example, the use of procurement policies for the greening of business. Green agriculture requires natural and physical capital assets, knowledge and financial investments and enhance the capacity building in efficient and sustainable management of soil fertility, water use, farm mechanization, crop and livestock diversification; etc. the analysis of investments measures benefits and costs of green economy and green energy policies taking into account capacity building, management, operation, research and development, expenditures in infrastructure, incentives, etc.

Measures for business and farm diversification can support transition activities to green economy. The sectors considered to have green potential are energy renewable, water, waste/recycling, sustainable farming and forestry, fisheries, public transport, green buildings, tourism, health care, education and training (Rosenberg et al. 2018), green finances, etc.

The fishery sector requires strengthening the fisheries management and financing fishing activities to maintain sustainable stocks within biological limits limit the environmental impact. Assessment of the impact on the dynamics of the fishing and marine ecosystem and biodiversity using quantitative indicators for socio-economic factors is required for a more effective exploitation of fisheries.

The sector of forestry must be focused on reducing deforestation and increasing reforestation in accordance with economic and market mechanisms including payments for ecosystem services, certified benefit sharing and other schemes, community-based partnerships, sustainable forest management instruments aimed to carbon reduction, enhance protection of forests against fires and pollution, biodiversity and forest ecosystems, provision of environmental services, etc.

The industry and manufacturing sector implies design to extend the useful life of goods and recycling them to support the use of by-products and alternatives for substitution to achieve a circular economy with a close-loop manufacturing in eco-industrial parks. The building sector requires a policy framework with instruments for development of sustainable building capacities and standards, cost-efficiency and incentives. Building requires investment and incentives for energy supply and renewable performance for sustainability of new and renovated buildings.

Transport policies intend to integrate land use and transportation planning for more environmentally efficient modes shifting to non-motorized transport and improving vehicle and fuel technology, avoiding or reducing trips and using water and rail transport for freight. All these policies are aimed to reduce the negative environmental and social effects. A greening transport policy framework to enhance sustainability though greener and efficient roads includes a strategy focused on the cost of transport in terms of environmental damage to society and reducing noise pollution.

The tourism sector can be done by increasing the involvement of the local community in the tourism value chain and the interplay between internal factors and structural conditions.

The waste management sector requires decoupling waste from economic growth and addressing the challenges of increasing the recycle rate of electrical and electronic equipment or e-waste, turning bio mass waste into recovered energy and other valuable resources, reducing food waste in the food chain, etc. The treatment of waste in the whole cycle from waste generation to waste disposal, should emphasize recovery for reuse and recycling of waste materials. Strategy for the prevention of industrial waste are based on industrial symbiosis resulting of collaboration to facilitate the exchange of by-products, water, energy and other materials.

The water sector requires water management based on quality standards, to increase investments with better financial arrangements, to achieve a more efficient water supply, to improve the institutional arrangements and allocation systems, entitlements and use pf payments for ecosystem services.

Green economy transition is relevant to all economic sectors and requires a change in their economic activities perspective. Green economic transformation should be supported by new institutional forms for organization and decision making supported by participation and collaboration structures between public, private and community agents and actors for sharing resources and knowledge in green economic activities. One of these economic transformations into green economy is the circular economy.

5 Circular Economy (CE)

Since its creation, the Circular Economy has been defined in different ways, in order to make it more understandable and easy to transmit for its application. Here are three ways to define this concept.

Claudia García Caicedo, in her publication Circular economy and its role in sustainable design and innovation, mentions that the Circular Economy aims to achieve product designs that reduce or completely eliminate waste, and also seeks to ensure that products are simple to dismantle-disassemble for reuse in new products. The CB is also responsible for defining business models that are exclusively dedicated to companies that apply the Circular Economy in their processes to achieve sustainable innovation and consequently feel economically motivated to recover their product after fulfilling its main function, use it again in manufacturing and repeating the cycle (Caceido García 2017).

Catalina Balboa and Manuel Domínguez, in their work Circular economy as a framework for ecodesign: the ECO-3 model, define CS as a philosophy of systems organization inspired by living beings, which pursued the change of an economy linear (produce, use and throw) increasingly difficult to implement due to the depletion of resources towards a circular and regenerative model, as occurs in nature and which also represents a great opportunity in the business world (Balboa and Somonte 2014). The interesting thing in the definition of Balboa and Dominguez is that they mention it as a way to try to solve the problem of scarcity of resources.

The Ellen MacArthur Foundation mentions that the Circular Economy usually has other names related to the schools of thought on which it is based, and that were already mentioned above, for example: economy of the cradle to the cradle or economy of closed loop (Ellen MacArthur Foundation (2) 2019). It is important to mention that this foundation declares that the defenders of the theory do not consider the Circular Economy as part of an ecological movement, but as a form of improvement of the design.

From the three definitions given above, it can be said that circular economy decouples gradually the economic activities from the consumption of resources and energy that are finite to underpinning a transition towards renewable energy and natural resources while keeping materials, goods and services in use, designing out waste and pollution out of the system and regenerating the natural systems.

6 Conceptual Background

6.1 Strategy

To define the strategy, the present work was based on the works of Michael Eugene Porter. This author defines the strategy as a differentiator that is created by making choices about several options that in the end would generate a unique value combination. Based on the conclusion by Porter, what really defines a strategy are the activities to which it specializes, that is, decision making is conditioned by the company's interest in differentiating itself from the competition. If the above is not respected, competitive advantage would not work as a differentiator but as an idea of marketing (Porter 1996).

Porter talks about how a strategy can scale a company in the market, and become a strategic position. This position comes from three sources that sometimes work together:

(1) Positioning based on the variety of products or services that exist in the area. The companies that use this positioning are usually those that have a better possibility of producing some good or service due to certain special characteristics that only they handle in the process.
(2) Positioning based on needs, is one that is responsible for trying to meet the needs, or most of these, of a certain group of people. In a market there are many types of customers who request special products, which require certain characteristics in the good they want to buy, with different tastes or preferences, so they usually need guidance, support or very specific services.
(3) Positioning by customer segmentation according to the way to access them or positioning based on access. Normally, this type of positioning is determined by the position or geographic location in which the client of interest is located. It can also be determined by the dimension or some specific situation that would hinder or hinder easy access to the client.

According to Porter, usually more threats come from sources outside the organization. When a strategy becomes part of a competitive advantage, it is likely to be threatened by changes in areas such as technology or the actions of competitors. It is mentioned that the event that can cause a certain strategy to fail is internal to the firm, and is mainly due to the underestimation of rival companies in the industry, poor planning, lack of information, or great ambition to grow without finalizing details.

6.2 Sustainability

The term Sustainability does not have a precise definition by itself, it is a rather ambiguous term that derives from the word sustainable, an adjective that implies "that can be maintained for a long time without exhausting resources or causing serious damage to the environment" (RAE 2017, page sp). We can also find that Sustainability comes from Latin etymologies such as sustenance, sustenance, sustentare, sustentavi, sustentatum, which mean: sustain, maintain in good condition, care, conserve, support, favor (Ecología UNAM 2015). In practice, we define Sustainability as a process that aims to ensure the satisfaction of the needs of the current and future generations.

6.3 Competitive Advantage

Competitive advantage can be defined as the essential aspect that demonstrates the performance of markets that are competitive, over the years the focus on competitive advantage has been lost to focus on the diversification and growth of organizations (Porter 2015). Porter declares that the source or origin of the competitive advantage is the value that the firm generates in its products or services to satisfy the clientele, in other words, it is considered as a plus that manages to surpass the competition, even when the rivals try reach the company that has an advantage of this kind. In Porter's book "Competitive Strategy" he describes three general strategies for achieving competitive advantage: cost leadership, differentiation and concentration.

7 Theoretical Background

The main objective of any firm is to generate high rates of return, in other words, obtain profits. Because of this, in the research work Sustainable Competitive Advantage: Combining Institutional and Resource-Based Views, Christine Oliver decided to create a hybrid model that would include the Approaches Based on Resources and Institutions, in this way the Model of the Advantage was born. Competitive Sustainable This author mentions that the reason why the resource-based approach is not only used is due to its limitations (Oliver 1998):

(1) Explain the heterogeneity of companies through the properties of resources and the markets of resources.
(2) Does not worry about including the social context, which affects the decision making about the use of resources.
(3) It does not talk about how the selection of resources is made.

By including the Institutional Approach, we seek to complete the spaces left, in the social sphere, by the Resource-based Approach, so that the Institutional will contribute (Oliver 1998):

(1) A study on how social influence affects decision making within a company.
(2) It will show us the close relationship that exists between the selection of resources and sustainable competitive advantage in relation to decision making.
(3) The importance of having an "institutional context" in the three levels of the company.
(4) Individual level, are the normal and individual values.
(5) Company level, are represented by the organizational culture and politics.
(6) Inter-company level, as an example: public relations and its pressure on the market, regulations and standards faced by firms.

Oliver also inspected the general notions of the new hybrid approach of the Sustainable Competitive Advantage, which turned out to be the following:

(1) The model divides decision making into three levels, as does the institutional one: Individual or managerial choice, company level, and inter-company.
(2) Includes the way in which managers select resources and capacities, that is, the decision of which resources and capacities to implement.

(3) Determines that in order to create and apply strategies, resources and institutional regulations must be taken into account.
(4) It defines what is a capacity, resources and its idea of sustainability of an advantage that must be competitive.
(5) Its analytical model consists of three determinants that assure us a sustainable competitive advantage, if it is integrated in the right way: those based on resources, such as managerial decisions; the selection of resources, the heterogeneity of the company; the institutional determinants: rational/individual regulation, institutional/business factors, and isomorphic pressure/between companies.

In relation to the selection of resources for its application, this new approach mentions that there are three cases in which it is more likely that a company is willing to acquire them or use the ones they have in reserve (Oliver 1998):

(1) When resources are acquired that are not major for the company, in this way we managers do not feel that they are risking their main activity.
(2) Companies tend to be traditional, they are affected by their institutional part, so the acquisition of resources must belong to the same item that the company manages.
(3) When a resource of the company is no longer considered productive, it is time to acquire new, since the organization does not feel so threatened by the change. The opportunity cost will have less impact.
(4) The accumulated resources should be periodically monitored to know what we have and what we can use.
(5) The training of the assets of the company helps them to know how to use potential resources.
(6) From the moment of hiring, people with attitudes that have a notion of the use of resources for an optimal management should be chosen.

From the point of view of the Institutions approach, there are certain assumptions that would allow achieving a sustainable competitive advantage, in relation to the use of valuable resources (Oliver 1998):

(1) The acquisition of a valuable resource will be accepted by the company when it does not violate the regulations or the corporate culture.
(2) The acquisition will be accepted if senior management gives it political support.
(3) For reasons of power struggle, certain valuable resources will be acquired if with it the power of a decisive voter increases or strengthens its place in the company.

The creation of a model that unites the two approaches was necessary, since all approaches have deficiencies that tend to focus too much on production, resources or institutions and their regulations at different levels. This theory also shows concern for the factors that affect the individual as such and the impact that this would have on the decision-making process when selecting resources and applying them in the correct manner. The Focus of the Sustainable Competitive Advantage can be considered as an advance in the evolution of the strategies, since not only is oriented to obtain an objective, but also it is in charge of making known how to achieve it.

8 Research Methods

Although the theoretical perspective that will be used in the present work is the theory of resources and capacities slightly influenced by the approach of the institutions, an analysis will be carried out through the VRIO Framework (Value, rarity, inimitability, and organization) to determine if The Circular Economy model could be considered as a viable resource as a competitive advantage that allows the company to achieve its objectives and position itself in the market, which would generate a competitive strategy.

VRIO is a method used to analyze the resources and capabilities of firms with the aim to determine the sources of sustainable competitive advantages. The tool was developed by Barney (1995). The VRIO analysis begins questioning if the resources and capabilities are valuable and if the result is negative, there is a competitive disadvantage. The second question is if the resources and capabilities are rare. If the answer is not, then there is a competitive parity. The next question is related with the cost to imitate and if the answer is not, there is a temporary competitive advantage. Finally, the four question is related with the issue if resources and capabilities are organized to capture value and if the answer is not, there is also a temporary competitive advantage. However, if the answer is yes for the four questions, it can be concluded that there is a competitive advantage in resources and capabilities.

9 Analysis of Results

As already mentioned before, the tool that will be used to analyze the situation of the Circular Economy as a competitive resource within an organization, is through an internal analysis called Marco VRIO. First the analysis will be shown in the form of a table and after the explanation will be made by means of the answer to the four questions that this frame generates.

To obtain the results, the research work matrix was used as a base: ICT as a source of competitive advantage in SMEs (Moncada Niño and Oviedo Franco 2013). The pertinent changes were made to be able to apply it in this specific case, shown in Table 1.

9.1 VRIO Analysis (Table)

See Table 2.

Table 1 Matrix VRIO—competitive Implications

Valuable?	Rarity?	Expensive to imitate?	Exploited by the organization?	Strength or weakness	Competitive implication
NO			NO	Weakness	Competitive disadvantage
SI	NO			Strength	Competitive parity
SI	SI	NO		Strength	Temporary competitive advantage
SI	SI	SI	SI	Strength	Sustainable competitive advantage

Source Moncada Niño and Oviedo Franco (2013)

Table 2 VRIO analysis: the Circular Economy as a business resource

Concept	Answer
Valuable: is it a valuable resource to achieve an advantage?	Yes
Rarity: is it being used by a small group of firms?	Yes
Inimitability: is it expensive to imitate?	Yes
Organization: Is the organization prepared in its policies and procedures to use this resource?	NO
Strength or weakness: is it considered a strength or a weakness	Strength
Competitive implication: results	Temporary competitive advantage

Source Own elaboration

9.2 VRIO Analysis (By Concept)

Next, the explanation of the previous table, concept by concept, in addition to the specific interpretation to this particular case.

Valuable: answer the following question: is it a valuable resource to gain an advantage? According to Álvaro Fernando Moncada Niño and Martha Lucía Oviedo Franco in their work, valuable resources are considered those that can be used as a response to external threats, and in turn, help take advantage of opportunities. "The definition of the value of the resource or capacity is related to its possibility to exploit an opportunity or mitigate a threat in the market. If one of those two things is done, it can be considered as a strength of the company; otherwise, it is a weakness. When these are properly exploited, they generally lead to an increase in income or a decrease in costs or both" (Moncada Niño and Oviedo Franco 2013, page 129).

Due to the above, the Circular economy can be considered a valuable asset, by exploiting the opportunity to reduce costs and reduce the waste generated by the company through a reengineering of processes and design.

The VRIO framework considers that resources must be rare, limited or unique, that is, very few companies are using it in their activities, otherwise the resource would not serve as a competitive advantage. If the rarity remains, and few companies manage to acquire it this would mean that the resource would remain scarce, which would give it the characteristic of sustainable competitive advantage (Moncada Niño and Oviedo Franco 2013). The Circular Economy is a scarce resource that has not been applied in a large number of companies due to its complexity, but it is very likely that this rarity is not held too long, because of the changes in the policies related to sustainability and its derivatives, so it is considered as a temporary competitive advantage.

Inimitability: is it expensive to imitate? "… resources are inimitable when the possibility for competitors to analyze and duplicate them makes their acquisition or acquisition costly or takes too long to replicate" (Moncada Niño and Oviedo Franco 2013, page 129). This characteristic of the VRIO Framework is usually related to the previous two, since the cost of use or application directly affects the rarity and its value. The CE is considered an inimitable resource, because when applied it would be considered quite expensive, since the generation of totally new designs thinking about the reduction of waste is expensive at the beginning, not any company can achieve it.

Organization: Is the organization prepared in its policies and procedures to use this resource? "Relating to the fact that the company has certain organizational aspects, such as the organizational structure, processes and systems, as well as the business culture itself, to exploit the full competitive potential of its resources and capabilities. Therefore, the resources and capacities have to be exploited efficiently by the company" (Moncada Niño and Oviedo Franco 2013, page 130).

When referring to this characteristic, it is deduced that the companies are not prepared, in their great majority, to install and use the Circular Economy, due to the great complexity changes and improvements necessary for an optimal operation.

10 Conclusions and Recommendations

At the beginning, it was mentioned that this research aimed to determine if the circular and green economy is a resource contributing to transformational transition to become a competitive advantage that has an impact on sustainable development based on the focus of resources and capabilities. In this specific case, it turns out that, if it has the majority of the features of the VRIO Framework that allow it to be a competitive advantage, but of a temporary nature, because in the rarity concept it is considered that in the coming years the business political demands will change, forcing economies and companies to look for ways to include this type of model, regardless of the cost.

As a recommendation, it can be highlighted that organizations that wish to implement this transformational transition model in their sustainable development activities will need to have a good economic position. The initial costs of application are usually very high as a result of the changes that must be made from the root of the product: the design of this.

According to the results obtained in the present investigation, the CGE is considered as one of the main factors that SMEs should consider carrying out re-engineering in the products they develop as in those that they generate through the innovation they generate. In this sense, the results are focused on those proposed by Ünal et al. (2018).

Likewise, for the business sector, the CGE turns out to be a competitive advantage that can be used to develop transformational transitions towards sustainable development through the use of resources and capabilities, being a strategy that can position not only in national markets but also international markets. In conclusion, it was proved that the Circular Economy can be considered as a competitive advantage in transformational transitions towards sustainable development, but of a temporary nature according to the characteristics of the VRIO Framework.

References

Alvial Muñoz A (Abril de 2015). Economía Azul: Una revisión en el marco de nuevas tendencias en Economía. Retrieved from Bioeconomía Argentina http://www.bioeconomia.mincyt.gob.ar/wp-content/uploads/2014/12/1.-Econom%C3%ADa-azul-A.-Alvial.pdf

Balboa CH, Domínguez Somonte M (25 de Mayo de 2014) Economía circular como marco para el ecodiseño: el modelo ECO-3. Retrieved from Universidad Nacional de Educación a Distancia. https://www2.uned.es/egi/publicaciones/articulos/Economia_circular_como_marco_para_el_ecodiseno_el_modelo_ECO-3.pdf

Ballie J, Woods M (2018) Circular by design: a model for engaging fashion/textile SMEs with strategies for designed reuse. In: Crocker R, Saint C, Chen G, Tong Y (eds) Unmaking waste in production and consumption: towards the circular economy. Emerald Publishing Limited, pp 103–121. https://doi.org/10.1108/978-1-78714-619-820181010

Barney JB (1995) Looking inside for competitive advantage. Acad Manag Exec 9(4):49–61

Barry J (2010) Chapter 6 Towards a model of green political economy: from ecological modernisation to economic security. In Leonard LA (ed) Global ecological politics (Advances in ecopolitics), vol 5. Emerald Group Publishing Limited, Bingley, pp 109–128. https://doi.org/10.1108/S2041-806X(2010)0000005010

Bıçakcıoğlu N, Theoharakis V, Tanyeri M (2019) Green business strategy and export performance: an examination of boundary conditions from an emerging economy. Int Mark Rev 37(1):56–75. https://doi.org/10.1108/IMR-11-2018-0317

Bocken N, De Pauw I, Bakker C, Van der Grinten B (2016) Product design and business model strategies for a circular economy. J Ind Prod Eng 33(5):308–320. https://doi.org/10.1080/21681015.2016.1172124

Brundtland GH (1987) Our common future. Retrieved from UN http://www.un-documents.net/our-common-future.pdf

Caceido García CL (2017) Economía circular y su papel en el diseño e innovación sustentable. Obtained from Libros Editorial UNIMAR http://ojseditorialumariana.com/index.php/libroseditorialunimar/article/view/1154

Cezarino L, Liboni L, Oliveira Stefanelli N, Oliveira B, Stocco L (2019) Diving into emerging economies bottleneck: industry 4.0 and implications for circular economy. Manag Decis. https://doi.org/10.1108/MD-10-2018-1084

Ecointeligencia (27 de Enero de 2017) ¿En qué consiste la Economía del Rendimiento? Retrived on May of 2019, de Ecointeligencia https://www.ecointeligencia.com/2017/01/economia-rendimiento/

Ecología UNAM (02 de Junio de 2015). Fundación UNAM. Retrieved from UNAM http://www.fundacionunam.org.mx/ecologia/sostenibilidad-vs-sustentabilidad/

Ellen MacArthur Foundation (2) (2019) Economía Circular: Concepto. Retrieved on the 10 of May of 2019, from Ellen MacArthur Foundation https://www.ellenmacarthurfoundation.org/es/economia-circular/concepto

Ellen MacArthur Foundation (2019) Economía Circular: Escuelas de pensamiento. Retrieved on the 16 of May of 2019, from Ellen MacArthur Foundation https://www.ellenmacarthurfoundation.org/es/economia-circular/escuelas-de-pensamiento.

Ferguson P (2014) The green economy agenda: business as usual or transformational discourse? Environ Polit 24:17–37

Fioramonti L (2014) The world's most powerful number: an assessment of 80 years of GDP ideology. Anthropol Today 30:16–19

Gleason Espíndola J, Cordova F, Casiano Flores C (2018) The importance of urban rainwater harvesting in circular economy: the case of Guadalajara city. Manag Res Rev 41(5):533–553. https://doi.org/10.1108/MRR-02-2018-0064

Gliedt T, Parker P (2007) Green community entrepreneurship: creative destruction in the social economy. Int J Soc Econ 34(8):538–553. https://doi.org/10.1108/03068290710763053

Green D, McCann J (2011) Benchmarking a leadership model for the green economy. Benchmarking Int J 18(3):445–465. https://doi.org/10.1108/14635771111137804

Haldar S (2019) Green entrepreneurship in the renewable energy sector—a case study of Gujarat. J Sci Technol Pol Manag 10(1):234–250. https://doi.org/10.1108/JSTPM-12-2017-0070

Heaven Grown. (s.f.). Diseño Regenerativo. Retrieved on Mayo de 2019, from Heaven Grown http://heavengrown.com/arquitectura-regenerativa/

Jänicke M (2012) Green growth: from a growing eco-industry to economic sustainability. Energy Policy 48:13–21

Jones P, Wynn M (2019) The circular economy, natural capital and resilience in tourism and hospitality. Int J Contemp Hosp Manag 31(6):2544–2563. https://doi.org/10.1108/IJCHM-05-2018-0370

Kavinski H, De Souza-Lima JE, Maciel-Lima SM, Floriani D (2010) La apropiación del discurso de la sustentabilidad por las organizaciones empresariales brasileñas. Cultura y representaciones sociales 4(8):34–69

Klingenberg B, Kochanowski S (2015) Hiring for the green economy: employer perspectives on sustainability in the business curriculum. J Manag Dev 34(8):987–1003. https://doi.org/10.1108/JMD-06-2014-0058

Kumar V, Sezersan I, Garza-Reyes J, Gonzalez E, AL-Shboul M (2019) Circular economy in the manufacturing sector: benefits, opportunities and barriers. Manag Decis 57(4):1067–1086. https://doi.org/10.1108/MD-09-2018-1070

Liakos N, Kumar V, Pongsakornrungsilp S, Garza-Reyes J, Gupta B, Pongsakornrungsilp P (2019) Understanding circular economy awareness and practices in manufacturing firms. J Enterp Inf Manag 32(4):563–584. https://doi.org/10.1108/JEIM-02-2019-0058

Missé A, Moreno JA, Vázquez Oteo O, Escorsa P, Casado Cañeque F (2015) Responsabilidad Social de la Empresa: ¿RSE o RIP? Retrieved from jstor http://www.jstor.org/stable/26360524

Moncada Niño ÁF, Oviedo Franco ML (17 de Junio de 2013) Las TIC como fuente de ventaja competitiva en las PYMES. Sotavento M.B.A., 21, 126–134. Retrieved on May of 2019, from Universidad Externado de Colombia https://revistas.uexternado.edu.co/index.php/sotavento/article/view/3441/3128

Morlet A, Blériot J, Opsomer R, Linder M, Henggeler A, Bluhm A, Carrera A (2016) Intelligent assets: unlocking the circular economy potential. Ellen MacArthur Foundation, pp 1–25

Oliver C (04 de Diciembre de 1998) Sustainable competitive advantage: combining institutional and resource-based views. Retrieved on Mayo of 2019, from Strategic Management Journal http://www.jstor.org/stable/3088134

Pelling M, Manuel-Navarrete D (2011) From resilience to transformation: the adaptive cycle in two Mexican urban centers. Eco Soc 16:1–11

Pelling M, O'Brien K, Matyas D (2014) Adaptation and transformation. Climatic Change 133:113–27

Popkova E, Bogoviz A, Ragulina J (2018) Technological parks, "Green Economy," and sustainable development in Russia. In: Sergi B (ed) Exploring the future of Russia's economy and markets. Emerald Publishing Limited, pp 143–159. https://doi.org/10.1108/978-1-78769-397-520181008

Porter M (Diciembre de 1996). ¿Qué es la estrategia? Retrieved from el Mayo de 2019, de Harvard Business Review https://s3.amazonaws.com/academia.edu.documents/37851742/4_Que_es_Estrategia.pdf?AWSAccessKeyId=AKIAIWOWYYGZ2Y53UL3A&Expires=1558381861&Signature=132Q27yedrcb1MADZWvjsnckcC8%3D&response-content-disposition=inline%3B%20filename%3DQue_es_la_estrategia.pdf

Porter M (2015) Ventaja Competitiva: Creación y sostenimiento de un desempeño superior. Retrieved from el Mayo de 2019, de Grupo Editorial Patria https://books.google.com.mx/books?hl=es&lr=&id=wV4JDAAAQBAJ&oi=fnd&pg=PT3&dq=que+es+ventaja+competitiva&ots=mwvClbT58A&sig=O2eioD4ADfMxwU5PBfOy20qHSA#v=onepage&q=que%20es%20ventaja%20competitiva&f=false

Puentes-Poyatos R, Yebra-Rodríguez Á, Guerrero F (2016) Responsabilidad Social Corporativa: El compromiso de la Universidad con los ciudadanos. Retrieved from Revista de Antropología Experimental http://revistaselectronicas.ujaen.es/index.php/rae

RAE (Diciembre de 2017) Real Academia Española. Retrieved from DLE http://dle.rae.es/?w=diccionario

Rosenberg E, Lotz-Sisitka H, Ramsarup P (2018) The green economy learning assessment South Africa: lessons for higher education, skills and work-based learning. High Educ Skills Work-Based Learn 8(3):243–258. https://doi.org/10.1108/HESWBL-03-2018-0041

Saavedra García ML, Saavedra García ME (2014) La PYME como generadora de empleo en México. Revista Clío América, pp 153–172

Sbicca J (2019) Urban agriculture, revalorization, and green gentrification in Denver, Colorado. In: The politics of land (Research in political sociology), pp 149–170. https://doi.org/10.1108/S0895-993520190000026011

Secretaria Internacional de la Carta de la Tierra (2019) La Carta de la Tierra. Retrieved from http://cartadelatierra.org/descubra/la-carta-de-la-tierra/

Shimova O (2019) Belarus on the way to sustainable development: circular economy and green technologies. In: Sergi B Modeling economic growth in contemporary Belarus (Entrepreneurship and global economic growth). Emerald Publishing Limited, pp 89–106. https://doi.org/10.1108/978-1-83867-695-720191007

Shurrab J, Hussain M, Khan M (2019) Green and sustainable practices in the construction industry: a confirmatory factor analysis approach. Eng Constr Archit Manag 26(6):1063–1086. https://doi.org/10.1108/ECAM-02-2018-0056

Stiglitz JE, Sen A, Fitoussi J-P (2009) Report by the commission on the measurement of economic performance and social progress. Commission on the Measurement of Economic Performance and Social Progress, Paris

Ünal E, Urbinati A, Chiaroni D (2018) Managerial practices for designing circular economy business models. J Manuf Technol Manag. https://doi.org/10.1108/JMTM-02-2018-0061

UNAM (02 de Junio de 2015) Ecología UNAM. Retrieved from Fundación UNAM http://www.fundacionunam.org.mx/ecologia/sostenibilidad-vs-sustentabilidad/

Zsolnai L (2002) Green business or community economy? Int J Soc Econ 29(8):652–662. https://doi.org/10.1108/03068290210434198

The Ecosystem Approach to Assessing the Quality of the Urban Environment and Managing Urban Development

Nataliya V. Yakovenko and Igor V. Komov

Abstract The issues of the urban environment, the comfort of cities for life, their ability to act as generators of innovations in various fields of life and points of attraction for the creative class are being actively discussed among researchers and politicians. The essence of urbanization in both territorial and environmental-economic aspects lies in the formation of large urban settlements, including in their area an increasing number of rural settlements, which ultimately acquire the functional significance of an urban character. The proposed research considers the urban area as a complex integrated ecosystem with specific regulatory functions to manage these processes. The purpose of this research is to consider the area inhabited by people, including urban education, as a triune system of "nature—economy—society", to determine the basic conditions for the ecosystem approach implementation to assessing the quality of the urban environment and managing urban development.

The main contradictions characterizing the processes of environmental and economic interaction in urban areas are being formed in the following areas:

1. between the increasingly large-scale targets of municipal socio-economic development and the strengthening of limited natural resources, and the environmental degradation;
2. the level of reflection in the educational processes of the complexity of the problems of interaction with the environment is inadequate to existing realities;
3. between the increase in the parameters of the use of natural resources and the insufficient level of technological realization of knowledge about the features of natural processes and phenomena.

All of the above allows us to justify the approach to the interpretation of the essence of the modern city as an urban ecosystem, which is a set of interdependent factors of economic, technical, technological, social and natural origin, which are manifested in the urbanized area.

N. V. Yakovenko (✉)
Voronezh State University Forestry and Technologies named after G.F. Morozov, Voronezh, Russia

I. V. Komov
Voronezh State University, Voronezh, Russia
e-mail: igrkom@bk.ru

© The Author(s), under exclusive license to Springer Nature Switzerland AG 2021
W. Leal Filho et al. (eds.), *Innovations and Traditions for Sustainable Development*,
World Sustainability Series, https://doi.org/10.1007/978-3-030-78825-4_6

In this work, the urban area is considered as a difficult complex ecosystem with specific regulatory functions for managing these processes.

The purpose of this work is to develop the principles of environmentally-oriented management of urban development and substantiate approaches to creating methods for evaluating the effectiveness of decisions.

Keywords The ecosystem approach · The quality of the urban environment · Urban development

1 Introduction

Society and nature as two systems do not exist independently. In fact, they are always merged into cohesive ecosystem. Stability of the society and all its integral parts nowadays turns to be a burning issue. There is no any social system, anywhere from a state to a city or even more to a village, which is protected from social catastrophe of certain rate, intensity and severity. Such responsiveness of the state of the society can be surely determined as typical. However, the essence of such regularities is still understudied. Currently sustainable development appears as the strategic goal in social and economic political activity of almost all developed countries covering national, regional and local levels.

Initially the term "sustainable development" for the first time was used in 1987 in the report called "Our common future", made by the World Commission on Environment and Development (UN), and meant such type of development, which "satisfies requirements of the modern world and at the same time doesn't disrupt capacity of the next generation to satisfy their own demands" ("Our common future" 1987).

The beginning of the twenty-first century was marked by escalation of such essential global issues in economic, ecological and social spheres as population growth, urbanization and climate shifts (Yakovenko 2007, 2009, 2010; Yakovenko et at. 2013). According to the UN estimations by the year 2050 the population will increase to the level of 9 billion people and approximately 80% will be city residents. Modern cities can be characterized by immense concentration of people, equipment and related flows of substances and energy per unit area, which are transcendent. Natural constituents of urban area, which directly influence human health, security and accommodation comfort in metropolitan cities, suffer extremely high level of anthropogenic and industry-related pressure, to the extent of extinction or considerable losses of existing ecological functions and services.

Sustainable urban development represents one of the top-priority human responsibilities for the nearest future, which is enrolled to the UN programme, national and regional strategies, international consortiums and programmes (C 40 Alliance, U.S. Mayors Climate Protection Agreement). Sustainable urban development supposes rational use of urban nature capital and maximization of urban ecosystem services, which include ecological, social and economic benefits, accessible to the citizens.

Soils and green areas play crucial role in stable functioning of urban ecosystems and perform essential ecosystem services: carbon sequestration, microclimate establishment, reduction of pollution and dust level of atmosphere, water cycle management as well as provide recreational, educational, cultural and esthetic services.

Unfavorable state and irrational use of urban ecosystem components, in its turn, is related to high risks. Thus, polluted and degraded urban soils appear to be a key penetration source of high-density metals, fine dust and pathogens for the human body. Weakened trees amenable to the high level of blowdowns, which is directly connected with significant risks for human life and health, on the ground of extreme weather conditions. Objective analysis of ecosystem services of the urban green infrastructure and soil as well as the estimation of potential ecological, social and economic risks due to its irrational use represents overriding information for the development and management of modern cities.

City functioning is defined by interaction of natural and anthropogenic subsystems and directly depends on the impact of environmentally-dangerous crisscross anthropogenic influences, which are generated by connected in terms of location industrial and residential facilities, and surge capacity of natural subsystems which compose general urban biocommunities.

Menace for urban vital capacity can be caused by structural damage of the system and its operations and deterioration of existence conditions or deterioration of the potential capacity of the system towards goals achievement. We will consider, that there is a threshold of external influence for the system under consideration and the system loses its qualities and functions reaching this line. Environmental resistance represents the fundamental property of urban vital capacity.

The issues of sustainable development and introduction of green economy are nowadays studied extensively by different schools of economic thought. First of all, they include such conventional approaches in the framework of the mainstream of economic theory as economics of nature use, economics of natural resources, human capital theory, public choice theory and collective actions theory. In a framework of heterodox approach can be revealed several types of economy: ecological economics (Shmelev 2012; Daly 2007), evolutionary economics within the macroscale (reports for the Rome club) and green economics (Scott 2009), economics of ecosystems and biological diversity (Helm and Hepburn 2014).

Academic discussion in the sphere of sustainable development nowadays is far from termination. At the moment following issues appear to be the most debated: capability for unconditional stability (Daly 2007; Georgescu-Roegen 1971), methodological background of economics of nature use and ecological economics (Daly and Farley 2004; Illge and Schwarze 2006; Söderbaum 2000). An individual unit of research over questions of stability is represented by highly specialized practice-oriented multidisciplinary approaches, including sustainable city conception, which unites general ideas of environmental economics, economics of nature management and the latest institutional theory (terms of public goods and general resources).

2 Literature Review

Estimation of stability allows to determine terms of urban vitality as urban ecosystem. However, the city appears to be the system which is prone to changes and evolution. For the purpose of survival and stability the system should be able to respond to threats and adjust to the menaces until they cause serious harm (Bossel 2001). Thereby, the term "sustainable development" appears along with the term "stability of the urban ecosystem". As in the case of stability, there is no single approach towards the mentioned issue. The term "sustainable development" can be used relating to the planet in general as well as with regard to individual countries, regions and various business profiles. Extensive theoretical and methodological analysis of the issue of the sustainable development is represented in the scientific work of V.I. Danilov-Daniljyan (2003). According to his point of view sustainable development represents "such social development, which is harmless to its natural base, provided life conditions do not involve human degradation and socially destructive processes do not develop to the scale hazardous to the social security" (Danilov-Daniljyan 2003).

Voronin (2000) defines a problem of the sustainable development as "the occurrence of the balance between static and dynamic stability, i.e. rational degree of decentralization in the structure of the big open system which stimulates its self-development and at the same time provides cohesiveness and security in changing external conditions". Such interpretation, while considering sustainable development as the feature of the city, perceived as the open city, wholly corresponds to the principle of sustainability stated by Le Chatelier. Practical recommendations for the choice of indicators and evaluation methods of stability of the territorial development are represented in the scientific work of H. Bossel. Among the scientific works, directly related to the sustainable urban development, can be marked several books (Tetyor 2011).

3 Methodology

The methodological basis of the study was a systematic approach. General scientific research methods such as synthesis, analysis, generalization and comparison were also used. Periodical, statistical, analytical and translated literature was used as information resources, which investigated the problems solved in the work.

4 Results and Analysis

4.1 Ecological and Economic Approaches to the Formation of Urbanized Territories

The first scientist, who proposed to consider the city as the ecosystem, was an English geographer Ian Douglas. Ecosystem (from the Greek "oikos", which means a residential unit, residence and system, which means combination, corporation) represents an ecological system, community of jointly coexisting organisms and their life conditions, which reside in appropriate interconnection among each other and constitute system of interdependent biotic and non-biotic phenomena and processes.

The difference of the city from the natural ecosystems is evident. Natural ecosystems unite all organisms with one another and external environment. They include imbedded stabilizers, which maintain balance and stability. Changes in one part of the ecosystem affect all other elements. City represents a distinctive ecological system, which includes natural and anthropogenic subsystems. Environmental expert Raimers (1992) suggested to define "unstable anthropogenic system, which consists of architectural and engineering objects and hardly destroyed natural ecosystems" as the urban ecosystem.

Within the cities the natural stabilizers are destroyed. The city lives not only according to natural laws, but also appears as a concentration of social life and in this sense it is set against pristine nature. In order to protect themselves from elemental forces people construct buildings and complicated engineering structures. For the provision of survival needs of citizens water-supply and wastewater disposal facilities, roads and communications are being constructed and developed. High population density within the cities requires development of social institutions aimed at regulation of relations between people and maintenance of order. Nowadays it is increasingly evident that technical and institutional regulators, created by people, are not able to provide stable and persistent urban development. City as an ecosystem appears to be a source of disorder within the surrounding environment (Philipson 2013).

Russian scientist Vladimirov (1999) features the following crucial peculiarities of the urban ecosystem: polymorphy, dependence on allied ecosystems and instability of general structures.

Polymorphy of the urban ecosystem consists in its combination of natural (hydrosphere, atmosphere, etc.) and anthropogenic (buildings, elements of infrastructure, etc.) ecosystems. This feature is connected with the complicity to provide constructive interference within the urban communities aimed at optimization and makes impossible to enhance them by the reconstruction of only one type of structure.

4.2 Conditions that Determine the Effectiveness of the Implementation of the Concept of Sustainable Development at the City Level

Any ecosystem represents an open unit which interacts constantly with the external environment and exchange life sources. The peculiarity of the urban ecosystem resides in its over-openness. A modern city exists on account of other ecosystems and it is not able to feed all their population. It "breathes borrowed air", "drinks borrowed water" and "eats borrowed biomass". Apart from consumption of natural sources and energy, constricted from vast areas, the modern city produces the great amount of wastes, emitting into the environment its metabolic by-products.

Key factors, which define efficacy of the implementation of the sustainable city conception, can be represented in the following way:

1. Formation of harmonized social, ecological and economic system aimed at realization of the interests of urban population on the base of provision of the greatest possible homogeneity and equity in distribution of benefits.
2. Implementation of the environmentally aware approach in the framework of interaction between economic and ecological subsystems based on discussion of the last, not from the point of the required resources acquisition, but in the scope of maintenance of its role of the main sustenance institute.
3. Maintenance of natural ecosystems on the level, sufficient for provision of stable environmental parameters.
4. Inherent necessity to implement innovation-based resource saving technologies, transformation of the urban economy structure and promotion of environmentally-oriented behavior as components of business activities ecologisation (Karpova 2019).

Path finding for "sustainable city" creation was also embodied in a city-planning conception. Evolution of the general conceptual approaches formation towards city-construction with respect to the necessity to make it complement with the environment is represented in Table 1.

4.3 The City as a Dynamic Ecological-Economic System

Under study of the city as the urban ecosystem the system approach should be implemented. The system approach represents a general scientific concept, which implies consistency in approaches and methods in problem solution management. Under the system approach a complex problem should not be divided into parts. The question should be consciously entangled until all essential correlations are taken into consideration. Fundamentally the system approach is based on the general systems theory. The general systems theory (according to L. von Bertalanffy) is a fundamental science, which covers the whole complex of problems, connected with research and

Table 1 Key architectural theories, aimed at creation of "sustainable city" (Peshina 2013)

Theory	Author	Content
Garden-city (1898)	E. Howard	Concept of "satellite cities" Combination of industrial and agricultural production
Segmented city (1918)	G. E. Saarinen	Decentralization of large cities on the base of the urban structure dispersion
Green city (1930)	M. I. Ginzburg, M. O. Barsch	Dezurbanism (désurbanisme) and linear system of, advanced housekeeping and industrialization of house-building
Spacious city (1958)	F. L. Wright	Dezurbanisation. Allocation of functions on agricultural suburbs
Biotechnical city (1969)	P. Solery	Maximum efficient land use and close settlement of people. Building of constructions with self-sustainable infrastructure
Smart city (1970)	Scientists of the Intelligent Building Institution in Washington	Security, comfort and resource saving for all citizens. Productive and efficient use of workspace
Ecocity (1978)	R. Register	Harmony with nature. The city, formed with reference to impact on the environment
Ecopolice (1981)	A. A. Brudny, D. N. Kavtaradze	The city, where people and wild life have mutually supportive relationship

construction of systems. It forms a methodological ground of the system approach in general and a theoretic base for the urban system analysis. System analysis represents interconnected and sophisticated examination of issues, which refer not only to organization, functioning and elimination of the city, but also to the management approaches with reference to legal, economic, natural and geographical, political and other aspects. In terms of municipal administration, the tasks of the system analysis consist in development of methods for informal and formal description of objects of management, identification of their functioning patterns and trends of development, composition of the system theory and establishment of practical techniques aimed at management. In other words, the system analysis is meant to implementation of the general functions of the municipal administration theory in terms of science—descriptive (what is it?), interpretative (how does it work?) and predictive (how does it develop and in which way?).

Establishment of market relations, emergence of private property, formation and development of institutes for democracy and decentralization of public administration initially go hand-in-hand with emergence of a new urban economic and social system. System (from the Greek "composed from parts", "composition", from "connect, compose") represents an objective unity of consistently interrelated units, events and knowledge of nature and society. There are another functioning laws within the system, which differs from laws in the environment (Elohov 2011). According to the system paradigm, changes in external or internal relations, preferences, economic, cultural or other processes and institutes cause changes within the elements of municipal living facilities and municipality in general.

Cities can be classified as *dynamic systems*. Cities are influenced by progressive and regressive development, division, convergence and etc. Any changes within cities are described as processes. It is desirable to foresee all the processes, influence on them in a positive for the municipality direction.

Urban systems act as *probabilistic systems*—their functioning can be predicted with a certain level of probability reviewing the systems past. This peculiarity significantly increases the role of individual, creative initiatives within administration of municipalities and the role of monitoring over key factors, which make municipal development more predictable.

In the context of urban openness nature of its relationship with other systems (state, constituent unit of a federation), which are the integral parts of environment, as well as the city itself, appears to be a crucial point. The city is rather independent from the environment, which suggests that it reacts on environmental influence and it evolves, but at the same time the city maintains qualitative determinacy and quality (as local governance), which distinguishes it from other systems (public and regional governance).

The key notion of the systems theory and the defining feature of any system is its unity.

Unity can be defined as a quality to characterize sustainability of the municipal system within its minimal structural complexity and minimum required resources. Absence of necessity to add or remove its specific structural elements in order to enhance efficiency and stability of operation appears to be the criterion of the unity.

Joint operation of heterogeneous interrelated elements of the city municipality creates non-additivity—the new functional property, which means, that municipal property does not represent a sum of properties of its integral elements and elimination of properties of the entire system out of elements properties. Non-additivity lies in the fact that decomposition of the urban system causes the inevitable disruption of horizontal and cross links, which characterize unity of the system. Thus, due to the decomposition of the system, it unavoidably loses qualities, aimed at provision of social and economic stability for the city functioning.

Non-additivity describes urban system as the product of integration, mergence into the unity of several elements and their interaction, reflection of some common cooperative properties of the given set of elements and events. The criterion for the well-organized, coherent system is that the whole is always greater than the sum of its parts, while in the disorganized and fragmented system the whole is usually substantially less than the amount.

The essential characteristic of the city in terms of coherent system is its structure. The term structure means the combination of components and connections, which define internal constitution and organization of the municipality in terms of coherent system. Such consideration of the city defines structure as a way to describe its organisation (Fig. 1), subsystems connections are accompanied by the relevant information flows).

The leading role within the strategic planning and management of the city in terms of EES should be given to the hierarchy of ecological and economic urban development factors, connected with the essence of such Russian science as patoeconomics (Luzin and Pavlov 1995)—the science of issues in implementation of reforms performed within economic systems, which is directly influenced by the content and state of ecological and economic system. At the same time objects and subjects of ecological and economic system, such as economic entities, environment and society, coincide with the subjects of patoeconomics. The crisis processes appeared within the subjects are abnormal for the traditional in the twentieth century research of economic systems.

The quality of the system's diagnostics on any stage of strategic planning and administration of the territory could be enhanced in case we arrange the hierarchy of ecological and economic urban development factors in a way similar to Maslow's hierarchy of needs (Maslow 1999) and Kuznetsova's factors hierarchy of social and economic development within the regions (Kuznetsova 2013) (Fig. 2).

In interpretation of patoeconomics authors (G.P. Luzin, K.V. Pavlov) the principal explanation and justification of the existence of such science as patoeconomics is the history of human development, notably the fact, that any social and economic system regardless the type of social structure (capitalistic, socialistic) got into crisis states caused by numerous factors.

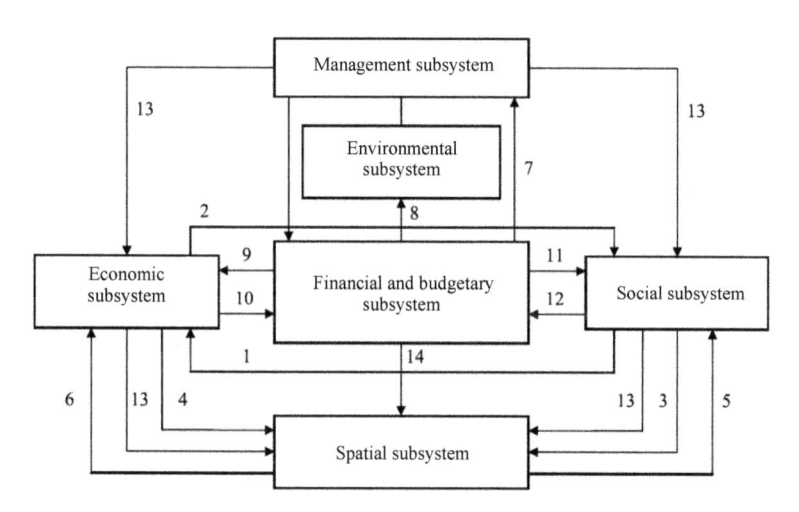

Fig. 1 Urban structure (Elohov 2007). Connections of subsystems: 1—labour, 2—living facilitis, 3—requirements, standarts, labour force, 4—production (goods), incomings, labour, 5—services, norms, requirements, 6—services, norms-values, 7—transfer and revenue of money, 8—funding of environmental conditions within the internal borders of the municipality, 9—funding of the local economy, 10—income of the local economy, 11—funding of the social sphere, 12—local social standarts, 13—organisational and governing interactions, 14—funding of the spatial development.

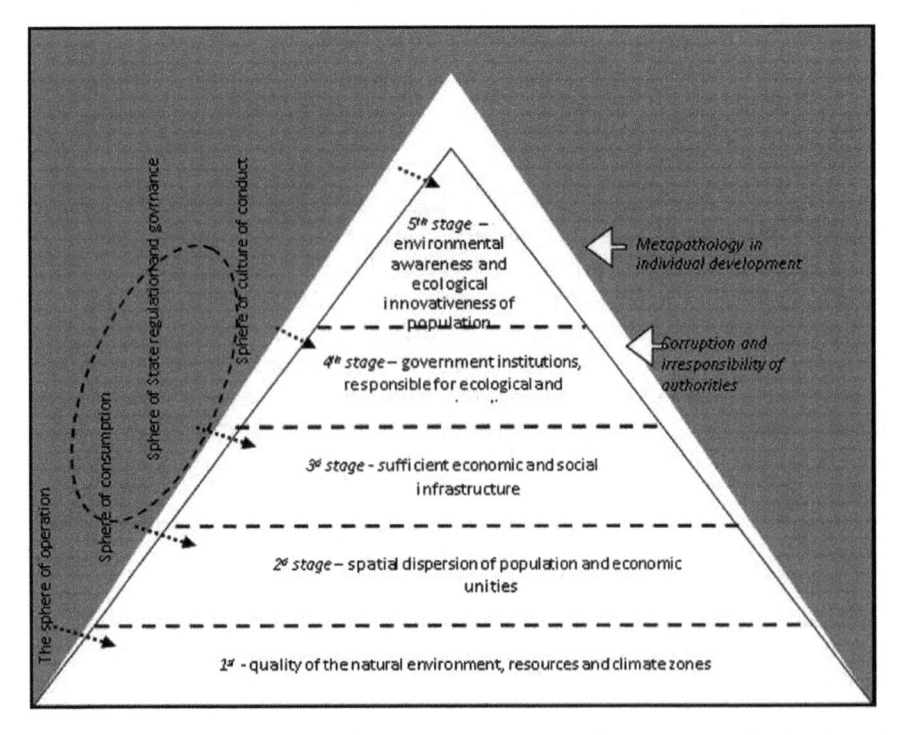

Fig. 2 Factors hierarchy of social and economic development within the regions (Zlochevsky and Buletova 2013)

Pavlov (2011) in order to describe the origins of the crises in post-socialist countries divides the reasons into the following separate groups:

- negative factors of the countries' economic development within previous, socialistic period (e.g. high level of functional and moral depreciation of the fixed capital);
- factors which had led to the development of crises already during the restructuring processes (lack of integral and comprehensive policy aimed at transition to a market economy—Gaidar's «shock therapy», misconceived tax system, lack of a general theory of crisis social and economic processes and states);
- external economic and political factors (such as changes in prices within the global market);
- environmental factors of economic development (substantial reduction of financial resources, allocated for environmental protection measures, an abrupt obsolescence of funds for of environmental purposes, substantial degradation of the environment in several areas of the country).

Under consideration of the hierarchy which divides factors of ecological and economic development into objective and subjective, it is necessary to specify that the main purpose of state regulation in terms of EES consists in provision of conditions for security and life-sustaining activity of the objects within the system, their symmetric development, which is aimed at balanced realization of interests within all components of the system, including a person, society, authority structures, economic subjects and natural environment itself. Attainment of the symmetry and bio-economic compatibility represents a complicated but necessary process, where special attention should be paid to the role of the state and municipal authority.

According to the represented hierarchy of the factors of social and economic development within the regions, three first stages include objective factors with the content inherent by nature due to unique character of the location area, history of the city development, its resource potential and attractiveness for people and enterprises. And most importantly, their content is entirely determined by the federal area development policy, which is forming at the level of federal conceptions-strategies-programmes. Capacities of municipal governments are limited by the potential of rule-making, low level of financial independence, absence of governance initiatives and responsible political and business community.

The fourth and the fifth levels of the hierarchy are also limited in the abilities of their performance and achievement of the state necessary for system's governance, conduction of its components by the high centralisation of adopted decisions over territorial development—financial, staff and legal centralization. However at the federal level it is also possible to make decisions and adopt programmes, capable of breaking the stalemate in the process of transformation within the Russian economy and its political, social and legal framework, exactly through provision of conditions for establishment of high-quality human resources in terms of labour resource and intellectual resource—by the system of environmental education (pre-school, school and higher), indication within the strategy of the state of the primary "national idea", which unite all social and age groups, in terms of the idea of building an environmentally-friendly economy, which is impossible under prevailing conditions, when not only population but also authorities and business are deprived of innovativeness, and when Russia could be considered as the a platform for introduction of innovative green technologies—the safe and attractive platform until the system approach is used for development of environmental awareness of an individual person, all social groups and groups of economic actors.

Significance of eco-systematicity is continually growing in the process of implementation of regulatory measures in environmentally oriented sphere, which could be defined as the necessity of explicit and attainable, systematically organized and internally consistent complex of legal, economic and organizational requirements towards priority of preventive measures in environmental sphere and prevention of harm to nature due to the use of a particular resource and to the environment in general (Kichigyn et al. 2009).

There is a variety of definitions for eco-system management depending on the scientific sphere of researchers. General and comprehensive definition was given by the American researcher Wood (1994): "Ecosystem management is an integration of environmental, economic and social intended to protect environmental sustainability, biodiversity and environmental productivity".

Its major advantage consists in determination of the underlying drivers (environmental, economic and social governance principles within biological and physical systems) and purposes of ecosystem management (environmental sustainability, biodiversity and environmental productivity). Another definition of ecosystem management was elaborated by the UNEP. In its definition ecosystem management refers to the approach of natural resource management, focused on the sustainability of ecosystems in order to provide environmental and humanitarian needs in the future. Sharing the UNEP position in definition of ecosystem management in terms of an approach towards natural resource management scientists Pavlikaki and Tsichrintzis (2000) highlighted that it represents new comprehensive approach.

Ecosystem management has the following features:

- focus on a long-term sustainability of resources;
- maintenance and enhancement of biodiversity;
- broad spatial and temporal scale thinking;
- integration of economy, sociology and environmental systems into the planning process;
- adjustment of management plans with regard to monitoring results and new information;
- recognition of the complexity and interaction of "ecosystems";
- recognition of people as a part of the ecosystem.

The foregoing reflects not only recognition of unity, interdependence and interaction of all environmental elements.

4.4 An Ecosystem Approach to the Management of Urban Development

Eco-systematicity, in its broad sense, is crucial due to significant ecological vulnerability and fragility of its balance.

Attention precisely to this side of the environmental issue was expressed in a caution and at the same time an objective requirement, set out by Academician of Academy of Sciences of the USSR and All-Union Academy of Agricultural Sciences Moiseev (1988) in 80s of the twentieth century: "There is an objective requirement for all those responsible for scientific and technological progress and even more responsible for use of its achievements with practical objectives: to take into account the vulnerability of the environment, to ensure that its "resistance limits" are not exceeded, gain a better understanding of its inherent complex interrelated phenomena, avoid conflicts with natural patterns in order to avert irreversible processes".

Such emphasis, highly commendable and worth to be recognized with respect to obvious wisdom and long-sightedness of the author, and confirms the increasing significance of several principles, which were updated during adoption of non-antagonistic environmental management in the twenty-first century. They include: supremacy of the system approach in terms of calculation and estimation of the environmental damage probability, including within the context of increasing man-made and natural disasters; comprehensive analysis of the data, formed within the results of environmental monitoring, and consideration of all of the circumstances in terms of the possible negative impact on nature in decision-making on business and other activities performance; recognition of the importance of prevention of environmental offences, including such measures as economic sanctions, educational measures, etc.; growing relevance of the precautionary principle, compulsory accounting of the probable environmental harm.

In order to reveal the commands of institutional environment, favourable for ecosystem approach towards environmental management, it is necessary to clarify the notion of institutions in this context. As it is known, there are several definitions of the term nowadays. From the one hand, institutions include superior norms, traditions and rules as well as the thinking stereotypes, enforcement "mechanisms", etc. On the other hand, institutions include different organizations, departments and agencies. Thus, one of the institutionalism's founding fathers—Veblen (1984) stipulated that an institution represents "a stereotype", "traditional ways of response on incentives", "common way of thinking towards relations".

North (1997) had also attempted to define an institution separately from an organization. In his works particular relevance is given to those institutions, which are embodied in customs, traditions and codes of conduct, and less sensitive to conscientious human.

A corresponding member of RAS Kleiner (2001) in the late 1990s offered a system of institutions classification based on the following criteria: type and form of subjects, their role within social and economic processes, the sphere of decision-making, period of emergence, sustainability, the degree of formalization, information and monitoring mechanisms. Its interpretation and account of "classic interpretations" form the ground for the next definition: the functional role of institutions, conducive to the implementation of the ecosystem approach towards environmental management and environmental protection consists in foundation and further use of the system of formal and informal rules, norms, incentives and restrictions, which influences on economic entities (including coercive measures) in the direction of increased responsibilities for resource conservation and general environmental welfare, handling of environmental issues within the appropriate level, rise of natural resource use culture. Academician Moiseev (1988) named such approach as collective decisions or "institutions of agreement".

Thus, creation of the institutional environment, favourable for ecosystem approach towards environmental management, suggests provision of convergence of environmental, economic and social interests harnessing the potential of formal and informal rules, norms, incentives and restrictions, which influences comprehensively on the activities of public authorities of the Russian Federation, local authorities and economic entities, including promotion of interaction amid them, stimulation, social contract and if necessary—enforcement, punishment and compensation for harm caused to the natural objects, implementation of the requirements of nature protection, reinforcement of environmental security.

Any urban settlement can be characterized as a stochastically determinate system, which is evolving under the influence of external and internal factors. Herewith the city by its nature represents a specific environmental and economic open type system which is stipulated by the constant energy and substance exchange between economic structures and environment. The purpose of the work is in the consideration of the populated territory, including urban area, as a triunique system "nature—economy—society" and definition of the key conditions for implementation of the system approach towards urban development.

Environmental and economic system represents a set of economic and environmental subsystems, which operates jointly and possesses emergence property. In this regard it should be noted that environmental subsystem is characterised by properties of the natural system, formed without human intervention, whereas economic subsystem possesses parameters of an artificial system, appeared as a result of human activities. Moreover, environmental subsystem in its essence represents the permanent entity with unlimited duration time, whereas economic subsystem has relatively temporary nature (Vergelis 2002).

A number of authors have suggested that it is possible to study the city not only from the point of view as a system formation, but also as a living organism, capable of self-stabilisation and self-development. Within this point urban system design should be implemented only in the context of consideration of the city as a process, but not as a complex of buildings and constructions, due to the fact that "objects obtain significance only as actors in the process".

Owing to the living components cities by their natural subsystem are involved into global biogeochemical cycles and thus, represent the elements of the biosphere as part of the Earth's geographical cover. Herewith city as a highly sophisticated system is functioning according to social laws of development, not on the base of biological laws. Social subsystem within such system performs strategic and executive functions (Vasyleva 2013). In this regard, the city as natural and man-made system, formed by architectural and engineering construction sites and undermined natural ecosystems, could be characterized as an urban ecosystem.

City consumes energy resources in the form of fossil fuel, nutrition and water and uses incoming external information resources. Performance results of the urban system are reflected not only in production of spiritual and material assets, but also in significant quantities of generated wastes, which are pollutants for environment, source of information "noise" and various types of climate change influences.

Interrelations in urban ecosystem can be divided into spatial and functional. The fist type is reflected in the territorial invasion of urban entities on natural ecosystems, which is defined by direction and intensity of development of the key constructive element of the economic subsystem (industrial, residential, communication, transport and engineering, etc.). At times such spontaneous invasion of nature is reflected in decrease of area and violation of the integrity of natural complexes, fragmentation of the structure and declining of ecosystem sustainability.

Ecosystem approach stipulates investigation of the functional relations of the living organisms among each other and with the environment, as well the factors that violate self-regulation of the ecosystem and lead to its imbalance. This approach is relevant for the development of an ecosystem economic use strategy. Within the ecosystem approach a prominent role belongs to the concept of homeostasis (self-regulation) in terms of estimation of anthropogenic impact on processes aimed at economic integrity.

Under investigation of community ecology, the main focus is given to the biotic components of the ecosystem, description of a species composition and factors that constraint their proliferation (biotic, abiotic and anthropogenic).

Ecosystem approach (EA) is connected with the influence of people's use of ecosystems on their functioning and productivity.

- Within the EA ecosystems are being studied comprehensively. Currently ecosystem management is generally connected with their use to derive a single dominant ecosystem good or service, e.g. fish, timber and hydroelectricity. At the same time there is no complete understanding how ecosystems react on such activities. Probably within the implementation of such approach we actually miss goods or services, which are more valuable than those received and underestimated by the market, such as biodiversity and flood control. In the framework of the EA the full range of possible goods and services is taken into consideration and covers and there is an attempt to rationalize multiple benefits, given by this system and other systems. The objective of the EA is to achieve the effective, clearly identified and inexhaustible (sustainable) functioning of ecosystems.
  - Within the EA there is a shift away there is a shift away in borders, which traditionally define our ecosystem management. The EA highlights the system approach, recognizing that ecosystems function as integral unities and it is necessary to manage them right in this way, not as separate parts. Thus, in the framework of the EA consideration of ecosystems is being given beyond traditional legally established boundaries as ecosystems frequently cross state borders.
  - The EA takes into account long-term processes. Herewith it focuses on the processes inside ecosystems within the microscale, but they are investigated in a broad perspective, which includes entire landscapes in space and decades in time.
- The EA includes people. Along with environmental information attention is given to social an economic data. In this way human needs explicitly linked to biological capability of ecosystems to satisfy the needs. Although in the framework of the EA processes in ecosystems and biological thresholds are under close attention, there is still space for human-made modification of ecosystems.
- The EA is aimed at maintenance of the productive capacity within ecosystems. It is not limited by operational functions of ecosystems only. Production of goods and services as a natural outcome of the health ecosystem operation is not considered as a final goal. In the framework of the approach ecosystems management is not

considered successful in case it does not preserve of even enhance abilities of the system to generate benefits in the future.
- The ecosystem approach does not mean rejection of other approaches towards management and maintenance of such ecosystems as biosphere reserves, protected areas and programmes, aimed at conservation of certain species, implemented in the framework of existing national policy and legislation. The EA could use all the approaches and other methodologies in complicated situations. There is no a single way of the EA implementation as it depends on local, district, national, regional and global conditions (Perelet 2006).

Sustainable urban development requires the use of a broad spectrum of tools, which affect various fields of the city and citizens life, including the following:

- urban agriculture;
- renewable sources of energy;
- urban planning and provision of urban amenities in order to reduce a "thermal spot" (tree planting, increase of water bodies, increase of greenfield lands);
- improvement of the public transport system, stimulation of cycling, increase of pedestrian zones to reduce car emissions;
- optimization of the building density;
- control over growth of the cities;
- sustainable urban design (green roofs, facades, etc.);
- sustainable transport;
- environmentally friendly and resource-efficient buildings, smart houses;
- development of a data collection and data monitoring system;
- effective urban management, elaboration of key performance indicators of urban development.

The given tools correspond to the such concepts of the spatial development as "the compact city" and "sustainable growth".

Russian experience in compact development is still quite often carried out in relatively perverse forms. Thus, pointlike, "infill" building and construction work entirely in the format of multistory buildings drastically increases pressures on all existing urban systems and creates strong resentment and opposition within the local residents in the spirit "Not in My Backyard" (or NYMBY).

Herewith peculiarities of the budgetary and urban management systems restrict authority's ability to experience the results of extensive development.

Conversely, "infill" development in its best sense and practice within foreign cities as one of the main policies of the urban compactness, is scrupulously planned and evaluated by the city; it is intended to integrate into gaps of the urban textures in a most possible delicate and multifunctional way in order to satisfy the urgent needs of the local people, provide coherence and accessibility of urban amenities while using power of the urban infrastructure wisely.

It is evident, that environmental economy of sustainable cities requires fundamentally different management practices, than the economy of resource base broadening. These are management systems, configured to break the barriers among economic sectors, jurisdictions and urban subsystems, provision of long-term investments within integral solutions, institutional and financial partnership-formation of different types (Administration of urban spatial and economic development: hidden reserves 2016).

There is no universal way of EA principles implementation, but actions are required in the following areas.

Eradication of the missing knowledge about ecosystems. Effective ecosystem management requires knowledge about their performance and current state. Detailed knowledge about ecosystems provides us with information about their productive capacity, influence of our activities and long-term impact.

- Conduction of a social dialogue over ambiguity of human activity towards ecosystems and ecosystem policy. Knowledge of the ecosystem processes and conditions is significant; however, it is just a ground for the policy, based on effective information and aimed at resource management. Within the EA objectives of the ecosystem management are set on the basis of discussion over society's needs and expectations from ecosystems, benefits sharing and actions we are ready to take within expenditures and compromises between our needs and possibilities with the informed citizens.
- Establishment of the quantitative assessment of ecosystem services. Underestimation of ecosystem services represented one of the key factors of short-sighted management practices in the past. Thus, an essential element of the EA is the opportunity to give to the community, governments and industry the ability to set up more accurate assessment of ecosystem services in such a way as to include the mentioned values into the activity planning process with ecosystems.
- Involvement of the local people in the management of ecosystems. Experience of many countries in the world shows that local communities often represent the most prudent managers of ecosystems. Their knowledge of ecosystems and direct interest in promotion of their health can be extremely valuable for provision of the long-term management within ecosystems. Involvement of the local communities in the management of ecosystems can also contribute to a more equitable distribution of benefits and costs brought by ecosystems.
- The assessment of capacity to ecosystem recovery. Ecosystem recovery represents not a brand-new idea, however for the past 20 years the scientific ground for ecosystem recovery had improved. Interest towards recovery has increased sharply as well as expenditures for this activity. However, there is lack of reliable assessment of the general degradation of ecosystems in the world and understanding of the part of degradation, which should be eliminated by efforts towards ecosystem recovery.
- Consideration of urban planning factors within ecosystem management. Urbanisation and the way of consumption used by citizens drastically affect ecosystems. Well-managed urban settlements are able to alleviate the pressure by savings through the scale of housing growth, transport systems and energy use. In the rational ecosystem management there is no any place for ignorance of significance of the cities or their review in terms of peripheral factor.

- Implementation of new approaches towards parks and protected areas. The EA is connected with adoption of new measures aimed at blend of human activities and objectives of ecosystem preserving. Parks and protected areas should be an integral part of an integrated landscape management strategy, which includes environmentally-friendly human activity. In some cases, the mentioned territories and objects can be psychically connected via landscape corridors in such a way as to provide further performance of the initial spatial nature (Analytical review of HELCOM materials towards Europe marine strategy 2019)

5 Conclusion

A sustainable urban environment can only be created if economic activities are combined with environmental protection measures in a rational way. For a modern Russian city, considered as an ecological and economic system and as an object of strategic management, a necessary condition for competitive development with an increasing conflict of economic and environmental interests of subjects and objects of the system is to achieve and maintain the state of ecological and economic security as the main system-forming factor. Diagnostics and monitoring of the state of the city as an ecological and economic system should be implemented within the framework of strategic planning under the condition of duality-the use of static and dynamic approaches at different stages and with different goals of diagnosing the ecological and economic security of the city. At the same time, the priority of the indicative approach also allows you to implement the duality of construction. The theory of ecosystem management of sustainable urban development is just beginning to take shape. Building and implementing sustainable development management models can be the most effective mechanism for studying the laws of city functioning and the best way to test ideas and technologies for sustainable development. The authors propose a spatial-economic ecosystem approach to urban development management. Assuming the provision of synergetic, integrative and co-evolutionary development of the economic, social, environmental, cultural and physical subsystems of the city, the model of sustainable urban development determines the fundamental relationship between the physical shell (space and infrastructure) of the city and the economic processes that arise in it. In addition, according to the model, the balanced development of all subsystems is possible if the local management style is as inclusive and participatory as possible, and horizontal partnerships are built as opposed to top-down directive administration.

Acknowledgments The reported study was funded by RFBR, project number 19-29-07400.

References

Administration of urban spatial and economic development: hidden reserves (2016) Moscow management school SKOLKOVO; INPO "Moscow urban forum". Center of urban research of the business-school SKOLKOVO, Moscow, p 192

Bossel H (2001) Indicators for sustainable development: theory, method, applications: a report to the Balaton Group (translated from English under the editorship of VR Tsybulsky). Institute of the North Development SB RAS, Tyuman, p 123

Daly HE (2007) Ecological economics and sustainable development, selected essays of Herman Daly. Edward Elgar, Cheltenham; Northampton, 270 p

Daly HE, Farley J (2004) Ecological economics: principles and applications. Island Press, Washington, DC, 454 p

Danilov-Daniljyan VI (2003) Sustainable development (theoretical and methodological analysis). Econ Math Models 39:123–135

Elohov AM (2007) Theoretical and methodological basis of strategic programming in municipal governance. Perm State University, Western-Urals Economics and Law Institute, Perm, p 547

Elohov AM (2011) System approach towards city as the subject of results-based management. ARS Administrandi 2:61–79

Georgescu-Roegen N (1971) The entropy law and the economic process. Harvard University Press, Cambridge

Helm D, Hepburn C (2014) Nature in the balance: the economics of biodiversity. Oxford University Press, Oxford, UK, 416 p

Illge L, Schwarze R (2006) A matter of opinion: how ecological and neoclassical environmental economists think about sustainability and economics. Berlin

Karpova NV (2019) Sustainable development of urban localities: theoretic postulates and practical implementation. Econ Ecol Territorial Entities. 3:64–70

Kichigyn NV, Ponomarev MV, Khludeneva NI (2009) Environmental law: lecture notes (2009). Higher education, Moscow, p 208

Kleiner GB (2001) Peculiarities of the formation processes and evolution of social and economic institutions in Russia. Preprint #WP/2001/126. M. p 16

Kuznetsova OV (2013) Factors hierarchy of social and economic development within the regions. Econ Questions 2:128–129

Luzin GP, Pavlov KV (1995) Patoeconomics: tasks, questions and areas of research. Published by RAS, Apatity, p 112

Maslow AG (1999) Motivation and personality (translated by AM Tatlybaeva). Eurasia, Saint-Petersburg, p 478

Moiseev NN (1988) Mankind ecology as seen by the mathematician. A person, nature and the future of the civilization. Moscow, p 133

North D (1997) Institutions, institutional change and economic performance. p 39

Our common future The report of the World Commission on Environment and Development (WCED) (1987) (translated from English by SA Evteev, RA Perelet). http://www.un.org/ru/ga/pdf/brundtland.pdf. Last accessed 27 March 2020

Pavlikaki GE, Tsihrintzis VA (2000) Ecosystem management: a review of a new concept and methodology. Water Resour Manage 14(14):257

Pavlov KV (2011) Essence and tasks of the general theory of ecological and economic policy. Int Collect Sci Res 3(21):284–289

Perelet RA (2006) Ecosystem approach for natural resources use and nature protection management. Econ Nat Resour Use. 3:3–19

Peshina EV (2013) Evolution of theoretical and methodological approaches towards understanding of "ideal city." Manager 4(44):32–40

Philipson YA (2013) City as an ecosystem. Youth and science: sourcebook of the IX Russia-wide science and technology conference of students, postgraduates and young scientists with

international participation, devoted to the 385th anniversary of the founding of Krasnoyarsk. http://conf.sfu-kras.ru/sites/mn2013/section025.html, free of charge. Last accessed 20 June 2020.
Raimers NF (1992) Hopes for human survival. Conceptional ecology. Young Russia, p 365
Scott CM (2009) Green economics, an introduction to theory, policy and practice. Earthscan, London, 224 p
Shmelev S (2012) Ecological economics: sustainability in practice. Springer, 248 p
Söderbaum P (2000) Ecological economics. A political economics approach to environment and development. Earthscan, London
Tetyor AN (2011) Sustainable urban development. Direct-Media, p 128
Vasyleva NA (2013) Modern city as a specific socio-ecosystem: challenges and prospects. Bull Irkutsk State Agrarian University 58:151–157
Veblen T (1984) The theory of the leisure class. pp 200–201
Vergelis YI (2002) Role of cities for Earth's biosphere. Public infrastructure of cities: research and technology collection of essays. Kharkov. 36:177–181
Vladimirov VV (1999) Urban ecology. The International Independent University of Environmental and Political Sciences, p 204
Voronin AA (2000) Sustainable development: fiction or reality. Math Educ. 1:59–67
Wood SA (1994) Ecosystem management: achieving the new land ethic. Renew Resour J 12:6–12
Yakovenko NV (2007) Quality of life of the population" as a scientific category of demography. Ecol Urbanized Territ 1:41–46
Yakovenko NV (2009) Hygiene of localities and its impact on the health of the population of the Ivanovo region. Probl Reg Ecol 2:123–130
Yakovenko NV (2010) Sustainable development model and socio-economic monitoring of the city. Probl Reg Ecol 3:118–126
Yakovenko NV, Markov DS, Molodzeva AA, Turkina EP (2013) Environmental factors in shaping the health of the population of the Ivanovo region (atmospheric air). Mod Probl Sci Educ 5:461–469
Zlochevsky IA, Buletova NE (2013) City as the ecological system: diagnostics and conditions to provide security. Manage Econ Syst Electron Acad J 7. http://www.uecs.ru/uecs55-552013/item/2251-2013-07-20-05-40-53. Last accessed 07 June 2020

Professor Nataliya Vladimirovna Yakovenko graduated from Melitopol State Pedagogical University (Ukraine, 1993), qualified as Teacher of Geography and German. In 2002, the author was awarded the Degree of Ph.D. of Geographic Sciences, in 2013—the Degree of Advanced Doctor in Geographic Sciences. At the present time author is the Director of the Research Institute of Innovative Technologies of the Forestry Complex of the Voronezh State University Forestry and Technologies named after G.F. Morozov (the Russian Federation). The author has over 600 publications. Research interests: sustainable development of cities and regions, environmental quality, regional differentiation of development.

Associate Professor Igor Vladimirovich Komov graduated from Voronezh State Pedagogical University (the Russian Federation, 1999), qualified as Teacher of Geography and Economics. In 2006, the author was awarded the Degree of Ph.D. of Geographic Sciences. At the present time author is Associate Professor at the Department of Socio-Economic Geography and Regional Studies at the Faculty of Geography, Geoecology and Tourism of the Voronezh State University (the Russian Federation). The author has over 300 publications. Research interests: sustainable development of cities and regions, environmental quality, greening of agricultural production.

Continuity of Technologies in Education, Science and Industry

The Next Steps for the Baltic Universities' Cooperation in Accordance with the Development of Education and the Upbringing of Students

Elena Kropinova and Eugene Krasnov

Abstract This article highlights the need and potential for more open systems of education, including the expanded mobility of students and teachers, a greater sense of international cooperation, and new teaching technologies in different spheres of university life. The authors of this paper look for new progress in this area across the Baltic Universities Programme—and other international activities—paying special attention to education and research in relation to the UN Sustainable Development Goals (at the global, regional and local level). The initial focus of this article is to understand human beings as an open yet self-developed system. This is an important lesson when it comes to the transformation of the educational process. On a larger scale, we also consider education in the context of one's upbringing. New steps towards the modernization of teaching are connected with the active participation of students. Self-assessment, self-respect, and self-actualization are the most acute components of the self-development of scholars. Nevertheless, we also need to complement our approach with the psychological instruments of a person's development (soft influence, communication, conviction, debate etc.). An interdisciplinary approach has been applied here, using international and personal sources on pedagogical development over the past 30 years. Many teaching materials and research outcomes were provided as common results of us and our students including peer-reviewed papers, final reports, Masters/PhD theses and dissertations. In practice and in curriculum, the findings of this paper might be used in modern psychological approaches including in the design of education systems.

Keywords International programmes · Education and upbringing · Baltic universities cooperation · Integrative facilities · New steps for sustainability · Problem-oriented learning · Combined effects

---

E. Kropinova (✉)
Institute of Economics, Management and Tourism, Immanuel Kant Baltic Federal University, 236041 Kaliningrad, Russia

E. Krasnov
Institute of Living Systems, Immanuel Kant Baltic Federal University, 236041 Kaliningrad, Russia
e-mail: ecogeography@rambler.ru

1 Introduction

The symbolic title *Man and the Biosphere* was suggested by UNESCO for some of their programmes—this is now set to be corrected since Man is also a part of the Biosphere and ecosystems of Earth. Humans play an essential role in living nature, and it is necessary to develop educational theory and practice to respond, including educational written texts.

Education and Sustainability Background. Following on from the definition of Sustainable Development by Brundtland in UN Commission, Sustainable Development has different foundations in the different countries and participants of the Baltic Universities Programme (BUP). Some nations focus upon Environmental Education (EE) and include additional issues of ethics, equity and new ways of thinking and learning. Others say that Education for Sustainable Development (ESD) should be a part of good EE and that there is no need to do away with EE as an umbrella. Another view is that EE is a part of ESD, because ESD includes development, cultural diversity, as well as social and environmental equity (Jutvik and Liepina 2020).

The series of printed materials *A Sustainable Baltic Region* from the *Baltic University Programme (BUP) and The Road towards Sustainability: a historical perspective (1997)*—was interesting for us and our students. In this booklet, the political history of sustainability is highlighted alongside the history of environmental science and scientists. Already in the 1990s the subject of paradigms in sustainability was being discussed: whether a society can become sustainable within the modern paradigm (the domination of nature through science and technology, growth and consumption), that is, through better technology and accurate pricing, or whether sustainability requires the transition to a post-modern world (The road towards Sustainability: a historical perspective 1997).

A number of teachers nowadays start their research on ESD by looking at the introduction of sustainable development ideas into education. Other papers provide the evolution in the relationship between education, nature and childhood, starting from Jean-Jacques Rousseau (Bengtsson et al. 2020) and Immanuel Kant (Allison 1971). Bengtsson has suggested that with the emergence of environmentally oriented education in the late 1960s "the process of reducing nature to processes of human production became an issue of problematization" (Bengtsson et al. 2020). This approach made people want to study nature in order to dominate it. "What human beings seek to learn from nature is how to use it to dominate wholly both it and human beings" (Bengtsson et al. 2020; Horkheimer and Adomo 2002).

Orienting Environmental Education (EE) towards Sustainable Development (SD) and capacity building has been suggested by Harvey (1995) and many other authors. In general, the study of SD—as well as ESD—developed together with the BUP. The combined effects of such activity and its influence on the development of ESD in Kaliningrad was observed in 2017 in our paper (Krasnov and Kropinova 2017).

The Baltic University Programme (BUP) is the network of Universities around the Baltic Sea. In the beginning of the 1990s, the main focus was on the topic of the

"Ecology of the Baltic Sea" (TV, courses and 10 books in Russian and English) and many others, for example, *A Sustainable Baltic Region course* (Sorlin 1997). More than 500 Kaliningrad students attended the BUP course "The Baltic Sea Environment" and the "Geoecology" Department was opened in the Immanuel Kant Baltic Federal University (IKBFU). Ecology became a compulsory subject in all educational programs. At this time, a subject oriented approach dominated, and the course "A Sustainable Baltic region" (1997) demonstrated this approach: included were booklets on sustainable energy, mobility, habitation, agriculture and forestry, industrial production etc. (Sorlin 1997). Since the mid-90 s, the role of the social sciences rose. The BUP course of lectures "The peoples of the Baltic" was introduced as a regionally compulsory subject in Kaliningrad universities. At this time, there was a need for further education on these issues. This improvement in the ecological education of municipal employees was the key to success in the implementation of sustainable development technologies at the level of urban management. As part of the BUP, the course "Sustainable Community Development" was designed to demonstrate a problem-oriented approach. However, sustainable development is possible only through an integrated approach. As a part of the BUP, a unique monograph "Environmental science" (Rydén et al. 2003) has been created and a new Masters Course was opened by Lars Ryden, Pawel Migula and Magnus Anderson (Eds.). A lot of educational material were published also in *A Handbook for Teaching and Learning Sustainable Development* (Jutvik and Liepina 2020). Since the very start of the programme's implementation, an emphasis has been placed on studying the opportunities of SD in the Baltic Sea region, with a particular focus on building joint resources for the purposes of SD, among which education is a priority. Today, the programme is an acknowledged world leader in education aimed at SD (Ionov and Kropinova 2010; Krasnov 2020).

The key events of the evolution of ESD in global politics started in 1972. ESD was one of the 26 principles of the Stockholm Declaration, and by 2015 it became one of the key UN Sustainable Development Goals (SDG), listed at No 4 for quality education with a target to ensure inclusive and equitable quality education and to promote life-long learning opportunities for all (UN 2015; Leder 2018). The modern reality offers new questions, such as whether Covid-19 jeopardises progress towards the Sustainable Development Goals, including No. 4, on the basis that remote learning remains out of reach for at least 500 million students (UN 2020).

Global aspects of Sustainable Development. The framework for the implementation of ESD was established during the 40th UN General Conference in 2019. The strategic objective for ESD by 2030 is connected with (a) individual transformation, (b) societal transformation and (c) technological advances (UNESCO 2019; Framework 2019). One possible way to solve environmental problems is to set up educational centers for university students and local people on the basis of leading environmental higher education. For example, in the Baltic University Programme (BUP) one of the teachers' seminars in Uppsala University completed its training course with the participation of one of the foremost prominent American scientists, Dennis Meadows. His open lecture (May 2015) was connected to the previously released Global Rome Club report *The Limits to Growth: The 30-Year Update by*

Donella H. Meadows, Dennis L. Meadows, Jørgen Randers (Meadows et al. 2004). In this report, and in his lectures, Dennis underlined the limits of natural resources including clean water, air, food and other great challenges for mankind in the near future. In a way, he announced a so called "zero-growth model". Significant environmental and social-economic problems were presented in Meadows' models as real challenges for both developed and developing countries. Similar challenges were highlighted by Moscow State University rector and academic Victor Sadovnichiy et al. in *Modeling and Forecasting World Dynamics* (Sadovnichiy et al. 2012). In sum, it was concluded that a global system crisis in all sectors of development will come soon if UN countries don't change the direction of their environmental, social and economic policy towards sustainability.

One of the examples of the advanced development of human potential is the results of a sociological survey, which was conducted in many countries across the world from 1996 to 2006. The respondents were asked questions to assess the scientific literacy of the population. The highest percentage of the correct answers (more than 70%) according to the integral indicator were found in the countries of Northern Europe: Sweden, Finland, Denmark, and Norway. Russia placed 32nd overall, while the USA was ranked 20th. Such differences in knowledge affect the behavior of people as consumers of information; it also affects the nature of the organization of production processes. Companies such as IKEA, Ericsson, and Volvo Group use matrix control systems organized according to a project principle. A lack of understanding of the need for basic knowledge is one of the most important factors hindering the creation and dissemination of innovation (Markelova 2007, Shuvalova 2007, World Intellectual Property Indicators 2011).

High hopes are often placed on innovation, in particular on *know-how* development, where regions move to more sustainable development models. The ability of younger generations to be able to generate innovations depends heavily on high-quality education systems being in place. It is no coincidence that in many educational programs, including grants, a focus on creative solutions is a priority. For example, according to an interview conducted in the Finnish technology sector with 45 employees, their views on both creativity and sustainability differ significantly (Lemmetty et al. 2020). Through an analysis of their responses, it was revealed that managers and employees of enterprises have different attitudes towards creativity as a tool for innovative development. While managers strive to innovate and "destroy the old," employees are more focused on the progressive improvement of existing technologies and their reuse. Both paths take place simultaneously. However, since creativity is present at various hierarchical levels of production, lifelong strategic approaches to sustainable education must be implemented to take into account the differences in the worldview and goals of the participants in the production process.

Regional aspects of Sustainable Development was summarized in one of Springer's proceedings *Baltic Region—the Region of Cooperation* (Fedorov et al. 2020). Evidently, for the authors, sustainable development is impossible at the same time as several economic activities (including harmful production, such as metallurgy) in small sea-side towns and regions like Kaliningrad Oblast. There is a need for new technologies and approaches without resulting in damage for the environment. Recreation and tourism, agriculture, and complex productions of amber are profitable, environmentally-friendly and a good alternative. In *Crisis Management Challenges in Kaliningrad*, (Krasnov et al. 2014) special attention was paid to the comprehensive analysis of different sectors of economic, social and environmental industries.

Local aspects of Sustainable Development. The Curonian Spit National Park in the Kaliningrad region of Russia is included in the UNESCO list of Cultural Landscapes due to its Outstanding Universal Value, and is a good example of the worldwide importance of local natural features. Knowledge of this area, produced by employees of Immanuel Kant Baltic Federal University (IKBFU)—and by international researchers—is valuable not just for Kaliningrad residents, but also for international tourists and scientists. That is why it is important to realize and support international projects such as *Ecotourism as a tool for preservation natural and cultural heritage* (Ecotour4Natur). The project is being implemented under the Cross-Border Cooperation Programme Lithuania-Russia 2014–2020. One of its activities is the development of educational course on SD for national park visitors and the revitalization of the historical exhibition pavilion of the first European ornithological station.

So the purpose of this study is to define the status quo of over 20 years of progress in Education for SD in the Baltic region taking the case for the contribution of the Baltic universities to this issue. The hypothesis was that BUP in particular and the other international activities contributed a lot in development of the teaching technologies, as well as to the improvement of researchers and teachers cooperation. So the objectives of this paper is to find out new steps that could facilitate the UN SDGs at the global, regional and local level. The authors' idea that taking into consideration that human beings is an open yet self-developed system, it is proper time to introduce into the education for SD not just knowledge on sustainability but to pay extra attention to the upbringing of students. So the objective is to find such components so as to complement the pedagogical approaches with the psychological instruments of a pupils' development.

2 Research Methodology

In this review, we used a systems approach to analyse the sources and experiences of both the authors of this paper and those of international academics over the past 30 years. We have applied the following searching strategy for sources: we used keywords in combination Education and Sustainability as well as names of researchers working in North-West Russia on the issues during last 10 years. There search sources were used as follows: Springer Links, Webofknowledge, Scopus as well as personal libraries collected during 20 year period working on the sustainability issues. Those analysed had a connection to and high involvement with the BUP, TEMPUS, INTERREG, TACIS and other international programmes; secondly, they had personal experience in teaching and transforming education within Immanuel Kant Baltic Federal University (IKBFU) in Kaliningrad, Russia. Many teaching materials and research outcomes were provided by our colleagues from Sweden, Germany, Denmark, Finland, France, UK and other EU countries. And, finally, there were some common results, including monographs, final reports, Masters and PhD theses and dissertations for a more comprehensive understanding of how to transform education systems. It is not possible, however, to make changes for teachers without involving and engaging students. It is essential that teachers and students work together. Different approaches in our curriculum might be used depending on the design of the Education System. Special attention must be paid to visual interpretations of integrative approaches for new curricula, designs and psychological instruments of ESD.

Transforming the process of education has a special role to play in really shaping more sustainable development. It is even viewed as a separate tool for sustainable development in its own right, by promoting efficiency; consistency; permanency and sufficiency. Education and social commitment, as presented in the work of Haubrich et al. shows that satisfaction and sustainable development can be achieved through education and a commitment to social justice (Leder 2018; Haubrich et al. 2007).

A socio-capital survey, conducted for Tanzania under the supervision of scientists from the Swedish University of Agricultural Sciences (Prof. Madeleine Granvik et al.) has shown that both bonding and bridging social capital are essential ingredients in building peri-urban resilience (Mngumi 2020). This and some other findings from this research underline the role of a social phenomenon including education for SD.

It might now be possible to add one more strategy: Networking and bringing all sectors together to address challenges associated with the SDGs. The combined effect of education and research for sustainable development facilitating this process could be of special importance not only for our University, but also for our neighbours (Polish, Lithuanian et al. educational establishments).

In the modern world, it is globally a very complicated time. When it comes to pure subject-oriented Education, we need to transform as fast as possible to problem-orient the teaching process, in order to directly find solutions to humanity's problems.

Many of the problems addressed above are of global, regional and local character. The solution for them is not possible just through educational projects. In order to realize these ideas in our own region we need to start upgrading the curriculum with an orientation towards mostly practical issues. For example, the study of Maritime subjects is a common research and education topic in Kaliningrad area. There are a lot of challenges for all organisations and companies. Fishery, aquaculture, mineral resources, recreation and others all have the possibility for more profitable development and self-efficiency in Kaliningrad. Such an approach is possible when there is coordination and a balance of interests between very different actors.

Some universities started the project-oriented work by developing for a sustainable future. For instance, one project was created in the University of Hamburg, Germany. The main idea is to link ESD and innovation processes (Schmitt et al. 2018).

Following the work of Leal et al. (2018), we would also like to use a more holistic approach to reform ESD. It is necessary to rethink the existing courses and programs and select the most effective practices with regard to the regional development of public relations. The analysis presented in the work of Leal shows that in many countries, campuses are well developed, enhancing student participation in the educational process. However, research in the field of sustainable development is still underdeveloped. The exception is Sweden and other Scandinavian countries.

We need to know more, and therefore, through our research, we must prevent society from over-exploiting natural resources. We must tackle an imbalance in their use and provide a biophysical estimate of Sustainable Development (death-birth rate etc.). There is a demand for a holistic approach to education, which might be more general, and not just focused upon didactics/pedagogy. When changing curriculum, one must not just investigate one problem (ex. rate of CO_2 in the air), but conduct research on the nature of humankind, and its potential to change. There is a need for a collaborative approach on a multi-subject basis. A multicultural vision must be applied. Transformation must serve as model for social justice and environmental stewardship.

When we speak about the penetration of sustainable thinking and ideas into education, we could evidence its success by looking at its acceptance by Education Authorities and its introduction into curricula. For example, the Dutch education system and the role of Ministries involved in the implementation of ESD in the curriculum of primary and secondary education is described in work by de Wolf and de Hamer (Wolf et al. 2015).

In some Russian universities, for example, new educational units and new curricula are discussed very actively in connection with the general priorities and changes that are made at the state level. A lot of attention is now paid by our leading teachers to inclusive education in this way (Simaeva et al. 2019). A fast transition from the previous education system to a new, much more modern one is in demand. The post-Covid-19 era creates new challenges: what is the "golden mean" between off-line and on-line education? Is distance learning itself a sustainable solution for education? Answers to these questions must be found rapidly, and at the least, the current environment provides a good opportunity to evaluate the pros and cons of distance learning in practice.

3 Geography for SD

Geography in many ways plays an important role as far as SD is concerned, since distance is important in producer-customer relations. Paul Krugmann (a Nobel Prize winner in economics) suggested that geographical knowledge (on relief, water-basins, climate and other factors) was essential in the realization of new infrastructure and other projects (Krugman 1999). Such a view introduced a new role of geography in geo-system development, and Butvilovsky says as much about "Geographical" matters, information and measures (Butvilovsky 2017). At the same time, student-geographers' practical activities are possible to connect with industrial innovations and their own campus reconstruction, etc.

From that point of view, we could agree with the ideas about the availability of geographical approaches towards the principles of ESD as suggested by Haubrich et al. (Haubrich et al. 2007) and Leder (2018). The most up-to-date of them are that ESD must: be relevant for everyone at all levels of learning, across formal and informal education; be an ongoing, continuous process; gain acceptance of processes of societal change; be a cross-sectorial task that has an integrative function; improve the contexts in which people live; create new opportunities for individuals, society and economic life; promote global responsibility; align with the principles and values that underline Sustainable Development; engender a healthy balance of all dimensions of sustainability, including environmental, social, cultural and economic sustainability; and it must fit local and global contexts. The recent findings by Demssie et al. on the role of indigenous knowledge in ESD confirm that it is important to link education to the cultural, traditional, and historical identity of learners and the context at large (Demssie et al. 2020). The scheme of ESD through the lens of Geography education is presented in Fig. 1.

We need to reconstruct the geographical units that are consisting of integrative multidisciplinary answers to many of the questions related to social, economic and

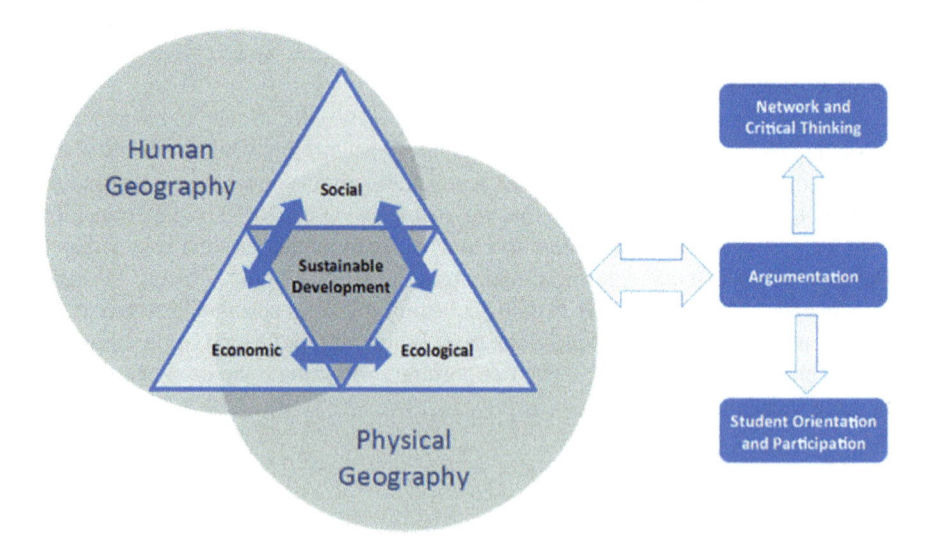

Fig. 1 ESD through argumentation for Geography (Leder 2018)

Fig. 2 Changing of environmental responsibility level in different age groups (Bovt and Buzova 1999)

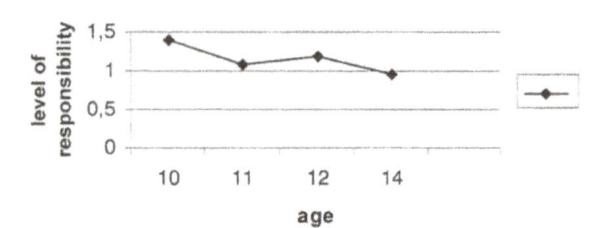

environmental challenges, first of all at a regional and local scale. This is very important for the Kaliningrad region, which is located as a Russian exclave area between foreign countries. In recent times our common research and collaboration with partners in Lithuania, Poland and others were very productive. Unfortunately, nowadays, the geopolitical situation is not as favorable. Nevertheless, some possibilities exist to develop cooperation across an EU project. For instance, the Kaliningrad region is a priority area for the operation of some important projects of the Cross-border Cooperation Programmes Lithuania–Russia and Poland-Russia 2014–2020 developed within the framework of the European Neighbourhood Instrument.

Participation in world events creates additional opportunities. The social effects of the World Cup in 2018 meant that for some Russian cities—including Kaliningrad—there was much more interest to sport, physical education and healthy lifestyles for all people, including our students. This opens new opportunities for a comparative geographical analysis of this effect in Russian cities in which the Championship took place.

4 Students' and Scholars' Visions

It is important to learn how our students and scholars understand the terms *Responsibility* and *Sustainability*. Research on this topic was conducted by IKBFU students during their stage in Sweden under the supervision of Prof. Krasnov from Kaliningrad and Swedish course leaders Ekborg and Malmberg from Malmo University. The children's most common explanations are presented in the Tables 1 and 2 and Fig. 2.

In Table 1 according to students respond there were 3 levels of responsibility for environment protection defined: starting from Level "0"— no responsibility at all—up to the 3d one— pupils act environmentally friendly.

Table 1 The levels of children's responsibility for environment protection

Level 0	Responsibility is not developed at all. Pupils do not behave in an environmentally friendly way and do not realise that they themselves are responsible for the environment
Level 1	Pupils still do not recognise themselves as responsible, but do behave an environmentally friendly way. The reasons of such a behaviour can be: the rules of their parents, habits, financial incentive, or a child simply cannot give the reason for their behavior
Level 2	Pupils behave in an environmentally friendly way, because they know that it is good for the environment, but do not give details in their explanations
Level 3	Pupils act independently and environmentally friendly, are well aware of the reasons for their behaviour and explain it according to their knowledge about the environment

Source Bovt and Buzova (1999)

Table 2 The children's most common responses and explanation of the term *Sustainability*

Question number	Do not behave in such a way	Behave in such a way
2	"Because it is cosier" "Because I'll come back into my room"	"The lamp can get hot and the fire will start" (1) "Not to waste electricity"(2) "To save energy" (3) "It takes current and you have to pay for it"(l) "Because my mother told me that (1)
3	"I need to rinse my mouth" "Teeth should be clean" "I don't want to brush my teeth if there is no water" "It's a habit, but it's stupid to do it like this, I should stop it"	"We should not waste the water"(2) "Because we can run out of water and it hurts the environment" (3) "Because of the sound and that we destroy nature"(2) "It's just a habit"(1)
4	"We live in the city and we just have a regular waste basket" "My Mom always works, so we don't have time" "It's easier to put everything into one waste-bin" "Our apartment is small and we throw everything in one and collect everything"	"Because you should take care of the environment" (2) "We recycle paper and glass and batteries" (1) "Not to destroy the environment" (2)
5	"We are too lazy to do it" "I want to, but my parents never help me, I forget it"	"We don't litter, don't throw plastic away" (1) "To save money" (1) "We don't waste plastic" (2) "You can use them for other things" (1)

Source Bovt and Buzova (1999)

Table 2 demonstrates how children explain the term Sustainability. As far as the question why people behave/not behave in sustainable way among the most common comments were ones dependent on the habits in the families. Although some comments reflected that not families behave environmentally friendly due to too much occupation of parents who do not have time to think about sustainability there were many more phrases really environmentally oriented, like: to save, not waste, not destroy etc (Table 2).

It is a good example of students' research, from which we could conclude that there is strong need to start environmental education at age 11–12.

The next example of the use of cognitive conflict in educational processes as a tool of problem-oriented learning was presented at Malmo University by our student-geoecologists Gouloubitskii and Belykh (1999). They explored ways to manage and direct cognitive conflict towards positive goals such as a better understanding and memorization of educational material. These young researchers also examined the attitudes of Swedish teachers and their pupils when it comes to cognitive conflict presentations.

Fig. 3 Approach towards combined effects to SD based on network thinking and new knowledge. *Source* (Leder 2018) with authors' modifications

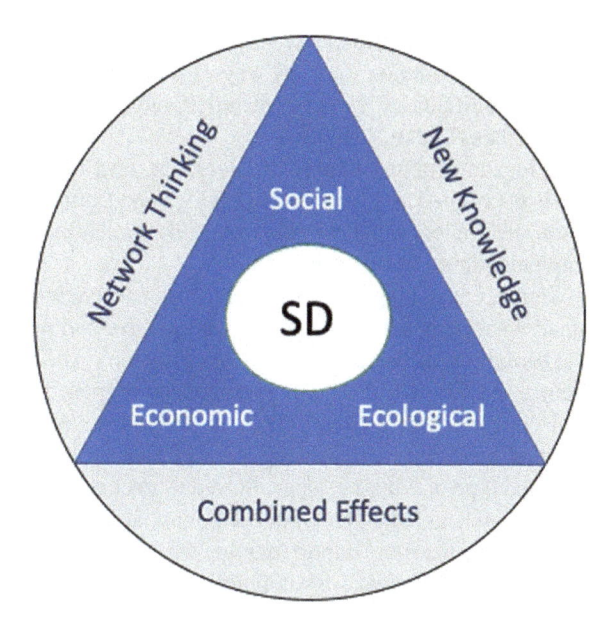

Another notable piece of research is one which was conducted on gender studies. Under the author's leadership about 200 students from geography and law departments took part in an opinion poll aiming to investigate their perception of regional issues. About 85% of male and 90% of female respondents outlined the importance of education for success in their career. Half of them regardless of gender suggested that ongoing technological upgrades were a necessary condition to ensure economic, social and environmental security at the local and national level. The same portion of respondents were sure that the knowledge and skills they obtained at university would help them to realize their future. When it comes to the most crucial problems which might affect future development, 25% of females considered environmental issues (placing in third place after problems caused by HIV (45%), and alcoholism and drugs (30%). Males did not consider ecological issues among the most acute (Krasnov 2001).

5 Collaboration and Combined Effects of ESD

Some authors criticize curricula which focus upon universalism and standardization (Kyle 2020). Indeed, education for SD must first of all develop the critical thinking and creative development of students. We share the understanding of the direction and development of pedagogical practices as set out in the work of Leder (2018). Curricula should be structured in such a way as to encourage learners to find the right arguments to facilitate decision-making, in which natural resources will be considered comprehensively—from the standpoint of their sustainable use, protection and reproduction. In many ways, the development of students' systems thinking will be facilitated by learner-centered problem-posing, and network focused teaching approaches (Leder 2018).

Education for sustainable development must be re-organized in such a way that SD is positioned as a worldview (Pavlova and Lomakina 2016). We also believe that the education policy of both the state and supranational structures is important in the implementation of this approach.

Before embarking on the formation of a worldview based on SD, one should understand what competencies a person should have and what methods will contribute to the formation of these competencies. This topic is covered in detail in the work of Yeung and Chow (2019). We analyzed the proposed approaches of various authors and compared them (Table 3). It is possible to conclude from this research that most of the experts are speaking about similar approaches towards development SD-oriented type of thinking. That allowed us to combine them into more general groups of competences, like mental development, actor interaction, strategic planning. For our research the most important issue was that all three experts were speaking about importance of personal development, involvement and responsibility, that pursued us to create such group as personal development approach.

We also visually developed the concepts of ESD (Fig. 3), which were earlier suggested by Leder (2018).

New knowledge is an important component as far as ecological technology and development of society are concerned. This issue was mostly addressed during Education process and it is a way of personal development as well. Network thinking is also important for psychological research with regards to "brain-based learning" and has been described by Spitzer (Hamburg University) as student-centered learning. It has been proven to be supportive as the number of transmitters between synapses is boosted and action potentials are increased in strength and frequency (Spitzer 2007). Therefore, we look at how these didactic principles and arguments can be used as a teaching methodology. Our findings from this very research, that these two (Network thinking and New knowledge) should be supplemented with Combined effects originated from the combination such approach with the psychological instruments of a person's development.

In very difficult and complicated global conditions, we need to start with radical education reforms in a much larger scale than before, because, the problems of "Man Quality" now are primarily that of character. In general terms, we need reform in a Didactic sense, related not only to pure teaching, but reforms which are also closely connected with the upbringing of our young generations of students in accordance with the triangle of ambitious ratios as seen in Fig. 4.

In more strictly way in Russia we need to develop didactic approaches to student's mentality (psychological) across several steps, which presented in a Table 4. As far as motivation as one of the important issue is concerned while speaking about pupils' approach to *Sustainability* (see Table 2) we much concentrate in our didactic aims on formation Environmental friendly attitude, Rational behavior, Positive actions towards nature, connecting all these issues to Personal Success. It might be possible,

Table 3 Conditions for the formation of an SD-oriented worldview

Groups of competences	Competences by Rieckmann (2012)	Competences by Lozano (2017)	Methods by Cotton and Winter (2010)
Mental development	systemic thinking and handling of complexity; critical thinking; anticipatory thinking	systems thinking; anticipatory thinking; critical thinking and analysis	debates and critical incidents (using critical events as background and asking student what they would/could/should do); critical reading and writing; problem-based learning
Actor interaction	cooperation in groups; participation; communication and use of media; interdisciplinary work	interdisciplinary work; interpersonal relations and collaboration; communication and use of media	role-plays and simulations and group discussions; stimulus activities (using videos, photos, poems, or newspaper for reflection and discussion); case studies and reflexive accounts; fieldwork; worldview and value research; action research
Strategic planning	planning and realizing innovative projects; evaluation	strategic action; assessment and evaluation	modeling good practice; future visioning
Personal development approach	ambiguity and frustration tolerance; acting fairly and ecologically; empathy and change of perspective	justice, responsibility, and ethics; personal involvement	personal development planning

Source Compiled by the authors after Yeung and Chow (2019)

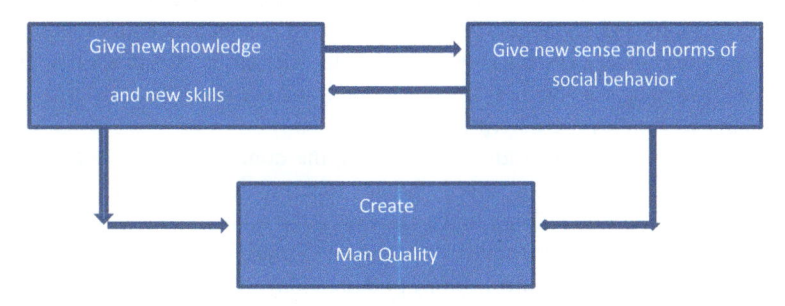

Fig. 4 The triangle of creative relations for SD (compiled by the authors)

Table 4 Upbringing spheres and aims for skills development

Psychology spheres	Examples of didactic aims
Intellect	Variability, flexibility, critical thinking
Motivation	Nature friendly, rational behaviour, positive actions, personal success
Emotions	Expressions, responsiveness, Sympathy, credulity, positive answers
Self-regulation	Self-criticism, self-control, overcoming the barriers, creativity
Will	Adherence to principles, determination, perseverance, sustainability
Legislation	Co-agreement, respect for applicable laws, compliance with legal regulations
Existence	responsibility, cooperation, collaboration

Source Glushkova and Krasnov (2001)

for instance, when make children acknowledged with famous persons—youth idols (actors, singers ect.) who are involved in actions supporting the environment and nature protection.

6 Conclusion

So, new steps towards the modernization of teaching are connected with the active participation of students. For the more effective realization of ideas and concepts related to ESD in a global, regional and local context, we need to integrate educational programmes and courses with a problem-oriented approach to find the solution of conflicts between society and nature with project-oriented activity. Such kind of activity calls for such acute components of scholars self-development as self-assessment, self-respect, and self-actualization. Environmental education and its design at all levels are to be transformed not only directly pedagogically, but also through a holistic approach. In this case, we agree with Leal (Leal Filho et al. 2018), because it is only the active interaction between teachers and students that could change the current situation to one that is more sustainable in our common localities, regions and countries. Only this approach has a chance to integrate ESD at these different levels. We also agree with Kyle, who says: "Youth are demanding action; science educators ought to enable learners and communities to transform and reinvent the world they are inheriting" (Kyle 2020).

Involving external experts and popularizing the concepts of SD is a chance to upgrade the value of education and improve outcomes in many real sectors of the economy and social life not only in Russia, but in the other countries too.

This programme falls under the educational initiative named "understandable science", which is becoming increasingly popular today (at open and stand-up conferences, festivals and another forums). There are Quantoriums and Techno-parks, techno-rooms and also interactive spaces in museums, youth science festivals etc.

all over the world. An upgraded BUP has to use all of these attributes to promote ESD. From our point of view, however, it will be much more useful and there will be far more opportunities if we use ESD across long-term education (BA, MA, PhD), summer schools, conferences, online learning and modern technologies. Research conducted by students and young scientists in the sphere of SD for the Baltic Sea region ought to be of interdisciplinary character and cover the same research across different life spheres.

Immanuel Kant Baltic Federal University participated in such activities within the BUP-network and in many other international programmes and projects on Educational development as well. For example, the result of the Project "Strengthening and Developing Business and Administrative Education in Kaliningrad region"—KalEdu of the EU-Russia Cooperation Programme TACIS is the new MBA/MPA (Kaliningrad oblast 2007), which includes the development of new courses on ESD with these authors' participation. A great number of the SDGs nowadays need to be ranged, systemized and matrix-classified through the use of IT and technologies which are connected with Europe Open Science Clouds, the Solution for Education Integrated Data System and other modern possibilities.

So while participating in all the innovation forms of education no one should forget about complementing our pedagogical approach with the psychological instruments of a person's development such as soft influence, communication, conviction, debate etc. The facilitation of such kind of components will bring visual results in near future.

And finally, let us conclude with Eastern wisdom, which says: if your plan is for one year—plant rice; if your plan is for ten years—plant a tree; if your plan is for one hundred years—educate students and teachers. This is the most important Goal of sustainable development.

Acknowledgments Authors Contributions
All parts of the article were developed by both authors in strong cooperation. Both authors have read and agreed to the published version of the manuscript.

Conflicts of Interest
The authors declare no conflict of interest.

References

Allison HE (1971) Kant's transcendental humanism. Monist 55(2):182–207
Bengtsson SL (2020) Outlining an Education Without Nature and Object-Oriented Learning. In: Cutter-Mackenzie-Knowles A, Malone K, Barratt Hacking E (eds) Research handbook on childhoodnature : assemblages of childhood and nature research. Springer International Publishing, Cham, pp 129–150

Bovt A, Buzova M (1999) Pupils' understanding of the responsibility for the environment. Malmo University, School of Teacher Educations, Sweden-Kaliningrad, Final Paper

Butvilovsky VV (2017) On "Geographical" matter. Info Measure Bullet Kemerovo State Univer 3(3):31–34

Cotton D, Winter J (2010) It's not just bits of paper and light bulbs: a review of sustainability pedagogies and their potential for use in higher education. In: Jones P, Selby D, Sterling S (Eds) Sustainability education: perspectives and practice across higher education, Earthscan London, UK, New York, NY, USA

Demssie Y, Biemans H, Wesselink R, Mulder M (2020) Combining indigenous knowledge and modern education to foster sustainability competencies: towards a set of learning design principles. Sustainability 12:6823. https://doi.org/10.3390/su12176823

de Wolf M, de Hamer A (2015) Education for sustainable development in the Netherlands. In: Jucker R, Mathar R (eds) Schooling for sustainable development in Europe: concepts, policies and educational experiences at the end of the UN decade of education for sustainable development. Springer International Publishing, Cham, pp 361–380

Fedorov G, Druzhinin A, Golubeva E, Subetto D, Palmowski T (eds) Baltic region—the region of cooperation. In: Springer proceedings in earth and environmental sciences. Springer, Cham. https://doi.org/10.1007/978-3-030-14519-4_10

Glushkova LS, Krasnov EV (2001) Ecologicheskoe vospitanie shkolnikov I studentov kak factor ustoichivogo razvitija regiona. [Environmental education of schoolchildren and students as a factor of sustainable development of the region] In: Orlenok V (ed) Geography the turn of the century, Kaliningrad, KSU, pp 236–243. (In Russian)

Gouloubitskii A, Belykh E (1999) Cognitive conflict as moving force of problem based learning in environmental education. Malmo University, School of Teacher Educations, Sweden-Kaliningrad, Final Paper

Harvey T (1995) An education 21 programme: orienting environmental education towards sustainable development and capacity building for rio. Environmentalist 15(3):202–210

Haubrich H, Reifried S, Schleicher Y (2007) In: Lucerne declaration of geographical education for sustainable development. Switzerland: Geographiedidaktische Forschungen

Horkheimer M, Adomo TW (2002) Dialectic of enlightenment. Sstanford University Press, Stanford

Ionov V, Kropinova Y (2010) The Baltic University programme on the eve of its 20th anniversary. The Baltic Region 3(5):70–74

Jutvik G, Liepina I (2020) In: Education for change: a handbook for teaching and learning sustainable development. Uppsala: Baltic University Press. Available Online http://www2.balticuniv.uu.se/bup-3/textbooks-course-materials/bup-course-materials/education-for-change (Accessed on 20th of June 2020)

Kaliningrad oblast (2007) Toward regional MBA/MPA programmes: Materials of the Integrated Cross-Modular Learning Programme in Business and Public Administration: in two books. KalEdu:. I.Kant State University of Russia Publishing House, Kaliningrad

Krasnov E (2001) New approaches to gender researches at the university level in Russia. In: World Wide Wisdom—socially responsible and gender inclusive science and technology (Gasat) 10 Book of Abstracts, July 16, 2001. Copenhagen, Denmark

Krasnov EV (2020) From nature conservation to sustainable development. In: Fedorov G et al. (ed) Baltic region—the region of cooperation. Springer, pp. 111–119

Krasnov E, Kropinova E (2017) The combined effects of education and research on sustainable development in the Immanuel Kant Baltic Federal University (Russia, Kaliningrad). In: Leal Filho W, Skanavis C, do Paço A, Rogers J, Kuznetsova O, Castro P (eds) Handbook of theory and practice of sustainable development in higher education: volume 2. Springer International Publishing, Cham, pp 413–423

Krasnov E, Karpenko A, Simons G (eds) (2020) In: Crisis management challenges in Kaliningrad. 1st edn. Farnham, UK. Available Online http://urn.kb.se/resolve?urn=urn:nbn:se:fhs:diva-5154 (Accessed on 20th of June 2020)

Krugman P (1999) The role of geography in development. Int Reg Sci Rev 22(2):142–161

Kyle WC (2020) Expanding our views of science education to address sustainable development empowerment and social transformation. Discipl Interdiscipl Sci Educ Res 2(1):2

Leal Filho W, Raath S, Lazzarini B, Vargas VR, de Souza L, Anholon R, Quelhas OLG, Haddad R, Klavins M, Orlovic VL (2018) The role of transformation in learning and education for sustainability. J Clean Prod 199:286–295

Leder S (2018) Education for sustainable development and argumentation. In: Leder S (ed) Transformative pedagogic practice: education for sustainable development and water conflicts in indian geography education, Springer Singapore, pp 55–88

Lemmetty S, Glăveanu VP, Collin K, Forsman P (2020) (Un)Sustainable creativity? different manager-employee perspectives in the finnish technology sector. Sustainability 12(9):3605

Lozano R, Merrill MY, Sammalisto K, Ceulemans K, Lozano FJ (2017) Connecting competences and pedagogical approaches for sustainable development in higher education: a literature review and framework proposal. Sustain 9:1889. https://doi.org/10.3390/su9101889

Markelova EY (2007) Strany Severnoy Evropy: factory uspeha. [Nordic countries: factors of success.] St. Petersburg, pp 18–21 (in Russian)

Meadows D, Randers J, Meadows D (2020) Limits to growth: the 30-year update. Chelsea Green Publishing Company, White River Junction, VT, Vermont, 2004. Available Online http://www.unice.fr/sg/resources/docs/Meadows-limits_summary.pdf (Accessed on 20th of June 2020)

Mngumi LE (2020) Exploring the contribution of social capital in building resilience for climate change effects in peri-urban areas, Dar es Salaam, Tanzania. GeoJ. https://doi.org/10.1007/s10708-020-10214-3

Pavlova M, Lomakina T (2016) Sustainable development as a world-view: implications for education. In: Lam C-M, Park J (eds) Sociological and philosophical perspectives on education in the Asia-Pacific Region. Springer Singapore, Singapore, pp 37–50

Rieckmann M (2012) Future-oriented higher education: which key competencies should be fostered through university teaching and learning? Future 44(2):127–135

Rydén L, Migula P, Andersson M (eds) (2003) In: Environmental science. understanding, protecting and managing the environment in the Baltic Sea region. Uppsala, The Baltic University Programme—Uppsala University, pp 824

Sadovnichiy VA, Akaev AA, Korotaev AV, Malkov SYu et al (2012) Modeling and forecasting world dynamics. ISPI RAN, Moscow

Schmitt CT, Palm S (2018) Sustainability at German Universities: The University of Hamburg as a case study for sustainability-oriented organizational development. In: Leal Filho W (ed) Handbook of sustainability science and research. Springer International Publishing, Cham, pp 629–645

Shuvalova OR (2007) Image of science: public perception of the results of scientific activity Foresight 2:50–59

Simaeva IN, Budarina AO, Sundh S (2019) Inclusive education in Russia and the baltic countries: a comparative analysis. Baltic Region 11(1).

Sorlin S (ed) (1997) In: A sustainable Baltic region. A series of booklets from the Baltic University Programme, Uppsala, The Baltic University Programme—Uppsala University. Available Oline http://www2.balticuniv.uu.se/bup-3/textbooks-course-materials/bup-course-materials/a-sustainable-baltic-region (Accessed on 20th of June 2020)

Spitzer M (2007) Lernen. Spektrum Akademischer Verlag, Gehirnforschung und die Schule des Lebens. Berlin

The road towards sustainability: a historical perspective. In: Sorlin S (ed) A sustainable Baltic region. A series of booklets from the Baltic University Programme, Uppsala, The Baltic University Programme—Uppsala University, Booklet 1. 1997 Available Online http://www2.balticuniv.uu.se/bup-3/textbooks-course-materials/bup-course-materials/a-sustainable-baltic-region (Accessed on 20th of June 2020)

UN (2015) About a sustainable development goals. Available Online https://www.un.org/sustainabledevelopment/sustainable-development-goals/ (Accessed on 20th of June 2020)

UN (2020) Covid-19 implications. Available Online https://sdgs.un.org/goals/goal4 (Accessed on 20th of June 2020)

UNESCO (2019) Framework for the implementation of education for sustainable development (ESD) beyond 2019. Available Online https://unesdoc.unesco.org/ark:/48223/pf0000370215 (Accessed on 20th of June 2020)

World Intellectual Property Indicators (2011) Geneva, pp 65 (in Russian)

Yeung S-K, Chow C-F (2019) Applied education for sustainable development: a case study with plastic resource education. In: Leal Filho W, Azul AM, Brandli L, Özuyar PG, Wall T (eds) Quality education. Springer International Publishing, Cham, pp 1–13

Continuing Education at Faculty of Geography (Lomonosov Moscow State University): Trend for Sustainable Development

Valentina Toporina and Alexandra Goretskaya

Abstract Current situation shows that education is enough for two or three years after graduating, and then there is a need for additional training to meet the rapidly changing requirements of the developing reality. The aim of the paper is to focus on concept of continuing education at the Faculty of Geography. It includes several stages. The School of Young Geographers (SYG) provides knowledge of geography that is not covered by school programme, introduces methods of field research. Bachelor's educational programme forms an understanding and framework of rational nature use as a result of human activity and way to solve adverse situations, gives a set of tools to improve land use and protect landscapes, choose the purpose of land use, etc. Masters are trained to conduct original researches at different levels, to propose recommendations on improving situation by providing principles of land adaptation, introducing various research methods and skills of mapping, training the ecological networks, learning to rank the priorities of use in accordance with the ecological capacity. Further educational programme develop a unique way for specialists graduated from different professional fields to receive or improve environmental knowledge and skills. At the end of each stage students demonstrate knowledge and skills in graduating works and make their scientific reports. In paper we stress mostly the bachelor's educational programme as it forms the majority of specialists in ecology and nature use.

1 Introduction

Present environmental situation in our country calls to solve tension between human load and ecosystems, locally there is an urgent need in decisions. So one of the main tasks is to develop and expand environmental education, since unskilled interference of economic and other activities in natural processes can further complicate the situation.

V. Toporina (✉) · A. Goretskaya
Lomonosov Moscow State University, Leninskie Gory, 117209 Moscow, Russia

© The Author(s), under exclusive license to Springer Nature Switzerland AG 2021
W. Leal Filho et al. (eds.), *Innovations and Traditions for Sustainable Development*,
World Sustainability Series, https://doi.org/10.1007/978-3-030-78825-4_8

In our paper we pose a research questions whether educational system brings people to understand sustainability and if educational system has undergone changes towards sustainability.

Research goal is to analyze the present situation of the environmental education system in Russia.

Before we reveal the essence of continuing environmental education at Lomonosov Moscow State University, as one of the leading universities in the field of fundamental and partly applied ecology, we will focus on some historical milestones in the formation of the system of continuous education.

In general, the idea of continuing education can be found in ancient papers among philosophers (Plato, Confucius, Socrates, Aristotle, Seneca) (Zhurakovskiy 1963) as well as scientists of the Renaissance (Voltaire, Goethe, Rousseau, etc.) (Kuznetsov 1965; Konradi 1987; Cassirer 1945). Traces of practical application can be found in the XIII–XIV centuries in the cities of Europe on the basis of craft workshops.

Jan Amos Komenski (Czech humanist, seventeenth century) is considered to be the founder of systematic and integral ideas about continuing education (Komenski 2009).

The idea of «continuity» in education received was reinterpreted in our country after 1917, which contributed to the formation of a new system of education in Russia. There were new types of educational institutions, including adult education and professional development for employees. However, by the end of the 60s of the last century, the concept of continuing education in the Soviet Union, stated in government documents, was not introduced, and did not become a component of the state educational system. Occasional references to this problem were based on the innovative practices of individual teachers and scientists.

Initially, in our country, continuing education was considered as an education for adults. Its purpose was to compensate the shortcomings or omissions of previous training, or to advance or enrich professional knowledge.

In the current situation, continuing education provides education at all stages of human life, which will help to improve its quality and attractiveness. Thus, a person has the opportunity to improve their educational level throughout their life. This is stated in the National doctrine of education in the Russian Federation until 2025.

The term "continuing education" was firstly used in 1968 in UNESCO's papers and after the report of the Commission under the leadership of Faure was published, continuing education has been recognized as the main principle, the "guidebook" for innovations or educational reforms in all countries of the world.

The adoption of the concept of sustainable development by the world community in 1992 at the UN conference in Rio de Janeiro confirmed the status of education as a «decisive factor of change». Since the transition to sustainable development is impossible without addressing the problems that have arisen as a result of inadequate use of natural resources and services, i.e. conflicts of nature management, continuing environmental education has become an organic part of continuing education in general.

The educational system in Russia was modernized in accordance with the challenges of the time, which led to large-scale transformations in Russian higher education, the most noticeable manifestation was the adaptation of domestic traditions to the experience of the Bologna system.

It should be noted that the Russian higher education system has always been characterized by high quality of students' training. Anticipating the concept of continuity in environmental education, we note that at our Faculty «environment» is regarded closely to practical nature use and social life, so our educational programme covers environment, ecology and land use.

Environmentalism has come into education in western countries much earlier than in our country. Simple reason explains such a state of affairs: unlike many countries, Russia has been an agricultural country for a long time, so majority of the population had clear understanding of natural occurring processes and phenomena. Evidently, increased economic activity has influenced negatively both the environmental situation and people. Now understanding of the environment and natural processes are again in demand, but on a different level. Humans already need the ability to manage ecosystems.

We will not cover every single aspect of education system transforming in Russia during the XXth and XXIst centuries, because the decisions were sometimes ambiguous. To date, the education system is similar to the global one and includes primary, secondary, and higher stages. Following trends in environmental education have clearly emerged in Russia:

(1) until the middle of the XIX th century environmental education was cognitive due to a rapid development of natural sciences.

An important prerequisite for environmental education in Russia was gathering of many natural facts by Russian naturalists of the XVIII th century (I. G. Gmelin, G. V. Steller, S. P. Krashennikov, I. I. Lepekhin, P. S. Pallas, V. F. Zuev, and others) (Kuzmina et al. 2014).

(2) From the second half of the XIX century people have become aware of the unreasonable nature use. At this time, the issues of moral, aesthetic, and careful attitude to nature were summarized and discussed in the research papers of a number of teachers (Kuzmina et al. 2014). Thus, the requirements for the environmental content for natural science teaching were determined, and a careful, moral attitude to nature was reasoned.

In the second half of the XIXth century V. G. Belinsky, A. G. Herzen, N. A. Dobrolyubov, D. I. Pisarev, N. G. Chernyshevsky, L. N. Tolstoy, K. D. Ushinsky, and others opposed to formal study of nature, narrow pragmatism, and utilitarianism in relation to it (Kuzmina et al. 2014).

As a result of this situation, special scientific and administrative organizations began to be created starting in the 1850s. This made it possible to introduce environmental issues into the curriculum. Till the 1950s environmental knowledge had biological character.

Substantial knowledge was given to the students of rural schools, where progressive text books prevailed. In the Ministry schools, such textbooks appeared only at the beginning of the XX century. Students of church schools received a minimal amount of natural science knowledge, but in this type of school a lot of attention was paid to bringing to students careful attitude to all life on earth (Reymers 1990).

Later ideas of teachers of the past about education and communication of the child with nature were summarized by Russian teacher A. Gerd (interpreter of Darwin's works in Russia), A. Pavlov, and many other naturalist teachers. They created a number of original manuals on the methodology of natural science, which put the issue of environmental education in primary school at the first place.

Issues of nature conservation and resource-saving technologies were gradually included in school curricula and textbooks as part of the teaching of natural science subjects, in particular biology and geography.

(3) Environmentalism has come into education due to the works of Johansen, Slastenina (1984), and others.

It has become recognized that in order to achieve systemic knowledge on nature, environmental issues must be included in both natural and human science.

Currently, theoretical researches and observations are being conducted in Russia in secondary and high environmental education. The Concept of General environmental education is being developed. But so far, only 30% of schoolchildren in Russia have received environmental education. It comes down to awareness of environmental issues. It includes discipline «Ecology»; environmental issues in «Chemistry», «Physics», «Geography», «Biology», «Chemistry Course for Humanities classes», «Nature and literature», etc. basic and specialized courses.

In schools environmental education is also presented in several adopted disciplines such as «Social Studies». Also, in various regions of our country, students and children get acquainted and information about their region from textbooks on regional studies or local history or from expeditions.

Today, environmental education in state and non-state organizations is supported by State educational standards, which implies that environmental education should be carried out at all levels of education through regular and extracurricular activities within the framework of independent educational program.

In the 2020–2021 academic years, the discipline «Ecology» is planned to be introduced in last school years, but some students complete their education before them, therefore, this discipline will never be able to study. In addition, it is important to consider the fact school state examination, and «Ecology» in the exam is not included. From the above, we can conclude that the implementation of various additional education programs and the development of individual research activities are very relevant at the school.

In Russia, there are centers for additional environmental education, where children of different ages can observe and study aspects of human-nature interaction, immerse themselves in the surrounding natural space. An even more effective way to solve this problem is to organize children's outdoor camps, where parents can also stay, which makes it possible to participate together in solving urgent environmental problems. Looking at how children relate to the environment, parents also change their behavior, in this way two generations are brought up at once.

In the last decade, non-profit organizations (NPOs) and individual initiatives of society have been most active. We can say that these organizations raised a «fashion on ecology».

In secondary professional educational institutions (colleges) the main activity in education is formation of eco-culture and a healthy lifestyle.

Much attention is paid to universities. Students should be able to gain knowledge on interaction between society and nature, and be scientifically and practically trained people in this field. One of the main signs of professionalism for many specialties in higher education is the presence of environmental training.

The higher school provides training in the field of ecology and environmental protection in more than 200 higher education institutions of the Russian Federation in bachelor's and master's degrees in the fields of ecology and nature management, environmental science and land use, safety, water use, and others. Today, universities can independently approve a list of specific disciplines that are included in educational programs based on the competence approach.

Trends for further development, for sustainable development, are certainly raised (Bobylev and Solovyeva 2017). In our opinion, this is no more than a tribute to some fashion. Of course, the trend towards interdisciplinarity has become obvious, disciplines have appeared in the curriculum and students are actively working on issues of living comfort, human ecology, environmental economics, and alternative energy.

The problem of higher education is that in Russia, biological, geological, geographical and engineering specialties are being captured by environmentalism. In this regard, there is often an «overlapping» of specialties. At the same time, environmental knowledge differs greatly depending on the fundamental faculty. And the education that a student or college graduate receives is also determined by the foundation of the teacher or tutor who is responsible for environmental education.

It should be noted here that in the USA and Europe, education for sustainable development is more often referred to (Barth and Rieckmann 2012; Amaral et al. 2015) a concept that includes related environmental, economic, and social issues (Education for Sustainable Development Goals 2017). It is important to emphasize that in the 1990s, environmental education in our country also followed this way.

In general, the problem of Russian education is the lack of a clearly defined model.

Let's focus on environmental education in western countries. In Western Europe, environmental education often begins before school (Savvateyeva et al. 2019).

Environmental education in Germany is aimed seriously at improving the knowledge of young people about the environment, developing their readiness to protect nature, and forming models of environmentally responsible behavior in the younger generations. There is also a vision of environmental education as political education, so specialists try to investigate current regional environmental problems, to bring them to the public attention, to work out alternatives and to actively bring the issues into the municipality of town council (Environmental Education in Germany 2000).

Environmental education in the United States has a rich history and has embraced huge pedagogical experience, which is actively used in the XXIst century. A profound contribution to the development of environmental education in the United States was made by H. T. Odum, E. P. Odum, A. M. Lucas, C. E. Roth, D. R. Hammermann, W. M. Hammermann, R. A. Wilson, and others.

Environmental education in the United States has two levels: environmental and nature protection. They are closely related to and complement each other. The first level covers issues of natural dynamic balance, organization of the biosphere, heredity, adaptation, and changes in nature. And at the other level, issues of nature management and nature protection are seized. Ability and skills are components of education. They are divided into individual and research. The first contribute to the effective understanding of environmental knowledge; the second are aimed at mastering the methods of independent creative scientific research by children and students including outdoor activities (Chawla 2003; Rivkin 2000).

Environmental issues are included in the programs of many academic disciplines. Environmental education is provided in the form of special courses, special training within scientific and technical organizations, higher educational institutions of industrial enterprises, local councils, publishing houses, as well as adult education during popular science programs on radio, television and other modern media. Experimental ecological stations and national parks are also operating successfully.

In France, disciplines on nature and environmental protection are included in the curriculum of secondary schools, lyceums, and colleges.

In some countries, a cognitive-value- model of environmental education is widely spread, according to which the latest knowledge about nature and its protection is combined with the traditional values of society developed in the discipline of the ethnic history. Particular success this model has been achieved in some Asian countries: China, Korea, Thailand, and especially Japan.

Since the authors work mainly in higher education, we emphasize the obvious similarities and differences between Russian and others universities (institutions).

Countries differ in the labor intensity and duration of training. We will not talk about this in detail, because what is important here is not the duration, but the worldview with which a graduate of a school, college or higher school will come out.

Note that in Russia and Italy, number of basic disciplines is very high (up to 50%). In contrast, students study fewer disciplines that form instrumental skills (GIS technologies, analytical and statistical research methods) but spend more time at field researches than in any European University. Abroad, long-term internships (2 weeks) have been preserved in British and French universities. Often the field researches are removed at the senior courses by academic exchanges with foreign universities.

In European universities, increased attention is paid to the study of territorial planning and management, and environmental protection.

We also differ in the specialization at the bachelor's degree. For example, in the Netherlands, different courses at the same faculty do not have common disciplines from the very beginning, in the UK and Russia—from the 2nd year, but the strategies are different. For example, in the UK, with the help of a scientific supervisor competencies, modules of interest and topics of graduation papers are selected.

Analysis of the schedule in universities showed that the volume of lecture hours varies, but is lower than the Russian. For example, in the School of Geography and the Environment at the University of Oxford (Great Britain) at the 1st year students listen to only 7–8, in the 2nd year 4–5 h of lectures per week; in Spain, in the 2nd year-10 h; in Russia, in the 2nd year of the Faculty of Geography—at least 18 h per week.

Abroad, great importance is attached to students «independent work»: searching for information, writing essays, performing practical tasks, working in laboratories. From the tutor, such student's work requires much time to prepare materials, tasks, and check the work. In the case of a large number of students, several teachers are often involved in teaching the disciplines.

This allows tutors to be engaged in scientific research, participate in grant work, and speak at scientific conferences. A completely different time is distributed among tutors in Russia. Supporting student's independent work and checking essays, other papers are almost devalued, but on the contrary, participation in administrative work is encouraged. Therefore, the content of disciplines does not find a response among students, and there is very little time left for working in grants.

2 Environmental Education

Constant threats of large-scale environmental crises that cause social conflicts at the regional, national, and recently at the international levels require educational paradigm that is adequate to new challenges.

The importance the prospects of further greening of education has been seen as act for creating the Department of Environmental management at the Faculty of geography at Lomonosov Moscow State University in 1987. The initiative belonged to corresponding member of the Russian Academy of Sciences, Professor Andrey Kapitsa. Subsequently, similar departments were created in the natural science faculties of many other universities in the country. Mentioned department and few others train specialists in the field of «Environment, ecology and land use».

Over the past twenty-five years, the curriculum of the Department has been repeatedly revised and enriched with new disciplines. Thus, at the very beginning of the Department's existence, the curriculum included disciplines of socio and economic character (Economics and environmental management, regional environmental management, natural and cultural heritage, etc.), strengthened links between the educational process and practice (expertise, management and audit in environmental management, technologies for improving land use, etc.).

Classical geographical education, like no other, contributes to the formation of an environmental worldview, which, together with training in the basics of land use, forms the foundation of a new paradigm of education.

The signing of the Bologna Declaration by our country has led to significant changes in the system of training in higher education. The implementation of the main principles of the Declaration, in particular the transition to a two—level system of training specialists-bachelors and masters—required a review of educational standards and curricula. Thus, interactive teaching methods are actively used, the independent work of students has increased, and computer technologies have been more widely introduced into the educational process.

In addition, the work with "future geographers" continued at the School of young geographers, which started its working before the II World War and organically fit into the system of continuing environmental education. The evening-classes at Faculty were partly transformed into additional education which provides knowledge in several fields.

Thus, at the Faculty, continuing education consists of several stages: School of young geographers, bachelor's/master's degree, postgraduate studies, and additional education. However compulsory high education is bachelor's degree, young geographers are prepared to pass exam to enter faculty, to graduate as master or PhD depends on desire.

Within the framework of continuing education, methods and forms of education have become specific. The tutors, scientific supervisors now moderate of the study and provide not only an individual approach to learning, but also effective teamwork.

The result of continuing environmental education is a constantly developing person who owns universal ways of working and therefore is able to adapt to the new working conditions.

3 Methodological Concept

The research is based on classical papers in the field of education, the results of investigations of different authors in recent year (grant of Russian Geographical Society) (Ratsionalnoye prirodopolzovaniye: perspektivy innovatsionnogo razvitiya 2016; Alekseyeva et al. 2015), as well as the authors 'experience over the past 15 years. We also widely used information from the websites of universities in Belgium, Germany, Great Britain, Italy, Spain, the Netherlands, Poland, Finland, France, and Sweden, including curricula and course programs. Additional sources were monographs, papers in scientific and methodological journals (Kholina 2004; Blumhof and Holmes 2008; Ottens 2013), programs (Programmy distsiplin professionalnoy podgotovki po napravleniyu 2013; Benchmark Statement, Geography 2019). As research method we preferred existing data, this type of research is done on the available data.

Analysis of the labor market and employers ' requirements for training specialists allows us to conclude that the role of: (a) economic approaches and criteria for assessing anthropogenic impacts and their consequences has significantly increased; (b) a wide range of remote methods and GIS technologies for analyzing the environmental condition of territories and making management decisions. In accordance with these requirements, both the curriculum and the educational program were changed. The variability of training forms has significantly increased. However, the core of the educational program covers the approaches and principles of environmental education.

The conceptual basis of this work, which reveals some aspects of the educational system, is based on approaches to understand the ecological state:

- natural science-the reason for ecological state is the lack of knowledge about nature, human influence on the environment. So the solution is seen in teaching as many people as possible providing with environmental knowledge;
- naturalistic—the main idea is " the study of nature is in nature (in the field), not through abstract theoretical knowledge";
- global-biosphere-considers the environmental crisis as a global planetary phenomenon; the solution is seen in people's understanding of global environmental problems and solutions that concentrate the efforts of the world community;
- cultural considers the problem as a crisis of civilizations, to solve which it is necessary to form centers of a new culture; it promotes norms of behavior that cause the least damage to nature (saving water, recycling, etc.); states that the solution of environmental problems is impossible without world peace, respect for human rights and social justice.

Based on these approaches, the goal and tasks of continuing environmental education were formed. The goal is to form a certain environmental worldview—responsible and positive attitude to their health, ecological state, improving the quality of life, meeting human needs.

The educational process is focused on:

- forming the necessary knowledge for understanding the processes occurring in the system «man-society- nature», contributing to the solution of local social and environmental problems;
- bringing up respect for nature and developing an active citizenship based on a sense of responsibility for the state of the environment;
- developing ability to analyze environmental problems and predict the consequences of human activities in nature, the ability to independently and jointly make environmentally significant decisions.

The following principles of continuing education are taken into account when forming the basic scheme of students-bachelors masters, and postgraduates training:

(1) «greening» considers each object from the point of view of interrelation and mutual influence. It means reflecting the ideas and concepts of sustainable development in the goals, content, methods, means and forms of environmental problems of the present. This process is aimed at developing an interdisciplinary content of training;
(2) «culture influences». The current environmental situation cannot be considered in isolation from the social and cultural background. That is why it is important to rise general culture among the youth: responsible and conscious attitude to nature, society;

(3) inheritance and continuity. This principle means that all blocks of the educational program, or rather disciplines, follow to a certain concept, and their content is not repeated;
(4) integrity provides transformation of environmental science into an interdisciplinary science. Its provisions are widely studied in philosophy, economics, psychology, and other social and natural sciences. According to academician Nikita Moiseyev (1997), a comprehensive approach to environmental issues takes it beyond the science of wildlife. This view meets all the needs of society, so an interdisciplinary approach helps to overcome the separation of academic subjects and form a clearer picture of environmental reality;
(5) interrelation within global, regional and local levels of understanding of environmental problems. Implementation of this principle will contribute to a broader understanding of environmental issues and processes, as well as spread the eco-friendly views to local conditions and activities in different regions. The knowledge of school leavers, and then bachelors, specialists who receive a new qualification in the field of «Environment and land use» will not remain formal if they can specify global environmental problems, as well as summarize specific phenomena of a local nature and submit them to the level of national and global environmental problems;
(6) developing creative, analytical, and critical thinking. The importance of this principle has roots from need to educate people capable of analyzing and predicting the environmental consequences of human activities;
(7) practical application favors to gain experience for students of educational institutions for their subsequent environmental activities. In accordance with this principle, continuing environmental education should be based on activities such as experimental research, modeling the environmental processes, as well as organizing practical activities for the protection and restoration of the environment, in addition to verbal teaching methods;
(8) interactivity introduces special technologies in environmental education, aimed at increasing personal responsibility and independence, favoring creative inspiration, free communication, joint activities. Interactive training has a complex effect on a person, who no longer just knows and can, but also wants to be involved in environmental activities.

4 Results and Analysis

In accordance with the approaches and principles outlined above, the goals and objectives formed at the faculty as a whole, and at the Department, in particular, within the framework of «Environment and land use», an educational program has been formed. It includes the basic and variable parts, as well as practices and state final certification (state exam and final qualifying work). Before focusing a little more on implementation of mentioned approaches, we will devote few words to the School of the young geographer. Without going into details about the history of this school, which has existed since 1948, we note that the disciplines that form the basic knowledge of future environmentalists and land users are already taught to applicants and school leavers. These include geomorphology, hydrology, earth science, etc. Students-applicants also go to field works and defend their qualification papers. Students are studying the same scheme, but the content of the educational program is somewhat more complicated (Fig. 1), since the basic and variable parts also include modules, and practices and certification are held not once, but at each course of both bachelor's and master's degrees.

The schedule is designed in such a way that the content of one discipline, for example, "Natural and anthropogenic landscapes", serves as a basis for the disciplines involved in the arrangement and improving of the landscape ("Fundamentals of landscape planning", etc.), so we favor the principle of inheritance and continuity.

The principles discussed in Sect. 3 are implemented in the educational program not directly, but through the development of certain competencies. Both undergraduate and graduate students, also post-graduate students have developed universal, common professional, professional and special professional competencies at the end of their studies. If we compare the competencies (Fig. 2), there is an obvious trend in increasing the number of disciplines that allow you to reveal the principle of practical orientation—that is, to give skills to solve the problem. While universal, aimed at the development of thinking, culture, understanding of the relationship between

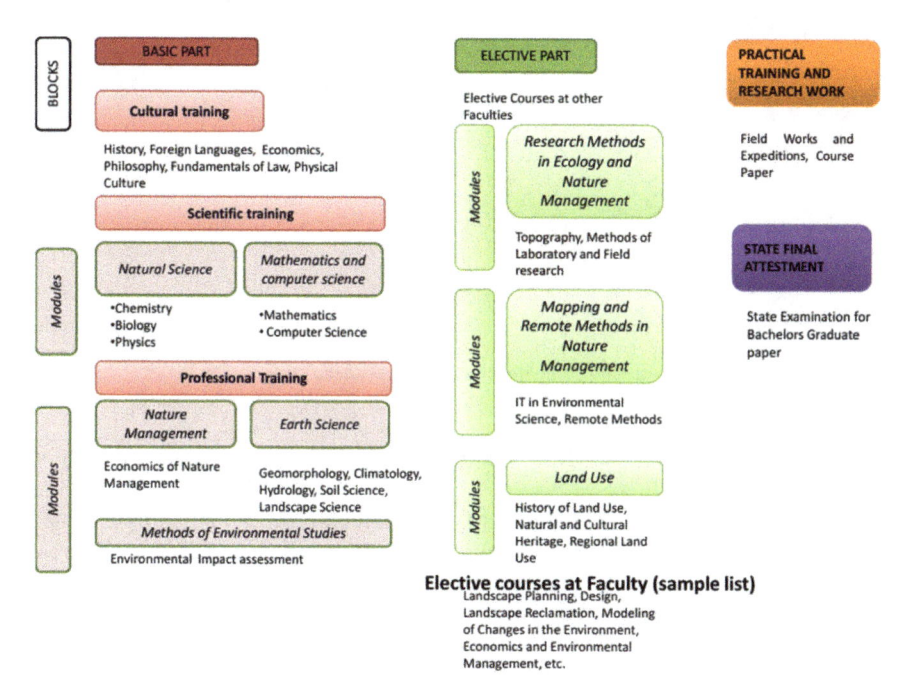

Fig. 1 The basic scheme of the educational program of the bachelors («Environment and land use») (sample)

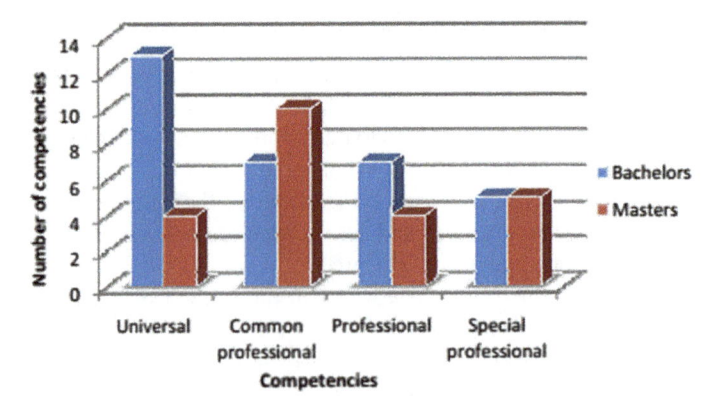

Fig. 2 Competencies developed by bachelors and masters

Disciplines / Training Period	Field Researches and Expeditions	Course paper Pre-graduate research	Land use	Environmental management	IT in land use	Remote Methods	Landscape Reclamation
2 semester	UC 1						
3 semester							SP 3
4 semester					SP 4		
5 semester					SP 4		
6 semester		UC 1　SP 4	CP 5				
7 semester				CP 5		SP 4	

Fig. 3 Coverage of disciplines by competencies (sample) (Bachelors)

global, regional and local levels of environmental problems, are formed mainly in the bachelor's degree.

The principle of greening and integrity of continuing education is implemented quite simply—one competence (universal or professional) is mastered in different disciplines (Fig. 3).

UC 1 Universal Competency 1	Ability to search, analyze critically and synthesis of information
CP 5 Common Professional Competency 5	Ability to use knowledge of nature management, sustainable development, environmental impact assessment, including legislation
SP 3 Special Professional Competency 3	Ability to analyze the relationship between society and environment, assess the consequences of resource use and human induced landscape transformation; management skills
SP 4 Special Professional Competency 4	Mapping skills, ability to use computer technologies and remote sensing data to analyze the landscape pattern, resource distribution and the dynamics of ecosystems caused by natural and anthropogenic factors

It is necessary to briefly describe the special professional competencies that graduates (bachelor's, master's, partly post-graduate and additional education) acquire, among them: knowledge of laboratory and field methods of research; methods of engineering ecological surveys and monitoring; the ability to apply them to assess the state of the environment; knowledge of the basic properties of the pollutants, the nature of their migration and accumulation in natural environment components; understanding of the laws of biosphere evolution as the basis for environmental protection, reduction of risk and industrial ecology; assessment of the degree of anthropogenic transformation of landscapes; ability to apply knowledge in the field of environmental regulation for natural resource management; theoretical and practical skills in identifying heritage sites at the regional level, knowledge of experience in their protection and use in management practice and many others.

Thus, knowledge is acquired comprehensively in the educational process of active and interactive forms (computer simulations, business and role-playing games, analysis of specific situations), and is fixed in independent student research—course and research papers (Golubeva et al. 2012).

As an example, we will give the general topics of research and course papers: environmental and social programs for regional development, environmental problems of a region/small rivers/ river basins, eco-tourism, eco-economic efficiency, environmental paradigm of education, protected natural areas, alternative energy, environmental management and its problems (sectoral or regional), improving the cultural landscape, etc.

When preparing a future specialist, as well as retraining, special attention is paid to the landscape criteria, which must necessarily be met, including: (1) proper use of natural resources and natural diversity; (2) reproduction and protection of natural resources—the presence of protected areas; (3) implementation of construction (and other engineering) measures that do not come into conflict with nature; (4) sanitary and hygienic conditions—constantly ongoing work to improve the landscape and

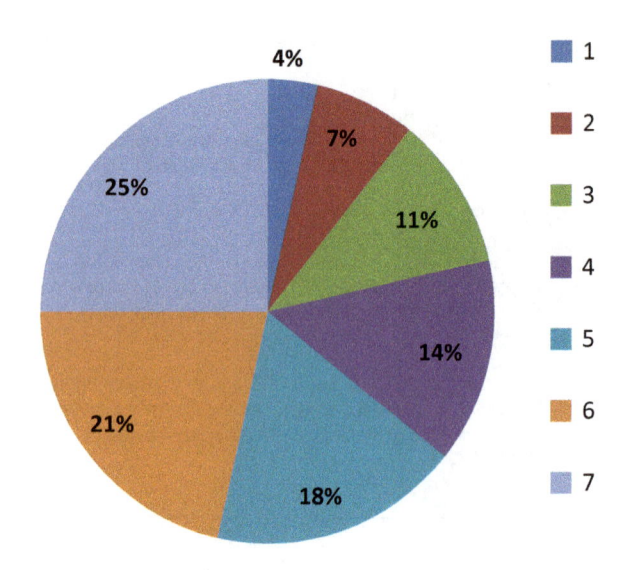

Fig. 4. Areas of professional activity of graduates of the educational program (according to 2015–2019). 1—ministries/departments (nature protection, regional development). 2—companies (consulting work: PriceWaterHouseCoopers, Zoï environment network, GazProm etc.). 3—Remote Sensing and IT. 4—Institutions and Universities. 5—Nature protection, urban parks. 6—Tourism. 7—Teaching in schools

its conservation; (5) absence of monotony and anthropogenic wasteland (wasteland, landfills, quarries); (6) functional zoning of the area.

As a result, bachelors receive theoretical knowledge, master the basic methods of conducting field research and get an idea of land improving.

If we talk about the results of practical implementation of the Department's program, most of the graduates apply their knowledge in specialized departments and consulting companies. In addition, graduates continue to improve their knowledge at the next stages (postgraduate studies, for example) both at their Alma Mater and other Universities (Fig. 4).

The next stages of education-master's, postgraduate and additional education-are operating in a similar way. Competencies also include universal, general professional, professional and special professional characteristics. The educational program also includes a basic variable part, practice and state examination.

5 Conclusion

Despite the fact that the theory of environmental education is generally the same for all countries, however, practical implementation significantly depends on the historical background, mentality and social and economic state of a particular state.

In Russia a pronounced structure of education has not yet been formed. It should be taken into account that the experience of environmental education in Russia is significant, but there is a number of problems that do not allow young people to fully realize their potential and form an environmentally literate person. The main reason, as we see it, is the lack of integrity, consistency of environmental education, as well as basic educational disciplines lack environmentalism.

In Russian universities, the curriculum does not pay enough attention to independent research work; we are talking about a weak focus on strengthening research skills both in the entire training and in individual disciplines.

The most successful systems of University geography are built on a strong foundation of school knowledge. Many disciplines from the curricula of Russian universities abroad are studied in specialized schools.

Russia, like the leading European countries (Great Britain, Germany, France), has an established system of higher l education, built on classical traditions, which is its undoubted advantage and a good basis for further development. At the same time, Russian universities are characterized by more conservative forms of teaching (lectures prevail) and still weak respond to the needs of society and students. At the same time, our education has documents that specify a significant percentage of elective disciplines, but there is a lack of simple and transparent algorithms for their implementation.

References

Alekseyeva NN, Klimanova OA, Naumov AS (2015) Sravnitelnyy analiz vysshego geograficheskogo i ekologo-geograficheskogo obrazovaniya v Rossii i stranakh zarubezhnoy Evropy, vol 1. Vestnik Moskovskogo universiteta. Seriya 5: Geografiya, pp 3–11

Amaral LP, Martins N, Gouveia JB (2015) Quest for a sustainable university: a review. Int J Sustain High Educ 16(2):155–172

Barth M, Rieckmann M (2012) Academic staff development as a catalyst for curriculum change towards education for sustainable development: an output perspective. J Clean Prod 26:28–36

Benchmark Statement, Geography (2019). The quality assurance agency for higher education. https://www.qaa.ac.uk/docs/qaa/subject-benchmark-statements/subject-benchmark-statement-geography.pdf?sfvrsn=4ae2cb81_4. Last Accessed 11 Sept 2020

Blumhof J, Holmes P (2008) Mapping the environmental science landscape—an investigation into the state of the environmental science subject in higher education. Planet 20:2–5

Bobylev SN, Solovyeva SV (2017) Sustainable development goals for the future of Russia. Studies on Russian Economic Development 28(3):259–265

Cassirer E (1945) Rousseau. Princeton University Press, Kant Goethe, p 118

Chawla L (2003) Bonding with the natural world: the roots of environmental awareness. NAMTA Journal 28(1):133–154

Education for Sustainable Development Goals (2017) Learning objectives. Paris, UNESCO. https://unesdoc.unesco.org/ark:/48223/pf0000247444. Last Accessed 10 Sept 2020

Environmental Education in Germany. Concepts. History. Projects. Visions (2000). Inter Nationes, Bonn (Germany), p 32

Golubeva E, Ignatieva M, Korol T, Toporina V (2012) Eco-geographical approach to investigation of stability of cultural landscape. Geogr Environ Sustain 4(5):63–83

Kholina VN (2004) Geografiya v sisteme universitetskogo obrazovaniya. Standarty i Monitoring v Obrazovanii 3:35–40

Komenski Ya A (2009) Uchitel uchiteley («Materinskaya shkola». «Velikaya didaktika»). M: Karapuz, p 288

Konradi KO (1987) Gete: zhizn i tvorchestvo. T.2: Itog zhizni. M.: Raduga, p 646

Kuzmina SA, Lubennikova SA, Popova LA (2014) Istoriya ekologicheskogo obrazovaniya v Rossii. Elektronnyy zhurnal Eksternat.RF. http://ext.spb.ru/2011-03-29-09-03-14/122-raznoe/5892-2014-09-05-13-39-50.html, Last Accessed 08 Aug 2020

Kuznetsov VN (1965) Volter i filosofiya frantsuzskogo Prosveshcheniya XVIII v. M. Izdatelstvo MGU, p 275

Moiseyev N (1997) Ekologiya v sovremennom mire, vol 12. https://www.nkj.ru/archive/articles/10376/, Last Accessed 11 Sept 2020

Ottens H (2013) Reflections on geography education in Europe. J Res Didactics Geogr (J-Reading) 2(2):97–100

Programmy distsiplin professionalnoy podgotovki po napravleniyu «Ekologiya i prirodopolzovaniye» kafedry ratsionalnogo prirodopolzovaniya. Uchebno-metodicheskiye materialy (2013). Geograficheskiy f-t MGU imeni M.V. Lomonosova, p 268

Ratsionalnoye prirodopolzovaniye: perspektivy innovatsionnogo razvitiya (2016). NIU VShE Moskva, p 172

Reymers NF (1990) Prirodopolzovaniye: slovar-spravochnik. M, p 318

Rivkin MS (2000) Outdoor experiences for young children. ERIC Digest ED448(013):8

Savvateyeva OA, Spiridonova AB, Lebedeva EG (2019) Sovremennoyeye ekologicheskoye obrazovaniye: rossiyskiy I mezhdunarodnyy opyt. Sovremennyye problemy nauki i obrazovaniya, vol 5. http://science-education.ru/ru/article/view?id=29188. Last Accessed 08 Sept 2020

Slastenina ES (1984) Ekologicheskoye obrazovaniye v podgotovke uchitelya: Vopr. teorii i praktiki. M.: Pedagogika, p 104

Zhurakovskiy GE (1963) Ocherki po istorii antichnoy pedagogiki. M. Akad. ped. nauk RSFSR, p 510

Valentina Toporina is graduated in Land Use (2008), PhD in Geography (2011). She is currently Senior Scientist at Lomonosov Moscow State University (Faculty of Geography). Her present research interest includes nature protection, environmental education, aesthetical and ecological state of cities.

Alexandra Goretskaya is graduated in Nature protection (1987). She is currently Assistant Professor at Lomonosov Moscow State University (Faculty of Geography). Her present research interest includes urban ecosystems, pollution, urban green areas.

The Culture of Learning in Organisations: What is the Current Perspective for Sustainable Development?

Orlando Petiz Pereira, Maria João Raposo, Miloš Krstić, and Oleksii Goncharenko

Abstract This study focuses on the organisational culture oriented to learning in the context of achieving the goals of sustainable development. Its objective is to perceive the positioning of values in the culture of organisations, in learning processes, and their influence on performance. This organisational typology facilitates the acquisition, use, and dissemination of knowledge. They learn on a basis of harmony, convergence, companionship, commitment, responsibility and focused on objectives. Which supports the sustainability of the organisation and its creative innovation processes and Sustainable development of society in general. As a methodology, it resorts to secondary data from several empirical works. It assumes that the organisational culture determines performance. However, this philosophy of learning cannot be viewed incoherently. This is due to the determination of behaviors which establish relationships of interaction, sharing, communication and commitment. It uses the model of contrasting values of the organisational culture by Cameron and Quinn to investigate the typology of organisational culture in Higher Education. Evidence largely confirms that the organisational culture, "Adhocracy" is prevalent in the institutions under analysis and that the "Clan" typology is the second most representative

---

O. P. Pereira
Economics and Management School, University of Minho, Campus de Gualtar, 4710-057 Braga, Portugal
e-mail: orlandop@eeg.uminho.pt

M. J. Raposo
Polytecnic Institute of Porto, Rua Dr. António Bernardino de Almeida, 431 4249-015 Porto, Portugal
e-mail: mrp@isep.ipp.pt

M. Krstić
Faculty of Sciences and Mathematics, University of Niš, Višegradska 33, 18000 Niš, Serbia
e-mail: milos.krstic1@pmf.edu.rs

O. Goncharenko (✉)
Department of Economics, Entrepreneurship and Business Administration, Sumy State University, Rimskogo-Korsakova str, 2, Sumy 40007, Ukraine
e-mail: o.goncharenko@econ.sumdu.edu.ua

one. As the following analysis confirms, this can help achieve more sustainable development goals. These results illustrate that in learning organisations, not only a single typology is adopted, but various others support a central one.

Keywords Organisational culture · Learning · Performance · Transformational values · Sustainable development goals

JEL Classification D23 · L25 · M14

1 Introduction

Organisations are living beings. They provide their employees with an identity, a meaning for work and enable them to defend themselves from disruptive social and organisational movements (Cullen 2008). They are tools to neutralise violence against people and the organisational environment (Capra and Jakobsen 2017). In recent years, the internal environment of organizations has been developing rapidly. This was facilitated by the UN in 2015 with its Sustainable Development Goals (SDGs). Not only were the SDGs the beginning of new relations of development but also led to organisations being producers of economic and social benefits. This can be considered a transition from a declaration to practically oriented activities (although so far mostly at the level of political decisions) to achieve them. This document recognizes the inseparable link between social, economic and environmental problems of the developed global world (Skliar 2018). When an organisation adopts a culture of centralised and bureaucratic power, the trampling on the performance of its employees can arise and have consequences for all stakeholders involved. This seems to happen less in more decentralised and less bureaucratic organisations. These appear to be more agile in the dissemination of knowledge and the motivation for quick and efficient decision-making. They help to improve the administrative efficiency and the intimate relationships between man, nature and institutions. They facilitate and motivate the development of the collaborator and respect their cognitive, emotional and social dimensions (Petriglieri and Petriglieri 2010). However, organisations, companies and social relations have been neglecting the most human and conciliatory aspect. In such an atmosphere, emotional instability has captured the hearts of organisations and their stakeholders, a reality that has been isolating and changing people's patterns of behavior (Pereira and Berta 2015; Sulkowski 2013). This reality is preventing access to higher levels of organisational intelligence, whereas the focus is placed more on things rather than on people. In this context, people feel undervalued and underappreciated and react in the presence of behaviours that are more hostile with the organisational objectives.

Against this reality, this work focuses on the organisational culture oriented to learning. It utilizes to the works of Wiewiora et al. (2013) and Cameron and Quinn (2006) to highlight the four typologies of organisational culture and the soul of its values. Also used the work of Raposo (2017), whose results validate the assumptions of these authors. Our objective is to understand the positioning of values in the culture of authenticated organisations and their influence on the performance of the employee, the organisation and the implementation of an organisational climate to stimulate and motivate learning attitudes in order to achieve SDGs. The 2030 Agenda for Sustainable Development reflected 17 goals and 169 objectives: 1. no poverty; 2. zero hunger, improving nutrition and promoting sustainable agricultural development; 3. good health and well-being, ensuring a healthy lifestyle and promoting well-being for all at any age; 4. quality education, ensuring comprehensive and equitable quality education and encouraging lifelong learning opportunities for all; 5. gender equality, ensuring gender equality, empowerment of all women and girls; 6. clean water and sanitation, ensuring the availability and sustainable management of water resources and sanitation; 7. affordable and clean energy, providing access to inexpensive, reliable, sustainable and modern energy sources for all; 8. decent work and economic growth, promoting progressive, inclusive and sustainable economic growth, full and productive employment and decent work for all; 9. industry, innovation and infrastructure, creating sustainable infrastructure, promoting inclusive and sustainable industrialization and innovation; 10. reduce inequalities; 11. sustainable cities and communities, ensuring openness, security, vitality and environmental sustainability of cities and other settlements; 12. responsible consumption and production, ensuring the transition to rational models of consumption and production; 13. climate action, taking urgent measures to combat climate change and its consequences; 14. life below water, conservation and sustainable use of the oceans, seas and marine resources for sustainable development; 15. life on land, protection and restoration of terrestrial ecosystems and promotion of their rational use; 16. promoting the building of a peaceful and open society; 17. Strengthening the means of implementation and intensifying work in the framework of the Global Partnership for Sustainable Development. In the following sections, only the numbers of the SDGs listed above will be used. It should be noted that not all of them will equally apply to one or another. Organizations need to first assess the positive and negative consequences of achieving them, and then build a plan to prevent negative consequences. In many respects, it will depend on the organizational culture.

With this perception, this study places emphasis on the following: the strategic value of a culture oriented to learning, supported by transformational leadership because it gives consistency to the structure of innovation and entrepreneurship processes within the organisation.

This study is structured in 5 sections. In the second one, the authors of this study design a theoretical contextualization of the climate and learning culture of organisations in addition to typologies of their culture and potential success in achieving the SDGs. The third section addresses the motivations to learn within the organisation. The fourth section is dedicated to empirical work. This section is divided into two parts. The first summarizes the results of the organisational culture of some empirical works. The second is based on the work of Raposo (2017). By Utilzing Raposo, an application of the Cameron and Quinn model (2006) in higher education to comparing the main component analysis procedure to discuss organisations roles in achieving the SDGs. Lastly, the study concludes with results and final remarks before a list of sources used.

2 Literature Review (1): Perspectives for a Learning Culture within the Organisation

The adoption of a culture of transformational values is a condition for the sustainable development of organisations and the Sustainable development of society as a whole. Just as a human being needs to look at the world from a positive, confident and humanising perspective, organisations also need sensitive collaborators who are aware of the "other" and "new" feelings (Costa and Pereira 2019). Like humans, organisations also have five senses—sight, hearing, smell, taste and touch—that underpin their performance and allow them to structure happiness and organisational peace. Hence, the importance of a more flexible, attentive and agile organisational culture and system in order to welcome, motivate and develop individual and group intelligence. This orientation as opposed to the traditional organisational values of obedience and dependence because the new organisational structures tend to be more thoughtful, critical, dialoguing, cooperative and interdependent (Pereira et al. 2019). There are more open organisations, eager to escape the society of ignorance and bureaucracy and materialise the so-called society of collective intelligence (Ramírez 2015). For the implementation of these structures, organisations need to assume new attitudes, feelings, emotions and employees should be more open to novelty and new organisational paradigms (Salas-Vallina et al 2017; Krstić 2014). This approach will allow organisations to confront and prevent new millennium challenges for sustainable development in connection with the emergence of new forms of crisis (pandemics, famine, inequality, etc.).

As happy people are more satisfied with their jobs (Gupta 2012), organisational happiness has an impact on performance and productivity (Frey and Stutzeer 2002). This reality underlines the need to bring behavioural, relational, ethical, emotional and spiritual competencies into organisations, as complementary and not as rivals of the technical-scientific and instrumental competences. It is a guideline for humanising feelings within organisations, which is based on shared values of their culture, beliefs, practices, rituals, customs, norms, traditions and leadership (Kefela 2010; Hoag et al. 2002; Robbins and Coulter 2005; Schein 1990; Krstić 2014, 2012; Costa and Pereira 2019). The combination of such feelings supports performance and organisational developments of reference (Tsai 2011; Tahir et al. 2011). In this sense, learning within the organisation is imperative (Pereira et al 2014). It opens the organisation to continuous flows of knowledge and the organisation is willing to learn in spaces of participation, interaction and knowledge sharing (Curado 2006). When this process is accompanied by instruments to reduce stress and anxiety, it enhances the rates of well-being, self-esteem and joy within personal and professional life. It gives relevance to the integrative evolution of different knowledge and leads employees to the learning of socio-humanist transversality (Paymal 2016). It enhances awareness of cooperative work, skills and intrapersonal and interpersonal values. It aims to improve relationships and focuses on building the maturity of social intelligence, which brings together skills dedicated to relationships (Costa and Pereira, 2019; King et al. 2012; Pereira 2013a, b; Pereira and Costa 2017; Pereira and Raposo 2019). It is interesting to mention that positive, enthusiastic and confident feelings and attitudes, are accompanied by the feeling of self-responsibility are the cornerstones for better performance. Nevertheless, the organisational culture cannot be seen ambiguously or as a pre-established menu. It is the result of the combination of variables and environments, which makes it a complex phenomenon (Ghinea and Brătianu 2012; Prugsamatz 2010; Pilat 2016). For a better understanding of this rationale, the emotional and spiritual capitals of employees and the organisation are indispensable in the equation of success because they compromise the collaborator with the notion of sharing, the exchange of ideas and dialogue, which makes the collaborator feel indispensable and active within the organisation. This position is also confirmed by the analysis of the structure of SDGs and their distribution by clusters. Therefore, the majority is set aside for the purposes of a conditional "social bloc". Therefore, improving the quality of life of people both today and in the future can be seen as a motivation to implement SDGs. The authors conducted further research in (Skliar 2018) on the hypothesis of "quality of life depends on the degree of achievement of SDGs" based on human development indices and composite index SDGs. As a result, it was determined that the goals of the conditionally social block $g1$, $g2$, $g3$, $g11$ and $g16$ showed a statistically significant effect on HDI (Skliar 2018). At the same time, it should be noted that the authors in (Skliar 2018) did not identify a statistically significant impact of SDGs on the quality of education $g4$.

Culture and knowledge play a relevant role in organisational performance (Hsu 2014; Wiewiora et al. 2013). According to Wiewiora et al. (2013), the organisational culture establishes a specific social context of interaction, sharing, and communication, marked by the conceptions of "wrong" and "right". Despite their importance, these authors affirm that organisational subcultures have not been the object of expressive reflection, a fact that may hinder the understanding of learning culture and organisational performance. They consider that organisational cultural differences influence behaviours and knowledge sharing. They use the organisational culture typologies clan, adhocracy, hierarchy and market, as shown in Fig. 1, to highlight the influence these values have on the typologies. We distribute the SDGs in Fig. 1 within each quadrant according to their role and priority for their reach for the types of organizational culture of a particular type.

The typologies unfold a diversity of attributes which explains why the organisational culture shapes learning within the organisation (Rebelo 2006; Santos 2011). Hence, there is the possibility of identifying different paths to learn in organisations, as opposed to the unique and standardised typology. This diversity transforms companies into practice spaces and facilitates learning and personal development for their employees. In this context, Pilat (2016: 88–9) considers that organisational culture creates a climate of stimuli to learning in the organisation, which involves the combination of the following four vectors:

CLAN		ADHOCRACY	
Mentoring	g1	Dynamic	g4
Extended family, nurturing	g2	Entrepreneurial	g5
Participation	g3	Risk-taking	g7
Teamwork	g8	Rapid change	g9
Employee involvement	g9	Innovation	g10
Corporate commitment to employees	g11	Creativity	g16
Rewards based on teams not individuals	g12	Temporary structure	g 17
Loyalty	g14	Power is not centralised, it flows from individual to individual or team to team	
Informality	g15	Sometimes exist in large organisations that have a dominant culture of a different type	
Job rotation	g16		
Consensus	g 17		
HIERARCHY		**MARKET**	
Structure	g1	Results-oriented	g5
Control	g2	Gets job done	g7
Coordination	g3	Competition and achievement	g9
Efficiency	g4	Focus on a transaction with external suppliers, customers, contractors	g10
Stability	g6		g16
Procedures govern what people do	g 7	Productivity	g 17
Formal rules and policies	g8	Tough and demanding leaders	
	g9	Emphasis on winning	
	g11	Success is defined in terms of market share and penetration	
	g16		
	g17		

Fig. 1 Attributes of clan, adhocracy, hierarchy and market cultures. *Source* Wiewiora et al. (2013: 1166). SDGs (g) supplemented by the authors (gray cells)

(i) *socialization process (teamwork, arranging social spaces in enterprises, integration and training trips, and so on),*
(ii) *key values, convictions and beliefs (initiative, learning, flexibility, self-development, qualify…)*
(iii) *management systems, policies, norms (avoiding red tape, employee empowerment, a down-top initiative of employees, system of awards and bonuses, …),*
(iv) *artefacts, daily practices (informal dress, common language, mentors, rituals, …).*

Cameron and Quinn (1999) are also enthusiastic advocates of the organisational culture because they believe it gives organisations a competitive edge. They consider that the most important objective of an adhocratic culture lies in its characteristics of adaptability, flexibility and creativity, in harsh environments for these conditions, where the excess of insecurity, ambiguity and overload of information is customary. Unlike market cultures and hierarchical cultures, adhocratic culture does not present a centralised power. Not relations of authority, but contact between individuals that foster relations and the work environment. Notwithstanding this notion, they consider that it all depends on the answer or problem one wishes to solve. Cameron and Quinn

(2006) argue, effectiveness indicators represent what people value about the organisational performance. Therefore, they define what is seen as good, and appropriate. In other words, the four types of culture define the core values where judgments are made about organisations. The authors stress that the determining aspect of the four sets of the core values is that they represent opposite or contrasting assumptions. Each continuum illustrates a set of values opposed to values that are at the opposite end of it—flexibility versus stability, internal versus external. The upper left quadrant, for example, identifies values that emphasise an internal and organic approach, while the lower right quadrant emphasises values from an external and controlled approach. Similarly, the upper right quadrant identifies values that emphasise an external and organic approach, while the lower-left quadrant underlines internal and controlled values. The opposite or contrasting values give rise to the designation of the Contrasting Values Model (Cameron and Quinn 2006), as shown in Fig. 2.

These authors consider that each quadrant was classified to distinguish its predominant characteristics—clan, adhocracy, market and hierarchy—by stressing that these

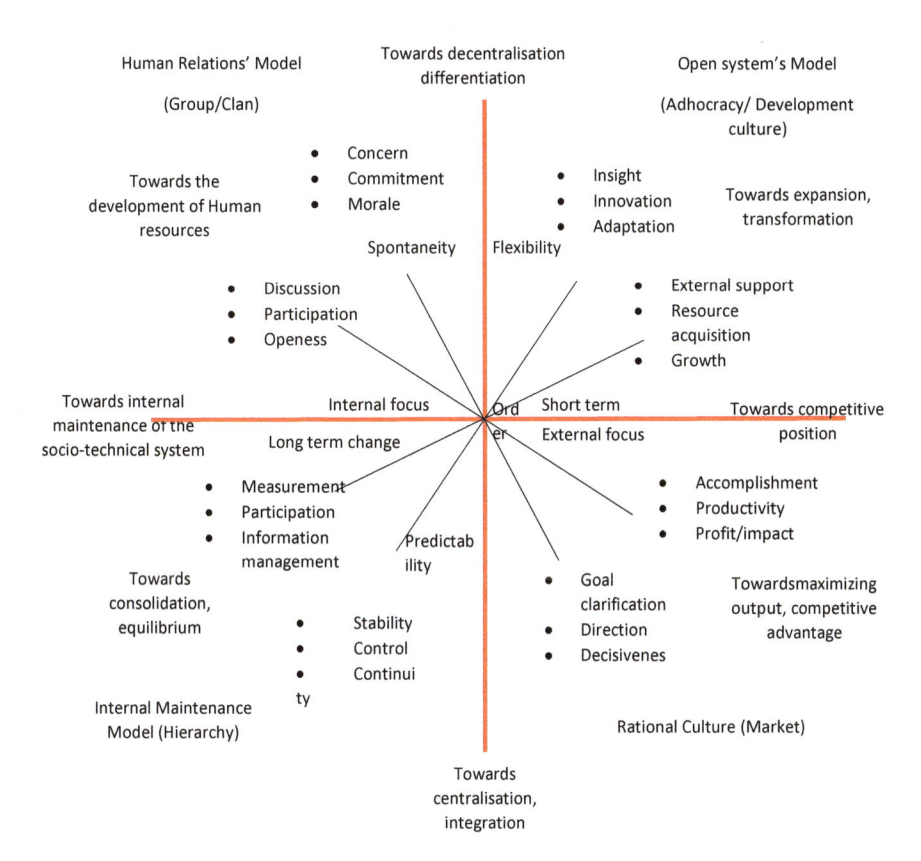

Fig. 2 Contrasting values model. Adapted from: Quinn and Rohrbaugh 1983; Denison and Spritzer 1991; Cameron and Freeman 1991 and Cameron and Quinn 1999, 2011

designations stem from the literature review, which explains how different organisational values associated with different forms of organisation. The authors found that these quadrants correspond precisely to those developed in the field of organisational sciences, as well as in key management theories, namely on organisational success, on quality issues, leadership and management skills. The consistency of these dimensions led the authors to identify each quadrant as a type of culture and to affirm that each quadrant represents basic assumptions, orientations and values, which are the elements that constitute the organisational culture (Cameron and Quinn 2006).

3 Literature Review (2): Motivation to Learn in the Organisation

According to Sallas-Vallina et al. (2017), there is a positive relationship between motivation to learn and the interactions established within an organisation. An organisational environment based on a culture of proximity to employees increases the circulation of tacit knowledge and becomes a source of productivity and longevity (Pedler and Burgoyne 2017). In the this evolutionary process, unlearning is an efficient form of learning (Martins et al. 2017a; b, c, d; Krstić 2012; Krstić and Krstić 2016). This is because in the process of unlearning tacit knowledge spreads more easily and allows more intensive and efficient use of knowledge, both coded and uncoded. This motivation to learn also feeds learning economies and learning societies, which are characterised by the intensity, novelty and volatility of knowledge (Stiglitz and Greenwald, 2014, cited in Pedler and Burgoyne 2017). This organisational spirit draws on goals of happiness at work and focuses on the emotional balance of the employee. In this organisational atmosphere, the predisposition to learn is greater, especially in learning organisations, as is the example of authentic organisations. These realise a work that "[…] it will be an antidote to stress, provide a healthy existence, enhance the human imagination and contribute to a more fulfilling life. They are organisations that help their employees balance their personal and organisational lives" (Kets de Vries 2001: 110). The term "authentizotic" is a neologism resulting from two Greek words "authenteekos" and "zoteekos". The first means that the organisation is authentic and deserving of trust. The second means that it is vital for people to provide meaning for their lives. According to the author, international organisations help employees establish a balance between personal and organisational life, providing people with the opportunity to meet the seven meanings that profile the human and professional existence: (i) purpose, (ii) self-determination, (iii) impact, (iv) competence, (v) belonging, (vi) pleasure and joy, and (vii) meaning. Kets de Vries (2001: 107) argues that in these companies the organisational culture is imbued with values that translate into behaviours of "joy, trust, frankness, empowerment, respect for the individual, justice, teamwork, innovation, customer orientation, responsibility, continuous learning and openness to change". Given these principles, the international organisations are seen by collaborators as good places to work, because there is a sense of confidence in their peers. They feel proud of their work and accomplishments in addition, they find pleasure in working with their colleagues. (Rego et al. 2006).

As with authentizotic organisations, other organisational typologies accept learning as a challenge to explore. This is the example of learning organisations, organisations learning and Organisational Citizenship Behaviour (Organ 1997). All these organisations are aligned with the objectives of learning and personal, group and organisational development. Their main goals are to increase organisational capital, grant longevity to organisations and to exceed the expectations of all investors. These organisational typologies are also sensitive to the issues of organisational citizenship. It is in this sense that an OCB (organisational citizenship behaviour) is centred on continuous improvement of performance and social and psychological context (Organ 1997). This author classifies a CBO by resorting to five dimensions: altruism, courtesy, righteousness, civic virtues and consciousness, which are part of the culture of organisations that are willing to learn by learning.

4 Methodological Route

The methodological path of this work, by the authors' strategy, is based on secondary data presented in the literature. Such methodological orientation is related to the central objective of the work, which is to reflect on the learning oriented organisational culture. This is in order to be able to frame the typology of culture with the new organisational capitals, which embrace emotional and spiritual aspects and which structure the conquest of competitive advantages, but which very often business organisations have difficulty in seeing and valuing their importance. Therefore, it was our purpose to reflect on the business reality in this field in order to, at a later moment, project new perspectives for business organisations. To this end, two works are considered structuring: one is the work of Wiewiora et al. (2013) and the other is that of Cameron and Quinn (1999, 2006), which assisted in highlighting and understanding the four typologies of organizational culture and the soul of its most expressive values. In this follow-up, Raposo (2017) is also utilized, which is in line with those authors and validates their assumptions. These three authors have assisted in understanding the strategic importance of values in the culture of autocratic organizations and their influence on the performance of the individual employee and of the organisation together, particularly in organisational climate cultures that encourage learning. Furthermore, by adding the variable of transformational leadership, the study is further enhanced as transformational leadership has the capacity to structure and feed the processes of innovation and organisational entrepreneurship. When such processes are consolidated, organisations gain more performance, more attractive longevity and therefore gain competitive advantages.

After the maturation of the above aspects, a focus on learning and organisational performance in the daily practice of business organisations is key. Thus, the study is expanded by drawing on the works of Dymock (2003), Egan et al. (2004), Joo (2010),

Pilat (2016), Kotter and Heskett (2011), Yang (2003), Khandekar and Sharma (2006), Tahir et al. (2011), Beliveau et al. (2011), Liker (2004). These authors illustrated that the happiness and success of the organisation depend on the positioning of the employee. This is because the culture of the organisation correlates positively with the satisfaction of the employees. Additionally, this satisfaction is influenced by the spiritual capital that is lived within the living being that is the organisation. With such an atmosphere, all the elements of the organisation are more sensitive to intra-personal and inter-personal development, to the diffusion of knowledge and to the creation of affective bonds which, on the one hand, avoids the migratory flow of the collaborators and, on the other hand, increases their performance and connection with the organisation. Thus, the adoption of this methodology has illustrated the day to day life of business and the designing of new guidelines for organisational culture learning.

5 Empirical Works (1): Learning and Organisational Performance in Practice: Empirical Work in Corporate Organisations

Organisations must be aware of the role that culture has in their development and performance not to violate guiding principles of the efficiency of its human, relational and intellectual capital. Each type of culture is associated with specific values that must be included in the equation of happiness and organisational success. Although the perception of the learning culture depends on the position the employee holds in the organisation (Dymock 2003). As per Egan et al. (2004), the culture of the organisation is positively correlated with the satisfaction of the collaborator in the workplace, which is attuned to the spiritual capital that is lived within it. This perspective converges to the guidelines of intra and interpersonal development, to the retention of the employee in the organisation and the reduction of the migratory flow, because when the organisation is absent it takes with him/her a relevant heritage which is no longer available (Joo 2010). Pilat (2016) also supports these fundamentals and states that contemporary organisations are increasingly looking at their human resources as the most valuable asset, stating that they are the basis of their competitive advantage. This is a necessary condition and tool in building Sustainable development.

It is interesting to mention that the organisational culture is the critical element of companies to influence the behaviour of their members, the relationships they establish with each other, as well as how to make decisions and establish priorities (Kotter and Heskett 2011). It has positive effects on the performance, results and longevity of the organisation (Yang 2003). To leverage and launch its performance, success and growth, organisational learning takes on a strategic role (Khandekar and Sharma 2006). These authors conducted a study that focused on 100 senior managers of Indian companies and their results show the importance of the role of learning in the organisational performance. These results are confirmed in the work of Tahir et al (2011), which shows the effect of organisational learning on the performance of employees. Moreover, Beliveau et al. (2011) and Liker (2004) reaffirm the importance of learning for the organisational performance and in its competitive position and consider it as "knowledge sharing". Hsu (2014) reflects on the effects of culture, learning and information technologies on the management and efficiency of knowledge. He constructed a model that equates those three domains. The results of his work highlight that these variables are essential to achieve high levels of performance, as found in Liker's study (2004). Martin and Siehl (1983) state that in the recruitment phase of employees it is important to be sensitive to personal values. Besides, personal values must be in line with the values as recommended by the organisation, bearing in mind that the organisational culture is a determining factor of its success and values of Sustainable development. This orientation is shared by Hoffmann and Pilat (2010), to whom the objectives of the organisational culture are to create a climate of stimulus to the learning process, taking into account its effects on performance.

The influence of culture on organisational performance is at the heart of the models of Cameron and Quinn (1999, 2006) and Wiewiora et al. (2013). The Cameron & Quinn model (1999, 2006), known as the Competing Values Framework (CVF), ascertains to evaluate the organisational culture. This model has four subcultures: clan, adhocracy, market, and hierarchy, as adopted in the study by Wiewiora et al. (2013). These authors, relate the organisational culture to the sharing of knowledge. They conclude that organisations with a "market" culture, which are based on specific values of competitiveness and achievement of organisational objectives, present a greater difficulty in sharing knowledge. They understand that in organisations with Clan-like characteristics, knowledge sharing is more natural and intense because these organisations tend to value a more collaborative, friendly and less competitive atmosphere. Hence, Wiewiora et al. (2013) show that the projects to implement and develop within organisations depend on the type of culture that exists for knowledge sharing. Therefore, when organisational culture is market-type, it tends to underline the values of competitiveness and overcoming objectives. Also, knowledge sharing is more cautious and less willful, because the values of competition exceed the values of cooperation. The opposite happens with the Clan typology, where the collaborative atmosphere outshines and where interaction and knowledge sharing is more common. In this type of culture, it is consensual to implement several projects with interactivity because of the organisation nature and the feeling of sharing cooperation and common goals within. The authors conclude that the dominant culture in the organisation inspires specific behaviours and values with different performances. Dymock (2003) and Egan et al. (2004) do also consider that the perception of the learning culture depends on the position of the employee within the organisation and his/her satisfaction in the workplace. This feeling is in tune with the spiritual capital lived within the organisation, because it points to goals of harmony and well-being, inside and outside the organisation. This perspective converges to the guidelines of the combination of intrapersonal and interpersonal development.

6 Empirical Works (2): The Cameron and Quinn Model of Organisational Culture: Application to Higher Education Institutions

Raposo (2017) applies Cameron and Quinn's Organisational Culture Model and adapts it to Higher Education Institutions. It focused on 6 (six) institutions, all belonging to the Polytechnic Institute of Porto, in Portugal. The sample consists of 585 respondents, 83% of whom are faculty members and 17% were staff collaborators.

In Raposo's study, it was found that the model of contrasting values of the organisational culture of Cameron and Quinn (1999) becomes more consistent when converted into three factors (Adhocracy Culture, Clan Culture and Hierarchy Culture), eliminating the typology of Market Culture, according to results obtained through the analysis of the main components (see Table 1).

Factor 1, which corresponds to the culture "Adhocracy", in addition to including the usual items of this cultural typology, also includes the variables "results-oriented", "dare and win" and "achieve efficiency in a quiet way", which usually belong to the Market culture type.

Factor 2 corresponds to clan culture and Factor 3 to Hierarchy Culture. The results validate the organisational culture model proposed by Cameron and Quinn. In Raposo's study, it is verified that for the specificity of higher education organisations, as with other corporate organisations, the dominant cultural typology points to the Adhoctacy culture, although it also brings together characteristics of Clan and Hierarchy cultures. That is, combining the results of the analysis of Fig. 1, we can also conclude that the distribution of types of organizational culture to achieve 17 SDG also occurs in the university. This confirms the importance of promoting the inclusion and promotion of the study of SR in university programs at all levels.

7 Discussion

In the study by Raposo (2017), despite validating the Cameron and Quinn model, the results unfold some specificity because the Market Culture is not expressive. However, in the organisational culture model under analysis, the four cultural typologies must be present, where one of them is dominant. For this rationale contributes to the temporal dimension and the internal and external factors that condition the assumption of the model variables. Therefore, organisations should be attentive to cultural fluctuations in markets, social, technological and behavioural fluctuations, inter alia, to define a cultural typology that best suits their objectives and performance. However, as the Market typology is not validated in this work, some of its variables are included in the typology of Adhocratic cultures, such as the variables of competition between people to achieve the objectives and competition for

Table 1 Analysis of main components of organizational culture

Variables/description	Components		
	Factor 1	Factor 2	Factor 3
2 Set goals and lead the market	0.848		
3 Innovation and development	0.833		
12 Leadership by market competency	0.814		
19 Product differentiation	0.768		
1 Efficiency	0.749		
21 Dare and win	0.727		
20 Assumption of risks	0.694		
6 Results-oriented	0.686		
16 Innovation and new opportunities	0.683		
5 Leadership results-oriented	0.675		
8 Cohesion through trust and loyalty	0.570		
7 Human development		0.800	
10 Follow the leadership		0.747	
4 Participative management		0.738	
17 Human development, care about people		0.731	
13 Take risks, just one person		0.637	
24 Leadership supports innovation and risk-taking		0.575	
14 Regulations and formal policies			0.754
22 Structured organization			0.682
23 Stability and efficiency			0.612
9 Job security, previsibility			0.550
Eigen value	6.66	4.13	2.34
% of variance	30.29	18.75	10.62
α Cronbach	0.93	0.83	0.69

Source Raposo (2017): 69

market leadership. Therefore, in today's markets, we understand that an organisational culture of learning should not be held hostage to the guidelines of a single typology. On the contrary, the trend seems to be a combination of the four typologies, because all of them have variables that are fundamental to the organisational balance. If new capital flows from within the employee and spreads throughout the organisation, the equation of organisational happiness should contain variables from the motivational, emotional, spiritual, personal intrapersonal development and shared leadership domains. Only in this way will it be possible to lead in the light of the sensitivity and the most congregating feelings for the objectives of the organisation.

8 Conclusion

An organisation whose dominant culture is the "Adhocratic Culture" indicates that the organisation is essentially focused on the foreign market. It easily adapts to the identified opportunities, accepts them naturally and implements innovative processes. In turn, what characterises the culture of "Market" is reputation. In this instance, people are competitive and more focused on achieving goals. The adhocratic typology is the most representative of the typologies, with evidence for the variables of setting ambitious goals, leading the market, implementing innovation processes and entrepreneurial capacities, the achievement of the market share. It also highlights the efficiency in the use of resources, the differentiation of its products and the assumption of risks, as underlined by the work of Raposo. In turn, what typifies the "Market" culture is reputation and success. Here people are competitive and more focused on achieving their goals. In this environment, internal competition prevents cooperation. As a consequence, the dissemination of knowledge within the organisation is hindered, which may have consequences on its performance. The Clan typology is the second most representative in the Raposo study. It emphasises human development, trust, openness and participation. It is a culture that points to open-minded human capital. Contrary to expectations, the perception obtained by the results of the study points to a more isolated (and not in a group) work as expected, as with participation and consensus. However, as it is a specific type of product—teaching and research—the results indicate that its success and development do not dependent on human resources, teamwork and commitment. Besides, Hierarchical typology is the least significant in the study. All its variables include job security, compliance, predictability and stability of relationships and institutions.

Given the results obtained by the studies, and the specificity of the type of institutions under analysis, it should be adopted conciliation strategies of the two main typologies (Adhocratic and Clan). As human development and on-the-spot capabilities are a priority, one should make an effort to increase such human development. This development will have to be looked at in light of new organisational and social capitals. Thus, technical and instrumental skills are powerless to leverage organisational success. Hence, such skills must be accompanied by emotional and spiritual capital. Thus, the construction and use of appropriate types of organizational culture will accelerate the formation and achievement of SDGs. Within these capitals are variables such as tranquillity, transparency, cooperation, help, empathy, respect for other, which leads to facing human development from a holistic perspective and providing reference performances. Therefore, for employees to follow leaders, they must set an example, they should inspire and motivate employees to participate collectively and actively and make them feel that they are an integral part of the organisation. It is within these assumptions that one can understand human development and the focus on human capital, as advocated by the clan typology culture. Failure to follow such orientations, the sustainability of the variables of the adhocratic typology is conditioned. Even if the orientation of organisations is in line with continuous learning, as is the case of authentizotic organisations, it will make little sense to continue to implement typologies that do not properly contemplate the human factor.

References

Beliveau B, Bernstein E, Hsieh H J (2011) Knowledge management strategy, enablers, and process capability in U.S. software companies. J Multi Res 3(1):25–46
Cameron K, Freeman S (1991) Cultural congruence, strength, and type: relationships to effectiveness. Res Organ Chang Dev 5:23–58
Cameron KS, Quinn RE (1999) Diagnosing and changing organizational culture based on the competing values framework. Addison Wesley Longman Inc., Reading, Massachusetts
Cameron KS, Quinn RE (2006) Diagnosing and changing organizational culture: based on the competing values framework, Revised ed. Jossey-Bass, San Francisco
Cameron KS, Quinn RE (2011) Diagnosing and changing organizational culture: based on the competing values framework, 3rd edn. Jossey-Bass, San Francisco
Capra F, Jakobsen OD (2017) A conceptual framework for ecological economics based on systemic principles of life. Int J Soc Econ 44(6):831–844
Costa CA, Pereira OP (2019) Values and trust in human capital: University students' perceptions, 2015–2017. Rev Galega Economía 28(1):102–116, ISSN-e 2255-5951. https://doi.org/10.15304/rge.28.1.6164
Cullen JG (2008) Self, soul and management learning: constructing the genre of the spiritualized manager. J Manage Spirituality Relig 5(3):264–292
Curado C (2006) Organizational learning and organizational design. Learn Organ 13(1):25–48
Deninson DR, Spreitzer GM (1991) Organizational culture and organization development—a competing values approach to research. In Woodman RW, Pasmore WA (eds) Organizational change and development, vol 5. JAI Press Inc., London, pp 1–21
Dymock D (2003) Developing a culture of learning in a changing industrial climate: an Australian case study. Adv Dev Hum Resour 5(2):182–195
Egan T, Yang B, Bartlett K (2004) The effects of organizational learning culture and job satisfaction on motivation to transfer learning and turnover intention. Hum Resour Dev Q 15(3):279–301
Frey BS, Stutzer A (2002) What can economists learn from happiness research? J Econ Lit 40(2):402–435
Ghinea VM, Brătianu C (2012) Organizational culture modelling. Manage Mark 7(2):257–276
Gupta V (2012) Importance of happy at work. Int J Res Dev-A Manage Rev 1(1):2319–5479
Hoag BG, Ritschard HV, Cooper CL (2002) Obstacles to effective organizational change: the underlying reasons. Leadersh Org Dev J 23(1):6–15
Hoffmann K, Piłat M (2010) The role of organizational culture in motivating innovative behavior. In: Bieniok H, Kraśnicka T (Eds) Innowacje w zarządzaniu przedsiębiorstwem oraz instytucjami sektora publicznego, Katowice: Wydawnictwo Akademii Ekonomicznej w Katowicach
Hsu S-H (2014) Effects of organization culture, organizational learning and IT strategy on knowledge management and performance. J Manage Stud 9(1):50–58
Joo B (2010) Organizational commitment for knowledge workers: the roles of a perceived organizational learning culture, leader-member exchange quality, and turnover intention. Hum Resour Dev Q 21(1):69–85
Kefela GT (2010) Understanding organizational culture and leadership—enhance efficiency and productivity. PM World Today XII(I):1–14
Kets de Vries MFR (2001) Creating authentizotic organizations: well-functioning individuals in vibrant companies. Hum Relat 1(54):101–111
Khandekar A, Sharma A (2006) Organizational learning and performance: understanding Indian scenario in the present global context. Emerald Group Publishing Limited 48(8/9):682–692

King DB, Mara CA, DeCicco TL (2012) Connecting the spiritual and emotional intelligence: confirming an intelligence criterion and assessing the role of empathy. Int J Transpersonal Stud 31(1):11–20

Kotter J, Heskett J (2011) Corporate culture and performance. Free Press, New York

Krstić M (2012) The role of rules in the evolution of the market system: Hayek's concept of evolutionary epistemology. Econ Ann 57(194):123–140

Krstić M (2014) Rational choice theory and addiction behaviour. Market-Tržište 26(2):163–177

Krstić B, Krstić M (2016) Teorija racionalnog izbora i društvena istraživanja. Sociologija 58(4):598–611

Liker J (2004) The Toyota way: 14 management principles from the world's greatest manufacturer. McGraw-Hill, New York

Martin J, Siehl C (1983) Organizational culture and counterculture: an uneasy symbiosis. Organ Dyn 12(2):52–64

Martins A, Martins I, Pereira OP (2017a) Shift in Paradigm: organizational values and performance. In: Afolayan GE, Akinwale AA (eds) Handbook of research on global perspectives on development administration and cultural change, USA, IGI Global, pp 158–172

Martins A, Martins I, Pereira OP (2017b) Embracing innovation and creativity through the capacity of unlearning. In: Ordóñez de Pablos P, Tennyson RD (eds) Handbook of research on human resources strategies for the new millennial workforce, USA, IGI Global, pp 128–147

Martins A, Martins I, Pereira OP (2017c) Challenges enhancing social and organisational performance. In: Ordóñez de Pablos P, Tennyson RD (eds) Handbook of research on human resources strategies for the new millennial workforce, IGI Global publisher of the Business Science Reference, pp 28–46

Martins A, Martins I, Pereira OP(2017d) Feedback and feedforward dynamics: nexus of organizational learning and leadership self-efficacy. In: Wang VCX (ed) Encyclopedia of strategic leadership and management, USA, IGI Global, Publisher of the Business Science Reference, pp 207–232

Organ DW (1997) Organizational citizenship behavior: it's construct clean-up time. Hum Perform 10(2):85–97

Paymal N (2016) *La Escuela de los 7 Pétalas*. La Paz, Bolívia; Editorial Ox-La-Hun/P3000

Pedler M, Burgoyne J (2017) Is the learning organisations still alive? Learn Organ 24(2):119–126

Pereira OP (2013a) Metacompetences: How important for organizations? Analysis of a survey in Portugal. Reg Sectoral Econ Stud 13(2):73–88

Pereira OP (2013b) Soft skills: from university to the work environment. Analysis of a Survey of Graduates in Portugal. Reg Sectoral Econ Stud 13(1):105–118

Pereira T, Berta W (2015) Increasing OCB: the influence of commitment, organizational support and justice. Strateg HR Rev 14(1/2):13–21

Pereira OP, Costa CA (2017) The importance of soft skills in the university academic curriculum: the perceptions of the students in the new society of knowledge. Int J Bus Soc Res 7(6):25–34. https://doi.org/10.18533/ijbsr.v7i6.1052

Pereira OP, Raposo MJ (2019) Soft Skills in knowledge-based economics. Mark Manage Innovations 1:182–195. https://doi.org/10.21272/mmi.2019.1-15

Pereira OP, Martins A, Martins I (2014) How important is learning on the firm's performance? Int J Soc Sustain Econ Soc Cult Context 9(2):30–40

Pereira O, Gomes D, Martins A, Martins I (2019) Education: instrument for socio-economic and humanizing development. In: Wang VCX (ed) Encyclopedia of transdisciplinarity in theory, research and practice, pp 284–295. https://doi.org/10.4018/978-1-5225-95311.ch020

Petriglieri G, Petriglieri J (2010) Identity workplaces: the case of business schools. Acad Manage Learn Edu 9(1):44–60

Pilat M (2016) How organizational culture influences building a learning organizational. Forum Scientiae Oeconomia 4(1):83–92

Prugsamatz R (2010) Factors that influence organization learning sustainability in non-profit organizations. Learning Organ 17(3):243–267

Quinn RE, Rohrbaugh J (1983) A spatial model of effectiveness criteria towards a competing values approach to organizational analysis. Manage Sci 29(3):363–377

Ramírez J (2015) Conocimiento que se genera en proyectos de aprendizaje servicio. In: Herrero, M, Tapia M (eds) Actas de la III Jornada de Investigadores sobre aprendizaje-servicio, CLAYSS-Red Iberoamericana de aprendizaje-servicio, Buenos Aires, Argentina, pp 81–84

Raposo MJ (2017) Las Instituciones de Enseñanza Superior como Organizaciones Aprendientes. Tesis Doctoral, Cádiz, Universidad de Cádiz, Un estudio de caso en Portugal

Rebelo T (2006) Orientação Cultural para a Aprendizagem nas Organizações: Condicionantes e Consequentes. Faculdade de Psicologia e Ciências da Educação, Coimbra, Universidade de Coimbra, Tese de Doutoramento em Psicologia do Trabalho e das Organizações

Rego A, Cunha M, Souto S (2006) Workplace Spirituality Climate, Commitment and Performance: an Empirical Study. Documentos de Trabalho em Gestão: WorkingPapers in Management, Área Científica de Gestão, G/n° 1/2006, Universidade de Aveiro, Departamento de Economia, Gestão e Engenharia Industrial

Robbins SP, Coulter M (2005) Management, 8th edn. Pearson Prentice Hall, New Jersey

Salas-Vallina A, Alegre J, Fernandez R (2017) Happiness at work and organisational citizenship behaviour: is organisational learning a missing link? Int J Manpow 38(3):470–488

Santos A (2011) Cultura de Aprendizagem em Organizações: revisão crítica de pesquisas nacionais e estrangeiras. *Gestão Contemporânea*, ano 8, No 9, pp 35–61. Disponível em: http://seer2.fapa.com.br/index.php/arquivo, consultado em o7 de abril de 2019

Schein EH (1990) Organizational Culture. Am Psychol 454(2):109–119

Skliar I, Kostel M, Petrushenko Y (2018) Sustainable development goals: ecologic and economic contradictions and challenges. Mech Econ Regul (3):9–18

Sulkowski L (2013) Strategic Management as the ideology of power. J Intercultural Manage 5(3):5–11

Tahir A, Naeem H, Sarfraz N, Javed A, Ali R (2011) Organizational learning and employee performance. Interdisc J Contemp Res Bus 3(2):1506–1514

Tsai Y (2011) Relationship between organizational culture, leadership behaviour and job satisfaction. BMC Health Services

Wiewiora A, Trigunarsyah B, Murphy G, Coffey V (2013) Organizational culture and willingness to share knowledge: a competing values perspective in an Australian context. Int J Project Manage 31(8):1163–1174

Yang B (2003) Identifying valid and reliable measures for dimensions of learning culture. Adv Dev Hum Resour 5(2):152–162

Orlando Petiz Pereira, PhD in Economics and Business Sciences by the University of Santiago de Compostela (Spain), Master degree in Social & Economics Studies by the University of Minho (Portugal), Bachelor degree in Economics by the University of Porto (Portugal) and Professional Coach by the ICU (International Coaching University). He is professor of economics at the University of Minho. The main area of research is Social Economy, Economics of Knowledge and Innovation, Economics of Education, Economics of Labor. He is developing the "Service Learning" project for application in teaching at different levels and degrees.

Maria João Raposo, PhD in Social and Legal Sciences by the University of Cádiz (Spain), Master degree in Human Resources Management by the University of Minho (Portugal), Bachelor degree in Human Resources Management by European University of Lisbon (Portugal). She is a professor of organizational behavior and human resources management at Higher Engineering Institute of Porto (Portugal). The main area of research is Organizational Culture and Leadership.

Miloš Krstić, PhD in Faculty of Economics by the University of Niš (Serbia), Bachelor degree in Faculty of Economics by the University of Niš (Serbia). He is research associate at the University of Niš. The main areas of research are:contemporary economic paradigms (rational choice theory, game theory, social choice theory, rational expectations theory), and macroeconomics. He is an author or coauthor of one book and more than 50 scientific papers.

Oleksii Goncharenko PhD. in Economics of natural resources and environmental protection (Sumy, Ukraine, 2019). Nowadays, he is a senior lecturer in the Economic, Entrepreneurship, and Business-Administration department at Sumy State University, Sumy (Ukraine). His current research interests include sustainable development, dematerialization of products production and consumption, social solidarity economy. He has experience in managing the department of work with international students at Odessa National Polytechnic University.

Interplay of Traditions and Innovations in Teaching Sustainability Issues: National and Global Discourses

Dzintra Iliško

Abstract The article discusses the necessity of gradual and meaningful integration of innovations in education by respecting cultural and educational traditions of the local context. Sometimes, the crises require urgent action to be taken to sustain the educational process, like digitalization of the educational process caused by the Covid-19. The international praxis shows that the most recent models of educational development lead to innovations and change for achieving sustainable forms of development. The author argues that Latvia needs to adopt sustainable discourses and practices for the local context. Currently Latvia is undergoing transitions towards competence-based curriculum at all levels in education by placing higher emphases on quality education. The transition is taking place towards deep learning, higher autonomy of students, transdisciplinarity, transversal skills, and integration of ITC technologies in order to meet the reality of a contemporary learner by building their competence as responsible global citizens. The author analyses teachers' perspective: Their concerns and difficulties while designing and implementing a new competence-based curriculum oriented towards quality education and other innovations while teaching sustainability issues. The methodology employed in this study are the focus group interviews supported by the in-depth interviews with the individual staff members who are integrating sustainability issues in their teaching. The focus is on how innovations and traditions can be understood from the perspective of the competence-based approach while teaching sustainability issues. It is concluded that Latvia needs evolutionary policies and strategies for implementing sustainable development based on flexible management and risk minimising approach for implementing innovative practices in education.

Keywords Tradition and innovation · Competence-based approach · Quality education

D. Iliško (✉)
Daugavpils University, Vienibas Street 13, Daugavpils 5401, Latvia
e-mail: dzintra.ilisko@du.lv

1 Introduction

All governments are encouraging scientists to build innovative practice. Higher educational institutions aim at educating innovative individuals (Keinänen and Kairisto-Mertanen 2019). Reform process towards transforming the curriculum towards a competence-based curriculum requires teachers to bring innovation in redesigning their teaching. Traditional ways of teaching in universities are contradicting the needs of a contemporary economy. The curriculum is being reformed in higher education in Latvia by taking into account the tendencies brought in by the globalization. There is quite a wide discourse in the scientific literature about the need to develop students' critical thinking, creativity, IT skills, problem solving and collaboration skills, however less attention has been paid to developing innovative competence (Edwards-Schachter et al. 2015). The new economy requires sustainable and innovative solutions; therefore, universities need to train innovative specialists for the future job markets.

Mhamed et al. (2018) argue that higher education in Soviet Latvia was "under a full state control and serve the needs of the centrally planned economy and the soviet ideology. Since Latvia has regained its independence in 1991, The Europeanization and internalization took place also in higher education. In 2004 Latvia has become a member of the EU. It was a transition from the centrally planned to democratically governed educational system. HE has undergone modernization and Europeanisation of educational systems, that brough huge transformations and innovations in both content and the ways of teaching. Educational reforms took place in line with the Bologna processes aimed at creating a European Higher Education Area (EHEA) with the intensive academic mobility. The system of quality assurance was introduced in 2011 in line with the Bologna processes as well (Saiema 1995).

Today higher educational institutions are placed in a very complex, unpredictable, changing social, political and cultural environment. Reforms have already started in 1991, since Latvia has regained its independence characterized by democratization and decentralization processes. Depolarization and democratization of education brought the end of strict ideological control of the state over what was taught in Universities and schools. Higher institutions have gained autonomy in designing their educational programs. The reforms were followed by stage of implementing legislative bases in education, like the adoption of *the Law on Education in 1991* as an initial bases of the reforms, followed by further developments. The emphases were put on gaining quality education, accessibility of education, and institutional developments. The reforms were rapid and changes were introduced without involving grassroots participants, and sometimes lacking a critical view.

Unfortunately, today the educational reforms have a tendency to turn back to centralization by diminishing the autonomy of regional Universities. The Universities are engaged in transforming the curriculum towards the competence-based approach, initiated by the project: *School 2030*. The theoretical framework of the reform was supported by significant international links, networking, cooperation and study trips of experts involved in this project to the countries whose educational systems have received international recognition.

Today reforms in higher education lead to internalization, mutual recognition of Higher Education Degrees between the Baltic and Benelux countries that was signed in Brussels in 2019, the possibility to obtain teacher qualification within one year for the professionals with higher education. This can be interpreted as a sign of diminishing of higher education. Further steps have been made for the internalization of higher educational system. The cabinet of Ministers has initiated further consolidation actions of regional Universities that cannot be defined as sustainable and will lead to further devaluations of education and diminishing the capacity of regional development. The current paper seeks to explore the opinion of the staff members employed in higher education on the following issues: To what extent are they competent to introduce innovations in their teaching in the reform processes in higher education? How are these innovations reflected in organizing their teaching? What are the driving forces and obstacles for introducing innovations is higher education in relation to teaching sustainability issues? How does innovations meet the needs of a contemporary learner? The paper intends to seek teachers' perspective: Their concerns and difficulties while designing and implementing a new competence-based curriculum oriented towards quality education and other innovations while teaching sustainability issues.

2 Challenges for the Higher Education

Today one of the challenges that the higher educational establishments are facing is reorienting a curriculum towards the aim of a sustainable education (Sterling et al. 2013; Filho 2011) and implementing the sustainable development goals (UNESCO 2017). *The guidelines for the development of education 2014–2020* define the aim to promote a sustainable development of the society. The document puts an emphasis on improvement of an educational environment by modernization of educational environment and by developing students' sustainability competencies. Another binding document is the strategy (Europe 2020). This strategy defines three EU priorities: (1) innovation-based economy, (2) resource efficiency and (3) a socially inclusive growth. Therefore, introducing sustainability in higher education requires new innovative approaches to promote sustainability in practice. Higher education is recognized as a major player in addressing unsustainability of current ways of being through research and teaching (Djordjevic and Cotton 2011). Sustainability is seen as part of curriculum, research, and leadership. Both, top down and bottom up approaches are needed to foster institutional change; therefore, universities need to foster engagement of multiple stakeholders in promoting a sustainable change. Integration of sustainability in higher education involves a shift in pedagogical strategies from teacher to student cantered, by the use of inquiry and action research strategies. One of the topical issues that need to be integrated is a climate issue since the role of Universities is of great importance in order to make informed decisions of future citizens (Filho 2011). Universities are seen as one of the stakeholders who need to take an active role in providing climate change education, students' awareness and active position.

3 Conceptual Framework for Introducing Innovations

OECD (2018) defines innovation as a new product or process that differs significantly from the previous products or processes. Process innovation refers to changes in organisational processes. Product innovation include a new pedagogy or educational experiences. The capability refers to the 'agency,' the capacity of teachers to act as change makers in transforming the environment of higher education and to contribute to a technological and social change (Drucker 1998).

Innovative competence includes the following aspects: Creativity, critical thinking, initiative, teamwork and networking (Perez-Penalver et al. 2018). Innovative thinking involves creative problem-solving skills, system thinking, team work, networking and goal orientation (Keinänena et al. 2018). Vias and Rivera-Cruz (2019) suggest social constructivism as a conceptual framework for introducing innovations in higher education that involves program design, new competencies and collaboration that require a new conceptualization of an individual and the way they interact with the world. Nielsen and Stovang (2015) suggest the model for introducing innovation in higher education that includes six stages: Discover the present, envision the future, use novel tools by sensing the future, go to theory, interact with others and implement a learner-centered pedagogy. This also involves some elements of experimentation. As Kolb and Kolb (2009) argue, this experimental learning is a part of holistic process of changes which involves thinking, feeling, perceiving and behaving dimension of a person. Hoover and Whitehead (1975) describes experimental learning as involving cognitive, affective and behavioural aspects characterized by active involvement of both, the student and the teacher.

3.1 Innovation Versus Traditions: National and Global Discourses

Under the pressure of globalization and digitalization tendencies in the world, the universities are gradually changing their habitual ways of teaching to online learning and introducing other changes that lead towards quality education. The integration of ICT in Higher Educational Institutions is considered as one of the innovative practices worldwide (IAU 2020). This requires changes in attitudes and as well as stems from the paradigm changes in education. Introducing innovations involve changes in the content, methods and practice. Introducing innovations will make education more flexible, develop teachers' digital competencies and increase students' interest about the open access resources. Staff members need to have access to the resources and the students need availability and accessibility of those resources.

Introducing innovative forms of teaching require transformations in teachers' frames of references. Teachers in higher institutions are recognizing the need of transformations how they teach sustainability issues. Teaching need to lead to a higher students' autonomy and ability to deal with complex real-life issues (Wiek and Kay 2015). Graduates need to acquire the competence to deal with wicked issues by transforming unsustainable ways of being and doing towards more sustainable ones. Future thinking competence can be developed by engaging students with real life issues. Sustainability issues are complex and cannot be dealt with linear methods and lecture type of methodologies, whether they are online or taking place in the physical setting of the university. They need to engage in problem solving processes in small interdisciplinary teams whether modelling issues in an online setting or in a real-life circumstances. The sustainability issues need to be taught by engaging students in problem analyses, future scenario constructing and strategic thinking. They need to be engaged in team building activities with the involvement of various stakeholders. This will build students' accountability and value-based competency about the future work in the community (Wiek et al. 2011). While dealing with complex sustainability issues, universities require encouraging innovations, creative thinking, problem solving and transdisciplinary approaches (Ellis 2017). Contemporary global trends of academic development require universities to develop more socially and environmentally literate citizens who are engaged in cross-sectorial and multi-stakeholder networking (Yarime et al. 2012).

One of the transitions that universities need to manage is a transfer from disciplinary to transdisciplinary discourse in dealing with sustainability issues. Universities need to engage in a cross-sectorial and multi-stakeholder networking where they can develop global perspective on responsibility (Riekmann 2012). Complicity of sustainability issues requires looking for innovative solutions while navigating across various disciplines and dealing with multiple discourses and underlying epistemological, conceptual and methodological aspects of knowledge (Spencer et al. 2018).

4 Methodology

The scope of the present paper is to map and discuss innovative approaches and tools for teaching sustainability issues in higher education institution as reinforced by the Covid-19. The methodology employed in this study are focus group interviews and semi-structured interviews with the University staff members. Prior to conducting interviews with the individual staff members, I carried out two focus group interviews with eight staff members in each group in order to obtain general ideal about challenges and concerns related to innovation and innovative teaching approaches employed in higher education by the staff members while teaching sustainability issues. Focus group interviews helped to identify the questions for the semi structured interviews.

Semi-structured interviews with the five University staff members allowed to get a deeper understanding of the staff's everyday life, difficulties and struggles during the Covid-19. The interview participants were selected by a snow ball technique by the recommendation of the staff members themselves of the interview candidate who is successful in introducing innovations in current teaching practice. The interviews were carried out online via skype or by doing telephone interviews due to restrictions by the Covid determined circumstances. By doing a careful analyses of interview data, I assigned codes and organized data into meaningful units. The thematic analyses of interviews allowed to identify the patterns in the interview data by grouping it into clear and identifiable units. By reconstructing stories told by the research participants, I got richer and more condensed story of issues raised by the staff members while introducing changes in their daily routine this getting a rich tapestry.

The interview focuses on the following questions: To what extent are you competent to introduce innovations in your teaching in the reform processes in higher education? How are these innovations reflected in organizing your teaching? What are the driving forces and obstacles for introducing innovations is higher education in relation to teaching sustainability issues? How innovations serve the needs of the twenty-first century learners? Now do keep balance of keeping good traditions and introducing innovations required by current times?

5 Research Findings

As reported by the staff members, a number of innovations in teaching in higher education were introduced as a result of forced changes regarding the strategies and methodologies of teaching by the reform processes of school 2030. Due to reducing a fragmentation in higher education, the Universities designed new educational programs for all levels in line with the new reform requirements unified for all universities in Latvia. One of such examples, as mentioned by the staff members, were changes in strategies and modes of teaching caused by the Covid-19 that can be seen as example of significant organizational changes in education. This could even be called a digital revolution that took place overnight when the staff members were forced to make a transition to online teaching. The staff members in higher education are quite slow in changing their teaching patterns. The changes in higher education are taking place slowly and gradually. This takes time for the staff members to understand, to adjust and to transform their teaching practice. Digitalisation of education involves various aspects such as organisational issues, technological infrastructure to pedagogical approaches (Tømte et al. 2019).

The crises caused by the Covid-19 made the staff members to develop their competence in introducing innovative solutions in their teaching and at the sime time preserving the best traditions what they had so far. New changes were in line with the priorities set by the reform processes in higher education. Those innovative experiences are reflected in organizing teaching in a remote mode of teaching, thus providing students a higher degree of autonomy and engaging them in an inquiry type of learning. This innovative experience is related to teachers' attempts to engage their students into transdisciplinary experience by helping them to find interconnectedness among the different disciplines of sciences.

One question was: *"To what extent are you competent to introduce innovation in your teaching sustainability issues in the reform processes in higher education?"*.

I received different responses. Some staff members were quite successful in reacting to changes by implementing efficient crises management strategies, for some staff members this caused some confusion. As one of them commented: *"I was not teaching in any online platforms before, but the situation with the Covid-19 made me to learn new ways of reaching my students by the ICT tools in different platforms. It was very difficult to change from the traditional teaching to teaching with the digital technologies. My working day lasted more than eight hours."*

Staff members pointed to a number of advantages of introducing innovations in a regular university's routine. As one of them pointed out: *"I discovered new opportunities for teaching online. I have created lots of tasks for my students to encourage their autonomy for completing the assignment. They received a gradual feedback as well. I have discovered new digital solutions for teaching sustainability issues."* There is a lot of discussion about the need to engage students in setting their goals and promoting a more autonomous inquiry type of learning in a new competence-based model of teaching that has been introduced in all levels of education in Latvia. The situation with the Covid-19 caused many staff members to create the tasks that foster more autonomous type of learning. One of the examples of such innovative practice to be mentioned was teaching sustainability issues. As many of staff members responded, in an online situation both, students and the teachers have developed a number of new skills, like collaboration and networking in the online environment, providing a reflective feedback. The staff members learned to design new interactive materials, demonstrated a commitment and open mindedness towards exploring new opportunities of work in a digital environment.

Among the *advantages* of digital revolution caused by the Covid-19, staff members discussed their attempts to use new technological solutions that they have not used before or which they have postponed to learn. They have designed lots of support materials for their students in the Moodle platform that fostered students' autonomy, inquiry type of learning and reflective feedback. As many of the interviewed staff members commented, they have learned how to use Zoom, Webex, Microsoft team communication and other teaching tools, including Moodle platform.

One of the staff members reported about the possibility to foster self-directed learning of students of the use of digital tools and platforms: *"Digital learning is based on the principles of self-directed learning, when the students plan their learning, as well as which subject they focus on in a particular moment of the day. The student takes a full responsibility of how to achieve learning outcomes. Remote learning mode differs in a way that students do not sit in the classroom, but learning takes place by the use of modern technologies. Technologies play a very important role, but even more important are the student's skills to learn the content and their motivation to achieve certain learning outcomes."* Another response was related to the conscious need of changes in tradition approaches towards teaching: *"We need to give up out traditional ways of teaching separate disciplines because it does not work anymore. We schools have gone a way ahead in bringing changed into the learning process towards higher transdisciplinarity, we need to rethink and redesign the learning process in higher institutions as well".*

As the staff members reported, the students have learned to communicate their stories digitally about the surrounding unsustainability and have learned to work in digital teams in order to find solutions to issues by the use of verbal and nonverbal means of communication and behaviour.

Among the difficulties of introducing innovations, was the access of all participants to the resources and ICT technologies. Not all students had an availability and accessibility of all required resources. As one of the staff members commented: *"Unfortunately, some students did not have a camera and a microphone, they could not respond. But for others, the video suddenly turns off (or they turn it off), I don't see them anymore, so I can't understand what the reaction is, or everything is understood."* They complained about the absence of emotional reaction from the students that they could evidence.

Digitalization of teaching caused by Covid-19 left many staff members not fully prepared for the remote teaching, particularly, the elderly staff members. They were forced to learn the use of technologies in a rather compressed time and to apply them in teaching. As one of the staff members commented: *"we cannot assume that all students have equal access to technologies and if their technologies are sufficient enough for online learning and compatible with the teachers' technologies."*

The latest inquiries indicate that majority of teachers don't possess technologies that are necessary for online teaching and they were left unprepared. Among of problems of digitalization of education, the staff members have mentioned the following: The use of diverse learning resources, tools and communication platforms by the teachers leading to a fragmentation and a lack of a common approach in Universities and schools. The solution could be the use of single digital platforms for the whole institution. Students received the tasks from 5 to 6 different digital platforms. Many universities have not developed fully their own virtual environment, so, that both teachers and students worked in many digital platforms at the same time. The working day of staff members lasted much longer than eight hours.

Among the *challenges* while introducing innovations, the staff members have mentioned availability of technologies and low skills of staff members to use technologies.

As one of the staff members commented: "The *digital learning process is not easy, as it has many challenges related to changes to the usual rhythm of work. However, it offers an opportunity to grow professionally by the use of digital platforms, since the main task of education is not only to provide knowledge, but also to teach how to analyse critically a wide range of information, how to select information and relate it to real life situations.*"

To describe the challenges for digital solutions, G. Catlaks, the head of the State Education Content Centre, used a concept such as the *'digital ecosystem at the educational establishment'*, which is essential for the curriculum planning, data management, internal and external communication. He refers to a digital ecosystem of a staff member that is necessary for planning teaching, data management, access to a variety of learning resources, tools for creating learning tasks for implementation support (access to tasks, resources, feedback and for tracking students' performance), professional development opportunities, internal and external communication. The student's digital ecosystem is necessary for the students to have access to a variety of learning resources, tools and opportunities for communication, and for receiving feedback. To avoid overloading of students with the tasks, teachers need to design and to assign tasks that are more complex, therefore, they need not work individually but they must cooperate. Unfortunately, the teachers are still assigning the tasks in their individual mode.

Among the *driving forces for introducing innovations* is higher education in relation to teaching sustainability issues, the staff members have mentioned the need to meet the needs of the contemporary learners who represent a completely different online generation and who prefer to work in a completely different mode. These are the learners who prefer a non-linear mode of learning, fluid learning environment, flexible learning structures, open access communication tools, and a responsive learning. There is a huge transition that is taking place in learners' and teachers' roles. The learner acts in a role of innovative designer (Wagner 2012) who generates fresh ideas. This requires for the staff members to create virtual and physical environment where the students engage in fluid collaboration and playful thinking and seek opportunities outside the box. As Wagner (2012) asserts, the educators are encouraged to "not merely tolerate but to welcome and celebrate questioning, disruption, and even disobedience that comes with innovation" (p. 242). As the staff members commented, that by engaging in an online teaching, they have offered the students a chance to generate ideas and suggestions for solving local environmental issues, provided them flexible learning environment, learning tools of an open access, and a reflective feedback. The students became global learners who had immediate access to the world, therefore, the teachers designed the tasks in their study courses in a way that the ideas discussed were related to the issues of a global scale by developing their understanding about the interdependence of global and local issues.

This presents a challenge for the staff members to develop student's global competence, Jackson and Boix Mansilla (2011) define global competence as "the capacity and a disposition to understand and act on issues of global significance." Therefore, in their study courses, the staff members tried to make a link to global issues. Globally competent students are interested about global issues, they able to rethink their practices, make connections, and take actions. As one of the teachers' commented, *"I encourage my students to frame local problems make connection to global issues, collect and interpret data and build informed arguments. They consider the spheres of their action—family, community, nation, and the globe and consider specific actions how they can improve conditions.* The students also develop environmental awareness and planetary sustainability.

Another important change in students' role is placing a bigger autonomy on the student learners. A transfer to a distance mode of learning made the staff members reconsider their teaching by giving the students a compass to navigate in unknown territories, by making their own choices and finding viable solutions and discovering new roads. This strategy is in line with the vision outlined in the competency-based curriculum framework that has been introduced in the educational system of Latvia. In a journey of self-directed learning the students critically engaged with the self-selected websites and pools of information by critically questioning where teachers provide instant reflective feedback. Social networking become a part of an everyday reality of the learner by helping them use constant messaging and encouraging them to stay connected with other learners. They were eager to collaborate with their classmates in an online environment in designing projects related to sustainability issues and incorporating technologies in every aspect of their teaching. This needs to be taken into account while designing the tasks for the contemporary students in future.

While introducing digital learning opportunities, the staff members tried to develop the main skills of the twenty-first century, namely, cooperation, problem solving, creativity, action and accountability for one's actions. Dealing with unsustainability issues, staff members engage learners in solving complex problems in transdisciplinary mode: *"The students are engaged in blended environments and work creatively in finding solution to those complex issues. They also were encouraged to think analytically, by comparing, contrasting, evaluating and synthetizing information. They were required to collaborate virtually in diverse multimedia formats with the local and global partners,"* as one of the staff members commented. Developing those competencies goes in line with developing students' accountability, environmental local and global awareness and tolerance (Table 1).

6 Conclusions

Despite of attempts worldwide to foster transformations towards sustainability, there is still a space for improvements for introducing the idea of sustainability in all sectors of life, including politics, business, teacher education and media. Universities play a major role in integrating the idea of sustainable development in the curriculum. Still, traditional academic disciplines are not sufficient to meet the contemporary challenges and contribute efficiently to a sustainable development. Universities need to introduce transformational agenda and problem-based approaches in education and move beyond disciplinary boundaries (Wiek and Kay 2015).

It is not possible to judge about the sustainability of the educational reforms in the long term. Implementation of a new curriculum, digitalization and internalization of higher education is a necessary step for the Universities to become a part of contemporary world in responding to the reality of a current day. In the global economy this is not enough with the memorization of information. The contemporary person needs to develop a conceptual thinking, ability to deal with the complex issues of

Table 1 Data gained as the result of analyses of the semi structured interviews

Code	Description	Examples
Competence in introducing innovations as a part in the reform processes in higher education	Staff members are sharing their new learning experience about the competencies they acquired	"The transition to a remote mode of organizing learning environment made me to acquire competence to work in a new environment in a very compressed time" "To meet the needs of contemporary students, I need to practice more flexibility, innovativeness, techno pedagogical skills, and flexible teaching skills while teaching sustainability issues." "I have learned how to organize leaching by providing more independency to my students"
Organizing the learning process in a transdisciplinary mode	Teachers shared new experience transdisciplinary learning design while teaching sustainability issues	"The students are engaged in blended environments and work creatively in finding solution to those complex sustainability issues by the use of transdisciplinarity perspective.". They also were encouraged to think analytically, by comparing, contrasting, evaluating and synthetizing information. They were required to collaborate virtually in diverse multimedia formats with the local and global partners"
Innovative approaches in introducing sustainability in higher education	Staff members argue about the use of new innovative approaches to teach sustainability	"To achieve a better quality of teaching I paid more attention to implementing inquiry type of learning that is also in line with the priorities set by the reform processes of School 2030" "Digitalization of the learning process fostered by the pandemic speeded ap introducing more student centered and self-directed learning that made me to reshape and to redesign my teaching to a large extent" "I tried to engage students in problem analyses, future scenario constructing and strategic thinking while connecting theory to real world and community issues"
Gains while introducing innovation in HE practices	Staff members share their new learning experiences while introducing online learning; they talk about the benefits it brings for the students in organizing their self-directed learning	"I discovered new opportunities for teaching online. I have created lots of tasks for my students to encourage their autonomy for completing the assignment. They received a gradual feedback as well. I have discovered new digital solutions for teaching sustainability issues." "Digital learning is based on the principles of self-directed learning, when the students plan their learning, as well as which subject, they focus on in a particular moment of the day. The student takes a full responsibility of how to achieve learning outcomes" "Remote learning mode differs in a way that students do not sit in the classroom, but learning takes place by the use of modern technologies. Technologies play a very important role, but even more important are the student's skills to learn the content and their motivation to achieve certain learning outcomes."

(continued)

Table 1 (continued)

Code	Description	Examples
Challenges while introducing innovative practice in HE	Staff members refers to challenges of their practice and thinking while introducing ICT technologies and organizing work in an online environment. They provide examples of what challenges they come across while introducing innovative practice during three months of work, like lack of proper ICT tools and technologies and training	"The digital learning process is not easy, as it has many challenges related to changes to the usual rhythm of work. However, it offers an opportunity to grow professionally by the use of digital platforms, since the main task of education is not only to provide knowledge, but also to teach how to analyze critically a wide range of information, how to select information and relate it to real life situations" "We cannot assume that all students have equal access to technologies and if their technologies are sufficient enough for online learning and compatible with the teachers' technologies"
Difficulties while introducing innovative practice in HE	Teacher expresses concerns about the lack of clarity how the distance learning will be organized in their institutions and concerns about quality of teaching in an online environment and lack of necessary tools for online learning	"I was not teaching in any online platforms before, but the situation with the Covid-19 made me to learn new ways of reaching my students by the ICT tools. It was very difficult to change from the traditional teaching to teaching with the digital technologies. My working day lasted more than eight hours" "The time will show the quality of teaching in remote mode" "The difficulty causes the fact that the whole institution has no single learning platform and its up to individual staff member which platform or tools to choose; this causes some confusion among the students"
Administrative support in HE	Staff members refer to the importance of support of administration in the process of introducing innovative practice in HE and providing necessary support	"The support of admiration was evident in all stages from the beginning of introducing remote mode of learning, to begin with joint planning of the learning process, the choice of the modes of communication with the students up to training to acquire new learning platforms and tools."

(continued)

Table 1 (continued)

Code	Description	Examples
Future challenges responding to the needs of future generation in HE	Staff members give examples of what they believe they will deal with while teaching twenty-first century learners and make a reference to specific elements of students' culture background, needs socioeconomic status, language that influence organization of online learning process	"We need to put the needs of contemporary students at the center of teaching by integrating technologies more into teaching, our students need a quick access to information and they prefer new forms of learning" "The contemporary students have expectations that are difficult to meet in the current model of higher education" Contemporary student prefers inquiry type of learning, a mixture of synchronous and asynchronous technologies

life, to generate new and innovative ideas, and to develop integrated view of issues. Therefore, universities need to engage students in the transformational sustainability research: Complex problem-handling, transition governance, integrated research, and the transformational sustainability research (Wiek et al. 2015).

Innovations are not the end in itself. They aim is at improving educational outcomes and the organization of the learning process. Innovations are recognized as a driving force of the socio-economic development. Universities are gradually becoming aware that they need to educate the students who will be able to generate and develop new and innovative products and services, to understand complex social challenges and to advocate for them. They are seen as drivers of change towards a more sustainable future (Riekmann 2012).

Innovations in higher education do not necessarily need to be related to introducing technologies, but IT has influenced changes in the higher education to a large extend. The challenge of the Universities is to provide quality education and to promote digital technologies for a most successful acquisition of the content by a contemporary student.

To be successful in implementing innovations, universities need to monitor the implementation of innovations, to seek information about external innovations and discover how they are applicable in the institutions by reviewing priorities of introducing innovations, consulting the staff members about the process of introducing innovations, modifying existing structures, roles and procedures and providing an ongoing support for introducing innovations.

The interviews results with the staff members indicate the following challenges while introducing innovative ICT solutions in their teaching: New acquired technologies and digital platforms in a compressed timeframe developed trust in engaging students in setting their goals and promoting a more autonomous inquiry type of learning in a new competence-based model of teaching, thus making a transfer to a non-linear mode of learning, fluid learning environment and more flexible learning structures, to work more creatively by engaging students in finding solution to complex sustainability issues while doing inquiry.

Among the difficulties the staff members have mentioned: A lack of unified technological platforms for the universities, lack of ITC knowledge, lack of digital materials, facing a completely new unknown situation, lack of experience and tools of work in a digital learning environment as well as resistance to changes of habitual ways of being and doing.

The staff members shared their intention and understanding about the need to realign the content and teaching in higher education to the needs of contemporary learners by providing more flexibility in teaching, fostering a student cantered learning, and engaging them in the inquiry type of learning in a real world situation, by the use of mixture of synchronous and asynchronous online technologies, and by engaging students in both, face to-face and online learning contexts while teaching sustainability issues. They recognized the need of giving up habitual ways of teaching and doing things in academia.

References

Delors J (1996) Learning: The treasure within. Paris, UNESCO. http://www.unesco.org/education/pdf/15_62.pdf. Last Accessed June 22, 2020s

Djordjevic A, Cotton DRE (2011) Communicating the sustainability message in higher education institutions. Int J Sustain High Educ 12(4):381–394

Drucker PF (1998) The discipline of innovation. Harv Bus Rev 76(6):149–157

Edwards-Schachter M, García-Granero A, Sánchez-Barrioluengo M, Quesada-Pineda H, Amara N (2015) Disentangling competences: interrelationships on creativity, innovation and entrepreneurship. Think Skills Creat 16:27–39

Ellis N (2017) Transdisciplinary perspectives on responsible citizenship, corporate social responsibility and sustainability. In: Brueckner M, Spencer R, Paull M (eds) Disciplining the undisciplined? Perspectives from business, society and politics on responsible citizenship, corporate social responsibility and sustainability. Springer, Cham

Europe 2020. A strategy for smart, sustainable and inclusive growth. Retrieved from: https://www.efesme.org/europe-2020-a-strategy-for-smart-sustainable-and-inclusive-growth. Last Accessed June 22, 2020

Filho WL (2011) About the role of universities and their contribution to sustainable development. High Educ Pol 24:427–438

Hayes JH, Hubley AM (2017) Bold moves for schools: how we create remarkable learning environments. ASCD Arias publication e-book

Hilla MP, Worsfold N, Nagy GJ, Filho WL, Mifsud M (2019) Climate change education for universities: a conceptual framework from an international study. J Clean Prod 226(20):1092–1101. https://doi.org/10.1016/j.jclepro.2019.04.053(LastAccessedJune22,2020)

Hoover JD, Whitehead C (1975) An experiential-cognitive methodology in the first course in management: some preliminary results. Dev Bus Simulation Experiential Learns 2:25–30

IAU (2020) Impact of COVID-19 on Higher Education around the World (2020). https://www.iau aiu.net/IMG/pdf/iau_covid19_and_he_survey_report_final_may_2020.pdf. Last Accessed June 22, 2020

Jackson T, Boix Mansilla V (2011) Educating for global competence, Preparing our youth to engage the world. NY: Council of Chief State School Officers' Education Steps Initiative & Asia Society Partnership for Global Learning

Kasulea GW, Wesselinkb R, Noroozib O, Mulderb M (2015) The current status of teaching staff innovation competence in Ugandan universities: perceptions of managers, teachers, and students. J High Educ Policy Manag 37(3):330–343. https://doi.org/10.1080/1360080X.2015.1034425

Keinänen MM, Kairisto-Mertanen, L (2019) Researching learning environments and students' innovation competences. Education + Training 61(1): 17–30

Keinänena M, Ursinb J, Nissinenb K (2018) How to measure students' innovation competences in higher education: evaluation of an assessment tool in authentic learning environments. Stud Educ Eval 58:30–36

Mhamed AAS, Vārpiņa Z, Dedze I, Kaša R (2018) Latvia: a historical analysis of transformation and diversification of the higher education system. In: Huisman J et al. (eds), 25 years of transformations of higher education systems in post-soviet countries, Palgrave Studies in Global Higher Education, pp 259–283. https://doi.org/10.1007/978-3-319-52980-6_10

Nielsen S, Storvang P (2015) DesUni: University entrepreneurship education through design thinking. Education + Training 57(8/9): 977–991

Perez-Penalver MJ, Lourdes EAM, Montero-Fleta B (2018) Identification and classification of behavioural indicators to assess innovation competence. J Ind Eng Manag 11(1):87–115

Riekmann M (2012) Future-oriented higher education: which key competencies should be fostered through university teaching and learning? Futures 44:127–135

Saeima (1995) Augstskolu likums [The Law on Higher Education Establishments]. Latvijas Vēstnesis 179(462). Retrived from http://likumi.lv/doc.php?id=37967

Spencer R, Paull M, Brueckner M (2018) Towards epistemological pluralism and transdisciplinarity: responsible citizenship, CSR and sustainability revisited. In: Brueckner M et al (2018) (eds) Disciplining the Undisciplined? Sustainability, ethics & governance, pp 255–265. https://doi.org/10.1007/978-3-319-71449-3_16

Sterling S, Maxey L, Luna H (eds) (2013) The sustainable university: progress and prospects. Routledge, Abingdon

The guidelines on the development of education 2014–2020. https://rio.jrc.ec.europa.eu/library/guidelines-development-education-2014-2020. Last Accessed June 22, 2020

The Latvian Sustainable Development Strategy to 2030 (2015) http://baltadaba.lv/wpcontent/uploads/2013/04/latvija2030_lv.pdf. Last Accessed June 22, 2020

The Law on Education (1991) http://likumi.lv/doc.php?id=67960; http://likumi.lv/doc.php?id=50759 (In Latvian). Last Accessed June 22, 2020

Tømt CE, Fossland T, Aamodt PO, Degn L (2019) Digitalisation in higher education: mapping institutional approaches for teaching and learning. Qual High Educ 25(1):98–114. https://doi.org/10.1080/13538322.2019.1603611

UNESCO (2017) Education for sustainable development goals. Learning Objectives. Paris, Francia: UNESCO. http://unesdoc.unesco.org/images/0024/002474/247444e.pdf. Last Accessed June 22, 2020

Vias MC, Rivera-Cruz B (2019) Fostering innovation and entrepreneurial culture at the business school: a competency-based education framework. Ind High Educ 34(3):160–217

Vincent-Lancrin S et al (2019) Measuring innovation in education 2019: what has changed in the classroom? Educational Research and Innovation. Paris, OECD Publishing. Last Accessed June 22, 2020

Wagner T (2012) Creating innovators. Simon & Schuster Children's Publishing, NY

Wiek A, Kay B (2015) Learning while transforming: solution-oriented learning for urban sustainability in Phoenix. Arizona Curr Opin Environ Sustain 16:29–36

Wiek A, Withycombe L, Redman CL (2011) Key competencies in sustainability—a reference framework for academic program development. Sust Sci 6:203–218

Wiek A, Harlow J, Melnick R, van der Leeuw S, Fukushi K, Takeuchi K, Farioli F, Yamba F, Blake A, Geiger C, Kutter R (2015) Sustainability science in action—a review of the state of the field through case studies on disaster recovery, bioenergy, and precautionary purchasing. Sust Sci 10:17–31

Yarime M, Trencher G, Mino T, Scholz RW, Olsson L, Ness B, Frantzeskaki N, Rotmans J (2012) Establishing sustainability science in higher education institutions: towards an integration of academic development, institutionalisation, and stakeholder collaborations. Sust Sci 7(1):101–113

Dzintra Iliško since 2017 until now she is professor at the Institute of Humanities and social sciences, Centre of Sustainable Education, Daugavpils University. She holds a PhD from Fordham University, USA. Total number of publications: 57 (Scopus and Web of Science) Hirsh index: h-7 (SCOPUS). Her current research interests include sustainability in high education and sustainable management issue as well as worldview education. ORCID ID: https://orcid.org/0000-0002-2677-6005; Web of Science Researcher ID O-3090-2019. She was invited speaker in Austria, Canada, The Russian Federation, The United Kingdom, Lithuania, Italy, Lithuania, Czech Republic, Lesotho and other countries.

Towards a More Sustainable Transport Future—The Cases of Ferry Shipping Electrification in Denmark, Netherland, Norway and Sweden

Maciej Tarkowski

Abstract One of the key challenges of sustainable transitions is a deep reduction of greenhouse gas emissions in transport. Therefore, new propulsion systems are being improved—mainly hybrid and electric ones. In the last five years, they have been increasingly used on ferries in ports and on coastal waters. The short route and the use of the same ferry ports facilitate the use of hybrid or electric propulsion. The basis of this article is the assumption that the development in ferry shipping electrification is a result of interactions between technical and operational features as well as geographical and economic conditions. Its purpose, however, is to identify these features and conditions based on four case studies of hybrid or electric ferry implementations on port or coastal routes. The final solutions for each ferry crossing are the result of a specific configuration of mentioned earlier features and conditions. All analyzed case studies concerned Europe, which is why there is a need for further research into this type of investment. Along with their spread in other parts of the world, it will be possible to verify the identified catalog of conditions.

Keywords Sustainability transitions · Sustainable transport · Ferry shipping · Ferry electrification · E-ferry · Hybrid ferry · Drivers of transport electrification · Case study

1 Introduction

Popularized since the 1960s, the concept of sustainable development did not cause a profound change in the model of social functioning. This term has been defined in many ways, but the most frequently quoted definition has been proposed by UN World Commission on Environment and Development (1987). The core of the idea is the meeting of the present needs without compromising the ability of future generations to meet their own needs. The concept of sustainable development implies the existence of limitations—not absolute limits but obstacles imposed by

---

M. Tarkowski (✉)
Faculty of Oceanography and Geography, University of Gdańsk, Jana Bażyńskiego 4, 80-309 Gdańsk, Poland
e-mail: maciej.tarkowski@ug.edu.pl

the present state of technology, ways of resource management and by the ability of the biosphere to absorb the effects of human activities. However, diagnosed in the 1970s barriers to further growth of industrial civilization (Meadows et al. 1972) remain in force (Meadows et al. 2004; Turner 2008; Bardi 2011). The climate crisis, which is one of the symptoms of industrial civilization reaching the limits of growth, has contributed to the intensification of activities aimed at reducing greenhouse gas emissions. Successive rounds of international negotiations (initiated 40 years ago in Geneva) result in specifying obligations of individual countries in terms of limiting greenhouse gas emissions. Although these activities are considered by the scientific community to be insufficient to maintain the stability of the Earth's planetary system (Ripple et al. 2019), even their implementation requires significant changes not only in the functioning of the main sectors of the economy but also in lifestyle—travelling, leisure activities and even dieting. The entirety of these transformations, going beyond the technological sphere, is referred to as sustainability transitions (Hansen and Coenen 2015). Thus, the concept of sustainability transition focuses on the ways of achieving the goal, i.e. on the implementation of specific solutions to make the model of sustainable development come true. Such an approach seems to be more mobilizing than quite abstract considerations about the well-being of future generations. It may thus facilitate the overcoming of the "Anthropocene stagnation" (Bińczyk 2018), in part due to the importance, difficulty and complexity of the challenges that the industrial civilization has to face due to the climate crisis.

The scientific discourse on sustainability transition is conducted in two main perspectives—systems of technological innovation and a multi-level one (Coenen and Truffer 2012). Both approaches place technological progress aimed at climate protection in the context of social conditions. The first approach narrows this scope down to the relations between research centers, enterprises and authorities at various levels (the so-called triple helix) (Scalia et al. 2018). The second approach arises from the critique of the previous one and extends this context with the concept of a socio-technological regime which is a constellation of technical products, infrastructure, regulations and user practices (a model of a four- or even five-element spiral enriched with cultural and environmental conditions) (Carayannis et al. 2018). As Coenen and Truffer (2012) note, socio-technical regimes are characterized by great stability. They can be put off balance by technological progress or social pressure. These disturbances lead to achieving another state of equilibrium. Sustainability transition can be interpreted precisely in terms of the process of transition of a socio-technological regime from a certain state of equilibrium to another one.

According to Coenen and Truffer (2012), the weakness of both approaches lies in the shallow and simplified treatment of the role of territorial differentiation of the components of these concepts. The spatial context is treated as a passive variable that rather hinders formulating scientific and research generalizations. Meanwhile, the concepts formulated on the grounds of geographic and economic knowledge indicate that local conditions play an active role in the processes of sustainability transition and thus spatially differentiate its paths.

Although the sustainability transition affects many areas of the economy, transport is particularly challenged (Centobelli et al. 2017; Przybyłowski 2013). The energy consumption and mobility of means of transport cause particular difficulties in replacing conventional fuels with renewable energy sources. Their popularization also requires the development of infrastructure as well as organizational changes. While the progress in reducing emissions in the power industry, industrial production and construction is noticeable, it is difficult to achieve it in transport (Fig. 1) just for the reasons mentioned.

Sustainability transition results in systemic transformations that lie in the field of interest of transport geography. Following the proposal of Liu and Gui (2016), identifying scientific schools of transport geography, the issue of sustainability transition is particularly close to the approach derived from spatial planning as well as from the school of new mobility.

The aim of this article is to identify geo-economic conditions and factors of sustainability transition in transport based on the example of ferry shipping. As already stated, the course of this transition depends, among others, on from the spatially diverse natural and, above all, anthropogenic features of the geographical environment. In the case of ferry shipping, this dependence is particularly visible. The technical and operational parameters of vessels result from adopting the concept and application of technical solutions tailored to the nautical conditions of a given body of water (route length, waves, ice), the demand for transport work (capacity and frequency of trips), intensification of competition from alternative means of transport, availability of funds or legal regulations.

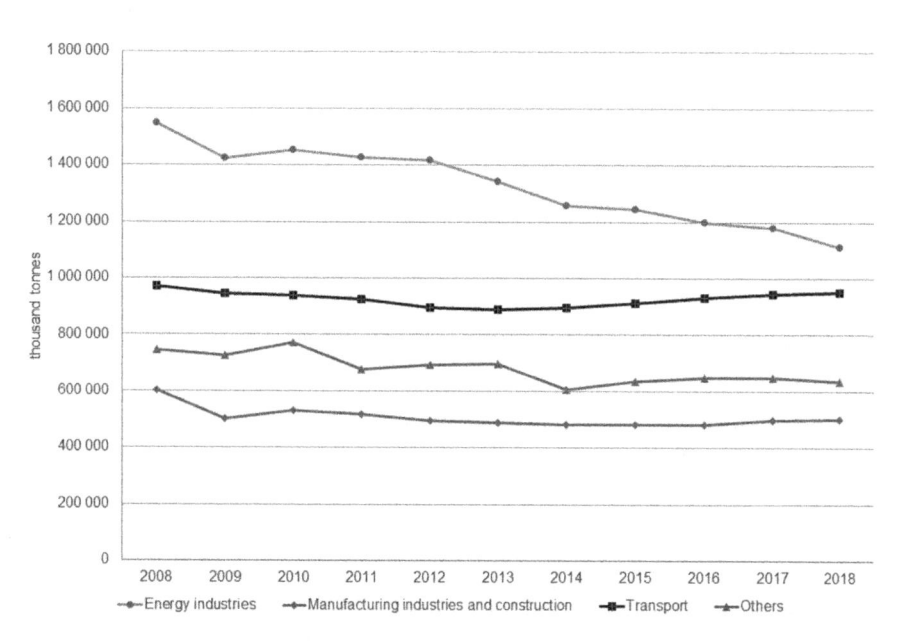

Fig. 1 Greenhouse gas emissions from fuel combustion by the source sector in the European Union (28 countries) in the years 2008–2018. *Source* author's own elaboration based on Eurostat data (https://ec.europa.eu/eurostat/web/environment/air-emissions)

Therefore, the material scope is limited to identifying the conditions with the greatest impact on the shape of the analyzed projects and at the same time characteristic of the area of ferry operation. The analysis was also limited to public passenger and car transport. *Strictly* tourism shipping was excluded, although electric drives are also implemented in them—from ships designed to explore cities using a network of canals to hybrid ocean units used e.g. for touristic exploration of the Antarctic. The implemented restriction is partially artificial, as public ferry transport also serves tourist traffic, to an extent depending on the presence and rank of tourist attractions in the vicinity (Le-Klaehn and Hall 2015). The crossing itself can also be such an attraction.

The last limitation of the material scope is related to the understanding of the concept of electrification, and thus to the evolutionary differentiation of the types of electric propulsion used on ships. Its use has a very long tradition—dating back to the nineteenth century—but its renaissance is observed in the second decade of the twenty-first century (Desmond 2017). The use of a hybrid drive system, in which electricity is generated by a classic diesel engine, has quite long traditions. It then goes directly to electromotors (diesel electric systems). This solution has become popular especially in the case of using AZ propellers. The electric transmission is cheaper to build, operate and more reliable than the diesel mechanical system. AZ propellers are mainly used to propel vessels with high maneuverability. The cases of ferries powered in this way are not discussed in this article, unless they have been modernized. Modernized ferries or built as hybrid ferries, whose propulsion system is additionally equipped with battery modules, are analyzed in this article, as are units equipped only with electric motors powered by battery modules charged while parked. At the same time, examples of the use of other alternative drives using hydrogen, biogas and liquid biogas as energy sources were omitted (Steen et al. 2019).

The spatial scope covers Northern Europe. It is Europe, especially the Nordic countries, that is among the leaders in the electrification of passenger transport. The spatial scope of the analyzed cases can be classified into two main categories: ferry navigation in ports and ferry navigation on coastal waters. The latter category includes the following subtypes: navigation in highly sheltered waters (in fjords and straits) and in archipelago—the degree of protection is lower, nautical conditions are more difficult, and the travel distance is longer.

The main temporal scope of the study covers the years 2015–2018. In justified cases, this time scope was slightly exceeded. In 2015, the first newly built fully electric ferry for coastal navigation was put into operation, contributing to the popularization of this solution.

The structure of the study reflects the course of the research procedure. Below, source materials have been characterized and research methods have been discussed. Then, a general characterization of four selected cases of implementation of ferries with hybrid and fully electric drives has been made. Synthesis of case studies enabled identifying the main geographic and economic conditions for electrification of ferry shipping. The article ends with conclusions, including proposals of directions for further research.

2 Materials and Methods

The study used an analysis of case studies as the main research method. The issue is so new that knowledge on this subject has not yet been accumulated and systematized. The analysis of scattered sources of information—sparse scientific literature, reports from specialist press, shipowners' websites, shipyards and equipment manufacturers, local and regional governments' statements as well as conference presentations—was used for the initial systematization of knowledge aimed at identifying the geographic and economic conditions of the transition of ferry shipping leading to reduction of emissions of greenhouse gases and other pollutants. Although the analysis of case studies, like any method, has its drawbacks (limited population range), in the context of the current state of knowledge and the nature of the discussed issues requiring a nuanced view of reality, its use seems to be justified. The analysis of the case studies is the first step in the systematic creation of model examples. As Flyvbjerg (2005) argues, without them, a scientific discipline is ineffective—it does not provide an empirical basis for the creation or verification of theories. When selecting case studies, a strategy of their differentiation was applied in order to obtain information about the significance of variable geographical conditions (Flyvbjerg 2005). The case study was used to find answers to the following questions: (i) which features of the natural and (ii) anthropogenic environment favor the electrification of ferry shipping? (iii) how can electromobility policy support the energy transition in ferry shipping? (iv) what role does the presence of local design and manufacturing capabilities play in the electrification process?

3 Results

As already mentioned, the strategy of selecting case studies served to show a wide range of conditions in which the electrification of ferry shipping takes place. The cases of ferry lines serving as urban transport in a big city (Amsterdam), a typical ferry connection across a fjord (Norway) and two coastal connections of a different nature, a different manner of service, operating in somewhat different geographical conditions—the South Funen Archipelago in Denmark and an international connection across the Øresund (Oresund) linking Denmark with Sweden—were analyzed.

3.1 City Ferry—Amsterdam (IJ Ferries 60 and 61)

Amsterdam, with nearly 870,000 inhabitants, is not only the constitutional capital of the Netherlands, but also one of the two largest economic centers of the country with a high rank in the EU settlement system (ESPON EGTC 2004). The city and its metropolitan area are served by a developed network of fully integrated public transport. In the Deloitte City Mobility Index 2020 ranking (Deloitte LLP 2020), the city is rated as a top performer in most categories. In the field of modal diversity and vision and strategy, it obtained the status of a global leader. The presence of numerous canals supports the widespread use of passenger ferries. The main hub for public transport, including the ferry, is the Central Station area (Fig. 2).

GVB—Amsterdam City Transport Authority—has adopted a strategy of mobility sustainability, one of whose goals is to replace the entire bus fleet with zero emission buses by 2025. This objective also applies to ferries, although hybrid ferries are allowed in addition to fully electrically powered vessels. The tool of change is the ELENA investment program implemented in 2017–2020. Investments in the ferries segment are estimated at 33 million euro. It is expected to result in a reduction in annual CO_2 emissions at the level of 4.5 thousand tons equivalent, nitrogen oxides at the level of 50 tons and a reduction in energy consumption of 13.5 GWh per year. Five over 80-year-old ferries operated by NZK Ferrylines will be replaced by fully electric vessels by 2022. In turn, 12 ferries operated by tIJ will be equipped with hybrid propulsion by 2025. Three new vessels with such a drive will also be built (Koek 2018). The use of fully electric ferries on the routes of this operator turns out to be impossible—the trip takes only four minutes and the stop takes two minutes. With the ferries operating 24/7, with the current technology, there is not enough time to recharge the batteries. In the case of a hybrid drive, the batteries are charged by a generator unit powered by a diesel engine during a stop or when the ferry is traveling

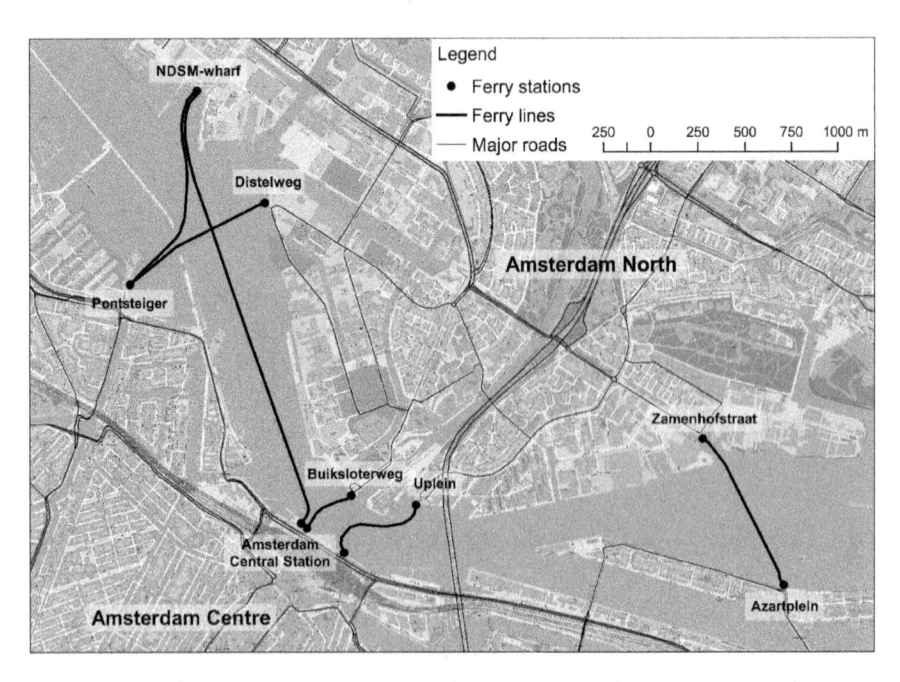

Fig. 2 Public ferry lines in the area of the main railway station in Amsterdam (2020). *Source* author's own elaboration based on OpenStreetMap data (https://www.openstreetmap.org)

at a slow speed. On the other hand, at peak power demand, batteries are the source of power. This solution allows the use of smaller diesel engines, whose speed does not change much. Thanks to this, fuel is saved and the emission of pollutants, noise and vibrations is reduced. The vessels also have better maneuverability. Two new vessels (IJ Ferries 60 and 61) entered service in 2016 and 2017. These are 34-m-long and 9-m-wide ferries. They are designed to carry 310 passengers as well as bicycles and mopeds. The power source consists of four power generators with a capacity of 133 kW each; lithium-polymer batteries have a capacity of 2×68 kWh, and two AZ propellers provide 250 kW of power each (Danfoss Drivers 2017).

3.2 Fjord Ferry—Sognefjord (MV Ampere)

The specific shape of the coastline becomes an important condition for the development of ferry navigation when it constitutes a transport barrier. Therefore, the coastline shape is vital as regards the spatial structure of the settlement network and the transport network. The longest Norwegian fjord—Sognefjord is an expressive example of this interdependence. It extends over 200 km inland, breaking the continuity of three roads of national importance. One of them is served by the first all-electric ferry in the world (MV *Ampere*) (Fig. 3).

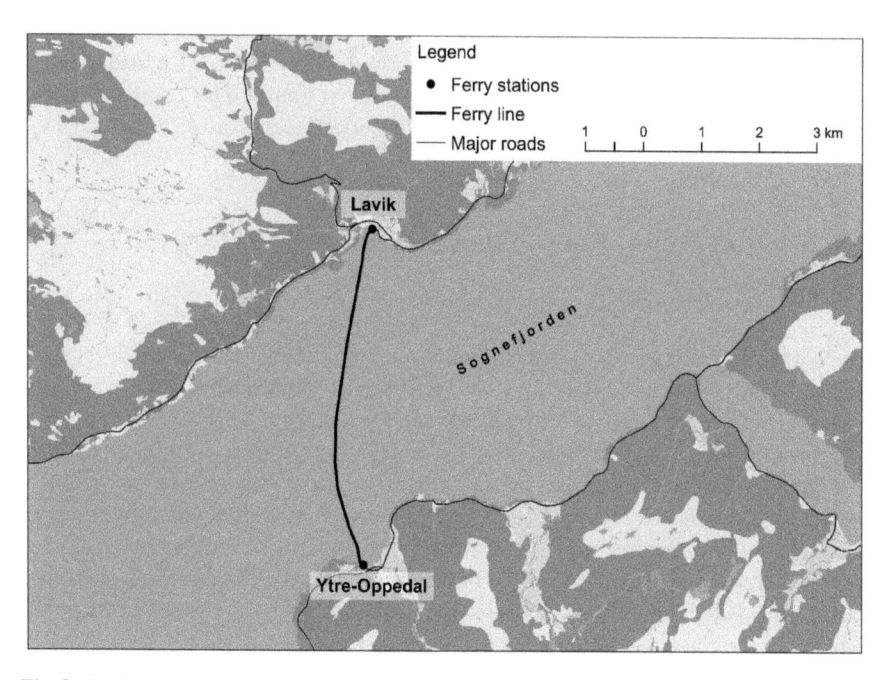

Fig. 3 Public ferry line Lavik—Oppedal in Norway (2020). *Source* author's own elaboration based on OpenStreetMap data (https://www.openstreetmap.org)

The vessel is a catamaran with an aluminum hull 80 m long and 20 m wide. The Nelton company from Pruszcz Gdański (near Gdańsk in Poland) participated in the design work, and the hull was built in the Aluship shipyard in Gdańsk. The ferry covers a distance of 6 km in 20 min up to 34 times a day. The stopover takes 10 min. It takes 120 cars and 360 passengers on board. The ferry was constructed at Norwegian Shipyard Fjellstrand in Omastrand in cooperation with Siemens and Norled—the ferry operator. The shipowner picked it up in October 201, and the transport began in May 2015. The aluminum hull structure reduces the total weight of the vessel and hence energy consumption. In order to save it, such solutions were used as LED lighting, photovoltaic panels, and the heating, ventilation and air conditioning systems with energy recovery. The drive system was integrated by Siemens using the BluDrive PlusC solution. It consists of two electric motors with a power of 450 kW each. They drive Rolls-Royce AZ propellers equipped with adjusting screws. The drive enables navigation at a speed of 18 km/h. The energy comes from lithium-ion batteries with a capacity of 1000 kWh and a weight of 10 tons. Charging stations are located on each shore. They are connected to the power grid, and at night, they use a local grid powered by hydroelectric power. They are also equipped with 260 kWh batteries that power the ferry when it is parked, when its batteries are being charged (Ship Technology 2015). Launching the ferry reduced CO_2 emissions by 95% and operating costs by 80% (Lambert 2018).

Putting the MV *Ampere* ferry into operation opened a new chapter in the development of ferry shipping in Norway. Siemens engineers involved in the project estimated in 2016 that 35% of 110 ferry lines could be electrified within 4 years (Siemens 2015). Similar data (50 ferries) was provided by the Swedish consul general in Hong Kong, when he was promoting this means of transport and persuading the city authorities to take advantage of the Swedish experience (Karacs and Ye 2016).

3.3 Coastal Ferry—Søby–Fynshav and Søby–Faaborg (EF Ellen)

Much of Denmark's population, economic actors and infrastructure are distributed around islands. The island town of Søby, located in the South Funen Archipelago, in 2019 began to be served by a fully electric ferry, EF *Ellen* (Fig. 4).

It is a one-sided ferry with one continuous deck for vehicles. The hull was built and partially equipped by a Szczecin (Poland) company Spawrem on commission by the Søby Shipyard. It is 60 m long and almost 13 m wide. On the main deck, equipped with bow and stern doors, there will be room for 31 cars or five trucks. In the summer season, it can also take 198 passengers, and in the winter—147. Two drive units provide a maximum power of 1500 kW, which allows for a speed of 28 km/h. However, when using only half the power, it can sail at a slightly lower speed—24 km/h, which will still shorten the travel time. The last element to be installed is two sets of batteries with 420 cells each. The batteries are characterized by greater

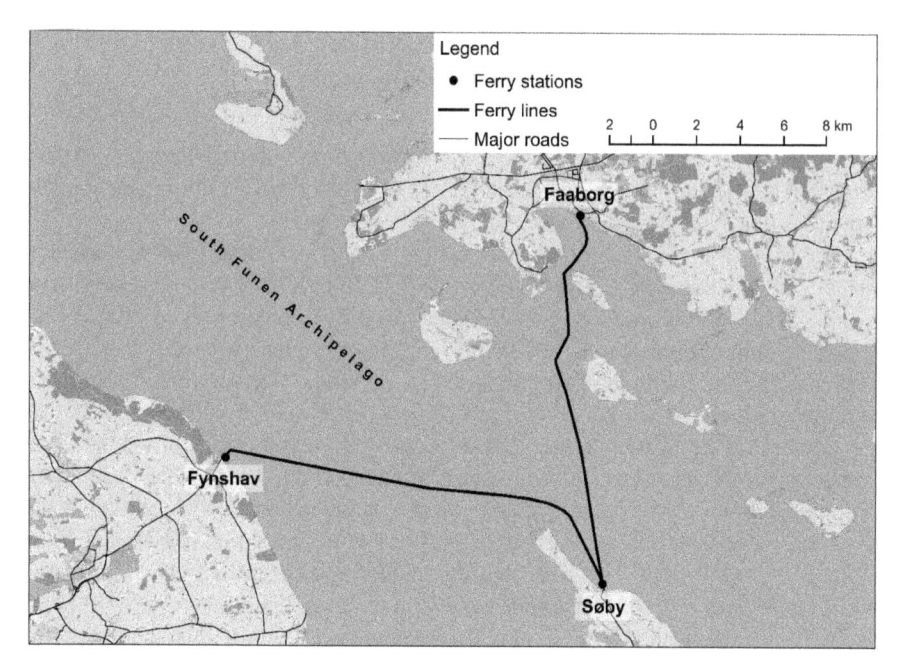

Fig. 4 Ferry lines Søby–Fynshav and Søby–Faaborg in Denmark (2020). *Source* author's own elaboration based on OpenStreetMap data (https://www.openstreetmap.org)

safety of use in marine conditions and a longer life cycle than the previously offered ones. Two sets of batteries provide the largest ever capacity installed on ships in the world. They allow the service of both lines, each about 18 km long, although their capacity will allow sailing for about 39 km. There is a chance that, under favorable circumstances, the ferry will only charge its batteries in one of the two marinas (E-Ferry Project 2019). The planned range clearly exceeds the one achieved in previous projects (usually no more than 9 km). The project is experimental and demonstrational in nature. It is financed by the EU program Horizon 2020. Its total value is EUR 21.3 million, of which EUR 15 million comes from EU funding. The aim of the project is to design, build and demonstrate the operation of a fully electric ferry with no emissions. It promotes water transport for communities in islands, coastal areas and inland waters in Europe and beyond that is energy-efficient and limits the emission of greenhouse gases and other air pollutants to zero. The direct effect of the project is to reduce travel time by 25% and to reduce CO_2 emissions by 2000 tons per year and 50% energy savings compared to the existing ferries. The four-year project is expected to end in mid-2019. The management of the Søby shipyard is very pleased with the investment, claiming that the construction of an electric ferry means jobs for approx. 40 people (Suchenek 2016).

3.4 Coastal Ferry—Helsingor–Helsingborg (MF Tycho Brache and MF Aurora Af Helsingborg)

Despite the operation of the bridge over the Oresund Strait, this ferry crossing is very popular. The four-kilometer wide strait (Fig. 5) is crossed by 7.4 million passengers and 1.9 million vehicles every year.

The crossing is served by four HH Ferries vessels, including two twin cruising ferries—MF *Tycho Brache* and MF *Aurora af Helsingborg*. These relatively large vessels—111 m long and 28 m wide—can carry 240 vehicles and 1250 passengers. They entered service in 1991. With four diesel engines with a total power of 9840 kW, they develop an operating speed of 26 km/h. The use of AZ propellers equipped with electric motors powered by energy generated by diesel engines was a good starting point for the modernization of both vessels. In addition, in 2013 the electricity management system was modernized, which was another facilitation. Original generators produce alternating current, which is used in motors used in AZ propellers. The batteries are charged with direct current, so the system had to be equipped with inverters. The essence of the modernization carried out in 2017 was the installation of a newer version of the energy management system, as well as an electricity source in the form of four 32-foot containers with 640 lithium batteries with a total capacity of 4160 kWh. The containers were placed on the upper deck

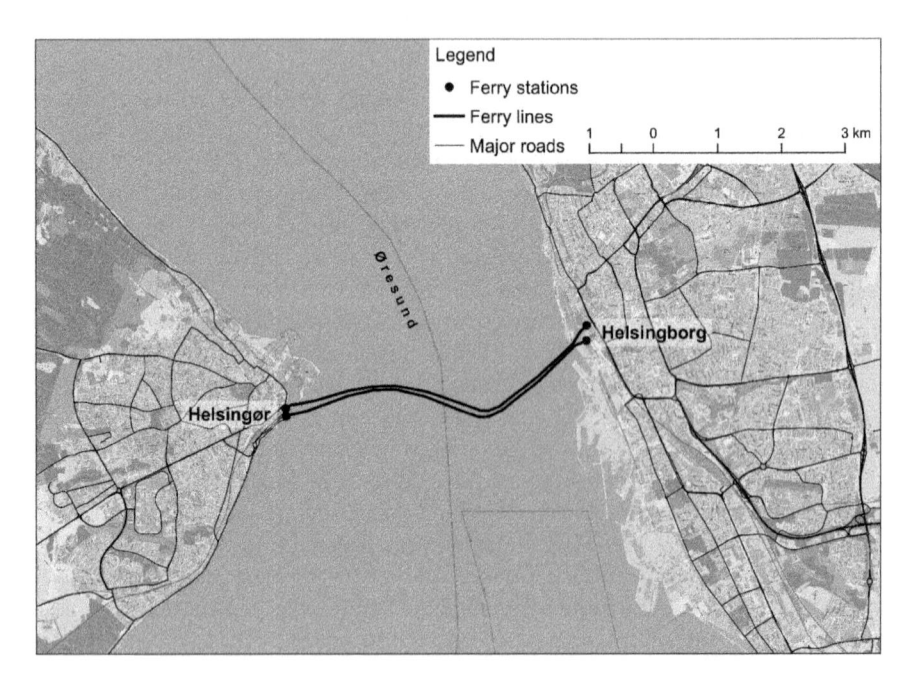

Fig. 5 Ferry line Helsingor–Helsingborg between Denmark and Sweden (2020). *Source* author's own elaboration based on OpenStreetMap data (https://www.openstreetmap.org)

and equipped with a water cooling system for the batteries. Ferries can be driven in three modes: as before—only with motors as the main energy source, a hybrid mode and a fully electric one. In the latter, about 29% of the battery capacity is used to cover the route (DEIF 2018). Recharging the approx. 1200 kWh used takes 5.5–9 min, depending on the port. Ensuring quick recharging of such high-capacity batteries was a great challenge. In each of the ports, ABB IRB 7600 robot connection charger is installed that enables charging with a power of up to 11 MW (220 kWh per min). Such intense charging required protecting the battery against overheating, which is ensured by a water cooling system. Each battery, consisting of 24 cells, is surrounded by a spiral conduit into which water is pumped. The modernization of two of the four ferries serving this line has made it possible to reduce CO_2 emissions by 65% per annum. The modernization was a very serious financial challenge for the shipowner—the cost of the project was SEK 300 million. The EU subsidy amounted to 13 million euro (equivalent to SEK 97 million) (Petersen 2016).

4 The Main Drivers of Ferry Shipping Electrification—A Summary of the Case Studies

An analysis of four case studies of the use of hybrid or electric ferries aimed at identifying the geographic conditions for the electrification of ferry transport. It was assumed that the final shape of each system, which consists of the ferry and the infrastructure enabling its operation, only partially results from the technical progress, i.e. the increasingly favorable operating parameters and costs of individual components: battery modules, chargers, energy management systems, electromotors or automatic mooring systems. It was also granted that the technical and operational characteristics of ferries and the infrastructure for their operation also result from taking into account the specific conditions of individual water bodies and coastal regions. Four categories of such conditions have been identified (Table 1).

The final shape of the solutions adopted for a specific ferry crossing is a result of a specific configuration of the above-mentioned conditions. It is important to emphasize their interdependence. The relationship between the shape of the coastline and the level of transport development determines the level of economic profitability of using electric or hybrid propulsion. On the one hand, short routes with a high frequency of journeys allow reducing investment costs (capacity of battery modules), and on the other hand, maximizing the rate of return on investment resulting from the difference in the costs of electric or hybrid ferries compared to diesel-powered ferries. The popularization of ferry transport and the consumption structure of primary energy carriers, in turn, affect the efficiency of the electromobility policy. From the point of view of reducing greenhouse gas emissions on a national scale, electrification of ferry shipping brings effects when it is possible to use renewable energy sources to power the battery modules of new ferries, and their very use will reduce emission. In addition, a local policy aimed at not only reducing greenhouse gas emissions but also

Table 1 The main drivers of ferry shipping electrification

Drivers of electrification	Explanation of the driver role in electrification
Features of the natural environment	• The shape of the shoreline (islands, bays, fjords, river mouths) requires the use of crossings to ensure adequate transport accessibility • Distance between marinas (a typical distance is approx. 6 km, the maximum one is approx. 40 km) will be increasing with the technical progress • Waves and wind determine the size of the ferries, the shape of the hull and other design solutions that affect the technical and economic aspects of the operation of the crossing • Low temperatures require energy-intensive heating of passenger compartments. Frozen waters require the use of a durable hull, greater power of the propulsion system and additional energy reserves
Features of the anthropogenic environment	• The spatial structure of the settlement and transport network determines the demand for ferry services • The operation of the conventionally powered ferry (diesel mechanical, diesel electric) provides knowledge and experience to facilitate electrification • The general level of wealth and the amount of budgetary expenses for the maintenance and modernization of the transport system
Electromobility policy	• The high priority to combat anthropogenic climate change creates the framework for the effective electrification of transport • Co-financing of R&D works – a cost-intensive element of the development of each technology • Co-financing of investments reduces the risk of the investor and lenders • The development of energy infrastructure ensures the supply of cheap renewable electricity • Including the time needed to recharge the batteries in the ferry timetable
Local design and manufacturing capacity	• The complexity and innovative nature of the construction of electric ferries requires constant communication with customers, which is easier when the main shipyard is close to the place of operation. Cooperators can be located further. For the discussed cases, Northern Europe seems to create a common market for this type of cooperation

Source Author's own elaboration

air pollution and noise emissions from urban transport may be a catalyst for electrification. However, the electromobility policy itself will not cause a real change without the design and manufacturing capabilities that will make it happen. The existence of local manufacturers also increases the effectiveness of the electromobility policy, which by creating a demand for innovative products helps to strengthen the manufacturers' competitive position. Moreover, successful projects, especially at the pioneering stage of electrification, have a very significant potential to promote a given country, region or city.

5 Conclusions

Technical progress in the field of electricity storage has created practical possibilities of using electric propulsion on ships. Ferry shipping has proved to be the most susceptible to this type of innovation. The short route and the use of permanent harbors facilitate the application of electric propulsion. The prior limits the necessary capacity of the battery modules, which reduces the costs of the propulsion system. The latter minimizes organizational problems and investment outlays necessary to provide infrastructure for charging batteries.

The analysis of four case studies shows that at the current level of technology development, electric ferries can be successfully used in harbor waters, bays and fjords as well as in archipelago waters. As shown by the example of South Funen Archipelago (Denmark), fully electrically powered ferries can handle connections of almost 40 km at a speed not inferior to vessels equipped with the traditional Diesel propulsion system. Technological progress is not the only factor determining the spread of full-electric drive systems in ferry shipping. As shown by the above analysis, four groups of drivers are of fundamental importance: features of the natural environment, attributes of the anthropogenic environment, electromobility policy as well as local design and manufacturing capacity. These groups of drivers are interdependent. In each of the analyzed cases, the specificity and role of the above-mentioned factors were different and resulted in the adoption of a solution corresponding to the transport needs, production capacity, available financial resources and the ability to organize the project in a given location. So while technical progress in the field of electric propulsion systems offers solutions available on the global market, their specific implementations are strongly determined by local factors.

The conducted research procedure was of a preliminary nature, which is largely due to the early stage of development of ferry navigation with a use of electric or hybrid propulsion. There is a need for further case studies. The variety of geographic conditions may result in application of solutions other than those presented, especially because the described cases, despite their diversity, come from areas with similar geographical conditions—natural ones (moderate climate) and anthropogenic ones (a high level of economic development). This can lead to excessive, and therefore distorted, generalizations regarding the regularities in development. The studies by Bandyopadhayay and Banerjee (2017) and Ghosh and Schot (2018) show that in less prosperous big cities of Southeast Asia organizational and structural transformations of urban transport systems can play a greater role in electrification of transport than technical progress.

References

Bandyopadhayay A, Banerjee A (2017) In pursuit of a sustainable traffic and transportation system: a case study of Kolkata. Int J Manag Pract 10(1):1–16

Bardi U (2011) The limits to growth revisited. Springer Science & Business Media, New York –Dordrecht – Heidelberg – London

Bińczyk E (2018) Epoka człowieka: retoryka i marazm antropocenu. Wydawnictwo Naukowe PWN, Warszawa

Carayannis EG, Grigoroudis E, Campbell DFJ, Meissner D, Stamati D (2018) The ecosystem as helix: an exploratory theory-building study of regional co-opetitive entrepreneurial ecosystems as Quadruple/Quintuple Helix Innovation Models. R&D Manage 48(1):148–162

Centobelli P, Cerechione R, Esposito E (2017) Environmental sustainability in the service industry of transportation and logistics service providers: Systematic literature review and research directions. Transp Res Part d: Transp Environ 53:454–470

Coenen L, Truffer B (2012) Places and spaces of sustainability transitions: geographical contributions to an emerging research and policy field. Eur Plan Stud 20(3):367–374

Danfoss Drivers (2017) Hybrid ferries connect the city of Amsterdam nonstop. Danfoss Drivers, Gorinchem. http://files.danfoss.com/download/Drives/DKDDPC939A402_IJFerry_LR.pdf. Last accessed 22 Jul 2020

Deloitte LLP (2020) Deloitte City Mobility Index 2020. Deloitte LLP, London. https://www2.deloitte.com/content/dam/insights/us/articles/4331_Deloitte-City-Mobility-Index/Amsterdam_GlobalCityMobility_WEB.pdf. Last accessed 22 Jul 2020

DEIF (2018) Tycho Brahe hybrid ferry case story. DEIF A/S, Skive.

Desmond K (2017) Electric boats and ships: a history. McFarland & Company Inc., Publishers, Jefferson

E-ferry Project (2019) E-ferry: prototype and full-scale demonstration of next generation 100% electrically powered ferry for passengers and vehicles. http://e-ferryproject.eu/. Last accessed 17 Jul 2020

ESPON ECTC (2004) ESPON 111 Potentials for polycentric development in Europe, ESPON EGTC, Luxembourg

Flyvbjerg B (2005) Pięć mitów o badaniach typu studium przypadku. Stud Socjol 2(177):41–69

Ghosh B, Schot J (2018) Mapping socio-technical change in mobility regimes: the case of Kolkata. SPRU working paper series, 2018–16:1–45

Hansen T, Coenen L (2015) The geography of sustainability transitions: Review, synthesis and reflections on an emergent research field. Environ Innov Soc Trans 17:92–109

Karacs S, Ye J (2016) Norwegian-made electric ferries 'could reduce pollution in Hong Kong's Victoria Harbour'. South China Morning Post, Jul 16th. https://www.scmp.com/news/hong-kong/health-environment/article/1990575/norwegian-made-electric-ferries-could-reduce. Last accessed 18 Jul 2020

Koek T (2018) Funding sustainable public transport in Amsterdam. GVB, Amsterdam. https://ec.europa.eu/energy/sites/ener/files/documents/2.2_smart_city_mobility_final.pdf. Last accessed 20 Jul 2020

Lambert F (2018) All-electric ferry cuts emission by 95% and costs by 80%, brings in 53 additional orders. Electrek, Feb. 3rd. https://electrek.co/2018/02/03/all-electric-ferry-cuts-emission-cost/. Last accessed 20 Jul 2020

Le-Klaehn DT, Hall MC (2015) Tourist use of public transport at destinations—a review. Curr Issue Tour 18(8):785–803
Liu Ch, Gui Q (2016) Mapping intellectual structures and dynamics of transport geography research: a scientometric overview from 1982 to 2014. Scientometrics 109(1):159–184
Meadows DH, Meadows DL, Randers J, Behrens WW (1972) The limits to growth. Potomac Associaties, Washington, DC
Meadows DH, Randers J, Meadows DL (2004) Limits to growth. The 30-year update. Chelsea Green Publishing Company, Vermont
Petersen SR (2016) Water-cooled batteries ensure fast charging of electric ferries Across Øresund. https://spbes.com/hh-ferries/. Last accessed 17 Jul 2020
Przybyłowski A (2013) Inwestycje transportowe jako czynnik zrównoważonego rozwoju regionów w Polsce. Wydawnictwo Akademii Morskiej, Gdynia
Ripple WJ, Wolf Ch, Newsome TM, Barnard P, Moomaw WR (2019) World scientists' warning of a climate emergency. BioScience biz088:1–5
Scalia M, Berile S, Saviano M, Farioli F (2018) Governance for sustainability: a triple-helix model. Sustain Sci 13(5):1235–1244
Ship Technology (2015) Ampere electric-powered ferry. Ship technology. https://www.ship-technology.com/projects/norled-zerocat-electric-powered-ferry/. Last accessed 20 Jul 2020
Siemens (2015) Electric operation makes seven out of ten ferries more profitable—a feasibility study. Siemens, Bellona, Oslo
Steen M, Bach H, Bjørgum Ø, Hansen T, Kenzhegaliyeva A (2019) Greening the fleet: a technological innovation system (TIS) analysis of hydrogen, battery electric, liquefied biogas, and biodiesel in the maritime sector. SINTEF Digital, Trondheim
Suchenek A (2016) Rozpoczęła się budowa nowego promu elektrycznego dla Danii. https://denmark.trade.gov.pl/pl/aktualnosci/200338,rozpoczela-sie-budowa-nowego-promu-elektrycznego-dla-danii.html. Last accessed 19 Jul 2020
Turner GM (2008) A comparison of the limits to growth with 30 years of reality. Glob Environ Chang 18(3):397–411
United Nations World Commission on Environment and Development (1987) Our Common Future. WCED, New York

Maciej Tarkowski Obtained a master degree (1999) and PhD (2007) in geography at the University of Gdańsk. He is Assistant Professor at Faculty of Oceanography and Geography in socio-economic geography and spatial organization. Transition studies are his main research scope. Initially, he researched the issues o local and regional dimensions of post-communist transition. He then directed his interests to sustainability transitions, especially in terms of urban transportation and mobility and maritime economy. In 1999-2015, he collaborated with Gdańsk Institute for Market Economics as a Regional and European Integration Research Team member. He was a co-author of numerous expert opinions for local, regional and national authorities.

Digitalisation and Communication for Sustainable Development

Overcoming Digital Inequality as a Condition for Sustainable Development

Vladimir Krivosheev

Abstract This paper stresses the need for overcoming one of the new forms of social disparity—digital inequality. It is largely a result of the institution and dominance of electronic culture and the widespread of digital technology. The acceleration of social change, which is a manifestation of social time compaction, gives rise to digital inequality. Rapidly proliferating digital inequality severs communication ties, disrupts normal interactions between people, social groups and different societies, and inhibits transition to sustainable development. The world community has to respond to these challenges and ensure equal access to information resources and technologies.

Keywords Sustainable development · Electronic society · Digital technologies · Digital inequality

1 Introduction

Transition to sustainable development urgently requires further development of social knowledge, which is believed to respond to current challenges and actual social processes. Sociology emerged as an independent science in the mid-nineteenth century when radical changes in society led to a new form of social disparity—class inequality. In other words, social factors prompted the rise of sociology as a scientific interpretation of all social processes and phenomena. Sustainable development was impossible when inequality was constantly causing conflicts and social tensions.

Today class inequality is playing a different role. Social partnership and social services have made it possible to smooth out many contradictions. The rise of information technology has brought about a new form of social disparity– digital inequality. This circumstance can interfere with transition to sustainable development

---

V. Krivosheev (✉)
Sociological Sciences, Institute of Humanities, Immanuel Kant Baltic Federal University, St. A. Nevsky, 14, Kaliningrad 236041, Russia
e-mail: VKrivosheev@kantiana.ru

at the global scale and cause numerous new conflicts. Many researchers and organizations have emphasized this circumstance. In particular, the UNCTAD report noted in September 2019 that a concerted effort was needed for a fairer distribution of the benefits of the digital economy and minimization of digital inequality (UNCTAD). For example, there are deep-seated ICT inequalities in the member countries of the Organization of the Islamic Conference (OIC), which brings together a very significant part of the world community (Siwar and Abdulai 2013). Digital inequality cannot be reduced to its technical components. Cut off from digital technology and high-speed Internet access, many people will be unable to assimilate common values (Ramade and Alvaro 2011). Digital inequality has also been viewed from a gender perspective (HayetKerras 2020). The role of digitalization in achieving sustainable development through transition to a green economy has been brought to the fore. Researchers from the International Organization for the Study of Environmental Problems and Pollution have considered the case of South Korea in the article "Environmental Friendliness: the Role of ICTs in Resource Management" to point out the connection between ICT, economic growth, energy consumption and carbon dioxide emissions (Batool et al. 2019). Digital inequality and poor access to IT hinder social development.

I consider inequality in terms of individual access to resources. Limiting access produces inequality. Post-industrial society has highlighted the importance of access to resources that sustain an entirely new, electronic, culture. A form of capital that is becoming dominant in the modern social world is information capital, as Pierre Bourdieu termed it (Bourdieu 2002).

In my understanding, information capital is rather an element of social capital expressed in the knowledge, skills and experience of an individual or a social group. Of course, I do not reject the material component of this form of capital. Yet the presence of computers, gadgets and accompanying infrastructure does not necessarily bring income or benefits. They remain nothing more than objects, albeit very modern, until an individual endowed with knowledge and competencies sets all these objects in motion.

It is important to identify the social component of information capital to understand how it is accessed and how the individual makes an informed effort to invest in it. The relevance of this study lies in the investigation of access to information capital or the lack of it. This problem has been addressed earlier (Bushuev and Lezhenina 2012; Korechkov 2018). It has been concluded that digital capital (or, more accurately, information capital) is a type of intellectual capital. I believe that this treatment fails to embrace the concept in question. Particularly, it does not make it possible to extend the term to attitudes, value preferences and strong-willed efforts, which cannot be considered as purely intellectual. Some researchers see the role of information capital in the formation of a new technological paradigm from a purely economic perspective. It is assumed to have penetrated all spheres of social production and thus to affect the dynamics of socio-economic development. Svetlana Mikhneva points to the emergence of information activities that rely on science-intensive technology (Mikhneva 2003). This means that a new form of relationships is emerging in the digital economy. The approach to information as a type of social capital makes this article a potential contribution to the corresponding stream of research.

Below I will discuss both various aspects of sustainable development and the need to advance sociological knowledge, which should not only adapt to changes but also offer effective tools to remedy the situation.

2 Forms of Inequality in Modern Society

Post-industrial society has not eliminated the old forms of inequality; moreover, it has created new ones. Below I will consider the forms of inequality that appeared in industrial or agrarian society and that has not been eliminated to this day.

A traditional form of inequality is income inequality. According to experts, in economically developed democratic countries, income inequality is registered when the highest-income decile of the population is about eight to ten times as rich as the lowest-income decile (Lapin 2020). In the richest country in the world, the US, there are sharp differences in incomes and living standards. "About 40 million live in poverty, 18.5 million in extreme poverty, and 5.3 million live in Third World conditions of absolute poverty", writes Philip Alston, the UN Special Rapporteur on extreme poverty and human rights. His report was presented to participants in a session of the UN Human Rights Council (The highest in the USA…). Researchers agree that inequality is out of control on the global scale: 1% of the wealthiest people are richer than the remaining 6.9 billion people (Kholyavko 2020).

Tax deductions for the rich and corporations, as well as tax evasion, have had a role in the accumulation of many huge fortunes, Oxfam experts say. After 1945, the maximum income tax rate in the US reached 90%; by 1980, it decreased to 70%. Now it does not exceed 40%. In developing countries, the maximum income tax rate is even lower, at 28%; the corporate tax rate is 25%. In some countries, the poor pay more taxes than the rich do. In Brazil, the effective rate for the poorest 10% is over 30%, and that for the wealthiest 10 is 20%. The billionaire Warren Buffet famously stated, "I pay less income tax than my secretary". The rich account for only 4% of global tax revenues, evading 30% of their tax liabilities (Kholyavko 2020). It turns out that inequality is enshrined in the law.

Another form of traditional inequality is poor accessibility of social institutions, such as healthcare and education, for various groups of people. The pandemic of the new dangerous infection, which became the scourge of many countries in 2020, has demonstrated the gravity of this problem. In the US, 9% of the population still do not have health insurance, which, even if available, does not guarantee access to high-quality medical care. The cost of COVID-19 treatment in the ICU averages USD 80,000 across the US (Kostyaev 2020). This does not mean that unequal access to health services is characteristic only of the richest country in the world. In varying degree, this type of inequality is present in all countries.

As to unequal access to education, many countries have only fee-paying higher education. This makes valuable professional knowledge inaccessible for many people. In Germany, tertiary education has been free for many years. In October 2011, however, three states imposed university fees (Komleva 2014). Since the academic year 1998/99, UK universities have charged at least GBP 3,000 to local students and much more to international ones (Rosenfeld 2006). At US top private universities and colleges, annual tuition and fees, which cover housing, dining, and books, often exceeds USD 30,000. Although public universities are much cheaper, they charge non-residents much more than the country's nationals. University fees are constantly growing. Since the early 2000s, it has been increasing by 5–7% a year. Paying tuition as well as housing and dining fees may be difficult for many students (Filipovich 2017). Although these examples do not exhaust the problem, they demonstrate that fee-paying education restricts access to knowledge.

Not everyone has equal access to a clean and safe environment. Particularly, transnational corporations have been moving environmentally hazardous production to developing countries for more than a decade.

All these forms of inequality preclude sustainable development of entire regions and many countries, even rich ones. Digital inequality complements and modifies other inequalities, making them increasingly tangible. It seems to be a new embodiment of class inequality. The COVID-19 pandemic has become a major challenge to sustainable development. It has been argued that the impact of COVID-19 on the world economy will be more intense and long-term than that of the global financial crisis of 2008–2009 (Behravesh and Rocha 2020).

3 Methodology

Methodologically, this article draws on the concept of the information society, which has been developed by many researchers, for instance, Manuel Castells. The concept holds that the key resource of modern society is knowledge and information. Insufficient access to either resource causes inequality. In this study, processes taking places in modern society at different levels of its organization are analyzed using this concept. The interdisciplinary approach is employed to consider social processes from the perspectives of sociology and economics. The idea of global development of humanity, which postulates the interconnectedness and interdependence of all social processes that have been taking place since the mid-twentieth century, aids in demonstrating that rapid and relatively even dissemination of knowledge and information is possible. Besides, the potential for innovations has not been fully realized. This has been convincingly shown by critics of an overly optimistic view of globalization. I also use my own concept of electronic culture. I define electronic culture as dominant and affecting all aspects of social life.

4 Results and Analysis

4.1 Electronic Culture and Its Characteristics

Digital inequality has become characteristic of post-industrial society. It is largely a result of the institution and dominance of electronic culture and the wide spread of digital technology. So far, researchers have not overcome the temptation to consider this phenomenon as part of other, more familiar and long-established forms of culture.

Written culture developed into print culture (McLuhan 2003a, b), and the "Guttenberg era" began. Printing made information, scientific knowledge, fiction and other literature accessible for huge masses of people. Nevertheless, the book-centric culture was infused with inequality.

Firstly, not everyone was literate when the printed book arrived. Secondly, some categories of people were forbidden to read the Bible. The historian Elena Brown, writes that, in the thirteenthcentury, the Catholic Church did not allow commoners to read the Holy Scriptures (Brown 2016). They could not even own a Bible. The Synod of Béziers decided in 1246 that neither the laity nor the clergy should be permitted to own the Scriptures in the popular tongue (Nosovsky and Fomenko 1997). Thirdly, books were very expensive, and very few could afford them.

It has been mentioned in the literature that the electronic form of culture was developing alongside of the traditional way of recording and saving information, scientific data, and images. These two technologies interacted and complemented each other, but sometimes competed or came into conflict.

The weakness of this approach is obvious. The appearance and spread of completely different media require new ways to work with, copy and transmit it. As a result, the new media evolve and incorporate new content. Human perception of information also changes.

Modern individuals are becoming increasingly focused on visual information. They both receive information in electronic form and see the worlds via the screens of electronic devices. People make videos of anything they find remarkable and upload them to social media. Many users never get off their gadgets, even when meeting people face-to-face (Busko 2009). They have a reduced ability for deep and meaningful communication: "When interchanges are reformatted for the small screen and reduced to the emotional shorthand of emoticons, there are necessary simplifications" (Turkle 2011).

Thus, electronic culture is an essential part of modern society. It permeates the content of social life and modifies it. Electronic culture gave rise to short life projects, i.e. extremely short-term social, economic, spiritual, and family plans (Krivosheev 2009). There is more to short projects: people strive to upload content to social media as quickly as possible to get instant approval. It does not matter what kind of content that is.

Because electronic culture dominates people's lives and creates a multifaceted reality, it has to be studied not as part of other cultures but as a separate phenomenon. Each culture represents the content of social life, and thus it should be considered along with its form. In other words, electronic culture has modified society and its structure as well as paved the way for electronic society, in which relationships and interactions between individuals and groups rely on completely new principles. Electronic culture has created a new form of disparity—digital inequality. From the very beginning, this new culture has developed in conditions of inequality. The mass distribution of radio has been identified as the starting point of electronic culture (McLuhan 2003a, b).

Radio access was limited only when the technology was in its infancy. Later, television became equally widespread. Since the late 1970s, the Internet has been penetrating social life. Microchips made possible the wide spread of mobile phones. Having acquired numerous additional functions, they became much more than a means of communication. Ironically, electronic versions of dictionaries and encyclopedias give a narrow definition of electronic culture. According to one of them, electronic culture is the use of ICTfor culture. The term applies to digitalized art, namely, fine arts (paintings, drawings, sculpture), performance arts (music, theater, dance, etc.), cultural heritage (architecture, cultural landscape), cinema, television, etc. Among objects of electronic culture are works that were originally created in electronic form; its scope includes electronic libraries, museums and archives (Encyclopedia...).

This approach does not seem to cover the essential characteristics of modern society. There is every reason to say that electronic culture has ceased to be an integral part of culture in general—it has become the foundation of the latter. Electronic culture has penetrated all aspects of social life and radically transformed the way individuals and social groups interact. It has also affected the government and authorities, main social institutions and the media. As Rezaev et al. put it, "interactions between people ... are becoming embedded in the communicative environment that is created and supported by relatively autonomous virtual agents and technical devices. The use of the Internet from computers or mobile devices, communication on social media and dating apps, content creation, time planning, geolocation, exchanging texts and images and receiving instant emotional feedback—all these phenomena have become commonplace".

Electronic culture is not the merespread of new information technology and its various applications. It modifies values, norms, and standards of behavior. The saying "if you aren't on the web, you don't exist" is nota joke at all.

As mentioned above, a special form of social capital is information capital. There are two types of information capital. The first one is individual information capital (IIC); the second public social capital. Individual information capital is an individual's knowledge and skills in an electronic society, the ability to use information for professional, educational, or recreational purposes. In its turn, there is professional, shared and massIIC. Professional IIC is that of individuals who create software, ensure information and social media security, etc. Shared IIC is the remit of professionals who teach others basic computer skills as well as how to use social media, e-mails, mobile banking, etc. In many countries, including Russia, special training centers have been established to teach retired members of communities computer skills (third-age universities…). Mass IIC is shared by everyone who regularly uses gadgets and personal computers to receive and process information, to communicate, to play video games, etc.

Like any other classification, the above is a simplification. For example, online courses and continuous communication make it possible for ordinary users, especially those from younger cohorts, to acquire programming skills. They can also learn how to download music and movies for free, bypassing the law. There is no insurmountable barrier between professional and massIIC. Still, the division of individual information capital into different categoriesis both reasonable and useful as long as digital inequalities in particular societies are concerned. It is important to understand how many and what IT specialists are trained in the country. This is becoming especially important as artificial intelligence is gaining momentum. Probably, even some IT specialists will be replaced by a powerful server or artificial intelligence.

4.2 Digital Inequality and the Global Community

Electronic culture, just like any other, inevitably generates certain forms of inequality. In this case, it is digital inequality, which has several manifestations.

The first one is age inequality. People of different generations do not familiarize themselves with the digital space at the same pace. This is an objective form of inequality, which also existed in the era of printed culture. Children, who could not read, were denied access to information. Today, on the contrary, the older generation may be less digitally literate than the younger ones. Before the era of electronic culture, the young acquired knowledge and learned skills, including that of using printed information sources, from their seniors. Now many adults learn how to use digital media from younger people. Remarkably, this happens not only at home.

The second type of digital inequality is that between societies and regions (Information Society…). In the Information Age, developed countries export research-intensive products: patents, technologies, complex devices. Developing countries and the Third World supply either raw materials for the produce of the First World or mass-production goods. Post-industrial countries can infinitely upscale production, whereas others need more materials and labor to produce more goods. At the same time, developed countries, for example, the US and the EU, outperform developing ones when it comes to industrial manufacturing and even agriculture. In bilateral trade with developed countries, the Third World has to follow the rules o the First. The West sells information products and technologies without forcing anyone to purchase them (Information Society…). Leading countries in the information and communication technology markets are trying to secure their superior positions. Regional digital inequality harms not only developing but also developed countries (Chen and Wellman 2020).

Regional digital inequality is closely linked to the monopoly on information trends. Information products are the keyto economies of scale. In combination with external network effects, this contributes to the monopoly over the information space

(Sagittarius 2003). The Western economic thought supports this view. In the information markets, monopolization is sustained by economies of scale related to both supply and demand (Shapiro and Varian 1999). The share of Google in the search engine market is 80%. The rapidly growing Chinese company Lenovo ranks first in the world in terms of PC production, yet its share is 16.9% (Kalustyan 2014). Information technology companies are growing through mergers and acquisitions.

The Digital Economy Report 2019 shows that the digital economy is concentrated in the US and China, whereas the rest of the world, especially Africa and Latin America, lags behind them. The US and China account for 90% of the market capitalization value of the world's 70 largest digital platforms, over 75% of the cloud computing market and 75% of all blockchain-related patents. By analyzing current rules and policies, the report predicts that this trend is likely to continue and contribute to increasing inequality (UNCTAD). The Sustainable Development Goals Report 2019 notes a huge gap between countries with access to the Internet: more than 80% of the population of developed countries have Internet access, compared with 45 and 20% in developing and least developed countries (Digital).

The third type of digital inequality is that in communication opportunities. There are two kinds of society in today's world: television society and Internet society. Members of the latter, especially young people and adolescents, have more opportunities to transmit and receive information than people in the former do. This type of disparity is similar in many respects to interregional inequality.

As to the fourth type, an information and financial oligarchy emerged in developed countries in the last quarter of the twentieth century to gain control over more than 90% of the planet's information and financial flows. (Information Society and Russia).

The fifth type of digital inequality stems from the power of public response. It is a wave destroying social space. Imaginary worlds of social reality are growing and multiplying to become real. Anomalies are becoming more numerous and "hotbeds of public response" are emerging. All this leads to conflicts and ruins the traditional space of social reality.

There are two perspectives on digital inequality. One holds that digital inequality is real, and a set of measures is needed to reduce it. The other treats digital inequality as a fait accompli (Global digital..., 2015). The World Bank recognizes digital inequality but stresses that it is being gradually overcome. I believe that digital inequality exists, and attempts to ignore it are nothing more than a manipulation aimed to conceal its scope. Something similar has always happened whenever inequality emerges.

For some time, Russia was a peripheral agent, an importer of information technologies and systems from other countries. Perestroika and the following uneasy decade almost destroyed the country's industrial potential and made Russia dependent on other countries for information technology.

As for digitalization, concerns about new types of inequality emerging in Russian society have been voiced since the early 2000s. Back in 2002, Vasily Kolesov wrote, "recent transformations in our society haveled to substantial social inequality. Concerns that new information technology will exacerbate this inequality merit special attention. Another urgent problem is the relationship between the state and society. It might be that the state in Russia is not in service of society but, regrettably, society is in service of the state. There are fears that information technology will cement this state of affairs" (Information Society and Russia).

Digitalization of society brings about the dispersion of values. Dispersed values are becoming the new norm and creating a "normal anomaly" (Kravchenko 2014). The fragmentation of society is reinforced by digital risks and the digital metamorphosis of society, which cause institutions to fail (Beck 2016). Institutional individualization is turning into a "radical form of individualization" (Beck 2002): young people seek to free themselves completely from institutional relationships, attachments to the local cultural context.

Under the influence of radical individualization, individualized forms of existence, which are passive in essence, become dominant. People are becoming increasingly isolated in the West and Russia (Puzanova 1998; Riesman 2001). A pronounced trend in the country is the individualization of leisure and unequal treatment of peers and seniors.

4.3 Transition to Sustainable Development and Social Technologies to Overcome Digital Inequality

Central to the progress of the world community is transition to sustainable development, which requires overcoming significant fluctuations in social development and reducing inter-country disparities, including digital inequalities. As early as June 2011, the UN recognized Internet access as an inalienable human right (Report of the Special Rapporteur... 2011).

The ways to overcome digital inequality and ensure sustainable development, which means economic progress and the use of artificial intelligence, are as follows.

Firstly, the need to reduce digital inequality has been emphasized by the UN and other international organizations. Certain steps have been taken to achieve that goal. For example, international funds have been established to support digitalization in the most deprived regions. International laws should be adopted to prevent the monopoly over the IT and IT-equipment markets.

Secondly, international efforts are needed to train IT professionals for developing countries. This also requires substantial financial support.

Thirdly, there is a need for targeted programs that combine IT development initiatives in developing countries with measures to support the local real sector.

5 Conclusion

A new form of disparity—digital inequality—has emerged. This inequality exists in each society. Its key manifestations include intergenerational and interregional inequality. Digitalization-related differences between countries are likely to grow.

Digital inequality is also observed in Russia. Since some other states faced the problem earlier than Russia, there is hope that the country will learn from their experience.

Concerted action to overcome digital inequality will ensure successful transition to sustainable development even amid the COVID-19 crisis.

References

Batool R, Sharif A, Islam T, Zaman K, Shoukry AM, Sharkawy MA, Gani S, Aamir A, Hishan SS (2019) Green is clean: the role of ICT in resource management. Environ Sci Pollut Res 26

Beck U (2016) The metamorphosis of the world. Polity Press, Cambridge, p 160

Behravesh N, Rocha EW (2020) Interim global economic forecast; IHS Markit: London. https://theinsiderstories.com/covid-19-recession-to-be-deeper-than-2008-2009/ (Date of visit: May 10, 2020)

Bell D (1974) The coming of post-industrial society. Harper colophon Books, New York

Bourdieu P (2002) Forms of capital. Econ Soc Electr Mag 3(5):60–74

Bushuev AV, Lezhenina LA (2012) Information capital in knowledge economics. Online J Sci 4

Busko M (2020) Cell phone crazy: anxiety linked with increased cell-phone dependence, abuse. Retrieved from http://cellphonecrazytech.blogspot.com/2009/08/effort-linked-withincreased-cell.html (Date of visit: May 10, 2020)

Chen W, Wellman B (2020) The global digital divide within and between countries. www.ITandSociety.org (Date of visit: April 1, 2020)

Digital watch obsterwatory. https://dig.watch/issues/sustainable-development (Date of visit: August, 10, 2020)

Encyclopedia of the Information Society: Electronic_Culture. http:// wiki.iis.ru/wiki/ (Date of visit: May 10, 2020)

Filipovich II (2017) The education system in the USA. Sci Bull SPM 4:99

Global digital divide "narrowing. http://news.bbc.co.uk/2/hi/technology/4296919.stm (Visit date: March 25, 2020)

In the US, the highest level of income inequality among Westesrn countries: UN News. https://news.un.org/ru/story/2018/06/1332802 (Date of visit: May 12, 2020)

Information Society (2004) Sat. Publishing House AST LLC, Moscow

Kerras H, Sánchez-Navarro JL, López-Becerra EI, de-Miguel Gomez MD (2020) The impact of the gender digital divide on sustainable development: comparative analysis between the european union and the Maghreb. https://www.mdpi.com/2071-1050/12/8/3347 (Date of visit: May 10, 2020)

Kolesova VP, Osmova MN (ed) (2002) Information society and Russia. Faculty of Economics, TEIS, 196p

Kalustyan DK (2014) The largest acquisitions in the ICT sector in early 2014. Bull Foreign Commercial Inf 1:126–127

Kholyavko A (2020) The state of 1% of the richest people in the world is more than the rest of the population. Vedomosti: https://www.vedomosti.ru/economics/articles/2020/01/20/821057-sostoyanie-bogateishih (Date of visit: May 10, 2020)

Kolesov VP (ed) Information society and Russia

Komleva NS, Shcherbakova EG (2014) Modern education system in European countries: problems and prospects. Bulletin of the Volga University named after V.N. Tatishchev 4(32):69–77

Kostyaev S (2020) Test for the insurance system: Novaya Gazeta. https://novayagazeta.ru/articles/2020/05/04/85211-ispytanie-dlya-strahovoy-sistemy (Date of visit: May 2, 2020)

Kravchenko SA (2014) Normal Anomia: outlines of the concept. Sociol Stud 8:3–10

Krivosheev VV (2009) Short life projects: manifestation of anomia in modern society. Sociol Res 3:57–67

Lapin NI, Ilyin VA, Morev MV (2020) Extreme inequalities and the social state (part 2). Sociol Res 46(2):20–30

La Rue F (2011) Report of the special rapporteur on the promotion and protection of the right to freedom of opinion and expression. United Antions General Assembley. Human Rights Council Seventeenth session. 16 May 2011. http://www2.ohchr.org/english/bodies/hrcouncil/docs/17session/A.HRC.17.27_en.pdf (Date of visit: August, 8, 2020)

McLuhan M (2003a) The Gutenberg galaxy. Dmitry Burago Publishing House, Kiev Nika CenterElga

McLuhan M (2003b) Understanding media: the extensions of man routledge. Zhukovsky, Kuchkovo Field, Moscow

Mikhneva SG (2003) Intellectualization of the economy: innovative production and human capital. Innovations 1

Nosovsky GV, Fomenko AT (1997) Mathematical chronology of biblical events/Ros. Science, Academy of Sciences

Puzanova JV (1998) The problem of loneliness: the sociological aspect. Publishing House of the Peoples' Friendship University of Russia, Moscow

Ramade AS, Cabrera A, Roberto MC (2011) The new digital divide: the confluence of broadband penetration, sustainable development, technology adoption and community participation. J Inf Technol Dev (Inf Tech Dev) 18(4):1–9

Rezaev AV, Starikov VS, Tregubova ND (2020) Sociology in the era of "artificial sociality": the search for new grounds. Sociol Res 46(2):4

Riesman D (2001) The lonely crowd: a study of the changing American character. Yale University Press, New Haven

Shapiro C, Varian H (1999) Information rules: a strategic guide to the network economy. Harvard Business School Press, Boston. lib.mexmat.ru/books/59593 (Date of visit: 25 Mar 2020)

Siwar C, Abdulai A-M (2013) Sustainable development and the digital divide among OIC countries: towards a collaborative digital approach. In: Digital literacy: concepts, methodologies, tools, and applications, p 13

Strelets IA (2003) New economy and information technology. Exam, p 144

Svyatlovsky VV (2019) History of socialism. Yurite Publishing House, Moscow, pp 67–68

Third-age universities in Russia and abroad. https://верити.рф/gerontologiya. (Date of visit: April 1, 2020)

Turkle S (2011) Alone together: why we expect more from technology and less from each other. Basic Books, New York, p 143

UNCTAD Report Recommends Actions to Reduce Digital Divide. https://sdg.iisd.org/news/unctad-report-recommends-actions-to-reduce-digital-divide/ (Date of visit: July 30, 2020)

Yu NR, Roshchupkin GV (2006) Features of the higher education system in the UK. Pedagogy Phys Cult Sports 12:141–143

Yadova VA (ed) (2001) Russia: a transforming society. Moscow

Vlalimir Venuaminovich Krivosheev Doctor of Sociological Sciences, Professor of the Institute of Humanities, Immanuel Kant *Baltic Federal University, st. A. Nevsky, 14,* Kaliningrad, 236,041, Russia. Author of more than 120 scientific publications. Research interests: problems of social anomaly, sustainable development, overcoming digital inequality. E-mail: VKrivosheev@kantiana.ru

Design Thinking and Collaborative Digital Platforms: Innovative Tools for Co-creating Sustainability Solutions

Diane Pruneau, Viktor Freiman, Michel T. Léger, Liliane Dionne, Vincent Richard, and Anne-Marie Laroche

Abstract Design thinking, which emphasizes users' needs and rapid prototyping, is a promising avenue in the field of sustainability. Used by international organizations such as IDEO.org and d.School, design thinking has yielded promising results in terms of sustainability solutions. Employed at strategic moments during the design thinking process, digital tools (ICT) can also facilitate the collaborative and co-constructive aspects of this approach to problem solving. Adults from four case studies worked on different problems, from adapting the University of Ottawa campus to international students' needs, to improving a drinking water problem in Quebec City, to adapting in the face of flooding in Morocco and to decontaminating of groundwater in a New Brunswick village. The researchers studied the affordances (perceived potentialities) of design thinking and collaborative digital tools (*Facebook*, *Realtime Board*, *Knowledge Forum*) during the process of solving of each case study problem. Interviews (individual and group) as well as evidence of participants' online work were used to identify several affordances of design thinking, namely broader understanding of a problem, awareness, identification of end users' real needs, caring towards users, production of multiple appropriate solutions and development of problem-solving skills. Different types of ICT have been employed

D. Pruneau (✉) · V. Freiman · M. T. Léger · A.-M. Laroche
Université de Moncton, Moncton, Canada
e-mail: diane.pruneau@umoncton.ca

V. Freiman
e-mail: viktor.freiman@umoncton.ca

M. T. Léger
e-mail: michel.leger@umoncton.ca

A.-M. Laroche
e-mail: anne-marie.laroche@umoncton.ca

L. Dionne
Université d'Ottawa, Ottawa, Canada
e-mail: ldionne@uottawa.ca

V. Richard
Université Laval, Quebec City, Canada
e-mail: vincent.richard@fse.ulaval.ca

to provide a remote support for participants at the various stages of design thinking: *Facebook* to share images of environmental problems and proposed solutions, *Realtime Board* to visualize a problem and to remember its details, and *Knowledge Forum* to analyze a problem more deeply and facilitate discussion around suggested solutions and prototypes.

Keywords Design thinking · Sustainable development · Collaborative digital tools · Environmental education

1 Introduction

Many environmental problems affect living environments and their residents: floods, lack of biodiversity, unsafe drinking water, etc. Faced with these problems, citizens are looking for solutions to improve their living conditions. Sometimes, they are accompanied by facilitators (teachers, animators) during their quest for solutions. What are the key interventions that promote effective support for these citizens as they search for solutions that are both feasible and adapted to their specific needs? Environmental problems are complex, potentially involving multiple causes, actors, health risks as well as social, economic, and ecological conditions that can be either unfavorable or critical. In this regard, numerous international organizations are currently using design thinking to address environmental problems. Design thinking is a creative problem-solving approach, structured in such a way as to pedagogically assist social groups in their efforts to analyze local problems and come up with realistic solutions. Similarly, in some cases, facilitators suggest the use of collaborative digital tools such as Information and Communication Technologies (ICT), which could help resolvers to think collectively, to represent situations, and to design and to discuss solutions (Innovation Training 2020). Examples of such ICT are *Realtime Board* (RTB[1]), *Facebook* (FB), or *Knowledge Forum* (KF).

An in-depth analysis of the origins of design-thinking conducted by Gamba (2017), provides references going back to 1960–70 with the emergence of the field of research on design as a way of thinking. In the 1980s, the field was enriched by the work of Cross (1982), who spoke of a 'designerly' way of knowing, later coined 'design thinking' by Rowe (1987), referring to the approaches used by architects and urbanists which focused on 'user-centered design'. In the early 2000s, design thinking was popularized in the fields of social innovation and education by Tim Brown and David Kelley, founders of the IDEO design and innovation firm. According to IDEO.org (2012), design thinking is a human-centered approach which relies on innovation, collaboration, and creativity to solve a multitude of social or environmental problems. It is a creative and collaborative way of working in which intuition matters, solutions are numerous, experimentation happens quickly, failures are given value, and, above all, the needs of users are considered (Brown 2009).

---

[1] There was a change in the product's name during our research; it is now called *Miro,* https://miro.com/.

A creative and analytical approach, design thinking is an amalgam of concepts in engineering, design, arts, social sciences, and business. By the end of 2010s, it is also considered as a collective intelligence approach that places humans and their needs at the center of a process of co-creativity, in which frequent feedback is invited from the end users of the solutions (Darbellay et al. 2017).

In organizations such as IDEO.org (in the United States), INDEX (in Denmark), Hasso Plattner Institute (in Germany) and *Design for Change* (in India), it is possible, with design thinking, to create products or experiences that improve the lives of communities. For example, in Colombia, as part of the *Cigabionica* project, inspired by *Design for Change*, teenagers made planes out of recycled materials to solve a problem of lack of vegetation in their city. Their aircraft transported and dropped seeds on arid lands, bringing local biodiversity back to the city (Design for Change 2018). There are many more such examples of sustainability projects on the web, all made successful thanks to design thinking. Examining these success stories, we wonder what, exactly, are the pedagogical impacts of design thinking. We also ask what type of tools could facilitate the design thinking process.

In design thinking, facilitators generally use traditional tools, such as Post-it Notes, whiteboards, role play, etc. (Evangelische Schule Berlin Zentrum 2017). However, nowadays, various collaborative digital tools (ICT) are widely available to facilitate the process of co-creating solutions during design thinking. Using ICT, solvers can share information (*Empathy Maps*), synthesize information (*Popplet*), propose and comment on ideas (*Padlet*), choose ideas (*Loomio*), draw prototypes (*Tinkercad*), plan (*Wrike*) and communicate (*Facebook*) (Pruneau and Langis 2015). The technological tools used in design thinking and their impact on the solution process have, however, been scarcely investigated (Gericke et al. 2012) in terms of their innovative potential for sustainability solutions.

Our research was aimed at answering the following question: What are the affordances of design thinking and of digital tools in the problem-solving processes? The concept of "affordances" corresponds here to the values and potentialities of action, perceived by a living organism, for a given object (Gibson 1979). Luyat and Regia-Corte (2009) describe affordances as a kind of adjustment between perception and action. Based on this concept, we will discuss four case studies which contributed to investigating the research question at hand. In the first case study, Moroccan women were accompanied by the research team as they were guided through the design thinking process and, on a FB group, while they sought solutions for adapting to the poor quality of water in their community because of flooding. In this first case, reaching Sustainable Development Goals (SDG) 3 and 6 (health and well-being, and water quality) were mostly targeted (United Nations 2015). In a second case, a group of students from the Faculty of Education at the University of Ottawa were accompanied by researchers as they used design thinking and RTB (a collaborative digital platform) to improve their campus environment in order to better meet international students' needs. Here, SDG 11 (sustainable cities and communities) was mainly targeted (United Nations 2015). In the third case study, university students from the Faculty of Education at Université Laval used design thinking and KF in a science education undergraduate course, as they tried to solve a drinking water problem in Quebec City (project focused on SDG 6, water quality). Finally, our fourth case involved engineering students from the Université de Moncton who used design thinking to solve a groundwater arsenic problem (SDGs 3 and 6). In this last case, students chose not to integrate ICT platform suggested by facilitators from our research team, preferring instead to use their usual communication and collaboration tools, such as *Google Docs*, *Google Drive* and *Messenger*.

This paper analyses the affordances (perceived potentialities) of design thinking and of three collaborative ICT (FB, RTB and KF), as perceived by resolvers and facilitators, in the four distinct case studies. We believe our research to be important because environmental problems are increasingly affecting the well-being of populations all over the world, and innovative and sustainable solutions are needed. Key interventions must be found to transform the environment, in collaboration with citizens, and to enhance the practices that shape its forms and functions. Could design thinking be one of these key interventions? If so, in what way? Can new technologies help resolvers achieve their goals regarding complex issues such as environmental problems? If so, what are the specific affordances of both, design thinking, and ICT? Globally, our research project was related to technology and innovation capacity-building mechanisms and looked at the ways to achieve SDG 17: Strengthen the means of implementation and revitalize the global partnership for sustainable development (United Nations 2015). Indeed, according to Salvia et al. (2019), SDG 17 is one of the most challenging and the least researched goals.

2 Design Thinking, ICT Tools, and Their Affordances

2.1 Design Thinking

Design thinking applies the designer's sensitivities and methods to solve complex problems. Indeed, it is used by designers to confront complex problems by generating various solutions that they test and gradually improve. Using a rigorous process and well-defined tools, design thinking, sometimes divergent and sometimes convergent, uses both creative and analytical modes of reasoning (Lietdka, 2015). Design thinking takes place according to definite but non-linear steps in which back-and-forth actions (iteration) intersect. The steps we used in our study (see Fig. 1) are inspired by Brown (2009) and Scheer et al. (2012):

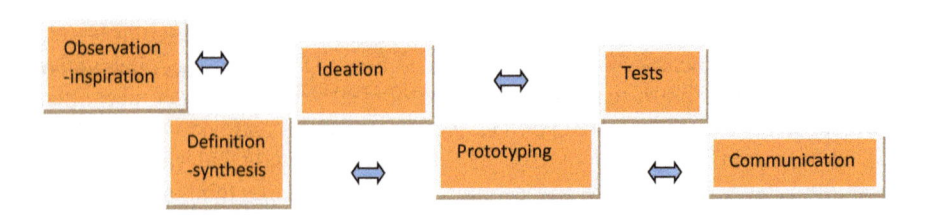

Fig. 1 The stages of design thinking

1. *Observation-inspiration*: an ethnographic survey is done to understand with empathy the people concerned by the problem (the users) and the situation.
2. *Definition-synthesis* : the problem is defined several times. Information and various perspectives related to the problem are researched. The information is synthesized to pose the conceptual challenge in a few statements.
3. *Ideation*: many ideas of solutions are formulated, and some are chosen for building prototypes.
4. *Prototyping*: prototypes are quickly constructed to illustrate the ideas, to share them with others and to evaluate their potential.
5. *Tests*: prototypes are assessed by collecting the opinions of users and experts. The winning prototypes are refined.
6. *Communication*: the final solution is made public.

2.2 ICT Tools

A literature review was conducted to look for possible benefits of ICT tools and digital environments in support of the design-thinking approach (Freiman et al 2020). The review showed considerable disparities between theoretical and conceptual frameworks as well as with the methods involved. We found that *Facebook*, *Twitter* and *YouTube* were the most frequently referenced digital tools used in design thinking studies. The problem-solving approaches mobilized by target populations were not always clearly explained by the authors. Several studies pointed at the benefits of digital technologies in problem-solving, especially related to environmental problems. Technologies were mainly used to promote sharing and dissemination of information, and to facilitate the emergence of shared and effective solutions within the community (Barborska-Narozny et al. 2016). Digital tools played a central role in project coordination, popularization and citizen engagement (Beche 2012). Finally, ICT allowed the development of competencies such as critical thinking, collaboration, and problem-solving skills, while helping people ask questions, review planning and materials and discuss problems (Pinzón and Nova 2018; Squires 2014).

2.3 Affordance

Affordance, a complex construct first introduced by Gibson in his works in 1960-70 s, is now widely used in different fields and contexts and could bear different meanings while still being difficult to define (Luyat and Regia-Corte 2009). We need to clarify this construct in the context of our study. Indeed, Luyat and Regia-Corte (2009) argue that affordance, in Gibson's sense, goes beyond direct perception of potentialities of actions by a human (or, in a larger sense, by all animals), while representing a rather complex and dynamic relationship between the properties of the human (animal) and the properties of the environment. From a different perspective, affordance could be viewed as 'disposition', as realized by a certain behavior—a sort of actualization of affordance.

Or, from a different perspective still, affordance could be seen as an 'opportunity for action' determined by the 'animal—environment' system, as a possibility of action (source). While recognizing the complexity of reconciling (or not) of different approaches related to a multitude of perspectives, we build our analysis around a vision of affordance as a means to interact with the environment, beyond 'action – as physical motion' and towards a larger ecological understanding of social interaction (interaction with others). Moreover, affordance is usually described in relation to the perceiver (Hammond 2010, p. 206). From the affordance's perspective, the possibility of action is historically and culturally embedded into the web of social activities of praxis (Bærentsen and Trettvik 2002; cited in Martinovic et al. 2013), shaped by past experience and context, while being "conceptually sophisticated, and might need to be signposted by peers and teachers" (Hammond 2010, cited by Martinovic et al. 2013, p. 220).

Furthermore, according to Walsh (2012), cited by Martinovic et al. (2013), affordance's landscape can be described as "co-constituting" and "commingled" by a reciprocity between organisms and their affordances. In any given context, there could be a variety of affordances (some could be directly perceived, others discovered, and/or constructed). For example, in the context of Computer-Supported Collaborative Learning, Jeong and Hmelo-Silver (2016, p. 247) suggest a framework built around the following seven affordances: (1) engage in a joint task; (2) communicate; (3) share resources; (4) engage in productive collaborative learning processes; (5) engage in co-construction; (6) monitor and regulate collaborative learning; and (7) find and build groups and communities. In our study, we adopt an inductive approach to describe affordances as being emerged in a complex process of solving a problem using design thinking approach and ICT tools.

3 Methodology

Considering the exploratory character of our research, a qualitative/constructivist methodology and a multi-cases method were adopted (Shkedi 2005). The purpose of the research was to identify affordances of design thinking and of some collaborative digital tools, in solving problems associated with the built environment. These affordances or perceived potentialities were sought from participants and facilitators in four case studies where adults were accompanied in their application of design thinking using at least one ICT to solve a local environmental problem. The stages of design thinking (Observation-inspiration, Definition-synthesis, Ideation, Prototyping, Testing and Communication) were carried out with the participants in each of the four cases, as the researchers invited participants to identify users' needs, to define a conceptual challenge, to propose and choose solutions, to prototype and test certain ideas and to communicate the best solution. Regarding the study of the affordances of design thinking and ICT, the resolvers and the facilitators in all four cases were asked to explain how they used the design thinking approach and the digital tools at their disposal, giving insight into what potentialities of actions or learning could result. The data sources included written and graphic traces left by the participants on the various ICT platforms used in the study (i.e. screenshots), as well as individual and group interviews with the solvers and with the facilitators at various moments in each study. The qualitative analysis of the data, carried out by two researchers in each site, mainly involved the extraction of themes as described by Paillé and Mucchielli (2008).

In the *first case*, a design thinking approach and Facebook were used to support ten Moroccan women with little education in their efforts to solve a problem of unsanitary drinking water caused by the flooding of the Ourika River (see Pruneau et al. 2017). The interventions with the women took place over a period of three years. Workshops followed the stages of design thinking. Electronic tablets and *Facebook* (FB) were used as networking tools when women were separated by long distances. Using videos and photos, the women shared their flood experience on FB, then chose collectively to focus and later work to solve the associated problem of unsafe drinking water. Along the way, the needs of the users collected consisted of the diseases contracted by the children and the elderly during the floods as the only source of potable water became contaminated by waste transported by these floods. The conceptual challenge was formulated as follows: How could we reduce the amount of waste on the ground and in the water in the region? In the workshops and on FB, the participant gradually created and implemented various solutions for flood adaptation, such as operating an electronic warning system, making homemade filters (to clean their water), making composters (to reduce their household waste), creating recycled jewelry with their plastic waste, and starting a waste reuse cooperative. All solutions were shared, assessed and further improved during the in-person workshops and the exchanges on FB.

The *second case* dealt with a group of 7 pre-service teachers at the University of Ottawa (the resolvers) working collaboratively on how to make their university campus more sustainable for the well-being of international students (the users). The collaborative digital platform RealtimeBoard (RTB) was chosen to support the participants in this process. The solvers identified various needs through talking with international students, specifically related to the university campus. Such needs included a lack of vegetation, works of art, spaces to work and socialize, places to meet people in their community, entertainment venues and signposts or places representing their culture. Following the steps of the design thinking process using RTB, participants met weekly on the virtual platform working to find ways in which to bring a better sense of well-being to the international student community. Participants proposed solutions such as the Megwetch park (*thank you* in the First Nation language of Anishinaabe) and a specific area on campus equipped with the following design elements: pergolas, trees, flowers, pools with fountain, swings, ping-pong tables, a giant chess game, tables for studying, an outdoor classroom, shady spots to socialize, trails, works of art and a garden displaying international flags.

The *third case* involved pre-service teachers (the resolvers) from Université Laval. These participants were given a water quality problem to solve within Quebec City.

Eighty-four students worked collaboratively in groups of 4 using the digital platform Knowledge Forum (KF) during an entire semester. As part of their homework, students made regular entries on KF. The main needs identified among users were twofold: on the one hand, students found that citizens lacked general information (especially farmers) as to the existence and seriousness of the problem at hand, the over-consumption of drinking water and, as a result, the need for better protection of watersheds supplying Quebec City with drinking water. Following the four classes devoted to design thinking as they made weekly entries on KF, students came up with several proposals attuned to users' concerns. For instance, some proposed a teaching guide to educate the children as to responsible use of drinking water resources, while others proposed an educational program to be deployed in Quebec City parks in order to raise awareness about the overconsumption of water. Yet another group suggested an educational kit to be used by a "beach squad" in order to educate navigators on the fragility of local streams. Some "political" solutions were also proposed as some teams stressed that government regulation may be part of the problem and should be adapted to protect sources of drinking water.

The *fourth case* looked at using the design thinking approach to solve a problem concerning high arsenic levels in the groundwater of a village near the university. Students from the civil engineering undergraduate program at the Université de Moncton were asked to address this environmental problem as part of their requirements for a compulsory water treatment class. In this case study, we divided all of the registered students ($N = 17$) into two distinct groups, the first tasked with finding solutions to the problem at hand through typical engineering procedures ($n = 8$), and the second charged with the same task while following a design thinking approach ($n = 9$). The design thinking group first met with citizens from the village to record their concerns around the water issue, a step unique to design thinking which the other group of students did not take. Despite knowing how the residents felt about the problem and what they thought would help from a consumer perspective, the conceptual challenge was to find a solution to treat the arsenic-laden water of wells in the Cap-de-Cocagne area. According to the students from the design thinking group, it was important to first verify whether the residents in fact had higher than acceptable concentrations of arsenic in their wells. Following testing, the students realized that arsenic concentrations varied greatly from one well to another. It was first recommended to each well owner to have a sample of the water in their well assessed to determine whether it was necessary to proceed with a technical solution. Talking with residents also helped to foster a better understanding of the problem, from both the consumer's side as well as the resolver's side, something the other group of students treating the problem did not experience. The design thinking students ended up proposing the following technical solutions: a reverse osmosis treatment system, a distillation system, the distribution of water bottles, an anion exchange system and an iron oxide absorption system. All of these solutions were feasible and offered the residents different approaches depending mainly on cost as well as on the priorities of each user.

4 Results

4.1 Affordances of Design Thinking

In the Moroccan case study, various affordances of design thinking were identified by participating women who used design thinking to solve a local environmental problem. University students who participated in the other three cases also demonstrated observable affordances while seeking to solve their respective problems through design thinking. In fact, design thinking was seen as a tool that provided a deeper (expanded) understanding and awareness of the problem at hand in each of the four study cases. For instance, Moroccan women mentioned that before using design thinking, they did not know the causes of the floods, nor did they fully appreciate the health risks incurred as a result of domestic waste being dumped in the river. For their part, Ottawa students did not suspect most of the needs of international students for university campus development before applying design thinking. In the case of Quebec City students, they were not aware of the drinking water issues in their city and did not feel concerned by the issue of contaminated drinking water because, among other things, they considered potable water to be abundant as a resource (observing a multitude of drinking fountains on campus as well as an abundant supply of bottled water).

Moroccan women, Ottawa students and future engineers also considered design thinking to be an effective tool for identifying the real needs of people in a given community. They felt that the process of design thinking helped them to identify authentic problems, which facilitated the formulation of appropriate solutions. For example, during the Observation-inspiration phase, the future engineers in the Moncton case study understood that all the users did not have the same problem regarding the presence of arsenic in their wells, which led them to the solution of initially evaluating the concentration of arsenic in the wells of citizens, before applying other engineering solutions. In Morocco, awareness of the needs of flood victims made it possible to prioritize issues with the quality of water as a sub-problem and to identify the local sources of contamination, that is to say the waste on the ground. Quebec City students, for their part, emphasized how the design thinking process allowed them to deal with the delicate issue of documenting the users' needs. For most students, the interviews with the users proved to be very demanding and, once the exercise was completed in the time available, many students had doubts about their ability to adequately report on documented needs. Design thinking allowed these same students to become aware of the complexity of the issue.

Moroccan women and Ottawa students also talked about mutual learning fostered by design thinking. Both groups indicated that knowledge consolidation occurred through the information received by and the questions asked of other solvers. The same is true for Quebec City students who pointed out that the sharing and organization of ideas throughout the process led to learning and improved understanding of the subject at hand.

Empathy is another affordance attributed to design thinking in the Moroccan case, as well as in Ottawa students and future engineers. A feeling of sympathy for the users' needs emerged. The solvers wanted to find solutions "for them and with them". For some Quebec City students, this empathy was coupled with a feeling of helplessness associated with the complexity of the problem.

Ottawa and Quebec City students, as well as future engineers, also believed that design thinking, led to better more numerous solutions, which were also more original given the goal of meeting users' needs. Design thinking seemed to lead to more involvement on the part of the solvers and a higher level of motivation to find solutions adapted to the users. According to these participants, this approach also allowed for complementarity and enrichment of ideas. One participant said: "Together, we did better!" In terms of creativity, design thinking seemed to support an open mind to new and innovative ideas, "[letting] the imagination go!", as another participant put it. Similarly, according to Ottawa students, prototyping brought ideas to life, contributing to the quality of the proposed solutions. On participant shared that "visual support [gave] us a better idea of what the space would look like". Finally, according to Quebec City students, the design thinking approach also led to valid solutions which highlighted their expertise as education specialists.

According to all research participants, from involved Moroccan women, Ottawa students, Quebec City students and future engineers from the Université de Moncton, design thinking seems to improve communication. More specifically, all four groups said that they learned to talk to people within the team and in the communities they were studying. The group of solvers becomes more cohesive, as knowledge of the people's needs and a desire for action develops. Team members gradually converged towards the common goal of solving the problem.

The following list of affordances emerged from some groups:

- development of general problem-solving skills (Moroccan women, Quebec City students and future engineers),
- development of collaborative skills, dialogue and discussion, critical thinking and analytical skills (Ottawa and Quebec City students),
- encouragement of attitudes of optimism and perseverance (future engineers),
- support of a learning process related to environmental issues (Quebec City students).

Referring to these affordances, Ottawa students, Quebec City students and future engineers all expressed a desire to reuse design thinking in their work.

4.2 Affordances of Technological Tools

Specific digital tools were used with three of the four cases presented: FB in the case of Moroccan women (on the flood problem), RTB in the case of Ottawa students (on sustainable campus development) and KF in the case of students from Quebec City (on drinking water contamination). In all three of these cases, solvers needed time at the beginning to familiarize themselves with the respective collaborative ICT used.

4.2.1 Facebook

Facebook (FB) is an online social network that allows users to post images, photos, videos, files and documents, exchange messages, create groups and use a variety of applications. The FB page provides a continuing list of members' entries composed of short text messages (news feed), illustrated by images, emoticons, videos, and Internet links. In the case of the Moroccan project, the news feed was created by entries provided jointly on the FB group by the facilitators and by the solvers. Along the way, FB's affordances were discovered and then gradually exploited by both parties, namely the facilitators and the participants. Thus, the various dimensions of the problem and the multiple solutions were documented and commented on, one at a time, by all members of the FB group.

More specifically, the FB group page was used to share and to comment on images of floods and waste in the project areas (see Fig. 2), images of solutions found by members (by themselves or on the Internet), pictures of the prototypes, solvers' questions and facilitators' advices during the testing of the prototypes (compost, jewels), planning suggestions (for the testing the prototypes) and posts of encouragement regarding the pursuit of the work.

According to participants (both solvers and facilitators), the Facebook group presented the following affordances:

- communication tool during floods,
- a tool for progressively defining the problem (one aspect at a time),
- learning tool (how to compost, how to craft, possible solutions towards sustainability),
- empowerment tool (encouragement to succeed),
- tool for sharing and critical assessment of solutions and prototypes
- planning tool (workshops, exhibitions, compost and jewelry sales, the management of the cooperative, etc.),
- decision-making tool (choice of workshops and products for sale),
- tool for building and maintaining solvers' team (to get to know each other personally to work together).

It should be mentioned that the FB page, growing over time with each added publication, did not allow for an overall view of the broader problem of flooding or the sub-problem of water contamination. FB, by its structure and its tools, made it possible to gradually develop information and opinions on each aspect of the complex problems at hand, addressing their causes, places, impacts, people involved and risks. Similarly, the solutions and prototypes were built and improved in person and then on Facebook, which allowed the participants to put the problem aside, to reflect and then to come back with new solutions. Facebook has also proven itself as a powerful communication tool, allowing participants to get to know each other

Fig. 2 Example from FB

and the facilitators. Online sharing, at every stage of the design thinking process, built upon the work done *in s*itu with participants, while regularly stimulating their interest for the problems they were studying. One of the advantages of Facebook for the participating Moroccan women was their ability to view real-time photographs or videos of the floods in the six villages as well as view representations of solutions and prototypes as they were developed. These images seem to bring the problems and solutions to life, making them real in the eyes of the participants and arousing their interest in both the problem and possible solutions.

4.2.2 Realtime Board

RealTime Board (RTB) is an online whiteboard allowing users to collaborate, write and draw on a shared screen (Squires 2014). An historical function keeps record of any change. Tables, concept networks, a chat, stickers, emoticons and comment bubbles are available. On RTB, resolvers were invited to represent each user's needs in a network of concepts (see Fig. 3), to collaborate online, to co-build an "empathy map", to synthesize all users' needs, to propose conceptual challenges, to vote for a particular challenge, to offer solutions, to generate new ideas, to evaluate and transform ideas, to vote for their conception of best solution, to share pictures of the prototypes, to assess the final solution and to compile users' comments about the final prototype. Even if RTB was reported by the participants as being initially difficult to master, it seemed that it gradually allowed everyone to converge towards a common goal in the search for solutions.

The main affordances students identified for RTB are:

- follow the progress of the project (get a general and specific looks of the problem);
- find and share information about the problem,
- propose, sort out and remove solutions,
- dvance the project in class and at a distance (hybrid approach, continuous contribution seen by each participant);
- learn what other colleagues do and say (find out others' viewpoint);
- react to colleagues' work (use of emoticons to assess suggested solutions).

RTB appears as a powerful collaborative working tool, allowing to visually represent and synthesize the various aspects of a problem (on the same page), the multiple solutions (on the same page), and the various prototypes (on the same page). Like

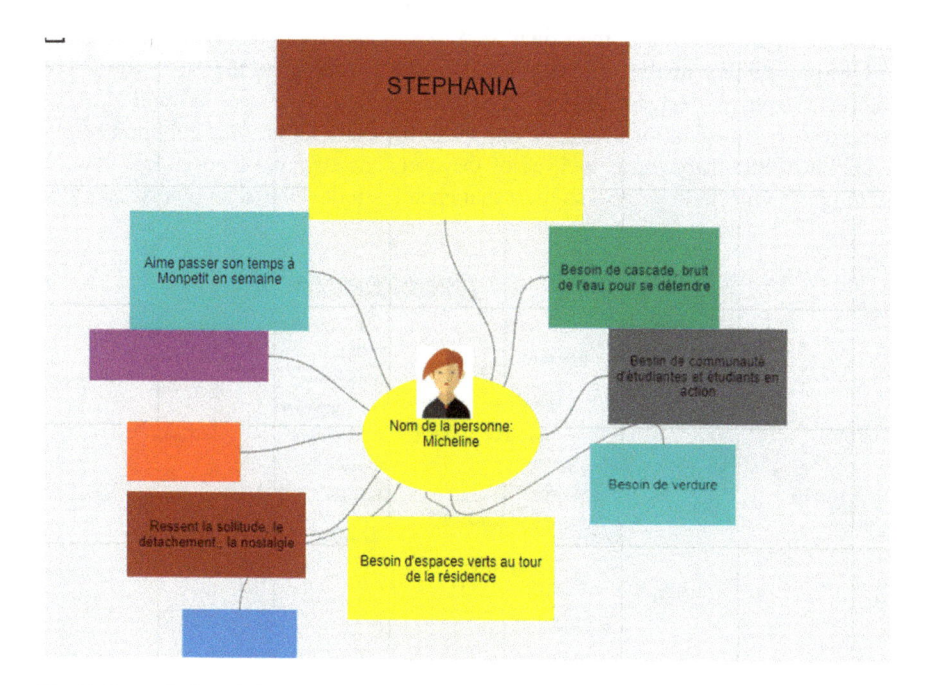

Fig. 3 Example from RTB

FB, it can be given the roles of progressive problem definition, sharing and criticism of solutions and prototypes, and building the team of solvers. Unlike FB, RTB is less of a spontaneous and ad hoc communication tool, but rather offers more of a systemic representation of a given problem. It allows information found about a problem and its proposed solutions to be gathered and remembered, which is essential because the complexity of environmental problems makes it difficult to recall all of the relevant information. By having a complete representation of a problem and its possible solutions on the screen, it is possible, outside of in situ meetings, to continue collaboratively working and reflecting on the issue. Thus, RTB relies less on images and more on conceptual tables and networks to capture and analyze problems.

4.2.3 Knowledge Forum

Knowledge Forum 6.0 (KF), a "discussion forum" tool, has been used with Quebec City students to share their perspectives, ideas and search result (information and data). Students also used KF to identify and develop possible solutions, discuss and assess them, and choose one for their prototype. The choice of this digital tool by the research team was motivated by its simplicity for the users, the availability of numerous writing aids (scaffolding, key words, feedback), as well as the quality of the representational tools for co-elaboration of ideas (structure of "conceptual map"). In addition, the tool is particularly flexible, allowing teams to easily exchange in a variety of ways (text tracking, hyperlinks, images, animations, videos) and to build a problem representation of their own.

Our preliminary analysis reveals that, for the participants, the KF affordances emerging from a design thinking approach seemed to evolve around the co-construction of discussions by each team and the elaboration of a common representation of the problem, as well as of the users' needs and of possible solutions (see Fig. 4). Our findings also suggest that each team developed its own strategy of

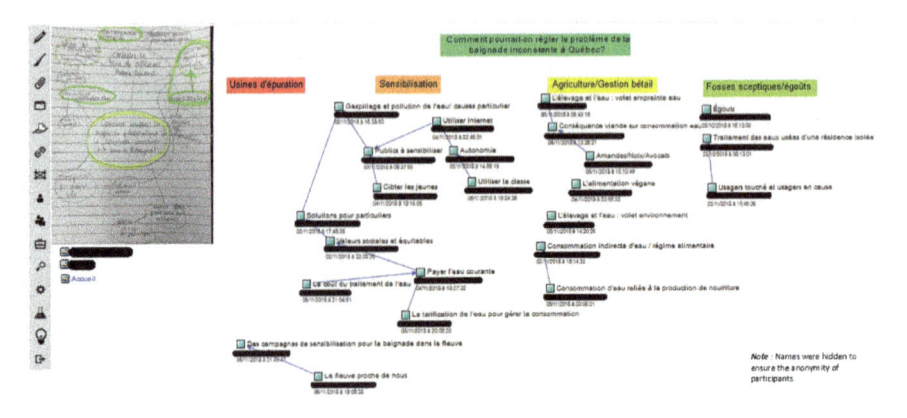

Fig. 4 Example from knowledge forum

exploring the problem and of suggesting solutions. At the beginning of the project, students were having difficulty grasping the nature of the problem. The problem being somewhat less accessible, the students did not have a clear idea of how to approach the development of possible solutions. However, in our view, a more explicit connection seemed to eventually emerge between each team member's contributions to the discussion and the stages of design thinking. Furthermore, this connection seemed to be afforded by the KF knowledge-building organization and structure.

It also seems that all the participants contributed, although in their own way, to moving the discussion forward, to organizing it in such a way that it went beyond the connections with the stages of design thinking, leading to more concrete and meaningful results rather than simply becoming sterile argumentation. Indeed, considering a) that the discussion on the forum took place over thirteen weeks, and b) that each member of the team had to contribute several times for each stage, we expected that the ideas addressed in the contributions made at the beginning of the approach would be very different from the ideas discussed in the last stages of the process. However, we found that, in general, the elements of discussion in the first stages of the design thinking process were again summoned when thinking about the solution of the problem. This leads us to believe that KF played a relatively strong structuring role in the team's thought process. In addition, our analysis suggests that the teams used what we might call "pivots", namely images, titles or pictograms that helped guide the contributions to the discussion. These pivots seemed both to serve as "summaries", allowing team members to navigate through the sometimes dense flow of the discussion, as well as markers for team members to adequately choose where they placed their contribution on the forum.

From the researchers' perspective, KF is a cognitive/collaborative tool aimed at enabling the development of a common representation of the problem, while being a sharing and discussion forum as well as a tool to keep tracks of the work-in-progress. From the affordances' perspective, KF seemed to serve as a tool for resolvers, offering them a variety of options to organize their ideas and a more holistic perspective in teamworking. However, following resolvers' comments, KF appeared to be quite challenging in terms of appropriation of its affordances.

5 Discussion

Concerning the use of design thinking as an approach to solving environmental problems, the following affordances seemed to emerge from our data across all four cases: broader understanding of the problems, increased awareness, clearer identification of the real users' needs, mutual learning, empathy towards users, better articulation of solutions which become more numerous and adequate. Because of its emphasis on empathy, design thinking seems to "humanize environmental problems", which may make them more tangible and more easily understood through the perspective of their impacts on people. The mobilization of participants to improve the studied situations increases as the problem space expands and as the teams actively collaborate to propose and improve solutions. The problem space widens to include not only scientific aspects, but also social aspects of problems (Pruneau et al. 2009). The problem space is also rich in terms of the users' emotions, of the perceived consequences and risks to humans, which in turn contributes in shaping the proposed solutions. Having a better understanding of the people's needs through the design thinking process, the participants really wanted to find effective solutions in order to genuinely help the people they talked to.

Based on our results from all four case studies, it seems that solutions emerging from design thinking process can be varied, but not necessarily original. The creativity and technical knowledge of the participants, the facilitators' interventions (which may more or less encourage novelty) and the strategies used during Ideation and Prototyping could influence the degree of innovation present in the solutions. Our analysis reveals that the solvers' final solutions very often meet the users' needs, and that they are not identical to the initial spontaneously emergent solutions. In-depth knowledge of the situations and of users' needs can lead to appropriate solutions, which are not always predictable, especially at the beginning of the study of problems. This emergence of new solutions along the way could be an element of transformational creativity. Boden (1999) defines creativity as the generation of ideas that are both novel and valuable. Boden (2009) also explains transformational creativity as the generation of new ideas by altering usual procedures and approaches, which is the case with design thinking.

Should collaborative digital tools be used during the stages of design thinking? Our study of collaborative digital platforms and social media (RTB, KF and FB) with adults, during design thinking, seems to demonstrate several beneficial affordances:

- visual, synthetic and organized representation of complex problems (in textual, pictorial and schematic format, using various functionalities available online, which serve as conceptual and collaborative illustrations and points of support);
- collection, analysis, processing and dynamic and efficient sharing of available information (structuring, restructuring, development, critical analysis, online backup and dissemination of collaboration between team members at different stages of the project; problem solving);
- effective management of the resolution process (consensual identification of the harmful elements of a problem and the conceptual challenge to be tackled);
- source of inspiration, reflection, support and motivation to continue working from a distance (between face-to-face meetings);
- support for creativity, since the various elements of the problem and the numerous solutions appear side by side on the screen and can therefore be moved more easily into the virtual space and then be modified or combined to reveal new ideas (innovation is often a question of connection);
- communication (the team gradually builds interpersonal relationships, which facilitates the ability to express themselves and work together).

In addition, each digital tool seems to make a specific contribution. The *Facebook* (FB) page, which was gradually expanded over the time, one entry at a time, did not seem to provide a global or holistic overview either of the problem of flooding, or the subproblem of water contamination. The use of FB, through its structure and its tools, as perceived by participants, allowed for gradual developmental flow of information and reactions / opinions from participants regarding aspects of each complex problem. In other words, participants were able to better familiarize themselves with the following aspects of a given problem through their use of FB: the causes, places, impacts and risks. In the same way, the solutions and the prototypes were built and improved in person and then shared on FB, which made it possible to leave the problem aside, to reflect on its various aspects, and then come back with new solutions. The use of FB also demonstrated its potential as an effective communication tool, allowing participants to get to know each other and the facilitators, facilitating the sharing of ideas and eventual results on a daily basis. Online sharing, at various stages of design thinking, enabled participants to continue the work done in situ, while regularly stimulating their interest in the problems at hand. Among other things, in the case of Morrocan participants, the interest in FB was linked to one's ability to watch photographs or videos of floods in the six villages, in real time, and of solutions and prototypes. These images seem to illustrate the problems and the solutions, increasing the interest of the part of participants.

Realtime Board (RTB) appears as a powerful collaborative tool, allowing for a visual representation of the problem solving process, including a view of the problem's various aspects (on the same page), the multiple solutions (on the same page), and the various prototypes (on the same page). Like FB, RTB seems to provide progressive support as a tool for defining a given problem. It could also allow the participants to share and criticize solutions and prototypes, while gradually learning to work as a problem-solving team. Unlike in the FB platform, RTB is not so much a tool for spontaneous and punctual communication, but rather a means for systemic representation and refinement of a problem in all varieties of its aspects. RTB can collect and gather information about a problem and its proposed solutions, which is helpful because the complexity of environmental problems makes it difficult to remember all the elements. By seeing all the aspects of a problem and its solutions simultaneously on the screen, it is possible, apart from the meetings in situ, to continue reflecting and working collaboratively. As such, RTB seemed to offer fewer opportunities to add illustrations (images), focusing more on conceptual maps and tables to capture and analyze problems.

Knowledge Forum (KF), from the researchers' perspective, is a cognitive/collaborative tool aimed at enabling the development of a common representation of the problem, while being a sharing and discussions forum and an effective tool for keeping tracks of the work-in-progress. In our opinion, this seems to combine FB's affordances in terms of everyday interaction among participants, while also allowing to better structure the problem even more holistically and consistently than with RTB. From an affordances' perspective, KF seemed to provide resolvers with a tool that offers a wider variety of options in order to organize their ideas, also offering a holistic view of teamwork. However, following resolvers' comments, as in the case of RTB, KF appeared to be quite challenging in terms of the appropriation of its affordances. Despite the "demanding" nature of the tool, several comments seem to indicate that the use of KF and its affordances remains a very positive experience. This was also the case of RTB and FB in the other study cases.

6 Conclusion

In concordance to the UN Sustainable Development Goals (United Nations 2015), our study on the application of design thinking among university students and with people from a small rural community demonstrates that this approach may offer a variety of affordances for the pedagogical accompaniment of people as they analyze local problems, propose and test solutions. The problem space constructed through the design thinking process seems vast, systemic, and analytical enough to thoroughly describe complex problems. A large and well-organized problem-space is conducive to proposing solutions adapted to end users' needs. According to Arras (2000) and Benkler and Nissenbaum (2006), relevant knowledge about a given problem is distributed among stakeholders and, as a result, by bringing together different perspectives, new ideas and different products can emerge. The point of view of a single participant can reframe the problem and thus lead to an unanticipated solution that can subsequently be shared and assessed by other participants (including end users). It is therefore essential to strive for a widening of both the problem-space and the solution-space, which would consequently seem to favor innovation (Paulini et al. 2012).

The affordances of technological tools in problem solving seem to depend on their nature (user-friendliness, available functions …), as well as on their potential uses as perceived by the solvers and the facilitators. It is interesting to see that *Facebook* makes it possible to exchange images of the problem as it unfolds (in different places), following a solution timeline of sorts, that *Realtime Board* allows for better visualization of a problem as a whole, and that *Knowledge Forum* is a powerful analytical tool, fostering discussion from a variety of perspectives and points of view. Finally, in light of these results, we believe that digital tools could play a potentially important role in the various stages of solving environmental problems. Our collective understanding of this role requires more investigation and experimentation. Nevertheless, it is already plausible to suggest that applying design thinking to solve environmental problems is an evolutive process, and one should take into account the extra time and techno-pedagogical support needed for resolvers to benefit fully from new digital tools, in order to make better use of associated affordances.

Acknowledgements The authors wish to thank the Social Sciences and Humanities Research Council of Canada (SSHRC) and the International Development Research Center (IDRC) for their assistance and financial contribution to this research.

References

Arras E (2000) Transcending the individual human mind-creating shared understanding through collaborative design. ACM Transactions on Computer-Human Interaction 7(1):84–113

Barborska-Narozny ME, Stirling E, Stevenson F (2016) Exploring the relationship between a 'Facebook Group' and face-to-face interactions in 'Weak-Tie' residential communities. In: Proceedings of the 7th international conference on social media & society, 17. London, UK: ACM, pp 1–8

Bærentsen KB, Trettvik, J (2002) An activity theory approach to affordance. In: Proceedings of NordiCHI 2002, New York, NY: ACM, pp 51–60

Beche E (2012) Les jeunes de l'Extrême-Nord/Cameroun, l'Internet et la participation au développement communautaire. Esquisse d'une stratégie de Net-développement. In: Oben A. and Zourmba E (eds), Enjeux et perspectives techniques, économiques et sociales pour le développement des régions septentrionales du Cameroun. http://hdl.handle.net/2268/133928

Benkler Y, Nissenbaum H (2006) Commons-based peer production and virtue. J Polit Philos 4(14):394–419

Boden MA (2009) Computer models of creativity. AI Magazine 30(3):23–34

Boden M (1999) Computer models of creativity. In: Sternberg RJ (ed) Handbook of creativity. Cambridge University Press, Cambridge, pp 351–372

Brown T (2009) Change by design: how design thinking transforms organizations and inspires innovation. Harper Collins, New York

Darbellay F, Moody Z, Lubart T (eds) (2017) Creativity, design thinking and interdisciplinarity. Springer, Singapore. https://doi.org/10.1007/978-981-10-7524-7

Design for Change (2018) Cigabionica, Columbia. https://dfcworld.com/VIDEO/ViewVideo/196

Evangelische Schule Berlin Zentrum (2017) Personal interview with Elias Barrash. Director, Berlin

Freiman V, Dionne L, Richard V, Pruneau D, Léger M, Laroche AM (2020) Exploring digital collaborative platforms in the context of environmental problem-solving using the design thinking approach. Paper presented (virtually) at the AERA Annual Meeting, San Francisco, Avril 202

Gamba T (2017) D'où vient la « pensée design » ? I2D – Information, données & documents 54(1):30–32. https://doi.org/10.3917/i2d.171.0030

Gibson JJ (1979) The ecological approach to visual perception. Psychology Press, New York

Gericke L, Gumienn RS, Meinel, C (2012) Tele-Board: Follow the traces of your design process history. In: Meinel C and Leifer L (Ed) Design thinking research. New York, Springer

Hammond M (2010) What is an affordance and can it help us understand the use of ICT in education? Educ Inf Technol 15(3):205–217

Innovation Training (2020) 5 Big advantages of remote design thinking. https://www.innovationtraining.org/5-big-advantages-of-remote-design-thinking/

Jeong H, Hmelo-Silver CE (2016) Seven affordances of computer-supported collaborative learning: How to support collaborative learning? How can technologies help? Educational Psychologist 51(2):247–265. https://doi.org/10.1080/00461520.2016.1158654

Luyat M, Regia-Corte T (2009) Les affordances : de James Jerome Gibson aux formalisations récentes du concept. L'année Psychologique 109(2):297–332. https://doi.org/10.4074/S000350330900205X

Martinovic D, Freiman V, Karadag Z (2013) Visual mathematics and cyberlearning in view of affordance and activity theories. In Martinovic D, Freiman V and Karadag Z (eds), Visual mathematics and cyberlearning. Mathematics education in the digital era (MEDEra) Book Series, vol 1, Springer Netherlands, pp 209–238

Paillé P, Mucchielli A (2008) L'analyse qualitative en sciences humaines et sociales. Armand Colin, Paris

Paulini M, Murty P, Maher ML (2012) Understanding collective design communication in open innovation communities. The University of Sydney, Sydney, Australia. http://maryloumaher.net/Pubs/2011pdf/codesign_Draft_Oct2011.pdf

Pinzón ANB, Nova YPM (2018) The articulation between project-based learning and the use of information and communication technologies in the foreign language teaching process. Revista Boletín Redipe 7(6):67–73

Pruneau D, El Jai B, Khattabi A, Benbrahim S, Langis J (2017) La pensée design et Facebook au service de la résolution d'un problème d'inondation : Une étude de cas au Maroc. Éducation relative à l'environnement : regards, recherches, réflexions, 14(1) http://journals.openedition.org/ere/2650; https://doi.org/10.4000/ere.2650

Pruneau D, Langis J (2015) Design thinking and ICT to create sustainable development actions. In: Proceedings of the 7th international conference on computer supported education, (vol 1) Lisbon, Portugal, pp 442–446

Pruneau D, Freiman V, Barbier PY, Langis J (2009) Helping young students to better pose and solve environmental problems. Appl Environ Educ Commun 8(2):105–113

Rowe P (1987) Design thinking. MIT Press, Cambridge, MA

Scheer A, Noweski C, Meinel C (2012) Transforming constructivist learning into action: Design thinking in education. Design and Technology Education: an International Journal 17(3):8–19

Shkedi A (2005) Multiple case narrative: A qualitative approach to studying multiple populations. John Benjamin Publishing Company, Amsterdam

Salvia AL, Leal Filho W, Brandli LL, Griebeler JS (2019) Assessing research trends related to sustainable development goals: local and global issues. J Clean Prod 208: 841–849

Squires A (2014) Group projects with technology: realtime board in a math classroom. Rising Tide, 7(3), http://www.smcm.edu/mat/wp-content/uploads/sites/73/2015/06/Amanda-Squires.pdf

United Nations (2015) Transforming our world: the 2030 agenda for sustainable development. Adoption resolution, 70th session. United Nations, New York, USA.

Walsh DM (2012) Situated adaptationism. In: Kabasenche W, O'Rourke M, Slater M (eds) The Environment: Philosophy, science, and ethics. MIT Press, Cambridge, pp 89–116

Hybridization of Time: Towards Temporal Sustainability of the Digital Economy

Bohdan Jung and Tadeusz Kowalski

Abstract This chapter briefly reviews a patchwork of newer business models that present themselves as attributes of the digital economy. Trust and attention seem to emerge as strategic resources in this economy. As both of them are in short supply, their management is of crucial importance to the sustainability of the digital economy. The specific new patterns of time use in this economy (multitasking), further in the chapter (referred to as time hybridization) point to the conceptual and methodological difficulties of the digital era and point to shortening human attention spans as one of the key weakness of digital consumption.

Keywords Digital economy · Time allocation · Temporal order · Hybrid time · Access economy · Sharing economy · Attention economy · Prosumption · Sustainability

1 Introduction

The last three decades have seen an unprecedented expansion of the digital economy, digitalization of many human activities and far-reaching changes in our lifestyles. The consequences of these have yet to be fully realized. Along with fast growing digitalization of the products and services, a number of hastily assembled theories[1] appeared in social sciences to explain new social and economic phenomena linked to success of the digital economy. This intellectual ferment included such concepts as access economy, 24/7 economy, creative economy, economics of attention and

---

[1] Even though these partial explanations were mainly referred to as 'economies', they were social in nature, deeply anchored in newly emerging social phenomena powerful enough to disrupt the logic of the (still mainly) industrial economics.

B. Jung (✉)
Faculty of Economic Analysis, Warsaw School of Economics (SGH), Warsaw University, Warsaw, Poland

T. Kowalski
Faculty of Journalism, Information and Book Studies, Warsaw University (UW), Warsaw, Poland
e-mail: t.s.kowalski@uw.edu.pl

© The Author(s), under exclusive license to Springer Nature Switzerland AG 2021
W. Leal Filho et al. (eds.), *Innovations and Traditions for Sustainable Development*,
World Sustainability Series, https://doi.org/10.1007/978-3-030-78825-4_14

wikinomics. These will be briefly discussed below. These efforts were often referring to examples of successful digital transformation, such as in the media sector, where design, production and distribution were fully digitized by turn of the century.[2] Regardless of the changes in the nature of products and services, which moved from the real to digital world, their immaterial character, weightlessness and their lack of anchoring in time and space (in the sense that they could not be physically located in a specific place and their date of production was irrelevant as in the case of updated software), yet another discourse summarized these changes from the perspective of the changing nature of work.

Remote online work (the scope of which was drastically widened during the COVID pandemics) could be carried out anywhere and at any time (De Wet and Koekemoer 2016; Carlson et al. 2018). The impact of constant online presence on the workplace and work's expansion outside of nominal place and time work was thoroughly analysed (Boswell and Olson-Buchanan 2007; Tennakoon et al. 2013; Thomas 2014). The growing number of employees carrying out their duties both at work and at home through the extension of new communication techniques has led some researchers to rise the point of reallocation of work to home (Halford 2005). Psychological aspects of work in work time and outside of its nominal framework were analysed for managerial staff, together with the provision of mobile equipment for them to extend work into private life (Vayre and Vonthron 2019). Also amply researched was the positive impact of new communication technologies on organization of work, communication within an organization, decision taking and flexibility of workplace and management styles (Bobillier-Chaumon 2003; Isaac et al. 2007; Scholz 2013), increase in the autonomy of employees (Lee and Sawyer 2010; Tremblay and Tremblay 2014). Research was also carried out on the employee overwork and fatigue caused by being overwhelmed with work outside of the workplace (Ninaus et al. 2015), difficulties experienced by employees trying to break away from professional duties outside of their work time (Fenner and Renn 2010), pressure and expectations from work on employee availability outside of working hours (Matusik and Mickel 2011).

This blurring of division between work and non-work time and the subsequent feeling of time squeeze (as studied by the attention economy) leads us to speculate on hybridization of time and a new temporal order of the digital economy, which breaks away from the traditional work/leisure division created by the industrial economy and still prevalent in the corporate world (Southerton 2006, Sullivan and Gershuny 2004). We conclude by voicing our concerns whether this hybrid temporal order is sustainable for human development. Referring to this notion of 'sustainability' with more precision, we use this term in line with goal 12 of UN's sustainable development

---

[2] The case of sectors that were inherently digital from their emergence (such as the computer or software companies) is not taken in consideration here, as they did not have to be digitized. To understand the potential challenges of this transformation, let's just invoke its remote work potential. (See: https://www.bls.gov/opub/mlr/2020/beyond-bls/the-number-of-people-who-can-telework-is-higher-than-was-estimated.htm. Accessed July 24, 2020).

goals (i.e. ensuring sustainable consumption and production patterns).[3] While UN focuses on Goal 12 in that it encourages doing more and better with less, it also refers to 'promoting sustainable lifestyles'.[4] It is in this area in particular that we see the linkage between long-term sustainability of digital economy's demand for human attention and sustainability of human lifestyles under time pressure.

2 Methodology and Objectives

Hybrid time and time hybridization are novel concepts yet awaiting for their own methodology to emerge. Prior to digitalization of human activity, use of time and temporal order were explored from the perspective of time budgets. 'Use of time' studies were conducted since 1950s on nationally representative samples by national statistical offices in most OECD countries and some international comparative studies were also made.[5] However, without an exception all these studies were based on the traditional time frames (aggregates) of the industrial era: division of time into work time, time of obligations and leisure/discretionary time.

The ability to carry various activities in a hybrid pattern, concurrently, via multi-tasking and across the three social time aggregates enumerated above has rendered these highly esteemed and methodologically sophisticated studies obsolete. More contemporary approach would call for use of mobile apps tracking respondents activities and also monitoring their location to provide context for whether (web-based?) leisure activities are taking place at work, work is conducted at home outside hours of work etc. So far such methodology for hybrid time has not been developed. In its absence, our chapter bases on a literature review and secondary analysis of statistics available in open access (such as: https://www.internetlivestats.com, https://wearesocial.com, http://hootsuite.com). At our level of conceptual exploration, we see this as sufficient for formulation of basic study objectives:

- establish the emergence of a new temporal framework (hybrid time) for human activity in which work and play are mixed in hitherto unknown forms,
- explore interaction between hybrid time and new forms of economic activity in the digital age,
- pinpoint human inadequacy in coping with availability and use of this time inexistent in the industrial era (i.e. the shortening of attention spans, life under constant time pressure),
- explore consequences of the collapse in traditionally perceived distinctions between work time and leisure time, validity of industrial epoch's temporal order in the digital era,

---

[3] https://www.un.org/sustainabledevelopment/sustainable-consumption-production/. Accessed August 9, 2020.

[4] Ibidem.

[5] International history of this research can be traced at https://www.timeuse.org/mtus. Accessed August 6, 2020.

Fig. 1 Patchwork of the digital economy. *Source* Own compilation

- signal the possible threats arising from the lack of long-term sustainability of such intense time use in the attention economy in the context of human wellbeing and development (some of the already diagnosed threats to this sustainability include depression, lack of sleep, inability to concentrate and focus (limited attention span), increased risk of overweight, back-pain complaints.[6]

3 Elements of the Digital Economy

The still tentative and partial theorization (in itself an innovation) on the digital economy leaves us at present with a somewhat patchwork picture. It offers a glimpse (rather than in-depth analysis) of its multi-dimensional nature and of a new (as compared to the industrial economy) mix of innovation and return to tradition (i.e. as in the gift, wiki, relationship and sharing economies). While 'digital' is generally less intensive in the use of material resources than 'industrial', which should appease worries about its sustainability, it is also raising new challenges to sustainability by making huge claims on one of our most rare and irreplaceable resources—human time (Fig. 1).

The idea of the gift economy by far predates other concepts. The term has been in use for over 200 years, but it was used in reference to archaic communities, which were ruled by culturally prescribed rituals. As researched by cultural anthropology, an exchange of gifts has preceded trade and it was taking place in scrupulously transparent conditions due to the obligatory reciprocity of gift-making and gift-taking.[7] The scope, context, ritual and cultural sense of gift-making was deeply changed in the new context of the digital economy with the emergence of User

---

[6] G. Gangwar, D. Suvidha, A Systematic Review of Literature on Effect of Internet Addiction in Adolescents, in: *International Journal of Scientific Research in Science and Technology*, Volume 2, 2016, Issue 5, pp. 172–79.

[7] Gifts had to be very stereotyped, since their value had to be clearly known to their beneficiary, who was obliged to offer in reciprocity a gift of approximately the same value unless he deliberately

Generated Content (UGC) freely available on the Web. It is estimated that a few billion people have contributed something to the Internet—a massive movement unprecedented in human history involving over a half of the world's population.[8]

The more contemporary practice of gift-making in the digital age was sanctioned by the legally defined concept of public domain and backward digitalization of old media content no longer protected by copyright laws and therefore made freely available to a global public.[9]

By now there is a body of freely available knowledge that can be interpreted as gift-type contributions of UGC, given to the public as an entrance fee to a career on the Web. These contributions are made to gain attention, create one's own brand and to establish a professional reputation. In other words, they appear as an inscription fee, an investment into one's future, a stepping-stone in one's work career. The long-term sustainability of genuine gifts thus seems to come under threat from a more pragmatic attitude of young content creators.

Slightly ahead of the digital age, Rifkin (2001) came up with a key concept, which proved to be so precious two decades later. Rifkin coined the term 'access economy' to discuss the implications of customers satisfying their needs not by buying goods and services, but merely by having access to them. This model is presently referred to as the subscription economy (as in Netflix) and is at the heart of key consumer services over many areas (health, culture, education, fitness etc.). At the heart of this model is a major (from the point of view of global sustainability) shift of an economic paradigm from ownership to use. This model of using something without owning it was of course known to humanity, but not on such a scale, now enabled by network-mediated communication. First of all, we must remind ourselves that all of the Internet and mobile services are run on a subscription mode, by paying a subscription (usually a flat rate) we gain access to the whole Web, within which some products/services require additional payment for premium content (as in the real world model). In respect to the subscription model, one of the key concepts is gatekeeper and gatekeeping becomes crucial to our access to global commerce (Rifkin 2014). From the point of view of this chapter transition from ownership to access is key to global social and economic sustainability since rare resources

---

wanted to offend the giftmaker (Mauss 2002). In those nonmonetary days the exchange of gifts was an early form of economic (possibly international) exchanges (Cheal 1988).

[8] These contributions range from trivial and mems to elaborate cultural and academic contributions. Research on motivations for contributions to social media shows that not all freely available UGC was made for altruistic reasons, but one thing it has in common is that these modern gifts carry no expectation of reciprocity, especially not of equivalence in value (Mauss 2002).

[9] Another kind of gift was made (especially in the early days of the Internet by hackers and pirates, who posted torrents giving access to music, films, software. In Sweden they were a *cause celebre*, giving rise to Pirate Party, which stressed in its agenda primacy of national copyright laws over international ones as part of the country's independence. The ease with which access to the global supply of culture and software could be obtained strengthened the users' feeling that Internet was free, i.e. its contents were freely available (like a gift) to anyone. It is often commented that this was reinforced by the fact that early hacker cultures were linked to the ageing hippies, whose attitude to private property was slack (Frayssé and O'Neil 2015; Lessig 2004). We could at this point speculate on global cultural sustainability of such open copyright systems and its impact on global creativity.

Transactions	Networks
Sales	Relationship
After-sales service	On-going feedback of info on consumer interests, preferences, habits
General-interest advertising	Targeting individualized ads
One-to-many communication	One-to-one relationship, with widespread use of influencers

Fig. 2 Moving from sales to relationships via access (from transactions to networks). *Source* After Rifkin 2001

can be shared. Use via acquirement of ownership (by transaction involving buying-selling, change of ownership) means that the main actors of this process are buyers and sellers. As their interests diverge, it is difficult to strike a relationship between them. This changes with access, where they change into operators/providers and users. Once again, it must be stressed that users are not owners. The operator is able to gather information on the users' behaviour and better match his services with interests of the user. Jointly used services (i.e. in transport pooled cars, bikes) create less pressure on the environment, need for a transport infrastructure etc. When user licenses replace contracts, user registration enables to establish an on-going relationship, which can be used to profit from user's feedback and suggestions for product improvement/development. In this sense access works both ways, giving to the user the ability to test at his ease the entire package of access provided at a flat rate by the operator (as in TV and radio programming) in exchange for tracking use patterns of the customer and his feedback.[10] While it may be speculated that access based business models were beneficial both to the economy and to the society, as well as being environmentally more sound than private ownership of everything, a quick scheme of the access-based model would accentuate its following features (Fig. 2).

The access mode has given rise to the R-economy (an economy of relationships), followed by R-marketing as specialists quickly realized that the full value of a relationship could be capitalized not through a single transaction but over the user's lifetime. A concept of Lifetime Value (LTV) was established to allow predicting

---

[10] This was seemingly a win–win situation for both sides, but soon there were growing concerns about user privacy as monitoring and tracking became more sophisticated. (see: L. Busca, L. Bertrandias, A Framework for Digital Marketing Research: Investigating the Four Cultural Eras of Digital Marketing, *Journal of Interactive Marketing, No.49, 2020, pp.1–19.*).

how much money could be made from a customer over his lifetime instead of a one-time profit gained from a single transaction. This gives a clue to the sustainability of business relations viewed as a lifetime relationship.[11]

Despite its seemingly humanistic and sustainable "R" focus, the age of access can be defined by an increasing commodification of all human experience—transforming every aspect of life into a purchased affair. In the propertied era of capitalism, the emphasis was on selling goods and services, while in the cyberspace economy, the commodification of goods and services becomes secondary to the commodification of human relationships. In what Rifkin (2001) sees as the new culture of hyper-capitalism, all of life becomes a paid-for experience. When even goods become platforms for managing services and services become the primary engine driving global commerce (as in the digital economy), establishing relationships with end users is critical, with access to customers as company's prime asset (as in Amazon). The company's standing in the digital world is viewed as a function of its ability to forge long-term commercial relationships with end users and seamless one-to-one relationship with customers, as well as its ability to customize production.

At this point, in the context of relationships it is opportune to evoke the idea of the trust economy, as evoked by Botsman (2011, 2012). Under the multitude of global service or product suppliers worldwide and fast changing nature on new businesses and with the bulk transactions being online, we can no longer use traditional means of assessing our customers' reliability, reputation and standing. Hence, we see the emergence of a new system of 'likes', consumers' reviews, recommendations all aggregated to provide a surrogate of trust and give us an impression on the reliability of our business contact.[12]

To carry out Rifkin's reasoning on access further, the contemporary economy is being transformed from a giant factory to a grand theatre (future for high-end global commerce) in which imagery and metaphors used to organize commerce change when industrious gives way to creative (Rifkin 2014). Can this 'culturalization' of service performance be interpreted as yet more evidence of a socially sustainable system of digital economic performance which has more in common with cultural production (as in the creative economy sector)? In our opinion it is but a newer and softer tool of increasing productivity and boosting efficiency, merely a window-dressing exercise as marketing mines culture for valuable potential meanings that can be transformed into commodified experiences. It is using the arts and communications technologies to ascribe cultural values to products, services and experiences and cultural value of designer labels. This process is deepened by appropriation and integration of countercultural trends as part of lifestyle and life-event. In terms of

---

[11] The actual groundwork for LTV strategy was laid by the automotive industry as car purchase was one of the key customer decisions made in that era. An honest car dealer could count to earn a few hundred dollars on a single transaction, but once he established himself in a relationship/position of a trusted family advisor on cars, his earnings from the customer and his family would be in tens of thousands of dollars (at US car prices of mid-70 s) (Pine and Gilmore 1999).

[12] Despite interesting efforts to aggregate these tokens of trust from different areas of activity i.e. selling on the Internet, doing small services in the gig economy, we are still far from a situation when these would be aggregated into a portfolio used by banks to assess their customers' status.

sustainability of this new (mainly economic) system real power lies in the hands of the new class of *"cultural intermediaries"* and their intangible assets, such as their knowledge and creativity, their artistic sensibilities and impresario skills, professional expertise and marketing sense. They hold the position of *"*n*ew tastemakers"*, *"cool hunters"* and cultural intermediaries (also influencers and bloggers) who act as gatekeepers in a world where access determines the parameters of lived experience for millions of people. In fact they are the key people to the mass understanding of what is new, innovative, creative and to what is sustainable in today's world.

Another much discussed change which is linked to digitalization is the concept of sharing economy. Due to the new opportunities offered by web-powered communication apps it became possible to find a market for dispersed occasional users of cars, bikes, spare rooms, social events, music etc. Once created, digital goods can be duplicated at no cost and their distribution comes at no cost (zero marginal cost economy). In contrast to the 'real' (i.e. physical) world, consuming a digital good does not limit access to it by others. A whole world of petty services can be opened to global customers (repair, personal services). This form was known and practiced in earlier pre-digital economies, but without a global information network it could not operate beyond the local level and word-of-mouth recommendation. This brings us to the 'gig economy' where millions of people make a living doing errands for others, rather than being formally employed. They rely heavily on opinions, recommendations and trust on the Web, on such a scale that would not be attainable prior to the digital age.[13] This also had consequence for a more environmentally sustainable where resources could be pooled and used jointly, instead of every individual striving to have his own.

The ease with which lines of code, digital images and sounds could be moved at no cost and with freely available software tools was also behind the sustainability of the *remix* (or cut-and-paste) *economy*. We can see this as intellectual recycling of pieces/quotes from our past and present which mixed together and recontextualized created a vast array of mems, new musical arrangements, software from trivia to little masterpieces of UGC.[14] This specific new mode of cultural production[15] brings us closer to yet another dimension of new sustainability of the digital age: the rise and prominence of *prosumption* (Toffler 1970), yet another step towards hybridization of seemingly opposed categories and concepts of the industrial era (Benkler 2006).

In the industrial (and corporate) economy based on private ownership economic activity was carried out in 'closed silos', in a secrecy needed to prevent competitors from being able to copy or counteract. New technologies and new products were among company's most valuable secrets and maintaining your dedicated staff of

---

[13] According to Upwork in 2020 57.3 million freelancers worked in the gig economy. Gallup estimates that 36% of all employees work within the gig economy while Mc Kinsey stipulates that 15% of the self-employed work through a digital platform: https://zety.com/blog/gig-economy-statistics?gclid=EAIaIQobChMIpNX8lqOW6wIVgbt3Ch3w9AKMEAAYASAAEgLM7PD_BwE. Accessed May 12, 2020.

[14] This leads us into a legal debate on the right to quote and its fair use (see L. Lessig below).

[15] *L. Lessig,* "Free Culture. How Big Media Uses Technology and the Law to Lock Down Culture and Control Creativity", *The Penguin Press, 2004.*

employed specialists was a means to achieving this goal. This logic came to be challenged when *wikinomics* came into existence in the early 1990s with the creation of such flagships as Wikipedia, detailed maps of the moon and outer space, the reading of the human genome (Tapscott and Williams 2006). In the world of games, the creation of Linden Labs' *Second Life* where only a fraction of the product was done by the producer and the rest was developed by users/gamers—a clear case of the power of prosumption—was showing the way for other developers on how to profit from collaboration between producers and customers. Feedback from the consumers/users, was technically made possible and easy in the digital era, when two-way communication for feedback channels became open. This process did not proceed without resistance. When LEGO users started to make motorized robots from company's components, its initial reaction was to take them to court for what was then considered an illegal product modification. It was soon realized that hobbyists, enthousiasts developing somebody's product out of intrinsic non-pecuniary motivation were a precious and free asset to the company. In line with supporting prosumers, this gave the company an advantage of establishing a network of relationships, take advantage of word-of-mouth recommendations and free publicity, forming a community around its products and services (a prelude to development of the relationship economy) (Allen et al. 2008). As would appear from these *wikinomic* projects, in the longer run cooperation paid off more than competition and required substantially less resources and effort (it was thus economically and socially more sustainable, with less resources wasted on fierce competition and holding back the competitors).[16]

Yet another dimension of changes is offered by the rapid expansion of a business model based on subscriptions (i.e., *the subscription economy*). This is closely related to the emergence of access and sharing economies.[17] From the perspective of sustainability of our development, they mark an important turning point—the passage from ownership of assets to these assets use through access, sharing and subscription.

4 The Attention Economy

Perhaps the most significant of these explanations related to the rapid expansion of the digital economy has been the attention economy and the acknowledgement of the availability of time as being key to the digital system's sustainability. The theoretical foundations for this concept were formulated at the beginning of the twenty-first century (Davenport and Beck 2001; Lanham 2006; Glorieux et al. 2008).

For the attention economy, the starting point is the assumption that information and knowledge are the most important resources for the valuation of capital. In the

---

[16] More on the 'preventive' use of patents to block competitors' expansion can be found in Lessig (2004) (op.cit.) and also in Tapscott and Williams (2006).

[17] *End of Ownership report,* (2018). Retrieved from https://www.zuora.com/press-release/new-international-survey-reports-on-the-end-of-ownership-and-the-rise-of-subscriptions/. Accessed February 10, 2020.

information society, knowledge and information resources are growing so fast that we can elaborate on an oversupply of information, which, in addition, is in multiform and can simultaneously be consumed in many communication channels. Large data sets are constantly analyzed and organized by various programming tools. Algorithmic selection has become a necessary element for the implementation of these processes, and the development of the semantic network seems to be a logical consequence of development. One of the special consequences of algorithm development is the transformation of large data sets, the so-called content macro into a large number of micro-content which is ever better adjusted to the preferences, expectations, interests and activities of virtual network users.

Algorithms increasingly affect everyday life, change our perception of reality, behaviour and everyday choices, and contribute to the shaping of social order. In other words, they've become an important player also in the area of development's sustainability. Programmed sequences of operations are able to distribute information (news), convey messages, create rankings of content interesting for users, assign to the activities of Internet users certain attributes or features and on this basis forecast future behaviours and interests. Navigating the web without algorithms is practically impossible, if only because searching for information in the conditions of rapidly growing amounts of data would be extremely time-consuming. Algorithms are therefore a necessity, a *sine qua non* condition for using the Internet, but also an element of risk, a potential field of manipulation, a threat to privacy and traditional forms of intellectual property protection. Algorithms create a technological and functional basis for new services and business models that are a challenge to the sustainability of traditional industries and business strategies (Bauer and Latzer 2016).

Growing interest in algorithmic selection on the part of both large broadcasters, content producers, as well as individuals (such as youtubers and bloggers) results from strong competition for the attention of users. Attention has become a basic element of competitive competition and a scarce/valuable resource.

> ...the richer a society is in terms of its production, distribution and consumption of information, the poorer it becomes in terms of human attention. Inventions such, as the Internet, e-mail, databases, digital television, social media, and so on, together with the radical informatization of the process of production of commodities have created both an abundance of information and a demand for new forms of organizaing and allocation attention... (Bueno, 2017, p. 1).

The basic measure in the attention economy is the time that the user is able to devote to the use of the offered resources (informational, entertainment content, advertising offers, products, services and others). Thus, time becomes a new field of economic exploitation and a new currency of business. In the attention economy, user time becomes one of the most important resources for monetization. The more time and attention the content offered on the web concentrates, the greater the chance to sell given products or services. Berardi (2009) points out that technological changes are much faster than the mental ability of a person to focus. Network technologies have significantly increased the length of time humans are exposed to information. Miniaturization and portability of devices allow access to information resources almost anywhere, anytime, and it does not require special technical skills.

Media use in the digital economy is pervasive and dominant (Webster 2014). At this point of digital economy's expansion not only media consumption time, but also place and context of their use seem to be under researched by traditional methods that ascribe media consumption to a given medium (text, audio, electronic etc.). Media users use and mix various information and communication techniques both in their leisure time and at/during work. The specificity of digital economy makes media use for professional use possible both within and outside work. This means that work time is absorbed by private time and vice versa. Thus, private time becomes a yet another dimension of economic exploitation. For many types of economic activity, the development of digital and information technologies has proven to be disruptive and digital media are right at the heart of the 'digital vortex', being one of the first sectors to have fully digitized design, production, distribution and consumption of their output (Yokoi et al. 2019). The development of media and communication contribute to time and space compression (as anticipated earlier by Lash and Urry 1994), which allows media consumers to communicate in real time and be immediately informed about events worldwide.

Compression of time and space, which is one of the consequences of the attention economy, is now (and will continue to be in the foreseeable future) of significant importance in digital adaptation of subsequent areas of human activity to technological progress. The consequences of these changes, although difficult to predict, can lead to a new organization of working and leisure time, development of remote production and service implementation systems (trade, education, telemedicine and others), which in turn will significantly reduce the consumption of some resources (e.g. mass environmental costs of commuting to the workplace, maintaining large office spaces, customization of production and services, new measures and standards of performance).[18]

5 Hybridization of Time

With development of the digital economy and its demanding claims on human attention, we are seeing the emergence of a new dimension of human development's sustainability—the hybridization of time in the digital era, resulting in a new temporal order of human activity.[19] The new concept of '*hybrid time*', in which much contemporary media consumption is nested, cannot be clearly identified as being either work, obligations or leisure time as it has characteristics of all three of them in a new mix not present in previously existing time allocation activities.

---

[18] Within the digital world both attention and trust seem to run in short supply, raising doubts about long-term sustainability of the system (Botsman 2012).

[19] We thus again make the reference to the concept of 'adaptive sustainability approach' and its recognition that 'sustainable development' refers to a holistic approach and temporal processes that lead us to the end point of sustainability.

There's a long line of postmodern deliberations on contemporary convergence of work and consumption, producers and consumers (prosumption), of time and space.[20] Hybrid time, tentatively defined here as flexible, multitasking, multi-threading mixed multidimensional time (a consequence of the digital vortex into which our life is pulled) is a product of the digital era, just like traditional division of social time into work, personal obligations and leisure was a consequence of the industrial era.[21] These categories matched the economic and social realities of the industrial era: work contracts stipulating daily hours of work, fixe organization of human life was basically confined within three aggregate categories (i.e. social times): work time, time of obligations and leisure (i.e. free, discretionary) time and working hours, days free from work, vacation entitlements. Work also meant it was a physical location different from home (as in: *going to work*). By shortening the obligatory work week, trade unions were struggling to claim back time 'stolen' during the early industrial revolution, when extremely long duration of work time was one of the tools for early capitalist accumulation. Work time was a primary driving factor in the organization of the system's temporal order as it predetermined the daily routines of commuting to and from work. Weekends that were free from work followed a visibly different time pattern especially with respect to sleep time, abundance and variety of consumption activities, peak media consumption, outdoor pursuits and physical activity (Fisher and Robinson 2009). Research on consumer behaviour has also pointed to distinctly different patterns of consumer strategies based on the availability of time.[22]

With the growing scope of remote work, expansion of the creative economy and its specific task-based work cycles, the development of multitasking, this clear division between the three social times is gradually beginning to lose its sense (Lee 2010). Each of these times still has its different function to fulfil (and these are still fully valid), but digital technologies allowed to pull down the walls between the hitherto separate and distinct silos of different social times, allowing for a range of activities to be carried *across* rather than *within* each of the three times. For example remote and mobile forms of media consumption (traditionally a part of leisure time) can now be squeezed into work time and time of obligations. The scope of multitasking is growing rapidly. In traditional time budget studies,[23] when multitasking was at its infancy, the problem of multiple activities carried over the same time was technically solved by using the distinction between 'primary' and 'secondary' activities. For example, to measure the allocation of time for then dominant leisure activity of TV watching, switching on of TV was considered the 'primary' activity, talking to family/friends

---

[20] As in: Toffler (1970), Bauman (2000), Lash and Urry (1994).

[21] Obligations were defined in terms of what was needed for our biological and social functioning (sleep, medical care, taking care of children and the eldely, religious practices, education).

[22] Money-rich/time poor vs time rich/money poor stategies (on the substitution of time and money in services). See also individual strategies for coping with time: front loading (of career), deferred life plan (postponing marriage and parenthood), life-shifting (mid-life change), time-deepening (through multitasking). See: Florida (2005).

[23] Complete and complex studies on the allocation of time across the population, conducted at long [usually: decade] intervals by many national statistical offices (see: https://www.bls.gov/tus/, https://www.timeuse.org/mtus).

while watching was the 'secondary' activity. Similar logic was applied to reading, walking etc. With the pervasiveness of mobile media, such distinctions have lost their sense as it became impossible to make the distinction between main and secondary activities in a patchwork of media consumption (Carlson et al. 2018). This has two dimensions: some leisure activities have moved into work and obligation time (i.e. browsing web pages), while some work activities have moved into our hitherto private and free time (checking and responding to emails, phone calls from work and work-related sources) (Tremblay 2012). The problem is not slight since some studies point that annually we spent on the whole one full month looking at the phone and scrolling pages.[24] If we add to this the idea of the 24/7 economy (Glorieux et al. 2008), we see a total disruption of time use patterns from the pre-digital era, which leads us to the question of its sustainability.[25]

Social communication became possible in places and situations that were unimaginable only a few years ago, in public places such as cafeterias, restaurants, cinemas, while watching TV or listening to the radio, during family gatherings etc. Research on FOMO (Fears of Missing Out) reveals that as much as 14% of the Polish respondents are highly dependent on the use of the Internet and social networks. Additionally, among all the respondents, 8% use the Internet while crossing the street, 27% while traveling by public transport, 7% in the theater or during a cinema screening, 10% during meetings with friends (phenomenon of so-called phubbing), as much as 8% use it while driving a car, 9% during flight, 6% during a business meeting and … 6% during a mass or church service (Jupowicz-Ginalska et al. 2019). In all of these cases, this media use can be both for work and non-work purposes. This was made possible by the invention and improvement of miniaturized and portable devices, which created a momentum for filling the chunks of time, which have so far been relatively "free" (i.e. between other activities).[26] Social time was thus emptied of little idle moments in between activities, probably contributing further to the feeling of 'time squeeze' (Clarkberg 2008).

The recently emerging studies of this new, more intense and varied use of hybrid time point to the somewhat artificial expansion of the media consumption time, as exemplified by total summary time of media and Internet use, which when combined with time of work and obligations exceed 24 h per day.[27] This is technically possible

---

[24] https://kobieta.onet.pl/w-ciagu-roku-spedzamy-30-dni-patrzac-w-telefon-i-scrollujac/mff6w44. Accessed July 18, 2020.

[25] This issue was raised earlier (see concerns about the overworked American (Schorr 1991) and the increasing feel of 'time squeeze' (Clarkberg 2008) and concerns voiced about the work schedules of the creative class (Florida 2005), but not in the context of digitalisation.

[26] In a certain sense this can be seen as an extension of an older process, where insertion of the radio in a car allowed for addressing the captive audience with media content and ads.

[27] According to ActivateInc study of 2020 time allocated to "media and consumer tech activities" (enumerated as video, audio, messaging and social media, gaming and other) is totalling 31 h per day and fills over half of US adults' daily time (12:07). Similar picture emerges from studies done in 2015–2019 period by Comscore eMarketer, Titbit, Gallup, GlobalWebIndex, Interactive Advertising Bureau, Nielsen, Pew Research, ResMed, U.S. Bureau of Labour Statistics. This is not

due to the multiple count resulting from users' multitasking. However, this multitasking is heavily paid for with dropping attention spans and what social psychology describes as "overstimulation," which further undermines the audiences' attention, information retention rates as well as falling attention spans and the feeling of stress as observed clinically among heavy internet users.

Furthermore, a study published by GlobalWebIndex in 2019 and done on a global sample in 2018 shows that on the average 4.6 billion Internet users were spending 6 h 42 min online per day. This means that annually people are spending 100 days online, just under one third of their lives! Again, this raises the issue of sustainability of such an intense and time-consuming hybrid.

To put things into context, the difficulty in conceptualizing and measuring 'hybrid time' is amplified by the widening scope of the "prosumption" phenomenon (Toffler 1980). The blurred line between production and consumption is highly reminiscent of hybrid time's positioning between work, leisure and obligations. The expansion of self-service activities such as self-service restaurants, filling stations, cash dispensers, check-in terminals, washing machines, vending machines or ticketing, has been around for several decades, but the switch to Internet communication has opened new areas for its expansion: accommodation and travel booking, medical diagnostics, online education and publishing, dating, even religious services (paradoxically, the COVID pandemic has made this process even more widespread and far-reaching during the lockdown period). From a purely economic perspective, this lead to further disintermediation, ruthlessly cutting out all intermediaries between the producer and the consumer. These prosumptive services (such as user-generated content dominant in the Web 2.0 era) are difficult to categorize within the existing conceptual framework of our temporal order as they are hybrid activities neither fully anchored in production nor in consumption. Yet another aspect of this mixed activity is conceptualized by some researchers as "serious leisure" or hobbies which through serious time commitments permit for amateurs to acquire professional competencies and expert status in the field of their hobby (Stebbins 2014). While satisfaction of amateurs reaching the pro status seems to originate from intrinsic motivation, their hobbies can be a steppingstone for their professional careers. This coincides with the monetization of special status, such as demonstrated in a spectacular way in apps such as Facebook, YouTube, Instagram or Tik-Tok.

6 Conclusions

Rapid digitalization of human activity calls for a new set of special skills and new types of sustainability to support them. In the first decades of XXI century we are seeing a rapid expansion of 'subeconomies' developing around the digital economy. We thus observe growing importance of economics based on traditional social skills,

---

a simple error of double counting, but the failure to design research tools capable to deal with such heavy multitasking within other activities.

such as and sharing (rather than ownership), collaboration (wikinomics), the building of relationships (R-economy).

On the side of innovation we also see the emergence and shortage of new key resources (substituting for capital) such as time, attention and trust. In particular, the pressure on time coupled with new multitasking communication technologies is creating new social patterns in the use of time, in which multitasking, hybrid time seems to prevail (already reported to engage 12:01 h per day, 30 days per year on mobile phones alone). We are particularly concerned with this hybridization of time in the consequence of which boundaries between work, obligations and leisure seems to collapse. For a variety of reasons (medical, social, cultural, economic) we see this a long-term threat to sustainability of human lifestyle.

This hybridization of time also impacts on the nature of work, its casualization, spill-over into non-work time (Independent Work 2016). This raises the question of sustainability of old distinctions between work/play time and between production and consumption (such as in prosumption).

One of the aspects of hybrid time is pressure from the workplace on employees to be always online, leading to symptoms of Internet addiction and strange behaviour (Salanova et al. 2011; Gangwar and Suvidha 2016). The relation between demands of the workplace and physical health of the employees and its consequences for work motivation is at heart of Job Demand Resources model, aspiring to offer an integrated approach to relationship of the needs of employers and of the employees (Bakker and Demerouti 2017).

Given the broad context of these changes at this point we could only speculate that in terms of human lifestyle and its sustainability of this digital world UN Goal 12, the overabundance of information, intrusion of work into our private time and the resulting pressure on our attention is such that much less time is available for satisfactory life.

References

Allen S, Deragon JT, Orem MG, Smith CF (2008) The emergence of the relationship economy. The new order of things to come. LLC, Silicon Valley

Bakker AB, Demerouti E (2016) Job demands–resources theory: taking stock and looking forward. J Occup Health Psychol 22(3)

Bauer J, Latzer M (eds) (2016) Handbook on the economics of the internet. Edward Elgar

Bauman Z (2000) Liquid modernity. Polity Press, Cambridge

Benkler Y (2006) The wealth of networks: how social production transforms markets and freedom. Yale University Press, New Haven and London

Berardi F (2009) The soul at work: from alienation to autonomy. Semiotext(e), California

Bertrandias L (2020) A framework for digital marketing research: investigating the four cultural eras of digital marketing. J Interact Mark 49:1–19

Bobillier-Chaumon ME (2003) Evolutions techniques et mutations du travail: emergence de nouveaux modèles d'activité. Le Travail Humain 66:163–194

Boswell WR, Olson-Buchanan JB (2007) The use of communication technologies after hours: the role of work attitudes and work-life conflict. J Manag 33:592–610

Botsman R, Rogers R (2011) What's mine is yours. The rise of collaborative consumption. Harper Collins Publishers, London

Botsman R (2012) Welcome to the new reputation economy. Wired Magazine

Bueno CC (2017) The attention economy. Labour, time and power in cognitive capitalism. Rowman Littlefield

Carlson DS, Thompson MJ, Crawford WS, Boswell WR, Whitten D (2018) Your job is messing with mine! The impact of mobile device use for work during family time on the spouse's work life. J Occup Health Psychol 23:471–482

Cheal D (1988) The gift economy. Routledge, London

Davenport TH, Beck JC (2001) The attention economy: understanding the new currency of business. Harvard Business School, Cambridge

De Wet W, Koekemoer E (2016) The increased use of information and communication technology (ICT) among employees for work-life interaction. South Afr J Econ Manag Sci 19

Fenner GH, Renn RW (2010) Technology-assisted supplemental work and work-family conflict: the role of instrumentality beliefs, organizational expectations and time management. Human Relat 63:63–82

Fisher K, Robinson JP (2009) Average weekly time spent in 30 basic activities across 17 countries. Soc Indic Res 93:249–254

Frayssé O, O'Neil M (2015) Hacked in the USA: prosumption and digital labour. MacMillan, Basingstoke

Florida R (2005) The flight of the creative class. The new global competition for talent. Harper Business/Harper-Collins Publishers, New York

Glorieux I, Mestdag I, Minnen J (2008) The coming of the 24 h economy? Changing work schedules in Belgium between 1966 and 1999. Time Soc 17(1):63–68

Goodin RE, Rice JM, Parpo A, Eriksson L (2009) Discretionary time. A new measure of freedom. Cambridge University Press, Cambridge

Halford S (2005) Hybrid workspace: re-spatialization of work, organization and management. N Technol Work Employ 20:19–33

Isaac H, Campoy E, Kalika M (2007) Surcharge informationelle, urgence et TIC: l'effet temporel des technologies de l'information, Management et Avenir 3(13):149–168

Lanham RA (2006) The Economics of attention: style and substance in the age of information. University of Chicago Press, Chicago

Latzer M et al (2016) The economics of algorithmic selection on the Internet. In: Bauer JM, Latzer M (2016) Handbook on the economics of the Internet. Chentelham

Lash S, Urry J (1994) Economies of signs and space. Sage, London

Lee H, Sawyer S (2010) Conceptualizing time, space and computing for work and organizing. Time Soc 19:293–317

Lessig L (2004) Free culture. How big media uses technology and the law to lock down culture and control creativity. The Penguin Press

Matusik SF, Mickel AE (2011) Embracing or embattled by converged mobile devices? User's experience with a contemporary connectivity technology. Human Relat 64:1001–1030

Mauss M (2002) The gift. The form and reason for exchange in archaic societies, Routledge Classics, London

Ninaus K, Diehl S, Terlutter R, Chan K, Huang A (2015) Benefits and stress-perceived effects of ICT use on employee health and work stress: an explanatory study from Austria and Hong Kong. Int J Qual Stud Health Well-being 10

Pańkowska M (ed) (2014) Frameworks of IT prosumption for business development, vol 2. Biuletyn Naukowego Towarzystwa Informatyki Ekonomicznej, pp 45–49

Pine R, Gilmore J (1999) The experience economy: work is theatre and every business is a stage. Harvard Business School Press, Boston

Qualman E (2011) Socialnomics. How social media transforms the way we live and do business. Wiley, Hoboken, New Jersey

Rifkin J (2001) The age of access: the new culture of hyper capitalism, where all of life is a paid-for experience. Penguin Putnam, New York

Rifkin J (2014) The zero marginal cost society. The Internet of Things, the collaborative commons, and the eclipse of capitalism. Macmillan, Palgrave, London

Salanova M, Llorens S, Cifre E (2013) Int J Psychol 48:422–436

Scholz T (ed) (2013) Digital labor. The internet as playground and factory. Routledge, New York

Schor J (1991) The overworked American. The unexpected decline of leisure. Basic Books, New York

Senarthne Tennakoon KIU, da Silveira GJC, Taras DG (2013) Drivers of context-specific ICT use across work and non work domains: a boundary theory perspective. Inf Organ 23:107–128

Southerton D (2006) Analysing the temporal organization of daily life: social constraints, practices and their allocation. Sociology 40:435–454

Sullivan O, Gershuny J (2004) Inconspicuous consumption: work-rich, time-poor in the liberal market economy. J Consum Cult

Stebbins R (2014) Careers in serious leisure. From dabbler to devotee in search of fulfillment. Palgrave Macmillan, Basingstoke

Tapscott D, Williams A (2006) Wikinomics. How mass collaboration changes everything. Portfolio, Penguin Group, New York

Thomas KJ (2014) Workplace technology and the creation of boundaries: the role of VHRD in a 24/7 work environment. Adv Dev Hum Resour 16:281–295

Tietze S, Musson G (2002) When 'Work' meets 'Home': temporal flexibility as lived experience. Time Soc 11:315–334

Toffler A (1970) The future shock. Bodley Head, London

Toffler A (1990) Third wave. The classic study of tomorrrow. Bantam Books, London

Tremblay DG (2012) Work-family balance; is the social economy sector more supportive and if so, is this because of a more democratic management? Rev Soc Econ LXX(2):200–232

Tremblay V, Tremblay CM (eds) (2014) Industry and firm studies. Routledge

Vayre E, Vonthron A-M (2019) Identifying work-related Internet's uses-at work and outside usual workplaces and hours and their relationships with work-home interface, work engagement and problematic Internet behavior. Frontiers Psychol 10

Yokoi T, Shan J, Wade M, Macaluay J (2019) Digital Vortex. continuous and connected change. IMD, Global Change For Digital Transformation

Webster JG (2014) The marketplace of attention. How audiences take shape in a digital Age, MIT Press Cambridge, Mass, London

Electronic Documents

Clarkberg M (2008) The time-squeeze in American families: from causes to solutions. Cornell University/US Department of Labour. Accessed 5 October 2008

Independent Work: Choice, Necessity, and the Gig Economy (2016) Mc Kinsey Global Institute. Accessed 13 December 2016

Jupowicz-Ginalska A, Kisilowska M, Jasiewicz J, Baran T, Wysocki A (2019) Fear of mising Out research study report (in Polish). https://kometa.edu.pl/biblioteka-cyfrowa/publikacja,824,fomo-2019-polacy-a-lek-przed-odlaczeniem-raport-z-badan. Accessed 5 May 2020

The New Digital Economy. How it will transform business (2011) Research paper produced in collaboration with AT&T, Cisco, Citi, PwC & SAP, June, Oxford Economics, accessed November 11, 2016

Bohdan Jung is professor at the Information Science and Digital Economy Institute at the Warsaw School of Economics (SGH). His interests are focused in the area of time use, attention economy, creative economy, socialnomics.

Tadeusz Stanisław Kowalski economist and media expert, a researcher at the University of Warsaw (as a professor since 2004), specializing in media economics and media management, the Internet's economics, social media. Co-author of the Broadcasting Act, expert of the National Broadcasting Council (1995–1997), expert of the Council of Europe (1995–2004), director of the National Film Archive (2008–2015), member of the supervisory board of public television (TVP, in 2009–2015). Author of several dozen scientific publications.

Impact of Uncompetitive Coexistence of Innovation and Tradition on Sustainable Development in McLuhan's Media Theory

Saulius Keturakis

Abstract The article discusses one of the most pressing problems of today's Western civilization—the harmonious interaction of innovation and tradition. The issue is presented from the perspective of media research by renowned media theorist Marshall McLuhan, the starting point of which is the almost absolute dependence of human behaviour on the media environment, its structure and characteristics. In McLuhan's view, so-called old media (e.g., writing) create an environment in which newer and more technologically advanced media inevitably discriminate the old ones. Therefore, from McLuhan's point of view, any harmonious state between innovation and tradition in, say, a book-based culture is impossible. However, based on the research of both primitive societies and the latest achievements of civilization, the media theorist proposes a solution—an electrical media-based human living environment, in which any inconsistent state is made impossible due to the information transmitted at the speed of light, ensuring an integrative rather than discriminatory approach to tradition. Based on less frequently discussed texts by McLuhan and data from the archives of his written legacy, the article reveals the development of such an approach, the main arguments, and the implications for areas of culture such as education or scientific cognition. It also shows how McLuhan's search for harmonious media interaction changed his general view towards the media and their subject, the human being.

Keywords Media · Environment · Sustainability · Innovation · Tradition

1 Introduction

In the famous computer game "Civilization" created by Sid Meier, the player finds himself in a situation that can be called the determinism of innovative technologies. Having received only a set of the most primitive technologies at the very beginning of the game, later the player, in order to maintain his empire, is forced to pursue

S. Keturakis (✉)
Kaunas University of Technology, K. Donelaičio St. 73, 44249 Kaunas, Lithuania
e-mail: saulius.keturakis@ktu.lt

more and more new technologies—writing, gunpowder, nuclear weapon. Stopping this constant pursuit of innovation in the game means an inevitable defeat.

Given that "Civilization" is called the most popular strategic game of all time, it could be argued that the approach embodied in the game is dominant in today's Western culture. It creates a situation where a harmonious interaction between innovation and tradition is not possible, it does not allow for the existence of several different cultures with equal rights; inevitably some become winners and others lose and are forgotten.

In order to understand the reasons for this approach and find possible solutions to change it so that innovation does not destroy tradition, it is worth remembering the ideas of the Canadian media theorist McLuhan about the media-conditioned human living environment and its impact on the nature of sustainable human relationship with the reality (Karakiewicz et al. 2015). In the world of new media, some of McLuhan's ideas today are very relevant to understanding how sustainability principles work in times of electronic communication (Bendor 2018). At a time when the latest technology is seen as the only way to develop and improve Western civilization, it is vital to bring back McLuhan's often-forgotten warning that neither innovation nor tradition alone creates a sustainable environment. On the contrary, the inability to find a way to synchronize the new and the old can undermine the ambitions of media ecology to become a meta discipline (Islas and Bernal 2016).

From McLuhan's point of view, the impact of the media on a person's ability to think or act on the reality is absolute. For this reason, uncompetitive, non-mutually exclusive interactions between tradition and innovation should be sought in McLuhan's media environment research, in which eco-friendly, i.e. the non-discriminatory human existence would be directly linked to the coherence of the media. And the most important condition of the latter, from McLuhan's point of view, is the speed of communication, on which a person's ability to react and restore the damaged harmony depends. Such a condition could be understood as the attitude that a person is inherently inclined to live by maintaining the harmonious interaction of all elements of his living environment, only some media are not opportune, creating a situation where the man plays only the role of a passive observer.

What media environment is the most harmonious and favourable, from McLuhan's point of view, to the interaction of tradition and innovation? What arguments have been presented to substantiate this?

2 Methodology

This article is based on a systematic historical analysis of McLuhan's texts from a methodological perspective on the problematization of the French sociologist and philosopher Michael Foucault (Gutting 2005). Of a rather rich set of features of problematization as a research methodology, two are used in this article. First, problematization as a methodology is characterized by special attention to those thinking and lifestyle events that are seen as problems at a certain historical moment. Second, problematization also raises the genealogical question of how the perceptions of certain thinking and action events as problems are interrelated and what the possible causes are (Rabinow 2003).

Based on the principles of problematization, this study reviewes both published and archived texts from the formation of McLuhan's media theory, looking for conditions for the emergence of a harmonious environment and their connections with cultural perspectives on tradition and innovation. The obtained data have been systematized in such a way that the assumptions of a sustainable environment in McLuhan's media theory as well as the arrangement of such assumptions in the history of media would be highlighted.

3 Results and Analysis

3.1 Environment is the Message

Media theorist McLuhan is usually remembered as an analyst of the impact of technology on a person's ability to experience, but it is much less often remembered that he saw his media theory as a study of a sustainable environment (Dale 1996). This McLuhan's perspective is usually attributed to marginal cases of his research and is left only to very curious investigators of McLuhan's heritage.

However, it is clear from the observations left by McLuhan himself in scientific texts and private letters that theorizing of the environment was an integral part of McLuhan's media theory. He had even said he wanted to use the term "environment" instead of "galaxy" in the title of his book "The Gutenberg Galaxy: The Making of Typographic Man" (McLuhan 1962, p. 1). Moreover, some researchers argue that the term "environment" was significantly more important in McLuhan's theory than "media" and is comparable in nature to the "épistémè" in Foucault's research on the history of thought (Cavell 2002). In private letters, McLuhan expressed the idea that he would even be eager to change one of his most famous sayings "the medium is the message" to another, in which "media" as a term would be replaced by "environment". The wording of the phrase should have been read as follows: "[t]o say that any new technology or extension of man creates a new environment is a much better way of saying the medium is the message" (McLuhan 1964a).

The word "environment" changes the content of one of McLuhan's most important concepts in media theory—extension. If, in the old media, the media extended a single sense, then the environment consists of an extended entire nervous system. In this way, instead of strengthening the capacity of any individual ability, a new type of situation emerges, which is called by McLuhan "A kind of involvement of the whole nervous system" (McLuhan 1967a). Involvement is a completely different way of relating to the environment than extension, because extension is related to the anaesthetic and amputation effects observed by McLuhan, where technology completely takes over some of the abilities from a person. In changing "extension" to "involvement", McLuhan was not stopped even by the inevitable situation that, in a cognitive sense, involvement creates an awkward position when the observer and the observed is the same. McLuhan initially liked this awkward situation, he called it "an invisible environment" (Marchessault 2005).

At this point, some preliminary generalizations can be made about McLuhan's hesitation on how to name the medium that determines a person's relationship to the reality and the decisions he makes. Clearly, in McLuhan's media theory, there is an opposition between the two relationships with the surrounding world.

McLuhan called the first relationship *the media* and described the connection it created with the reality as an exclusion—an amputation, in McLuhan's terms. This relationship is typical of so-called old, non-electrical communication-based media. They discriminate rather than integrate the elements of the reality, elevating some in relation to others. This is why McLuhan called the old media "anti-environment" (McLuhan 1997). Due to their anti-integrative nature, such media could be observed much more easily than the new, inclusive ones, which were described as invisible by McLuhan (Marchand 1989, p. 177).

Another human relationship with the reality is integrating and connecting rather than separating the person from the whole surrounding environment. McLuhan called it "the environment". The content of this term is basically the same as the media—it is about the communication environment that regulates all the relationship between the man and the reality. However, unlike in the case of the media, the environment is integrative rather than discriminatory in nature.

It could be said that McLuhan used two terms referring to the history of the relationship between the man and technology. One term, media, was defined as the period of the displacement of one media by another. It was during this period that McLuhan spoke of the so-called media laws, detailing how one media replaces another.

However, McLuhan sought to describe the period of electrical media in another term—environment—by giving it completely different characteristics. The most important of them is that the environment is the pursuit of a harmonious, ecological coexistence of the media.

While McLuhan's media theory is very widely discussed, much less attention is paid to so-called environment theory. What is the nature of this environment-integrating relationship?

3.2 Environment and Ecology

McLuhan's aspiration to change the concept of the media to the environment had an ecological implication. Like the media, the environment is not a passive medium. It is also a process that absolutely changes the content of communication (McLuhan and Zingrone 1995). However, if the evolution of the media is the replacement of one media by another (McLuhan 1964b, p. 158), then the principles of the environment are oriented towards organic evolution inherent in nature, which changes in the way of evolution rather than destroy it.

When thinking about the content of McLuhan's term "environment", it is essential to understand that the most important assumption is the communication speed resulting from electrical communication. Only because of the possibility to exchange data at the speed of light can the age of electricity be called, according to McLuhan, the age of ecology. Only during this period can the data on the entire biosphere be obtained at such a speed that harmony between them is finally possible (McLuhan 1969, p. 36). In a sense, the non-ecological nature of the old media, which ignores the principles of harmony, was, in McLuhan's view, a kind of trauma caused by ineffective communication.

In contrast, electric technology created the opportunity for the environment in which the media do not deny but support each other. McLuhan argued that from the old media perspective, it might be thought that the radio is a better medium for developing literacy than television, but television is perhaps much more effective in teaching languages. This approach to the media distributes various human activities to different media, identifying each one according to what it can do best. And ultimately, if effective communication enables an overview of the entire media sphere, there is no medium to be pushed out (McLuhan 2004).

As an example of such ecological approach, McLuhan saw changes in libraries that evolved from written media and book culture centres to multimedia centres. Now there are not only books, but also collections of sound recordings, films, and various events—concerts, art exhibitions take place there. The emergence of computer terminals in libraries seemed to McLuhan to be the most important feature of such integrative but not eliminative thinking. After all, it was digital communication that had to push books to the margins of the media sphere, but two symbols of different cultures, simultaneously representing tradition and innovation—the book and the computer—can coexist side by side thanks to integrative ecological thinking (Logan and McLuhan 2016).

3.3 Discipline Is the Message

Another important change brought about by electric communication is elimination of the fragmentation of human activity into disciplines (McLuhan 1964b, p. 243). The alphabet, the press, and mechanization in Western civilization created special zones to which certain cultural forms and functions were assigned. The role of history, let's say, was to interpret the past to the people of the present. The numbers were given to mathematicians, and philology studied linguistic signs. One of the features of electric media, McLuhan said, was abolition of classical dichotomies between old and new, culture and technology, art and business, work and leisure, along with creation of universal multi-disciplinarity (McLuhan 1964b, p. 301).

A closer look at the changes in terminology of McLuhan's media theory shows that the concept of the media—the medium of human presence and communication—was constantly modified in order to define the nature of that medium as precisely as possible. Now, the system of the media and the environment needs to be supplemented by a third term—discipline. Because McLuhan argued that the media can be called the discipline as well as the environment.

It is interesting that the discipline has its own hero—a specialist, an expert, and when electric communication eliminates the boundaries of disciplines, a completely new type of cultural character emerges—an amateur, non-professional or artist. It is the artist who represents integrated, undisciplined knowledge; he points out the most important directions of change in the multidisciplinary field (McLuhan 1964b, p. 71).

At this point, one of the biggest intrigues in McLuhan's media-environment-discipline theory emerges. It was mentioned at the beginning of this article that McLuhan believed that the environment, despite its positive ecological interaction with everything, has one feature—it is invisible. In cognitive terms, the anti-environment system created by the old media, in which a non-ecological, fragmented by individual disciplines world was visible and could therefore be thought of, was much more favourable.

After transforming the media into an ecological environment by means of electric communication and fast data flows, McLuhan had to look for a way to bring back cognition and, at the same time, education to the environment that could not be observed.

Paradoxically, McLuhan returns to the idea of anti-environment related to the discriminatory environment created by the old media, where the old and the traditional are doomed to be forgotten. At this point, the curious followers of McLuhan's intellectual activity had to shudder: didn't he abandon the environment project determined by eco-communication, and thus ecological thought?

In an environment where electrical media interconnects everything and where the disappearance of disciplines means that there is no room for a "different point of view", artists are taking over the creation of anti-environments, which are a condition of the perception of the environment (McLuhan 2005). Otherwise, an absolutely integrated environment could not be reflected, it would simply break out of the field of any intellectual activity (McLuhan and Carson 2003).

Further developing the conditions for cognitive and intellectual activity in a sustainable ecological environment, McLuhan complemented the character of an artist as a creator of anti-environments by a figure of an amateur. If the artist creates anti-environments that allow to think an invisible environment, then the role of the amateur is more concerned with preserving individuality. For in the cohesive environment, where all the boundaries between phenomena, whatever their nature, disappear, individuals who merge into a collective personality that does not yield to any definition, also vanish (McLuhan and Fiore 2001, p. 93).

It is easy to get lost in McLuhan's explanations how electric communication unites the separate media—the anti-environments—into a single environment in which discriminatory principles are abandoned, and then how anti-environments are again returned, because the environment is inaccessible for cognition without them. In McLuhan's case, however, this would not be a reason to call his deliberations incomplete or superficial, for this media theorist liked to demonstrate the operation of the ideas by various rhetorical strategies of the text immediately. If the conditions for cognitive activity are created by amateurs and artists, then the perception and analysis of the environment should be the opposite of the solidity and completeness inherent in classical scientific discourse—it must be playful and flimsy (McLuhan and Parker 1968). Such nature of cognition in an ecological environment should allow the abolition of both the departments of individual sciences and the descriptions of university subjects themselves, which emphasize the differences in the courses taught (McLuhan 1964b, p. 47). Cognition itself should lose the nature of a predetermined process and become a nomadic practice that does not oblige to any order (McLuhan 1964b, p. 311).

In McLuhan's environment theory, a character who can integrate various disciplines as well as tradition and innovation appears in an unexpected place—in "Playboy" magazine (McLuhan and Norden 1969). In a famous interview with this magazine, McLuhan both outlines the most important strategy for interdisciplinary environment education and points to the possibility of finally synthesizing innovation and tradition:

> PLAYBOY: Might it be possible for the "TV child" to make the adjustment to his educational environment by synthesizing traditional literate-visual forms with the insights of his own electric culture—or must the medium of print be totally unassimilable for him?
>
> McLuhan: Such a synthesis is entirely possible and could create a creative blend of the two cultures—if the educational establishment was aware that there is an electric culture. In the absence of such elementary awareness, I'm afraid that the television child has no future in our schools.

3.4 Genealogy of the Environment

What intellectual story lies behind McLuhan's shift from the reality to the solid electric media environment?

Before examining the traces of this change, it is important to remember that McLuhan became very popular after publication of his book "Understanding Media: The Extensions of Man" in 1964. His interest in the problem of the environment was taken for granted, as was the view of architecture as a particular form of communication. In a letter to his friend anthropologist Edward T. Hall, McLuhan wrote that if the new media environment is made up of electrical circuits and information, architecture also becomes the content of the new information environment (McLuhan 1965).

Along with the environment created by the electric media, McLuhan seemed to become obsessed with the topic of architecture and the environment. At first glance, this may seem coincidental, eccentric, like many of McLuhan's other activities, but a closer look at McLuhan's texts shows that his interest in the environment and architecture was indeed an interest in the changed conditions of modern human existence. The term "environment" in McLuhan's texts covered many changes in psychological, somatic, cultural, technical, and natural areas.

There is a very interesting document in the Edward T. Hall archives of the University of Arizona. It is a manuscript of the book "The Eternal Present: The Beginnings of Architecture" (Giedion 1964) by the historian of architecture Giedion (1963), based on the lectures given at the National Gallery of Art, Washington, in 1957. While reading the manuscript, McLuhan wrote down his remarks in the margins. They reveal the genesis and evolution of McLuhan's ideas about space and environment.

At the time the aforementioned Giedion text fell into McLuhan's hands, McLuhan's views on the media and the environment were just being shaped (Darroch 2008). McLuhan had begun to believe that the concept of space was historically changing. Therefore, it was very important for McLuhan that in Giedion's book he found the same historical model that he had constructed in his book "The Gutenberg Galaxy: The Making of Typographic Man", only related to the evolution of the perception of space.

In his book on the emergence of literary and print civilization, McLuhan argues that, by the electric media, the modern conception of society ceases to develop in a linear fashion and eventually returns to the original—true, modified—tribal social formations (McLuhan 1962, p. 268). While reading Giedion's book, he found the same idea, only related to the perception of space.

The most intriguing is one of McLuhan's remarks in the margins of Giedion's book. The book contained a description of one of the original limestone bas-reliefs, accompanied by the following commentary by Giedion: "With this first appearance of fully sculptural treatment of the human body, every effort was made to break free from the surface." (Giedion 1963, p. 162). This break of a man from the surface is accompanied by McLuhan's note: "non-ecological".

Giedion viewed the cultural break as an increase in human autonomy and emancipation in nature. But for McLuhan, this event was proof that the man is beginning to alienate himself from his environment. At the time he read the manuscript of Giedion's book, McLuhan was already interested in the interrelationships of all the elements that make up the whole of the environment.

Giedion's book strongly influenced McLuhan. As he later admitted, this book was one of the cornerstones that shaped his views on the environment where all the elements are interconnected and mean something. It is worth quoting McLuhan's longer remark in this regard:

> Space, Time, and Architecture was one of the great events of my lifetime. Giedion gave us a language for tackling the structural world of architecture and artefacts of many kinds in the ordinary environment. [...] He approached them not descriptively—not by classification—but structurally. Giedion began to study the environment as a structural, artistic work—he saw language in streets, buildings, the very texture of form. (McLuhan 1967b)

Giedion published some of his reflections on primitive space in the magazine "Explorations", founded by McLuhan's group. In it, he spoke of the concept of space of the primitive human, which was not guided by linearity, and where all directions were equally important (Giedion 1955). Moreover, this world was characterized by the unity of everything, the connections between the man, the beast, the plant and the rock.

McLuhan's remarks in the manuscript of Giedion's book indicate that he marked almost all the passages where "invisible forces" and cosmic unity were mentioned. Together with Giedion, McLuhan believed that the beginning of history was somehow connected to post-history. For this reason, the man of the electric age seems to return back to the state of the tribal human, at the same time associating himself very closely with the whole environment. From McLuhan's point of view, the architecture of the second half of twentieth century abandoned the linearity and perspective of writing and rediscovered the dynamic unity of the primitive man (McLuhan and Powers 1989).

"Reading" the interfaces of space was McLuhan's favourite creative pastime. According to McLuhan's biographer Philippe Marchand, McLuhan and his colleague Edmund Carpenter loved to walk down the streets of Toronto in search of connections: the clothes of the passers-by were associated with street advertising, etc. Importantly, it was the urban environment that McLuhan saw as a kind of form of consciousness in which everything was paranoidly connected (Marchand 1989, p. 116).

Even before McLuhan got his hands on the book by Giedion, he had come up with the idea of an alternative to visual (linear, logical, computational, and specialized) space—acoustic space (White and Logan 2016). In contrast to visual space, where objects can only be found by coordinates in three dimensions, in acoustic space the number of such dimensions is infinite. Acoustic space is not the capacity in which an object is stored, here the object itself creates its own space, so it is always dynamic, fluid, and changing. In the visual space, the eye finds an object at a specific point, separates it from other objects, and the ears—on the contrary—identify an object in relation to other objects. And most importantly—in acoustic space, objects are never far away, they are always close because there is no objective distance in it. It is a tactile, interactive, engaging space (Carpenter and McLuhan 1960).

An acoustic concept of McLuhan's space fits in perfectly with the primitive human space described by Giedion: the dusk of caves, the tactile orientation when light sources are unreliable and more trustworthy information is obtained by hearing and touching (Giedion 1955).

The story of the sense of space described in Giedion's book helped McLuhan take a step from the concept of acoustic space to the overall unity of the primitive man and the environment that electronic communication could recreate. It is important that this change in the age of electronic communication occurs in any case, whether a person consciously perceives it or not. Having entered the environment formed by electronic communication, a person breaks the cycle of linear history and, after trying to break away from the environment for almost two and a half thousand years, seeks connections with everything around him again thanks to the new media (Bornstein 1966).

Yet, despite the man's return to the primitive spatial experience, electronic communication changes something in the worldview of the primitive man. Yes, in McLuhan's understanding, electronic communication creates a person who cares about the environment and is connected to the whole world, but that environment is no longer random and the connection to the environment itself is not intuitive—it is managed and controlled (Powe 2014).

However, McLuhan sees no problem in this controlled ecological state, on the contrary, he sees it as a very effective tool. Let's say such an environment can be configured so that a second foreign language is learned as perfectly as a native language, because learning is not about constructing knowledge of grammar rules and vocabulary but engaging in the world of a new language at the experiential level (Gordon 2010).

McLuhan believed that the city should have become such a fully inclusive and controlled environment. In one letter to Jaqueline Tyrwhitt, McLuhan wrote (Darroch 2008) that the city should first be planned considering that it is an educational machine. If various stimuli of human senses are successfully placed in the urban space, such an educational medium would be, on one hand, ecological, and on the other hand, very effective (McLuhan 1987).

In the letter, McLuhan details his vision of the environment:

> It is possible to design a computer-controlled space in which the geometry of the room, as well as all its other sensory components, could be precisely varied. Groups of students could be taught various types of problems under these controlled conditions. Depending on their cultural and perceptual bias, one could discover exactly the focus for the various senses which would enable them to learn any given problem in math or biology or language at maximal speed. These levels would in turn reveal the sensory parameters of the culture. A Chinese could be provided with an environment which would enable him to see the West as if it were the East. (McLuhan 1964c)

It is important to remember that the artist becomes the creator of such a controlled environment (McLuhan 1966). Despite this frivolity of control, McLuhan felt that the aspect of global controllability of electronic communications was bringing the public close to a dangerous threshold beyond which George Orwell's "1984" Big Brother, field of global surveillance and monitoring, could be found (Massolin 2001). McLuhan sought to divert this problem of total control of human life from the field of consideration of the good and the evil—the philosophy of morality—by saying that technological change and its consequences for the man cannot be considered from a moral perspective (McLuhan 1962, p. 158). To do this, according to McLuhan, would be the same as to blame the saw for cutting fingers. The very problem of the interaction between technology and the man, McLuhan believes, should be taken away from the perspective of moral evaluation for the sake of moral philosophy itself (McLuhan 1962, p. 159).

Technology-driven environmental change is morally neutral (McLuhan 1969). And if someone were to decide to undertake an assessment of change caused by technology, such a person would behave in a non-ecological way—he would exchange life with all its dynamics and natural transformations into "troglodytic security of the ivory tower" (McLuhan 1962, p. 159).

Having spoken of evolutionary changes in the media sphere, compared it to the biosphere and the laws of natural Darwinian change, McLuhan speaks positively about total control, because change is inevitable and its control means being in the forefront of change, and not following it from behind (McLuhan 1964b, p. 179).

In McLuhan's honour, it must be said that he was a self-critical scientist. He acknowledges that in a medium of total electronic control, where the man can regulate all parameters of the reality, it is no longer possible to speak of classical nature. He said:

> Since Sputnik and the satellites, the planet is enclosed in a manmade environment that ends "Nature" and turns the globe into a repertory theatre to be programmed. (McLuhan and Watson 1970)

In this way, McLuhan's project of the harmonious media environment, created by electrical communications, fundamentally eliminates the classic notion of natural environment. Does this mean that sustainable development, ecological existence and elimination of oppositions such as tradition and innovation can only be artificially created today and would not survive without special, controlling intervention? McLuhan himself had anticipated this desire to criticize, blame, and moralize, warning that one should not be outraged by the changes brought about by technology. That is just the way it is, and all you need to do is live.

4 Conclusion

In search for ways to resolve the conflicting nature of the interaction between innovation and tradition in contemporary Western culture, the article explores less frequently remembered texts and archival material by the Canadian media theorist McLuhan.

First of all, from his point of view, there is a correlation between the structural features of the media sphere and human behaviour: the media sphere of a confrontational nature creates the same discriminatory interactions in human activity and its reflection. McLuhan sees the modern Western man as being fundamentally inclined to pursue harmonious interactions in the media sphere, while at the same time establishing ecological connections throughout human life and eliminating hierarchical divisions in dichotomies such as tradition and innovation, expert and amateur, culture and technology.

However, human tendency to create sustainability is not enough, as this requires communication technology of a very high performance and speed. Unfortunately, from McLuhan's point of view, all non-electric media condemns people to be separated from the environment and be non-ecological, because while creating a convenient distance for scientific analysis, they fail to provide the necessary communication speed to create an inclusive rather than discriminatory environment. For the first time in the history of mankind, such an opportunity is provided by electric media, which allow the modern man to feel like a primitive member of the tribe, whose ecological—integrative—connection with the environment was intuitive and non-reflexive.

Consequently, McLuhan draws attention to the potential for a hybrid presence of tradition and innovation, pointing to the benefits of such a cohesive, ecological environment by exploiting media from different historical times in the most efficient way. Unfortunately, delving into McLuhan's tools for creating a harmonious interaction between innovation and tradition, it becomes clear that this harmony can only emerge as a result of enormous control of the entire environment. This artificiality of the harmony of the media sphere, as well as the dependence of the whole ecological existence project, indicated by McLuhan, on the effectiveness of communication technologies can raise questions from the perspective of moral philosophy: how much ecological attitude, the harmonious interaction of oppositions such as innovation and tradition, depends on the man, and how much—on technology? Is it not the case that, by harmony and ecological existence, we define the level of technological development rather than human consciousness? McLuhan himself tried to anticipate intentions to raise such questions, acknowledging, on the one hand, that nature no longer existed in the classical sense, and, on the other hand, suggesting us to accept technological evolution as natural and simply live in the environment that technology offers us.

References

Bendor R (2018) Interactive media for sustainability. Palgrave McMillan, Delft, p 16
Bornstein E (1966) An interview with Marshall McLuhan. Structurist 6:61
Carpenter E, McLuhan M (1960) Acoustic space. In: Carpenter E, McLuhan M (eds) Explorations in communication: an anthology. Beacon Press, Boston, pp 65–70
Cavell R (2002) McLuhan in space. University of Toronto Press, Toronto, p 52
Dale S (1996) McLuhan's children. The Greenpeace message and the media. Between the Lines, Toronto, p 12
Darroch M (2008) Bridging urban and media studies: Jaqueline Tyrwhitt and the explorations group, 1951–57. Can J Commun 33:147–169
Giedion S (1955) Space conception in prehistoric art. Explorations 6:33–54
Giedion S (1963) The eternal present: the beginnings of architecture. In: Edward T (ed) Hall papers. University of Arizona, Box 39
Giedion S (1964) The eternal present: the beginnings of architecture. Oxford University Press, London
Gordon WT (2010) McLuhan. A guide for the perplexed. Continuum, New York, p 81
Gutting G (2005) Foucault: a very short introduction. Oxford University Press, New York, p 104
Islas O, Bernal JD (2016) Media ecology: a complex and systemic meta discipline. Philosophies 1:190–198
Karakiewicz J, Yue A, Paladino A (2015) Promoting sustainable living-sustainability as an object of desire. Routledge, London, p 16
Logan R, McLuhan M (2016) The future of the library: from electric media to digital media. Peter Lang Publishing, New York, pp 72–73
Marchand P (1989) The medium and the messenger: a biography. The MIT Press, Cambridge
Marchessault J (2005) Marshall McLuhan. Cosmic Media. SAGE Publications, London, p 202
Massolin PA (2001) Canadian intellectuals, the tory tradition, and the challenge of modernity 1939–1970. University of Toronto Press, Toronto, pp 58–59
McLuhan M (1962) The Gutenberg galaxy. University of Toronto Press, Toronto

McLuhan M (1964a) Letter to hall (September 16, 1964). In: Edward T (ed) Hall papers, special collections. University of Arizona, Box 8, Folder 28
McLuhan M (1964b) Understanding media: extensions of man. Mentor, New York
McLuhan M (1964c) Letter to hall (August 24, 1965). In: Edward T (ed) Hall papers, special collections. University of Arizona, Box 8, Folder 28
McLuhan M (1965) Letter to hall (June 22, 1965). In: Edward T (ed) Hall papers, special collections. University of Arizona, Box 8, Folder 28
McLuhan M (1966) The emperor's old clothes. In: Kepes G (ed) The man-made object. George Braziller, New York, pp 90–95
McLuhan M (1967a) Invisible environment. Perspecta 11:161–167
McLuhan M (1967b) A dialogue. In: Stearn G (ed) McLuhan, hot & cool. Signet Books, New York, pp 263–267
McLuhan M (1969) Counterblast. Rapp & Whiting Ltd., London
McLuhan M (1987) Letter to Jaqueline Tyrwhitt (May 11, 1964). In: Molinaro M et al (eds) Letters of Marshall McLuhan. Oxford University Press, Toronto, pp 298–299
McLuhan M (1997) Media research. Routledge, London, p 111
McLuhan M (2004) Understanding me: lectures and interviews. MIT Press, Cambridge, p 271
McLuhan M (2005) Marshall McLuhan unbound. Ginko Press, Berkeley, p 20
McLuhan M, Carson D (2003) The book of probes. Ginko Press, Berkeley, pp 30–33
McLuhan M, Norden E (1969) Playboy interview: Marshall McLuhan: a candid conversation with the high priest of popcult and metaphysician of media. HMH Publishing Co., Chicago, p 14
McLuhan M, Parker H (1968) Through the vanishing point: space in poetry and painting. HarperCollins, New York, p 243
McLuhan M, Powers BR (1989) The global village: transformations in world life and media in the 21st century. Oxford University Press, New York, p 31
McLuhan M, Watson W (1970) From cliché to archetype. Viking, New York, p 9
McLuhan E, Zingrone F (1995) Essential McLuhan. Routledge, London, p 275
Powe BW (2014) Marshall McLuhan and Northrop Frye—Apocalypse and Alchemy. University of Toronto Press, Toronto, p 102
Rabinow P (2003) Anthropology today: reflections on modern equipment. Princeton University Press, Princeton, pp 44–48
White EF, Logan RK (2016) Acoustic space, Marshall McLuhan and links to medieval philosophers and beyond: center everywhere and margin nowhere. Philosophies 1:162–169

Saulius Keturakis is a Professor of Media Studies at Kaunas University of Technology in Kaunas, Lithuania, where he teaches courses in Creative Writing and Digital Media. Saulius's current research interests include media philosophy and new media rhetoric, digital research methods, and computer-assisted text analysis.

Sustainable Event Management: New Perspectives for the Meeting Industry Through Innovation and Digitalisation?

Dirk Hagen

Abstract Through their growth in the postmodern and capitalist-oriented economies, "meetings", all forms of MICE events (Meetings, Incentives, Conventions, Exhibitions) represent purely quantitative, ever more comprehensive, resource-consuming gatherings. Such a development does not only increase the complexity of the business gatherings in terms of their global economic importance, but also in terms of increasing CO2 emissions. In recent years there has been a response to this development and new potentials for a sustainable meeting industry has been created, driven also by the megatrend of digitalisation, and the innovations related to it. Although sustainability of meetings have already been discussed more from the perspective of environmentally friendly or resource-conserving use, the aim of this paper is to discuss innovative examples like virtual or hybrid events, i.e. a kind of mixed event consisting of physical and digital meetings. The methodology approach is based on a theoretical background and in the gathering of empirical evidence, combined with a literature review. The results indicate positive opportunities, risks and challenges by virtual and hybrid meetings by a sustainable perspective that contribute to solutions that go beyond the general approach of sustainability in the debate.

Keywords Sustainability · Digitalisation · Event · Meeting Industry · Hybrid/Virtual Meetings

1 Introduction

In recent years, three trend topics have strongly influenced the public and scientific debate in the event and meeting industry worldwide: sustainability, digitalisation, and security. Additionally, with the Covid-19 virus, a disruptive development has taken place: The accompanying ban on the implementation of almost all event forms in

---

D. Hagen (✉)
Professor Of Business Administration Focus On Event Management, Berlin School of Management, SRH Berlin University of Applied Sciences, Ernst-Reuter-Platz 10, 10587 Berlin, Germany
e-mail: dirk.hagen@srh.de

the sense of socio-physical meetings since spring 2020 for hygienic security reasons has caused the number of virtual events carried out online to rise sharply and has inevitably led to an unexpected surge in digitalisation in the event industry. Although many digital tools and digital possibilities such as apps, streaming services or video conferences etc. have been available and used for many years (Boshnakova and Goldblatt 2017; Grosser 2017), the crisis has now led to a strong increase in virtual trade fairs, conferences, and conventions. The question rightly arises as to what prospects this could create in terms of sustainability. Particularly in view of the debate on sustainability in the events industry which has increased sharply in recent years (cf. e.g. Holmes et al. 2015; Jones 2017), this study, which has a focus on all types of meetings, examines this debate case under these current changed conditions: What developments in terms of sustainability potential and strategies can be derived from digitalisation or the increased use of online transmissions as a substitute for physical meetings? Does this create a separate, sector-oriented option in terms of sustainability, now driven by innovations in this global industry? In this text, it is assumed that this sector-oriented potential for sustainability differs from the general approaches to resource conservation and emission reduction for economic and service sectors or is sector-specific. In the context of this paper, the extremely heterogeneous event industry and event forms are also generally dealt with, but the focus is clearly on conventions, conferences, also understood or summarised as MICE (Meetings, Incentives, Conventions, Exhibitions) in the sense of an overarching international term—the so-called "Meeting Industry".

In the following, a first overview of the scientific discussion regarding the MICE industry and sustainability is provided by a brief lierature review: Due to a systematic literature review based on published articles from 2010 to 2018, Anas et al. (2020) identify accommodation, transportation, safety, security and facilities as the general key success factors of the MICE industry, however, without making a direct link to the sustainability approach. Nevertheless, events put the quality of its economic, social and natural environments at risk. For that reason, educators and researchers see a need to integrate sustainability into meeting and event management. In this context, Presbury and Edwards (2005) investigate the Best Education Network to identify the need of an incorporation of sustainability into meeting and event management. They conclude that information is not easily available and that there is only little published research about the incorporation of sustainability into texts and teaching materials about meetings and events. MICE (Meetings Incentives Conventions Exhibitions) professionals' perceptions of sustainability in meeting and event education is analysed by Millar and Park (2018) to identify the essential knowledge that should be taught to students to cope future challenges. In general, the professionals believed that sustainability is important for the industry and should be taught to the students. However, their definitions of sustainability varied and many of them are focused on the environmental impact, but not on the social and economic impacts.

Sustainable practices in convention facilities are summarized by Draper et al. (2011). They estimate the significance of these practices according to the meeting planners and develop three general categories: (1) energy efficiency, (2) recycling, and (3) sustainable policies. Whereas Jung et al. (2016) the perceptions and willingness of visitors to take part in sustainable practices of the MICE industry were investigated by doing a paper-based survey at the Consumer Electronic Show in Las Vegas. They conclude that the visitors' perceptions of the sustainable practices applied are not so positive and that the visitors have different opinions about the importance of the different sustainable practices. Furthermore, the results also suggest that the willingness of visitors to engage in sustainable practices can be traced back to perceived engagement efforts. Kim and Ko (2020) analyse the essential factors for a sustainable design of a MICE event that might be helpful in practice. They conclude that residents' willingness to participate in a MICE event increases and becomes relevant for the social identity if the name of a MICE location becomes more context-specific and MICE events have similar features. With mixed methods, Buathong and Lai (2017) explore in the metropolitan area of Bangkok, Thailand, the perceptions about sustainable event development and provide empirical evidence on important issues of event sustainability. They identify government agencies and the private sector as driving forces of the sustainable event development, but claim that also societal areas could support the development, e.g., like "environmental education and course curriculum, set standards for event sustainability, improved knowledge and understanding of sustainable event development, as well as related national policies and budget management" (ibidem. 15). In addition, Reshetnikova and Magomedov (2019) investigate the cooperation between local government and foreign private companies during development organisation of the industry associated to MICE in the Russian Federation. Due to their study, they link the concepts of sustainability and corporate social responsibility with social and financial results, showing the most important ways to develop public–privatepartnerships in tourism in Russia and present general methods and instruments to classify the different regions regarding the necessity of state investments for further development. In this context, Ramgulam et al. (2012) examine business tourism and its contribution to socio-cultural sustainability and the decisive theories regarding socio-cultural impacts and spill-off externalities. Moreover, they compare the tourism industry of Trinidad with key strategic areas of tourism of Australia. They come to the result that business tourism is socio-cultural sustainable and that it should be continued as a micro-sector to support the diversification of the economy of Trinidad and Tobago. Furthermore, they recognize potential benefits if the countries actively pursue business tourism that is in line with established socio-cultural practices. However, Hall (2012) argues that not all events are sustainable. Against the background of three different sustainability interpretations—economic sustainability, balanced sustainability, and steady-state sustainability, he (2012) concludes that mega-events are focused on economic or balanced sustainability and, thus, are not sustainable in their actual form. In comparison, sustainability is more likely to be found if the events are smaller, locally limited, community-based, and designed for the longer term or at least contribute that helps to maximize the use of the existing infrastructure. However, compared to mega-events, the problem with these events is that they are not so attractive for the self-presentation of political and corporate interests. The same applies to profit-oriented consulting firms and academics who gain from mega-events. To enable planners in the MICE industry to create an integrated sustainable event marketing plan, including the planning process, meeting and communication with participants, Tinnish and Mangal (2012) develop a framework by using a holistic view of sustainability, with a particular focus on marketing and communication practices, and a model for forestry marketing. On the basis of actual meeting planning priorities and sustainability values, Lee and Slocum (2015) explore the feasibility of sourcing local food for the meeting industry. The study results show that there exists a solid demand for locally sourced food by meeting planners. Meeting planners have contractual flexibility in their food choices and that there is a willingness to pay a higher price for local products. Nevertheless, from their perspective, there is a need to improve the knowledge of the benefits of using local food among meeting planners and participants. Obviously, the sector has a reference to the environmental dimension, while the social and economic pillar is hardly taken into account. This is indicated by the terms "Green Meetings" (cf. e.g. Green Meeting Industry Council w. D.). The term "Green Meeting" covers almost all forms of events that fall under the term "Meeting", i.e. from smaller meetings with only a few participants to conventions with over a hundred thousand participants. "Green Meetings" can thus be understood as an approach to the planning, implementation, documentation, and further development of environmentally compatible events. This reference in terms of dimension, however, is restricted primarily to environmental orientation. Getz and Page (2016: 395) see the term "green" as being associated with "reduce, reuse, recycle". In the above-mentioned descriptions, as a trend, sustainability in the sense of the environmental dimension is clear, but especially the economic dimension and also the social dimension has been given less consideration here.

2 Event and Meeting Industry: An Outline

Within the scope of the manuscript size, only a rough overview of the phenomenon "event" is offered with the aim of providing a first orientation. There is a wide range of literature on the most diverse forms of events and different (scientific) perspectives, including economic, sociological, ecological, and geographical views (cf. e.g. Getz and Page 2020). Over the last decades, there has been a continuous increase in the number of event formats carried out and the number of participants/visitors/guests, which can be understood on the one hand as a form of "eventisation" in the sense of a quantitative increase. The growth in Germany alone in terms of participants in all forms of events developed steadily and dynamically up to 2019 and is estimated at over 423 million, with thousands of venues, congress centers, or conference hotels and around 2.89 million events (EITW 2019). On the other hand, many more festivities, private celebrations, as well as more social and social areas are being professionally planned and managed. These range from organised food festivals over tattoo conventions to dinner events and can also be understood in terms of qualitative eventisation (cf. Hitzler 2011). While this increasing phenomenon of an experience-oriented, postmodern society can also be understood as forms of new community building, in these event-oriented societies, the orientation towards western economies, as in Europe and North America, is more likely to be observed, while Asian meeting formats in their heterogeneous forms are becoming increasingly dynamic in their development (ICCA 2019; cf. Bathelt et al. 2014).

Forms such as festivals, trade fairs and large congresses as well as the Olympic Games or FIFA World Cup are to be understood as mega-events in the sense of their number of visitors, economic significance, or even the media attention (Getz and Page 2016: 59). They are ultimately regarded by many neutral observers not involved in the discussion as "actual events" because they receive a lot of media attention. Especially the area of mega-events and larger festivals is the subject of papers from various scientific perspectives, especially with regard to their urban development and infrastructural consequences (cf. Hagen et al. 2021; Hiller 2000; Jago et al. 2010). In recent years, however, the topic of environmentally friendly sustainability has become the focus of attention with a large number of publications (Holmes and Mair 2018: 584; Case 2013; Holmes et al. 2015; Jones 2017). In the reality of the industry, however, by far the greater number can actually be attributed to smaller or smallest event forms. In the case of Germany, it is the largest meeting market in terms of time—after the USA (ICCA 2019)—according to the annual "Meeting & EventBarometer" (EITW 2019). It includes, for example, lists of over 60% of all events with less than 100 visitors/guests. A further distinguishing element—which is also relevant in terms of the sustainability debate—is the question of whether an event takes place outdoors, e.g. a festival on a meadow, or indoors in exhibition halls, congress centers, conference hotels or such attractive venues as castles, airport hangars or old factories.

The definition of these events is not uniform as an overarching description, but they can be understood as planned, targeted, one-off meetings with a clearly limited time frame. However, a core element is also to be seen in the emphasis on an experience orientation, multi-sensory appeal with characteristics such as staging and aestheticization. Getz and Page (2016) provide a good overview of the different types of events, presented in Fig. 1. In their typology of planned events, they differentiate between six columns, "Cultural celebrations", "Business and trade", "Arts and entertainment", "Sports and recreation", "Political and state", and "Private functions." Getz and Page mention for each type several subtypes. According to the main focus of this treatise, the column "Business & Trade" is highlighted, but it is to mention that the focus is on all kinds of meetings.

Within this framework, the focus is on the so-called meeting industry or MICE industry (Meetings, Incentives, Conventions, and Exhibitions). This can be understood as business or subject-oriented event forms, the importance of the "meeting industry", as only one part of the event industry. This is observed in Germany, as over 50% of all events are assigned to such meeting formats (EITW 2019) with a turnover of 114 billion euros (R.I.F.E.L. 2020). For a better international overview, the annual ICCA (International Congress and Convention Association) report can be used. The ICCA report provides due to the narrow definition of "international meetings" by this association. It is important to note, however, that the Western economies, in particular, are in the lead with the USA as first, followed by Europe with Germany in the

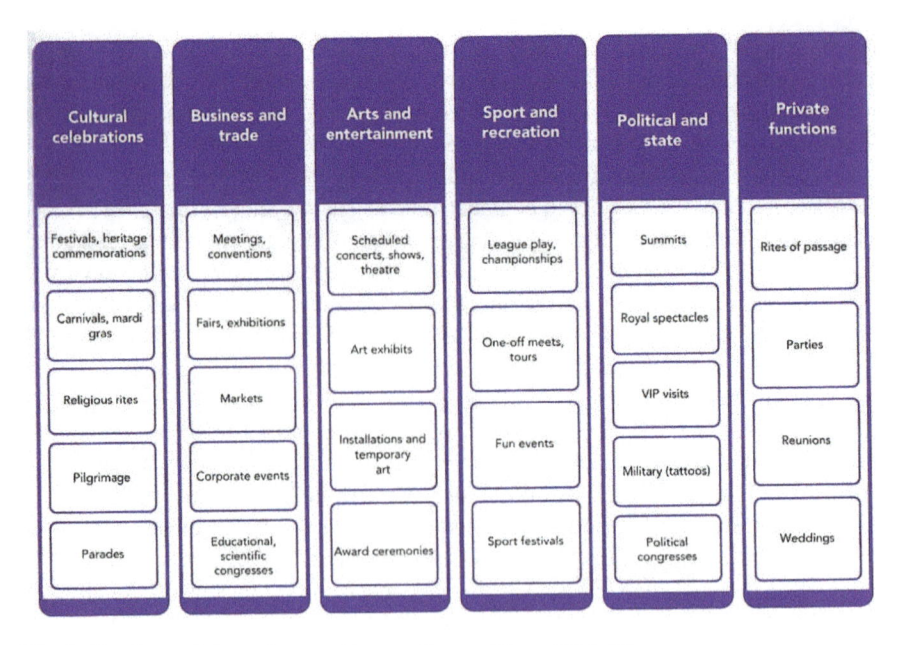

Cultural celebrations	Business and trade	Arts and entertainment	Sport and recreation	Political and state	Private functions
Festivals, heritage commemorations	Meetings, conventions	Scheduled concerts, shows, theatre	League play, championships	Summits	Rites of passage
Carnivals, mardi gras	Fairs, exhibitions	Art exhibits	One-off meets, tours	Royal spectacles	Parties
Religious rites	Markets	Installations and temporary art	Fun events	VIP visits	Reunions
Pilgrimage	Corporate events			Military (tattoos)	
Parades	Educational, scientific congresses	Award ceremonies	Sport festivals	Political congresses	Weddings

Fig. 1 Typology of planned events (Source: Getz and Page 2016: 53)

lead and with a strong increase in number and importance of the Asian region with China in the lead. Additionally, the report shows that central European destinations such as Paris, Berlin, London, Madrid, Barcelona, or Vienna, to name but a few, have benefited enormously from this growth for years or even decades (ICCA 2019).

A noteworthy feature of these event formats, apart from the provision of information, is the strongly increasing importance of networking, which is promoted by social contacts and promotes complementary knowledge. On the one hand, such personal contacts are responsible for project collaborations and even business transactions within the framework of meetings, conferences, conventions or trade fairs (Hagen and Luppold 2017: 254ff.; Zenk et al. 2014: 216ff.) Whereby here too, in the context of eventisation, more and more experience-oriented influences are increasing, up to a festivalization of congresses and trade fairs etc. (Getz and Page 2016: 36f, Hagen 2019: 281ff.). From an economic-geographical perspective, meetings can then also be understood as "temporary clusters" (cf. Henn and Bathelt 2015). Such business events then represent an accumulation of actors from industry, services, associations or politics, where knowledge is conveyed and exchanged. The personal contact that is established during such meetings, the so-called face-to-face contacts (cf. Storper and Venables 2004), is decisive at this point. Due to the spatially and temporally condensed structure and organisation, such events make such contacts possible. Networks are primarily understood as drivers of economic dynamics and innovation development (cf. Grabher 2004; Granovetter 1973). The networking or network formation between the various participants shows overall potential for innovative and production-promoting developments. It is crucial for the establishment of networks that social interactions between the various participants and stakeholders take place at all. For this purpose, coffee breaks, lunch, dinner, and get-togethers are usually available. Such relationships can be established here rather randomly and unstructured. For many participants, especially these contacts are of great importance for a "successful" participation in meetings, which is usually time-consuming and costly (cf. Hagen and Luppold 2017; Zenk et al. 2014).

3 Sustainability as a Mega-Trend in the Event and Meeting Industry

During the last years, sustainability has also developed into an important strategic and operational task in the event industry (cf. e.g. Bowdin et al. 2012; Case 2013, 2015; Holmes et al. 2015; Jones 2017). Various areas relating to the realisation of event formats will be included in this respect. These include (Hagen 2020: 80–82):

- sustainable mobility
- sustainable logistics
- sustainable procurement
- sustainable catering or regional /organic cuisine
- sustainable conference material
- sustainable waste management
- sustainable use of technology
- sustainable event/conference venues

The sustainable goals of event management can be based on

- Energy efficiency
- Conservation of resources
- Climate protection or neutrality
- Full compensation of the CO_2 footprint

A sustainable approach becomes particularly complex in the event industry because of the number of stakeholders such as venues, destinations, event cities, municipalities, and their authorities (e.g. security requirements). A large number of different service companies are involved in the realisation of even just one meeting or congress. Included in these complexities are event agencies, technical specialists, caterers, security companies, hostess companies, logistics providers, etc.

Case (2015) makes it clear that the basic idea of events, is to bring people together in a venue, uses up resources, and causes environmental damage. Although this idea is reduced through sustainability measures when central destinations or venues are chosen, it does not seem realistic for decentralised destinations or venues (Case 2015: 368; Holmes and Mair 2018). The costs of transport/logistics and personal mobility is also an attempt to reduce CO_2. This applies to all forms of outdoor events, but also to the meeting industry. This means that physical meetings remain a challenge especially, when the participants are dependent on CO_2-producing means of transport or their energy drives, such as aircraft, due to long distances.

Numerous national and international certifications have been developed, as well as international standards such as ISO 20121:2012 (Event Sustainability Management Systems) or EMAS (Eco Management and Audit Scheme). These are standards that at least indicate a development towards a more sustainable event management. In Europe/European Union, this has led to the publication of CSR (Corporate Social Responsibility) reports for larger forms of companies, including the promotion of event and meeting planners with sustainability certifications. However, it remains unclear to what extent such event sustainability standards are specifically geared to the particular challenges of events or meetings or are ultimately geared quite generally to organisational forms, companies etc. Ultimately, they do not go beyond general, less sector-specific approaches (Große Ophoff et al. 2020: 171).

4 New Technologies for the Meeting Industry

The general digital transformation has led to a surge of innovation in the meeting industry, including digital tools, but clearly spurred on by the pandemic crisis, there is a wide abandonment of physical meetings to purely online meetings, also understood as virtual meetings. Hybrid meetings, in contrast, are characterised by the fact that at least in part a physical meeting still takes place. In hybrid meeting forms, for example, parts of the keynotes and guests can be in a venue in the sense of being present on site, while other speakers or guests, e.g. during workshops, are then added online and are regarded as non-physical participants. The basis for the use of digital tools and digital applications is the technical or digital development during the last years, from which possible applications in the event industry can be seen or are already being used in the industry. These include among others (GBC 2020a):

- Artificial intelligence (AI)
- Big Data/Data Analytics
- Blockchain
- Cloud computing
- Robotic Process Automation
- Internet of Things
- 5G

These innovative developments have led to a variety of applications in the meeting industry, but in fact, these are new technological developments which, in the sense of a digital transformation, have reached all sectors of the economy and are not specifically used only in the meeting industry.

The specific use of certain, individual digital tools in recent years has certainly led to a surge in innovation, especially in the operational implementation in the meetings industry. However, the physical meetings were not questioned in terms of their breadth and mass. The digital tools were sometimes rather understood in terms of an improvement or modernization of the customer experience. The following is a selection of some possible digital tools and applications used currently in the meeting industry (cf. GCB 2020a,2020b):

- Virtual Reality (VR)
- Augmenting Reality (AR)
- Holograms
- IBeacon Technology (Location Contextual/ Location Specific)
- Navigation/GPS (Indoor/outdoor)
- Event platforms for participant recruitment and management
- Chat offers for customer communication
- Digital Twins
- Event Apps

Event apps are certainly one of the current focal points in terms of applications in the meeting industry. At this point, a distinction can be made between native apps, browser-based apps, shared apps and multi-event apps (Grosser 2017: 54). Ultimately, these offer general information on and about meetings and are already being used in many cases (Vogel and Thomas 2020: 54ff). In fact, they do indeed represent a sustainable component if the information on the meetings is only or mainly regulated by the event app. This would demonstrate a significant reduction in paper and material consumption. The programmes' information could be ideally offered as exclusive digital versions instead of being physically printed (ibidem: 62ff).

While keynotes, lectures or information of all kinds can also be transmitted as a stream via video, personal interpersonal contact with the aim of building a network is an increasingly important reason for physically attending meetings, as already described above. The following digital tools offer extensive technical support in this respect (cf. GCB 2020a, Hagen 2017):

- Event apps for interaction
- Matchmaking tools based on virtual assistant systems

While event apps are used, for example, for live voting or are increasingly being used as digital tools in workshops or interactive formats such as World Café, matchmaking tools represent a more complex form of interaction. In the meeting industry, the term "matchmaking" is the easiest to compare with the matching systems used in dating agencies (Hagen 2017: 752). The aim here is to enable a comparison between the content interests and topics of participants and providers (sponsors). In this process, large amounts of participants' data is collected virtually, e.g. on a landing page, and immediately compared with all existing participants' profiles on an AI-based basis and adapted at any time before a matching, i.e., a bringing together of participants on the one hand and solution providers, consultants or other participants on the other hand in "one-to-one meetings". In the run-up to the events, participants fill in multi-page, subject- and topic-oriented questionnaires and information sheets online and use the software-supported matching to set up face-to-face appointments with time and place details. The daily schedule or agenda can be updated at any time—almost in "real time" and displayed in the event app. In the context of the meeting industry, not only new potential for controlling interactions at meetings has been developed on the basis of such digital systems, but it has also led to a more efficient use of temporary meetings (Hagen 2017: 752f.).

5 Methodology

Since spring 2020, semi-structured, qualitative interviews with 10 German-speaking experts from practice were carried out. The expert survey was conducted in German. The selection of the experts was purposeful, and a focus was put on practitioners who were active as company representatives of relevant companies from the German-speaking countries, but also active as planers of the international MICE sector or meeting industry. Various questions regarding possible future sustainable perspectives of virtual/hybrid meetings potentials, virtual/hybrid meetings restrictions, challenges of virtual or hybrid event formats were asked. In addition, homepages and other corresponding information from relevant companies were studied with regard to the use of digital tools and the implementation of analogy or virtuality, as well as in the sense of hybrid events. These empirical results have been supplemented by a literature review, especially with regard to sustainability and the meeting industry, as already described above. Of course, like most qualitative studies, this does not provide any representativeness, but with regard to a scientific triangulation approach, meaningful results or trends can be provided. The transcriptions in the text were freely translated into English.

6 Results and Discussion

6.1 Sustainable Positive Effects of Virtual Meetings

From the experts' point of view, there are lasting positive effects for participants in such virtual events. The complete elimination of travel activities with simultaneously increased accessibility of—now virtual—attendees is listed. A further focus is on the time savings and lower costs resulting from the elimination of travel and hotel accommodation costs that result from virtual participation. It becomes clear that virtual meeting formats could also replace a considerable part of the physical meetings in the future, or at least some of the current participants could only join online. The ecological aspect, resulting from the almost complete elimination of travel and the associated lower CO_2 production is considered relevant. Whereby, this aspect is in the foreground compared to the economic sustainable positive effects for participants or for companies that cover the costs for such participation. It becomes obvious that—for example, in the form of hybrid meetings—hardly any solutions such as online participation for participants with high CO2 production due to travel with, e.g., aeroplanes have been considered in the past. Again, this impact is seen as a long-term perspective after the current pandemic limits.

As a respondent commented:

"Travel costs are completely eliminated, which takes the form of air travel, hotel accommodation, etc."

Furthermore:

"So I don't have to rent the hall, I don't need catering for the masses of people. I do not need transfers."

This is an important result, which points to a new sustainability option. Some authors have recently underlined once again the great importance of CO2 production through travel in the meeting industry. Within the framework of event realisation, travel activity represents the highest share of CO2 production or the largest share of carbon emissions (cf. Holmes and Mair 2018). This is certainly also true for the meeting industry (cf. Hischier and Hilty 2002; Neugebauer et al. 2020). The arrival and departure of participants for just one convention produces a large amount of CO2 as well as total emissions within the global meeting industry (cf. Civardi w. D.; Neugebauer et al. 2020). Publications indicate that the share of CO2-equivalent caused by travel, especially international or intercontinental air travel, for (international) meetings, for example, can be as high as 95%, depending on the share of air passengers (cf. Hischier and Hilty 2002). While new or renovated locations have made progress in terms of environmental sustainability (Große Ophoff et al. 2020: 178f.), as well as in the operational implementation (e.g. catering), and progress in logistics can also be seen to be moving forward (Hagen 2020).

6.2 *Sustainable Negative Effects of Virtual Meetings in the Business Chain of the Meeting Industry*

The lasting negative economic effects particularly affect planers of all previous physical forms of meetings, conventions, exhibitions, etc., as well as the venues that are places of realization. Massive losses in turnover can be expected and the fundamental question can be raised as how such venues can continue to exist economically. The effects due to the almost complete loss of local economic activities in virtual meetings have a lasting negative economic impact on all companies belonging to the production chain for the realization of such events. The economic sustainability or a large part of the MICE or meeting industry is questioned with such formats. The loss of jobs is also highlighted, and for example, caterers, event technicians, restaurateurs, hotel and service staff are listed. The negative impact on the travel industry as a whole is also listed, which has so far benefited from business events in particular. The prospects for the future are also assessed as economically negative, as a relevant number of participants and companies will not return to physical meetings in the future.

As a respondent commented:

"I believe that event companies, including hotels and restaurants, will have problems in the future because more will shift to the digital sector. We will never be as well off as we were before in my view."

6.3 Sustainable Effects of Virtual Meetings for Destinations

Compared to the direct effects on companies, the experts tend to treat or list the perspectives for cities, or destinations as secondary. The meeting industry has long been of great economic importance for destinations in terms of various positive economic effects (Rogers 2013: 251ff.). Cities and municipalities would lose a significant economic source of income without the significant business travel that results from conferences, meetings, etc., which often goes hand in hand with hotel accommodation, taxi rides, restaurant visits, etc. Venues, e.g. historic castles, museums, but also convention centers and meeting hotels, would suffer significant declines in turnover due to a sharp drop of physical events. In the same way, a significant proportion of the companies who concentrate in the destinations will suffer declines. As a result, jobs are threatened to a considerable extent, with correspondingly negative social consequences in these destinations who would suffer from losing tax income.

A recent study (R.I.F.E.L. 2020) uses the example of the German business events market to illustrate the consequences: The turnover generated by physical meetings to date is estimated at around 114 billion. Around one million jobs in Germany depend on this. In future, hardly any jobs would be created in destinations, cities and municipalities.

6.4 Technological Challenges for a Sustainable Successful Realisation of Virtual/Hybrid Meetings

In this context, the picture is divided. The experts mention the simplicity of video conferences or zoom meetings, but also the orientation to offer technically smooth conferences. The innovative development of related event technologies is seen as positive, but not yet sufficient. The creation of experiences and atmospheres, which often justifies high participation fees, can only be made possible by further innovations in event technologies (Wreford et al. 2019). The lack of personal contact opportunities is more difficult from the experts' point of view. However, personal contacts or network building, e.g. between speakers or participants and sponsors, are important elements in successful meetings and MICE formats. Without these aspects, one must also reckon with declining interest or declining economic prospects for the meetings industry as a whole.

As a respondent commented:

"The disadvantage definitely is the loss of networking opportunities during coffee breaks or during meals; these cannot be replaced in the same quality [...] digitally there is no possibility of a friendly smile or a personal greeting with a handshake."

Furthermore:

"You can't shake handshakes over the internet and experience shows that the acquisition and regeneration of leads is only physically possible. [...] I cannot simply get to know people spontaneously."

The importance of non-verbal communication and social interaction is already highlighted by Argyle (2013). Conferences, or almost all forms of meetings, are also to be understood as cumulative places for initiating business, which is important for personal interaction (Bathelt and Turi 2011; Hagen and Luppold 2017). With a virtual assistant system for matchmaking of one-to-one contacts, at least some (virtual) interactions can be facilitated, but clearly with less quality as in physical meetings.

6.5 Hybrid Meetings as a Perspective for the Meeting Industry

The expert survey clearly illustrates that—in addition to purely virtual meeting formats—hybrid forms, in particular, will be granted a sustainably successful future perspective. It is assumed that a certain number of physical participants will stay in the corresponding venues, and that hotel use and travel activities will then take place again. A larger number of participants will then participate virtually, in other words, online. Thus, on the one hand, it is assumed that a certain number of former physically present participants will only join virtually, while on the other hand new groups of participants will be opened up by the hybrid formats.

As a respondent commented:

"In the future, we will definitely go two-track and hold events hybrid, if not only digitally, because it offers the possibility of overcoming distance quickly and cost-effectively bringing together international guests."

Furthermore:

"So for us, the topic of hybrid events is actually the topic of the future."

Furthermore:

"We can reach a much larger audience with a format that we do not only plan live, i.e. that combines digital and presence. [...] We are convinced that we will certainly be able to present many formats in hybrid form that we originally only had a one-time presence."

Grosse Ophoff et al. (2020) point to the potential positive characteristics such as efficiency and eventfulness of hybrid conventions, although the approach of environmental sustainability remains since a larger proportion of participants do not travel. Nevertheless, virtual events, in turn, mean a significantly higher use of computers, tablets, data centers, communication networks, internet services, streaming services etc., which also have a CO2-producing effect. Thus, such ICTs (Information and Communication Technologies) have a carbon footprint as well (cf. Belkhir and Elmeligli 2018). The production of such technical devices, e.g. smartphones for event app use, is also associated with social distortions regarding poor working conditions in developing countries. Therefore, it seems relevant and reasonable to consider ICTs that are used in virtual meetings in terms of a life cycle assessment (LCA) or sustainability calculation. Despite such a consideration of such influences, the CO2-eq is many times higher in physical meetings compared to virtual events (Civardi w.D.), of course to an increasing extent with the increasing use of air transport; in the ecological sense, the advantage of virtual meetings is therefore beyond doubt. However, hybrid formats would still mean the need for a physical meeting but with fewer participants. This would continue to have a positive impact on destinations and service companies, and business chains of the meeting industry.

7 Conclusion

Although some efforts have been made in recent years to improve sustainability approaches, particularly with regard to the ecological dimension in the meetings industry, no significant solutions have been developed for CO2-producing travel, which is the main source of harmful emissions. The results of this paper clearly show that if these event formats take place for the most part only virtually, there is a danger that although the environmental dimension is very much satisfied by the elimination of travel, the social and economic dimensions in terms of declining turnover and job losses will develop in a clearly negative way, with negative consequences for destinations and businesses. Furthermore, the results of the expert interviews and the theoretical background of this paper show that the hybrid meeting form could represent a sector-specific sustainability option. Although meetings are held physically, some of the stakeholders and associated companies are still assured of maintaining existence, while at the same time using the technological, virtual or digital solutions which have little negative ecological impact. Furthermore, digital and innovative tools such as virtual assistant systems for matchmaking could at least partially secure or enable networking options between people who are not physically present. This would enable that the meeting industry could reach even broader groups and range of participants virtually. Overall, this could result in an industry-oriented sustainability approach for the meeting industry that takes into account all dimensions of the sustainability approach. However, more future research in the interface of sustainability and integration of virtual and hybrid meetings will be necessary.

References

Argyle M (2013) Bodily communication, 2nd edn. Routledge, London, New York

Anas MS, Maddiah NA, Eizamly NUEN, Sulaiman NA, Wee H (2020) Key success factors toward MICE industry: a systematic literature review. Special Issue Sustain Safety Secur (3S) JTHCA 12(1):188–221

Bathelt H, Golfetto F, Rinallo D (2014) Trade shows in the globalization knowledge economy. Oxford University Press, Oxford

Bathelt H, Turi P (2011) Local, global and virtual buzz. The importance of face-to-face contact in economic interaction and possibilities go beyond. Geoforum 42(5):520–529

Belkhir L, Elmegli A (2018) Assessing ICT global emission footprint: Trends to 2040 & recommendations. J Clean Prod 177:448–463

Boshnakova D, Goldblatt J (2017) The 21st century meeting and event Technologies: Powerful tools for better planning, marketing, and evaluation. Oakville

Bowdin G, Allen J, O'Toole W, Harris R, McDonnell I (2012) Events management. 3rd edn. Routledge, Oxon, New York

Buathong K, Lai P-C (2017) Perceived attributes of event sustainability in the MICE industry in Thailand: a viewpoint from governmental, academic venue and practitioner. Sustainability 9:1151

Case R (2015) Event impacts and environmental sustainability. The Routledge Handbook of Events. Abingdon, New York, pp 362–384

Case R (2013) Events and the Environment. Routledge, London

Civardi C (w. D.) (2020) Making industrial exhibitions green. A literature research on the LCA of physical virtual exhibitions. https://www.v-ex.com/wp-content/uploads/2020/06/Sustainability-Report_Whitepaper.pdf (Last Accessed 27 Oct 2020)

Draper J, Dawson M, Casey E (2011) An exploratory study of the importance of sustainable practices in the meeting and convention site selection process. J Con Event Tour 12(3):153–178

EITW (Europäisches Institut für TagungsWirtschaft GmbH) (2019) Meeting—& EventBarometer Deutschland 2018/19. Pressekonferenz 15. Mai 2019. Wernigerode. https://www.vdr-service.de/fileadmin/services-leistungen/fachmedien/fachliteratur_studien/GCB-EVVC-DZT-EITW_Meeting-Event-Barometer_2018-2019.pdf (Last Accessed 27 Oct 2020)

GCB (German Convention Bureau e. V.) (2020a) Innovation Catalogue 4.0. Future Meeting Space. Frankfurt/Main. https://gcb.de/de/trends-inspiration/future-meeting-space.html (Last Accessed 27 Oct 2020)

GCB (German Convention Bureau e. V.) (2020b) Digitale Tools für die Veranstaltung der Zukunft. Frankfurt/Main. https://www.gcb.de/de/germany-meetings-magazin/meetings-hands-on/2020/digitale-tools-fuer-die-veranstaltung-der-zukunft.html (Last Accessed 27 Oct 2020)

Getz D, Page SJ (2016) Event studies: Theory research and policy for planned events, 3rd edn. Routledge. New York

Getz D, Page SJ (2020) Event studies: Theory, research and policy for planned events, 4th edn. Routledge. New York

Grabher G (2004) Learning in projects remembering in networks? Communality sociality and connectivity in projects ecologies. Eur Urban Reg Stud 11(2):103–123

Granovetter MS (1973) The strength of weak ties. Am J Sociol 78(6):1360–1380

Green Meeting Industry Council (GMIC) (w. D.) http://www.gmicncn.org/ (Last Accessed 10 Oct 2020)

Grosser T (2017) Bits und Apps im Messe—und Tagungswesen. Veranstaltungen 4.0. Konferenzen, Messen und Events im digitalen Wandel. Springer Gabler. Wiesbaden, pp 37–65

Große Ophoff M, Griese KM, Werner K (2020) Event organisations at the interface between sustainabillity and digitalisation. Events—Future, Trends, Perspectives. UTB, München, pp 161–187

Hagen D (2020) Industrie 4.0/Logistik 4.0: Potenziale für eine nachhaltige Event-Transportlogistik—oder: Interdisziplinäre Ansätze zur Curricula-Entwicklung durch Logistik. Trends und Event Education. Gabler Springer, Wiesbaden, pp 73–87

Hagen D (2019) Festivalisierung in der Meeting Industry: Von der Kontaktanbahnung zur Spaßgesellschaft und zurück. Ausgesuchte empirische Ergebnisse zum Digitalfestival re:publica, Berlin. Eventforschung. Aktueller Stand und Perspektiven. Gabler Springer, Wiesbaden, pp 281–287

Hagen D (2017) Matchmaking. Innovative Ansätze zur Steuerung sozialer Interaktion, Netzwerkbildung und Geschäftsanbahnung in der Meetingbranche. Praxishandbuch Kongress-, Tagungs- und Konferenzmanagement. Gabler Springer, Wiesbaden, pp 749–754

Hagen D, Luppold S (2017) Matchmaking: Steuerungsinstrument für Interaktion und Netzwerkbildung—Ansatz zur Incentivierung und Emotionalisierung. Events und Erlebnis. Gabler Springer, Wiesbaden, pp 251–262

Hagen D, Gaeva D, Krasnov E, Barinova G (2021) Challenges by cultural and sport mega-events: Socio-economic and environmental effects: proceedings of international conference. Publishing office of Immanuel Kant Baltic Federal University [Electronic resource]

Hall CM (2012) Sustainable Mega-Events: Beyond the Myth of balanced Approaches to Mega-Event Sustainability. Event Manag 16:119–131

Henn S, Bathelt H (2015) Knowledge generation and field reproduction in temporary clusters and the role of business conferences. Geoforum 58:104–113

Hiller H (2000) Toward an urban sociology of mega-events. Res. Urban Sociol 5:181–205

Hischier R, Hilty L (2002) Environmental impacts of an international conference. Environ Impact Assess Rev 22:543–557

Hitzler R (2011) Eventisierung: Drei Fallstudien zum marketingstrategischen Massenspaß. Gabler Springer, Wiesbaden

Holmes K, Mair J (2018) Events, festivals and sustainability: The Woodford Folk festival, Australia. The Palgrave handbook of sustainability. Palgrave Macmillan and Springer Nature. Cham, pp 583–597

Holmes K, Hughes M, Mair J, Carlsen J (2015) Events and sustainability. Routledge. Abingdon, New York

ICCA (2020) The international association meetings market 2019. ICCA Statistics Report—Public Abstract. https://www.iccaworld.org/knowledge/article.cfm?artid=701 (Last Accessed 27 Oct 2020)

Jago L, Dwyer L, Lipman G, van Lill D, Vorster S (2010) Optimising the potential of mega-events: an overview. Int J Event Festival Mgmt

Jones ML (2017) Sustainable event management: A practical guide. Routledge, Abingdon, New York

Jung S, Kim Y-S, Malek K, Lee W (2016) Engaging attendees in environmental sustainability at trade shows: attendees' perceptions and willingness to participate. Anatolia 27(4):540–542

Kim K, Ko D (2020) How to build a sustainable MICE environment based on social identity theory. Sustainability 12:7166

Lee S, Slocum S (2015) Understanding the role of local food in the meeting industry: An exploratory study of meeting planners' perception of local food in sustainable meeting planning. J Con Event Tour 16(1):45–60

Millar M, Park S-Y (2018) Industry professionals' perceptions of sustainability in meeting and event education. J Teach Travel Tour 18(2):123–137

Neugebauer S, Bolz M, Mankaa R, Traverso M (2020) How sustainable are sustainable conferences?—Comprehensive Life Cycle Assessment of an international conference series in Europe. J Clean Prod 242:118516

Presbury R, Edwards D (2005) Incorporating sustainability in meetings and event management education. Int J Event Manag Res 1(1):30–45

Ramgulam N, Raghunandan-Mohammed K, Raghunandan M (2012) Exploring the Dynamics of Socio-Cultural Sustainability in Trinidad's Mice Market. Am Int J Contemp Res 2(6):44–56

Reshetnikova N, Magomedov M (2019) Influence strategic competitive advantage on the MICE industry and its sustainability. Studia Ekonomiczne. Zeszyty Naukowe. Uniwersytetu Ekonomicznego w Katowicach Nr 382:170–185

R.I.F.E.L. e. V. (Research Institute for Exhibition and Live-Communication) (2020) Die gesamtwirtschaftliche Bedeutung der Veranstaltungsbranche. Berlin https://rifel-institut.de/fileadmin/Rifel_upload/3.0_Forschung/Meta-Studie_gesamtwirtschaftliche_Bedeutung_der_Veranstaltungsbranche_RIFEL.pdf. (Last Accessed 20 Jun 2021)

Rogers T (2013) Conferences and conventions: A global industry. 3rd edn. Routledge. London, New York

Storper M, Venables AJ (2004) Buzz: face-to-face contact and the urban economy. J Econ Geogr 4(4):351–370

Tinnish SM, Mangal SM (2012) Sustainable event marketing in the MICE industry: A theoretical framework. J Con Event Tour 13(4):227–249

Vogel J, Thomas O (2020) Digitalisierung als Enabler nachhaltiger Veranstaltungen: Potenziale und Handlungsfelder durch neue Technologien. Trends in Event Education. Gabler Springer, Wiesbaden, pp 49–71

Wreford O, Williams NL, Ferdinand N (2019) Together alone: An exploration of the virtual event experience. Event Manage 23(4–5):721–732

Zenk L, Smuc M, Windhager F (2014) Beyond the name tag. Wissen nimmt Gestalt an—Beiträge zu den Kremser Wissensmanagement-Tagen 2013, Krems, pp 215–224

Global Trends and Regional Aspects of Environmental Management

Towards Sustainable and Responsible Regional Innovation Policy—the Case of Tampere Region

Anna Martikainen, Mika Kautonen, and Mika Raunio

Abstract The sustainability of regional development largely depends on the balance between social, economic and environmental processes. This paper presents the Responsible Research and Innovation (RRI) methodology as a regional policy tool that facilitates democratization of research and innovation processes and involves the public in bringing direction and more versatile insights into processes. In other words, among other things, it is supposed to help solve problems related to the pressing economic, social and environmental challenges of our time. This chapter uses data gathered from a development project conducted in Tampere region, Finland, to discuss benefits, challenges and good practices related to adopting the RRI methodology. The region has approximately 35 years of history with various innovation policies, from science parks and cluster policy to platform policy aimed at forming a sustainable and responsible innovation ecosystem. Sources of data include desk research, regional maturity mapping and an interactive workshop. The results show that maturity mapping, despite the challenges that arise, is useful for indicating areas of intervention to strengthen the RRI in regional innovation policy. As a part of a regional innovation policy strategy, Open Innovation Platforms can be a useful tool for implementing certain aspects of RRI in regional policy.

Keywords Regional innovation policy · Responsible research and innovation · Open innovation platforms · Sustainability

---

A. Martikainen (✉)
Institute of Rural and Agricultural Development, Polish Academy of Sciences, ul. Nowy Świat 72, 00-330 Warsaw, Poland
e-mail: aciechomska@irwirpan.waw.pl

M. Kautonen
Tampere University, Pinni B, Kanslerinrinne 1, P.O. Box 607, 33014 Tampere, Finland
e-mail: mika.kautonen@tuni.fi

M. Raunio
Migration Institute of Finland, Kampusranta 9C, 60200 Seinäjoki, Finland
e-mail: mika.raunio@migrationinstitute.fi

1 Introduction

Responsible Research and Innovation (RRI) is an approach that anticipates and assesses potential implications and societal expectations with regard to research and innovation, with the aim of fostering the design of inclusive and sustainable research and innovation (van den Hoven et al. 2013). RRI brings together topics of theoretical concern that appeared after the eighteenth century and crystallized during the nineteenth century around the expected and unexpected impacts of research and innovation, technology assessment and regulation of science (Monsonís-Payá et al. 2017). However, using the initialism RRI to address these concerns is much more recent (Zwart et al. 2014). The approach is included in the European Framework Programmes and is developed as an approach to governing research and innovation at the European Union level. It is an objective of the Europe 2020 strategy to create smart growth and is reflected in the Horizon 2020 programme, which defines tackling societal challenges as one of its main priorities (von Schomberg 2013). The commitment of the European Commission to promoting the RRI approach even resulted in the inclusion of specific calls for projects of research and coordination under this topic in Horizon 2020 in Part V "Science with and for Society Work Programmes". However, the approach was also investigated outside of Europe and has the potential to be included in policy frameworks globally, due to its relevance for supporting sustainable development goals and inclusive growth.

In recent years, the idea of RRI has started to be emphasized, especially by the European Commission in its research and innovation framework programmes (European Commission 2012, 2017; MoRRI consortium 2018). While research ethics has been a fundamental part of programmes before, and RRI has its forerunners, for example in technology assessment practices (Pellé and Reber 2015), RRI extends the idea of ethical behaviour and responsibility from the research process to societal impacts and consequences of innovation (Timmermans and Stahl 2013).

This corresponds with a recently conceived turn in framing innovation policy to deal with social and environmental challenges as identified in the Sustainable Development Goals (SDGs) and calling for sustainability transitions (Schot and Steinmueller 2018). For such an innovation policy, innovations are expected to address certain well-chosen societal objectives by contributing to, for example, a low-carbon and inclusive economy (Schot and Steinmueller 2018: 8). This contrasts with an innovation policy from the past where innovations were, in general, considered positive for the society at least as long as they contributed to employment and/or competitiveness.

In the following, we will briefly study the relationship between RRI and its regional dimensions. From the outset, it seems that a regional (or local) level is viable for RRI because of the geographical proximity that enables different kinds of stakeholders to be engaged in dialogue that is frequent and intense. Geographical proximity often comes with other types of proximity such as cultural proximity, which facilitates communication and thus mutual understanding between those various stakeholders (c.f. Torre and Rallet 2005). However, it should be noted that RRI so far has lacked the place-based approach and is not fully compatible with the prevailing EU approach to regional development, smart specialization (RIS3), due to its unclear concept of geographical scale (Fitjar et al. 2019).

The regional level also often forms an institutionally suitable environment when there are governmental and municipal authorities with resources and some autonomy. In addition, this is the level where usually most people operate daily and on which their sphere of life is focused. Bathelt and Glűckler (2011), in their overview of the relational turn of the school of economic geography, for example, stress the contextual nature of knowledge, i.e. processes of knowledge creation are cumulative and evolutionary, and different kinds of decision-making processes are the results of networks of social and institutional relations within which organizations and individuals operate. Therefore, the regional level provides a suitable environment for service and social innovations, for instance, as people tend to be committed to improving their immediate living environment. People engaged in co-creation processes may also get to see the results concretely in their vicinity, further increasing their motivation provided the results are perceived as positive (e.g. Needham 2007; c.f. Moulaert et al. 2007).

The extent to which, and how, RRI can be adopted as part of the regional policy toolbox depends on the regional system of innovation. There is always a question of resources, for example, but the regional institutional set-up, both in terms of formal and informal institutions, also alters the way RRI may be deployed. It seems likely, for example, that a regional culture characterized by a stark social hierarchy does not contribute to RRI measures as positively as a culture of equalitarianism. Other factors of importance include industrial structure and the level of education in a region. The composition of industries makes a considerable difference: RRI issues related to medicine and biotechnology, for example, may be much more complex than, say, a case in the mechanical wood industry. The educational level of the region's inhabitants may affect the ways in which RRI measures are conducted.

2 Open Innovation Platforms as a Tool to Increase Sustainability and Responsibility of Regional Innovations

During the last decade or so, the concept(s) of platforms has emerged to denote such modes of cooperation that unfold innovation processes for new actors as well as considering new types of value creation (Asheim et al. 2011; Cooke and de Laurentis 2010; c.f. Raunio et al. 2018). Under the broad umbrella concept, the use of the concept ranges from technological product platforms (e.g. the iPhone), value chains and industrial platforms (e.g. the car industry) to digital platforms (e.g. Űber and Facebook). Especially in the latter case, value creation is dependent on those platforms' ability to attract users and/or developers to achieve network effects (e.g. Choudary 2013; Haigu 2014).

The Open Innovation Platform shares some attributes with these platform concepts but is an independent concept that relates to innovation policy institutions, often on a regional/local level. Its roots can be seen to lie in a user-driven and open innovation approach (von Hippel 2005; Chesbrough 2003) as well as in the field of social and inclusive innovation approaches (George et al. 2012; Heeks et al. 2014; Avelino et al. 2017). All these approaches have contributed to widening the understanding of relevant stakeholders of innovation activities; they have helped to bring in, besides the traditional "Triple Helix" (Etzkowitz and Leydesdorff 2000) of industry, government and academia, many actor groups representing civil society. Early on, von Hippel used his well-known term "democratizing innovation", although at that point he was merely referring to users (or consumers) of a certain innovation. Inclusive innovation and related approaches have brought in a different horizon where it is stressed that there are many, often disenfranchised, societal groups (e.g. elderly people or people with disabilities) that cannot participate in innovation processes or cannot benefit from many innovations that are potentially vital for them (Schillo and Robinson 2017; Heeks et al. 2014). In addition, the concepts of frugal innovation and bottom of the pyramid have helped to include a perspective of poor people from the global south.

To date, there has been no generally accepted definition of an Open Innovation Platform (OIP). We have (see Kautonen et al. 2017: 6) preliminarily defined it elsewhere "…as a space for facilitation of co-creation processes that enables various third parties, in addition to producers and users, to participate in innovation processes, usually with the support of a peer-to-peer community". Thus, an OIP is an innovation hub that has some systematized facilitation methods and is at least relatively open to gathering insights from various stakeholders of its innovation processes.

OIPs provide a practical institution to ground RRI in the use of regional-level innovation policy. This does not mean there are not any other institutions of relevance; it is just the focus of the paper to study OIPs in relation to RRI. To begin with, there are some similarities in both: they are usually based on principles of openness and inclusion (at least relatively), and they enable the participation of new kinds of societal groups not traditionally associated with innovation activities. For companies that base their businesses on sustainability in broad terms an OIP is an attractive partner due to deploying RRI as an integral part of the facilitation process. Whenever an OIP is tasked with a social or environmental challenge of some kind, following the RRI methodology provides it with relevant guiding tools. For example, RRI puts emphasis on inclusivity, which guides an OIP to possibly broaden its partnerships, thereby potentially gaining more varied insights and feedback for an innovation process in question. This may also reduce the risk of lack of social acceptance in the early phase of development, contributing to the successful commercialization of an innovation. Finally, the regional level with its geographical proximity supports the processes of control and trust (e.g. Cooke 2004) that are needed for RRI.

Tampere region has approximately 518,000 inhabitants in 2020. The city of Tampere is the second-most important centre of research and development in the country, with research, development and innovation expenditure accounting for about 3.7% of the gross regional product (Council of Tampere Region 2020). Both the main industries of manufacturing and, more recently, information and communication technologies have recently been surpassed by service industries in terms of employment. A cornerstone of the regional innovation system is the higher education sector, which has approximately 4700 personnel and 32,000 students, as for 2018 (Statistics Finland 2020).

There is a long history of regional innovation system development in Tampere region. The main phases according to the main regional development programmes have included the following (c.f. Sotarauta and Kautonen 2007; Kautonen 2012; Kautonen et al. 2017; Raunio et al. 2018; Nordling 2019; see also Council of Tampere Region 2020):

- Prior to 1994, many institutions related to the regional innovation system had already been created, including the first science parks and technology transfer programmes at the end of the 1980s
- 1994–1998 (1st regional innovation programme): The First Centre of Expertise Programme where many localized networks among industry, local government and academia were created. This was the time when Nokia Corporation was rapidly growing and its R&D personnel grew almost exponentially in Finland, including Tampere.
- 1999–2006 (2nd regional innovation programme): the Second Centre of Expertise Programme during which the cluster policy was introduced and put into practice (particularly mechanical engineering and automation, ICT, and biomedical and health-care clusters)
- 2007–2013 (3rd regional innovation program): around 2010—the switch from cluster policy to platform policy (OIPs such as Demola, New Factory, Start-up accelerator, etc.)
- 2014–2020: The Six City Strategy (6Aika), which includes the six largest cities in Finland, effectively established Open Innovation Platforms as a mainstream of the regional innovation policy
- Currently: creation of an innovation ecosystem. Open Innovation Platforms as a centrepiece; a recent fusion of the two major universities in Tampere led to a search for new practices in regional innovation policy too.

3 Method

The main objectives of the research were to examine the role of regional policy practices (OIPs) in supporting RRI in the region, to identify the strong and weak points of the region in the context of RRI keys—needs for improvement and to indicate obstacles and difficulties in using the currently developed methodology in regional RRI assessment.

The research method used in this study consisted of a background analysis (with a review of the literature), complemented by the collection of empirical evidence using participatory active research. Due to the scope of the study and the nature of the methodological approach, the study cannot be regarded as representative. However, it builds a profile of regional policy tools to increase their sustainability and responsibility.

The research techniques used included desk research, regional maturity mapping and an interactive workshop. Desk research was conducted to identify the status quo of RRI in regional policies, as well as regional conditions and processes affecting RRI. Regional maturity mapping, supported by a focus group and interviews, was realized as part of the INTERREG Project MARIE—"Mainstreaming Responsible Innovation in European S3". It was aimed at auditing policy practices supporting RRI in the region and indicating obstacles in using the currently developed methodology to assess RRI on a regional level. An interactive workshop, based on the "World Café" methodology, was conducted to deepen the understanding of stakeholders of the role of OIPs in RRI and facilitate the co-learning process.

Not all regions have the same level of maturity in their awareness and implementation of RRI in their public policies on innovation and they do not have the same needs in terms of supporting RRI in their regional context. Therefore, an RRI maturity assessment was designed and implemented to assess the RRI maturity of Tampere region and identify its strengths and weaknesses in terms of RRI awareness and implementation. The assessment involved the self-assessment of the region in terms of its level of maturity in understanding the RRI concept, the regional conditions and processes affecting RRI, and the consideration and inclusion of RRI dimensions and elements in the regional policy for supporting research and innovation. The self-assessment featured 12 indicators chosen by Apospori and Tsanos (2018) (inspired by Strand et al. 2015; Ravn et al. 2015), which represented the six keys of RRI, namely Public Engagement, Open Access, Ethics, Gender Equality and Science Education, and Governance. The assessment leads to categorization of the region in terms of each indicator in three maturity rankings: "Substantial maturity", "Moderate maturity" and "Modest maturity".

An interactive workshop on the topic of responsibility in Open Innovation Platforms was conducted. To increase the effectiveness of the co-learning process, an international group of RRI professionals was invited to take part in the workshop. It was conducted based on the "World Café" methodology. Participants were divided into four groups. Each of them visited four different spots, where different topics related to the responsibility in Open Innovation Platforms were discussed. The topics included the RRI keys most relevant to OIPs, namely Engagement, Open Access and Ethics, as well as RRI Governance and Evaluation. The questions involved the benefits, challenges and good practices of incorporating RRI keys in Open Innovation Platforms. There were four facilitators, one per spot, who introduced to new groups the ideas of previous groups and helped the groups with the process of generating new ideas by collecting suggestions and writing them down. Each group spent 20 min at every spot. After the workshop, a short summary of the results was presented during the joint session by facilitators.

Fig. 1 RRI maturity of Tampere region according to maturity mapping results. *Source* Own study

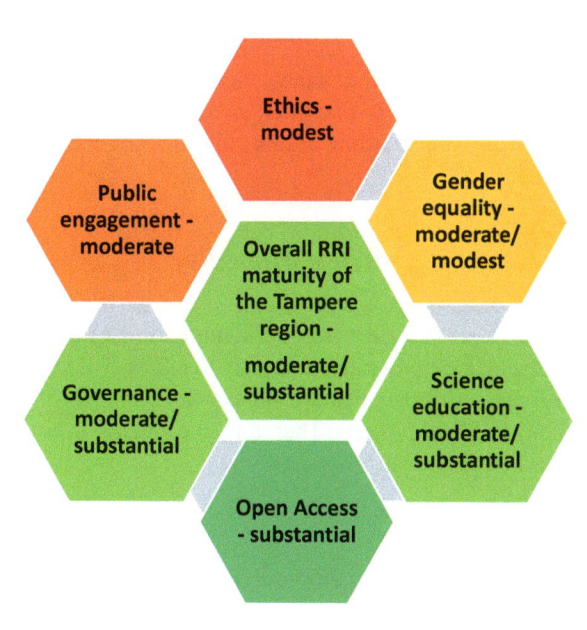

- Ethics - modest
- Public engagement - moderate
- Gender equality - moderate/modest
- Overall RRI maturity of the Tampere region - moderate/substantial
- Governance - moderate/substantial
- Science education - moderate/substantial
- Open Access - substantial

4 Results

4.1 Results of Maturity Mapping

Results show that the average maturity level of the region can be considered moderate to substantial, due to the results obtained in all RRI keys separately (see Fig. 1). Below, results for the assessment of Public Engagement, Open Access, Ethics, Gender Equality and Science Education and Governance are presented.

4.2 Public Engagement—Moderate Maturity

In terms of Public Engagement, one indicator was of a quantitative and one of a qualitative nature. According to Special Eurobarometer 340 (European Commission 2010), the percentage of respondents who stated "the public should be consulted and public opinion should be considered when making decisions about science and technology" was 47% for Finland, compared to 29% in the EU27. Therefore, the maturity level according to this indicator is considered substantial. The Regional Council has legal obligations to share and discuss the regional strategy and smart specialization strategy with citizens and give them the opportunity to comment on them. Formalized structures and mechanisms at the national level for involving citizens in regional science and technology decision-making, such as existing organizational bodies facilitating public involvement and legal frameworks mandating citizen participation in S&T decision-making, were classified as relatively "formalized". However, the degree to which citizens are de facto involved in making decisions in Tampere region was classified as relatively "low involvement", which is why the maturity level according to this indicator was considered moderate.

4.3 Ethics—Modest Maturity

The evaluation for the regional funding of R&I proposals and funded projects does not include criteria related directly to ethics. There were no R&I projects funded by regional authorities that are subject to evaluation of ethical concerns over the total number of R&I projects. Therefore, the maturity level according to this indicator was considered modest.

4.4 Gender Equality—Moderate/modest Maturity

According to EUROSTAT (2016) in Finland the percentage points difference between the share of the economically active population of women and that of men among scientists and engineers amounted to 43.48 percentage points. Among 260,800 scientists and engineers in Finland, 187,100 were male and 73,700 female. The maturity level according to this indicator is considered modest. Approximately 38.5% of regionally funded R&I projects support gender equality by creating RDI jobs that employ women, which is considered a moderate maturity level.

4.5 Science Education—Moderate/substantial Maturity

RRI-related training, and thus training in ethical, economic, environmental, legal and social aspects (EEELSA) in regional R&I strategies and projects, is not required. It is, though, a part of the education at universities at a substantial level. There are many events organized in the region that include RRI-related topics (for example: Corporate Responsibility Days at the University of Tampere, European Robotics Forum, etc.). There are many open lectures available at the universities that are related to thematic elements of RRI, such as ethics and gender equality. Therefore, stakeholders taking part in the focus group assessed the maturity level as substantial in science education.

4.6 Open Access—Substantial Maturity

The open access in the region is at a high level, especially in a broad sense. There are projects involving multiple actors that have the goal of sharing the experience and good practices of multiple actors, not only a MARIE project. For example, Tampere takes part in a project in which the six largest cities in Finland join forces to address their common urban challenges. The Six City Strategy (6Aika) is implemented in the form of cooperative projects. They engage different actors of the urban community to create smarter and more open cities. Two other projects involving multiple actors with the goal of sharing experience and good practices are Open Tampere and Open Innovation Platforms. Every regional funding call includes elements of open science/open innovation requirements. However, there are no specific open innovation calls available in regional scale. Most such projects have a national scope. The Council of Tampere Region does not fund projects of individual companies but funds different forms of cooperation, such as Open Innovation Platforms, which results in a much higher level of openness than in the regions that fund projects of separate companies. Stakeholders taking part in the focus group assessed the maturity level as substantial in open access.

4.7 Governance—Moderate/substantial Maturity

Tampere region is characterized by a high number of networks and high involvement of regional authorities in these networks, both of a formal and informal character. The region has a long tradition of establishing Open Innovation Platforms, such as Health HUB (promoting RRI in the medical industry), Demola (with a strong component of open access, as well as science education and equality), Inno-Oppiva (with a strong emphasis on ethics and science education) and Koklaamo (with a strong involvement of citizens). At the University of Tampere operates an Ethics Committee of Tampere region, which connects different fields of research and organizations to give opinions about ethical aspects of research. There are also committees related to particular topics, for example bioethics. RRI is supported by various projects, such as MARIE, 6Aika and various Open Innovation Platforms. There is also financing for projects related to responsible business models and sustainability, such as a circular economy. However, promoting RRI per se only started recently. Due to a lack of agreement between stakeholders taking part in the focus group, the governance maturity level was estimated as moderate/substantial.

4.8 Results of the Workshop

During workshop stakeholders indicated many challenges related to RRI, benefits that different aspects of RRI can bring and ways to implement RRI in Open Innovation Platforms.

4.9 RRI Governance and Evaluation

During the workshop, participants answered three questions related to governance and evaluation. The first was related to the most important keys (such as engagement, open science, open access, open innovation, science education, capacity building, equality, diversity, ethics and critical discussion) in the context of open innovation platforms. According to participants, engagement is a key for capacity building and increasing trust between stakeholders. It is a challenge to indicate which groups should be engaged and how. The transparency of the whole engagement process is important to avoid mistrust and lobbying. It is also important to ensure the highest possible level of equality among stakeholders. Everyone should know what the input is and what the outcome is, what is required or expected and what the benefit is. It increases engagement and thus enables value creation and capacity building. Participation facilitation is a useful tool for increasing the engagement level of stakeholders. Open access allows platform diversity and promotes critical discourse. Ethics is also very important in the context of innovation platforms.

The second question was why the implementation and development of the RRI approach should be evaluated and measured in the various innovation platforms. According to stakeholders, in the case of public policy, citizens should know how the public funds are used, which is why the evaluation of platforms is necessary. Change management also implies evaluation and measuring the implementation and development of RRI in innovation platforms, because management without measurement is dubious. A fact-based approach increases the credibility of the platform. Evaluation should be a standardized process to allow comparability of results. Measuring the implementation allows the quality of the platform to be improved and its effectiveness, efficacy and matching to stakeholders and societal needs increased. Among the measured dimensions, engagement, open access, communication, user experience, ethics and equality of accessibility to all Quadruple Helix actors should be included. Evaluation of platforms allows benchmarking among platforms, as well as the identification of weak points and areas for future improvements.

The third question was how the implementation and development of the RRI approach should be evaluated and measured in the various innovation platforms. Participants pointed out that the challenges related to measurement include possible measurement bias and finding a methodology that fits the needs of platform evaluation. Such a methodology should take into account the different objectives and interests of stakeholders, as well as different impacts of RRI. The methodology should be suitable for its goals, and the decision as to whether a quantitative or qualitative (or both) method should be used has to be made based on these goals. Impact assessment and results chain application can validate the activities of the platform.

5 Conclusion

Based on our case, we found maturity mapping as a potentially useful tool for indicating areas of intervention to strengthen the RRI in the regional innovation policy. However, it is a challenge to find data for quantitative indicators, and also cultural differences between the countries may cause serious challenges (e.g. share of women may be higher in science and engineering positions in the countries, where business sector is more exclusive for them). Relying on a qualitative assessment, as well as cultural differences, leads to difficulties in comparison between regions. In addition, simple indicators to complex issues lead easily to wrong conclusions, whereas lengthy lists make work laborious and not necessarily more comparable (see: MoRRI consortium 2018). However, we may consider that it is more important to conduct RRI without a possibility for comparisons than to not conduct RRI at the regional level at all. The maturity mapping tool is mostly useful for encouraging consideration of different aspects of RRI in the region and for communicating with stakeholders. Ongoing discourses on start-ups, open innovation, a circular economy, and so forth, provide many opportunities to add in RRI in those discourses. Future research could focus on different methods of assessing regional maturity.

There might be a risk that adopting RRI in a straightforward manner may generate rigid processes and discourage important innovators. Therefore, communication is important for showing what positive consequences may emerge. It might be useful to reflect the role of RRI in concepts like the Sustainable Development Goals (SDGs) and greening of the economy. Both of these concepts have a strong business dimension when defined goals are attempted to be reached by various actions. RRI seems to lack this dimension, possibly due to its science-based origins. In a creation of scientific value, practical actions are not always in focus, but the creation of new knowledge per se. Here, RRI with too rigid and narrowing keys may face another problem: while unfolding the unknown, it is difficult to anticipate the risks and systemic outcomes that the future knowledge provided by explorations will bring. Therefore, in interfaces within the field of science, technology and innovation (STI), one-size-fits-all RRI solutions are unlikely to be found. While simplicity is desirable, simple tools like maturity mapping must be cautiously applied when attempting to answer the complex and serious questions in the highly complex field of STI.

Currently, it is very difficult to find organizations with a strategic, comprehensive approach to RRI, including Open Innovation Platforms located at the intersections of STI, in Tampere region, as well as in other EU regions (see: Apospori et al. 2018). However, Open Innovation Platforms can be a useful tool for implementing certain aspects of RRI in regional policy, such as engagement. They also may support open access, although rather in the context of innovation than science. These qualities are parts of OIP practices as they are based on the idea of open innovation. However, it was difficult to find conscious attempts to introduce other aspects of the RRI from OIPs. Results of the workshop suggest that some of them, such as ethics or equality could be integrated to their existing practices, but would require conscious, targeted measures to achieve that. RRI provides also benefits for actors involved in the platform (see: Table 1). However, as stated in Table 1, OIPs have to face some challenges if they are aiming to apply the whole spectrum of RRI efficiently in their practices. This is not necessarily the purpose of the RRI framework as it represents more a high-level policy design or an academic exercise than a down-to-earth management tool for responsible business and innovation practices. As Fitjar et al. (2019) suggest, regional applications of RRI may be challenging or not even the most appropriate method for approaching ethics and responsibility in the regional innovation policy.

These limitations considered, the regional level might provide a space for experimentations on RRI, to integrate its keys to both regional innovation policy design and to individual instruments or projects (c.f. Fitjar et al. 2019). These experimentations would serve identifying and further developing RRI practices that can be embedded into everyday functioning of the regional innovation ecosystems. It is evident that regional innovation policies need to address grand societal challenges and in doing that, they need to integrate values and moral dimensions explicitly into their policies and practices.

Table 1 Benefits, challenges and good practices indicated during the interactive workshop *source* own study

	Benefits	Challenges	Good Practices
Open access	• new ideas due to sharing and combining existing knowledge • knowledge and experience accessible to the users • increased possibilities of new products and services • more options for unplanned and unexpected but desirable outcomes • avoiding duplication of innovations and increasing effectiveness • allowing benchmarking	• definition of openness of a platform • division of responsibility for managing the output • potential costs related to increasing openness • questions related to intellectual property rights (IPR) and data protection	• publishing research results • creative commons licences provide a flexible range of protection and freedom • Free and Open Source Software (FOSS) • legislation encouraging increased open access • public consultations increase legitimacy • code of conduct • organizing open events, projects and hackathons • egalitarian membership – even if companies pay for membership, it should be free for individual citizens • shared IPR between all involved in an innovation process • not utilized ideas made public after a given time to increase innovativeness also outside a platform
Engagement	• building common ground for discussion • enhancing collaboration, co-creation and research co-development and implementation • creating a culture of sharing and allowing the spreading of knowledge • promoting just innovation • speeding up the innovation process and making it easier due to access to users' feedback and diversity of ideas • transparency and increased awareness • better suitability for user needs • opportunity to test different concepts	• coordinating cooperation of organizations with varied specializations • ensuring engagement in the long term • management of IPR • decisions on whom to engage to ensure reaching the goals of the platform and desired level of maturity of solutions created • filtering the results of the process and choosing the best solutions • too many networks may limit the possibility of stakeholders engaging in activities	• communication beyond traditional networks • structuring of the process of engagement with clear expectations (input) and added value (output) • measuring, and supra-regional coordination of various platforms in different regions • effective internal and external partnerships with varied societal stakeholder groups

(continued)

Table 1 (continued)

	Benefits	Challenges	Good Practices
Ethics	• increase of consistency of values shared by stakeholders • increased trust and safe environment for cooperation, which especially attracts citizens and NGOs • stakeholders learn to consider ethics in their activities • increasing social benefits and regional impacts of innovation • ethical branding, which may attract consumers and increase profits of involved companies	• differences and inconsistencies in values • lack of ethical awareness • lack of explicit ethical guidelines may lead to misunderstanding and conflicts • risk of poor implementation of ethical rules and guidelines • concern about perceived costs • keeping shared values while increasing participation • power imbalance in creating ethical guidelines	• common values • introducing ethical committee or clear, written rules and guidelines created in participatory approach for a platform in question • ethical education within the platform • transparency • diversity ensures that heterogeneous stakeholder groups and their varied interests are represented

References

Apospori E, Tsanos CS (2018) MAinstreaming, Responsible innovation in European S3. Interregional comparison of regional RRI maturity and needs. www.interregeurope.eu/fileadmin/user_upload/tx_tevprojects/library/file_1548345177.pdf. Last Accessed 25 Sep 2020

Apospori E, Tsanos CS, Paraskevopoulou L (2018) MARIE MAinstreaming Responsible innovation in European S3. Enterprise survey. https://www.interregeurope.eu/fileadmin/user_upload/tx_tevprojects/library/file_1532507725.pdf. Last Accessed 29 Sep 2020

Asheim B, Boschma R, Cooke P (2011) Constructing regional advantage: platform policies based on related variety and differentiated knowledge bases. Reg Stud 45(6):1–22

Avelino F, Wittmayer JM, Kemp R, Haxeltine A (2017) Special issue on: transformative social innovation and game changers. Ecol Soc 22(4):41

Bathelt H, Glűckler J (2011) The relational economy. Oxford University Press, New York, Geographies of knowing and learning

Chesbrough H (2003) Open innovation: the new imperative for creating and profiting from technology. HBS Press, Boston

Choudary SP (2013) Platform power. Secrets of billion-dollar internet startups. http://platformed.info. Last Accessed 3 Apr 2018

Cooke P (2004) Introduction: Regional innovation systems—An evolutionary approach. In: Cooke P, Heidenreich M, Braczyk H-J (eds) Regional innovation systems, 2nd edn. Routledge, London

Cooke P, De Laurentis C (2010) Platforms of innovation: some examples. In: Cook P, De Laurentis C, MacNeill S, Collinge C (eds) Platforms of innovation: dynamics of new industrial knowledge flows. Edward Elgar Publishing, London

Council of Tampere Region (2020) Innovation Ecosystem. https://www.pirkanmaa.fi/innovation/?lang=en. Last Accessed 18 June 2020

Etzkowitz H, Leydesdorff L (2000) The dynamics of innovation: from national systems and "Mode 2" to a Triple Helix of university–industry–government relations. Res Policy 29:109–123

European Commission (2010) Special Eurobarometer 340 "Science and Technology Report". https://www.techylib.com. Last Accessed 20 Sep 2020

European Commission (2012) Responsible research and innovation. Europe's ability to respond to societal challenges. https://ec.europa.eu/research/swafs/pdf/pub_rri/KI0214595ENC.pdf. Last Accessed 23 Sep 2020

European Commission (2017) Strengthening innovation in Europe's regions: strategies for resilient, inclusive and sustainable growth. https://ec.europa.eu/regional_policy/en/information/publications/communications/. Last Accessed 24 Sep 2020

EUROSTAT (2016) Human Resources in Science & Technology (hrst). https://ec.europa.eu/eurostat/cache/metadata/en/hrst_esms.htm. Last Accessed 15 Sep 2020

Fitjar DR, Benneworth P, Asheim, BT (2019) Towards regional responsible research and innovation? Integrating RRI and RIS3 in European innovation policy. Science and Public Policy 1–12

George G, McGahan AM, Prabhu J (2012) Innovation for inclusive growth: towards a theoretical framework and a research agenda. J Manage Stud 49(4):661–683

Haigu A. (2014) Strategic decisions for multisided platforms. MIT Sloan Manag Rev 55(2)

Heeks R, Foster C, Nugroho Y (2014) New models of inclusive innovation for development. J Innov Dev 4(2):175–185

Kautonen M (2012) Balancing competitiveness and cohesion in regional innovation policy: the case of Finland. Eur Plan Stud 20(12):1925–1943

Kautonen M, Pugh R, Raunio M (2017) Transformation of regional innovation policies: from "traditional" to "next generation" models of incubation. Eur Plan Stud 25(4):620–637

Monsonís-Payá I, García-Melón M, Lozano JF (2017) Indicators for responsible research and innovation: a methodological proposal for context-based weighting. Sustainability 9(2168):1–29

MoRRI consortium (2018) The evolution of responsible research and innovation in Europe: the MoRRI indicators report. http://morri-project.eu/reports/ Last Accessed 21 Sep 2020

Moulaert F, Martinelli F, Gonzalez S, Swyngedouw E (2007) Introduction: social innovation and governance in European cities. Urban development between path-dependency and radical innovation. Eur Urban Reg Stud 14(3):195–209

Needham C (2007) Realizing the potential of co-production: negotiating improvements in public services. Soc Policy Soc 7(2):221–231

Nordling N (2019) Public policy's role and capability in fostering the emergence and evolution of entrepreneurial ecosystems: a case of ecosystem-based policy in Finland. Local Econ 34(8):807–824

Pellé S, Reber B (2015) Responsible innovation in the light of moral responsibility. J Chain Netw Sci 15(2):107–117

Raunio M, Nordling N, Kautonen M, Räsänen P (2018) Open innovation platforms as a knowledge triangle policy tool: evidence from Finland. Foresight and STI Governance 12(2):62–76

Ravn T, Nielsen MW, Mejlgaard N, Lindner R (2015) Metrics and indicators of responsible research and innovation. Progress report D3.2 Monitoring the Evolution and Benefits of Responsible Research and Innovation (MoRRI). https://www.researchgate.net/profile/Niels_Mejlgaard/publication/. Last Accessed 22 Sep 2020

Schillo RS, Robinson RM (2017) Inclusive innovation in developed countries: the who, what, why, and how. Telev New Media 7(7):34–46

Schot J, Steinmueller WE (2018) Three frames of innovation policy: R&D, system of innovation and transformative change. Res Policy 47(9):1554–1567

Sotarauta M, Kautonen M (2007) Co-evolution of the Finnish National and local innovation and science arenas: towards a dynamic understanding of multi-level governance. Reg Stud, Spec Issue Reg Gov Sci Policy 41(8):1–14

Statistics Finland (2020) PxWeb databases. Students and qualifications in education leading to a qualification by location of education and field of education (National classification of education), 2018

Strand R, Spaapen J, Bauer MW, Hogan E, Revuelta G, Stagl S, Paula L Guimarães Pereira A (2015) Indicators for promoting and monitoring Responsible Research and Innovation. http://ec.europa.eu/research/swafs/pdf/pub_rri/rri_indicators_final_version.pdf. Last Accessed 19 Sep 2020

Timmermans J, Stahl B (2013) Annual Report on the main trends of SiS, in particular the trends related to RRI. https://www.great-project.eu/deliverables_files/deliverables05/view. Last Accessed 18 Sep 2020

Torre A, Rallet A (2005) Proximity and localization. Reg Stud 39(1):47–59

van den Hoven J, Jacob K, Nielsen L, Roure F, Rudze L, Stilgoe J, Blind K, Guske A-L, Riera, CM (2013) Options for strengthening responsible research and innovation. https://ec.europa.eu/research/science-society/document_library/pdf_06/. Last Accessed 18 Sep 2020

von Hippel EA (2005) Democratizing innovation. MIT Press, Cambridge

von Schomberg R (2013) A vision of responsible research and innovation. In: Owen R, Bessant J, Heintz M (eds) Responsible innovation: managing the responsible emergence of science and innovation in society, 1st edn. John Wiley & Sons Ltd., Chichester, pp 51–74

Zwart H, Landeweerd L, van Rooij A (2014) Adapt or perish? Assessing the recent shift in the European research funding arena from "ELSA" to "RRI." Life Sci, Soc Policy 10(11):1–19

Sustainable Development of Russia's North-Western Border Areas and Their Neighbors: A Study of Landscape Effects on the Settlement Patterns of Villages and Towns

Elena A. Romanova

Abstract This paper explores the regional settlement patterns of the past 20 years (1998–2018) as seen in villages and towns. The study area is Russia's north-western border areas and their neighbors. The border areas are the Murmansk, Leningrad, Pskov and Kaliningrad regions and the Republic of Karelia. Their western neighbors are Norway, Finland, Estonia, Latvia, Lithuania and Poland. The Arkhangelsk, Vologda, Novgorod, Tver and Smolensk regions and the Republic Belarus comprise the eastern neighborhood. Settlement patterns of rural areas and towns are key to the conservation of landscape diversity as well as to sustainable spatial development, which rests on three pillars—the environment, the economy and society. This study of settlement system uses statistical materials to analyze population density, population dynamics in towns, settlement density and road infrastructure. It is also shown how settlement dynamics influence sustainable development in border areas.

Keywords Settlement · Landscape · Population density · Population dynamics

1 Introduction

A basic definition of sustainable development includes interactions between three pillars—the environment, the economy and society, which are closely related phenomena. Sustainable development of regions requires a sustainable settlement system in a favorable environment. The stability of settlement systems depends on population dynamics. Population growth has a negative effect on development in low and middle-income countries and a positive one in developed states (Güney 2017). The state of the landscape reflects the economic standing of a region and depends on the development of the settlement system. Border areas are of particular interest. In most cases, two regions divided by a border, share the same landscape but have different settlement systems and perform differently in economic terms.

E. A. Romanova (✉)
Nature division, Museum of the World Ocean, Petra Velikogo embankment, 236006 Kaliningrad, Kaliningrad Region, Russia

The border areas are the Murmansk, Leningrad, Pskov and Kaliningrad regions and the Republic of Karelia. Their western neighbors are Norway, Finland, Estonia, Latvia, Lithuania and Poland. The Arkhangelsk, Vologda, Novgorod, Tver and Smolensk regions and the Republic Belarus comprise the eastern neighborhood (Fig. 1).

Environmental conditions vary very much along Russia's over 2500 km-long border. The study area consists of regions lying in the temperate climate zone, although some northern parts of Russia, Norway and Finland (Lapland) belong to the subarctic. It has two distinct geomorphological units of different genesis: an ancient crystalline shield and the East European Platform. The terrain of the entire territory was formed by the events of the last Quaternary glaciation event, which is called Valdai in Russia, and the post-glacial period. The relief is mostly flat, although some glacial forms are present: moraine elevations and ridges, kames, eskers and drumlins. There are marshy lowlands of the ancient deltas of the rivers Vistula, Neman, and Neva. The north-western part of the Baltic region, which is a territory of Sweden, is more elevated. It is home to the eastern slopes of the Scandinavian Mountains. The zonal types of soil and vegetation are variably affected by human occupation because people are unevenly distributed within the Baltic region. The most similar landscapes in terms of origin are located astride national borders. Most of these areas

Fig.1 Russia's north-western border regions and their neighbors;
1—Pskov region,
2—Kaliningrad region,
3—Arkhangelsk region,
4—Vologda region,
5—Novgorod region,
6—Tver region,
7—Smolensk region

belong to the same domain or even province (Milkov and Gvozdetsky 1986). The modern appearance of landscapes is shaped not only by their natural genesis, but by the degree of human occupation, including residential development (Isachenko 2001).

In this article, I will consider how settlement dynamics influence sustainable development in border areas. A particular focus will be on villages and towns as well as on current residential pressure on local landscapes.

2 Theoretical Background

2.1 The Phenomenon of Villages and Town

The settlement system of any country is a spatially cohesive and functionally connected set of populated areas—cities, towns, and villages. Towns have a special place in this system. They represent, on the one hand, the lowest level of the urban settlement system and, on the other, the highest level of the rural settlements system (Vaishar et al. 2016). In Russia, urban settlements with a population of fewer than 50,000 people are subsumed under the category of towns, which comprises as a result significantly different objects (Smirnov 2019). Whether a settlement has the status of a town depends on its history, the role it plays in regional politics and municipal laws (Gunko 2015). Towns make it possible that the spaces of urban and rural Russia become integrated in terms of socio-economic development, transport, and infrastructure (Leksin 2012). The national village—town—city hierarchy contains one more element—the urbanized village. The latter is usually a settlement that has the population size of a town and develops as one but does not have a corresponding status.

2.2 Factors in the Transformation of Modern Landscapes and Landscape Classifications

The European Landscape Convention defines "landscape" as "an area, as perceived by people, whose character is the result of the action and interaction of natural and/or human factors". This definition reflects the idea that landscapes evolve influenced by both natural forces and human occupation. It also emphasizes that the landscape is a whole, whose natural and cultural components are taken together, not separately (Article 38). Article 42 states that landscapes have always been subject to change and will continue to change in response to natural factors and human activity. The aims set out in the Convention is to manage future changes in a way that recognizes the great diversity and quality of the landscapes we inherit and that seeks to preserve and even enrich this diversity and this quality instead of allowing them to decline

(European Landscape Convention 2000). Thus, the document postulates the idea of continuous development of modern landscapes under the influence of natural and socio-economic factors.

The Russian and Western schools of landscape geography treat the anthropogenic factors of landscape genesis differently. The Russian school of thought holds that landscapes should be studied using the tools of physical geography (Isachenko 1991), whereas Western geography pays greater attention to the factors of landscape transformation associated with human activity. Particularly, Bürgi et al. (2004) study the driving forces of modern landscape change. Lambin and Geist (2007) examine a wide range of natural, social and economic factors causing changes in land use and ensuing alterations in the appearance of landscapes. A collective monograph of the European Environment Agency, Landscape in transition (2017), contains a report covering 25 years of land cover change in Europe. Understanding that the main factors in the development of landscapes of populated areas are settlement and land-use patters makes it easier to predict the evolution of landscapes (Pedroli et al. 2016). Central to landscape typology are current landscape transformations. Most recent landscape typologies take account of land use (Lipský and Romportl 2015; Metzger 2018; Mücher et al. 2003; Zandena et al. 2016).

2.3 Settlement and its Role in the Development of Landscapes

The land-use system of any territory is inseparable from the settlement system, the latter affecting the former. This study into the relationship between the landscape development and settlement patterns builds on contributions of geographers, archaeologists and historians. These works pursue three main lines of research:
- the effects of landscape on settlement patterns (Isachenko 2008);
- "the landscape of settlements" (Roberts 1996);
- the impact of the settlement system on the landscape (Olišarová et al. 2018; Romanova 2015, 2017).

In recent decades, research has increasingly focused on transformations of settlement systems caused by various processes, including geopolitical changes. Divisions and unifications of territories always affect social phenomena and landscapes. Of particular interest is the work of the German researcher K. Waack (1995). Two decades ago, he described the differentiation of the cultural landscape divided by the Russian-Polish border. He links this process to not only economic and geopolitical factors, but also changes in settlement patterns.

2.4 Materials and Methods

This work uses statistical data on the populations of ten Russian regions (border areas and their eastern neighbors) and border regions of seven adjacent countries—Norway,

Finland, Estonia, Latvia, Lithuania, Poland and Belarus. Statistical data were used to calculate and analyze the settlement density, population density, rural and urban population size and highway density for territorial units of a lower hierarchical level (42 Russian and 21 international districts),

The first part of the work considers the development of settlements at the western border of Russia over in 1998–2018. The population dynamics and population density are investigated. The settlement density was also computed for areas with a significant percentage of the rural population.

The second part of identifies similarities and differences in the population and settlement patterns of Russia's border areas and their western and eastern neighbors. The study of the current state of residential development in the neighboring areas is based on a comparison of the situation in neighboring small border units, their density of population and settlement density. The third part of the work concentrates on the Kaliningrad region and the Warmian-Masurian voivodeship, where a state border serves as a new landscape boundary.

3 Results and Analysis

3.1 The Development of Settlement Patterns at the Western Border of Russia in 1998–2018

The Murmansk region. The Murmansk region, which lies entirely in the Far North, lost almost a quarter of its population (24.3%) over the study period. There is only one city—the capital of the region, Murmansk. Its population decreased by 23.7%. The largest town is Apatity (its population declined by 21.2%). Most of the remaining fourteen towns are military garrisons. The stability of population in a Murmansk town depends on the function this town performs. The populations of military garrisons either did not change at all (Gadzhievo, Severomorsk) or fell because of a planned reduction in the contingent (Ostrovnoy). The other towns were affected much more strongly over the 20 years. For example, the population of Kandalaksha decreased by 33.5%; of the single-industry town of Kovdor, by 31.1%; of Monchegorsk, by 28.3%. In 42.8% of 14 towns Murmansk towns, the population fell by more than a quarter (Fig. 2). There are twelve urbanized villages in the region, of which only four have a stable population.

The percentage of the rural population in the Murmansk is insignificant (7.9%) (2020). Most rural settlements are concentrated in five municipal districts. The average population density in these districts is very low. The most densely populated area is the Pechenga district (4.3 per km^2), and the most sparsely populated districts are Lovozersky (0.26 per km^2) and Tersky (0.27 per km^2). As of 2020, there were only 67 villages in the entire Murmansk region, of which nine are seats of rural administrations, twelve are associated with military garrisons, thirteen are railway stations and four service power plants. The population of most rural settlements

Fig. 2 20 years of population change in Murmansk towns (1998–2018) (stable: 0 ± 10%; declining: 10–25%; significantly declining: >25%)

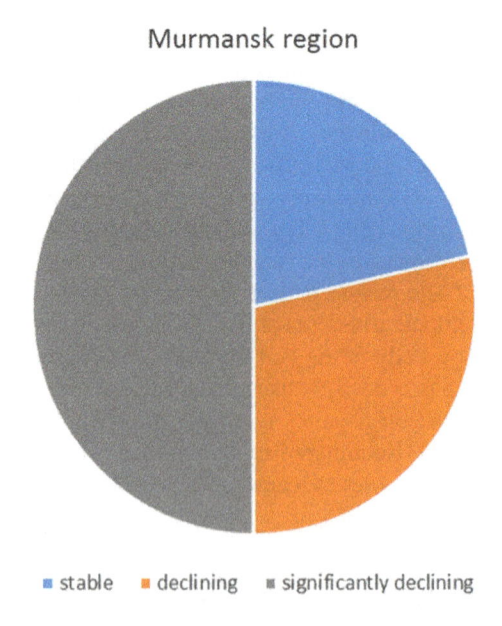

Murmansk region

■ stable   ■ declining   ■ significantly declining

decreased significantly over the past decade. Some others, such as Kuropta (Kovdor district), ceased to exist.

The settlement pattern of the Murmansk region is extremely uneven. There are large unpopulated areas. Since the trend towards a decline in the population continues in both towns and villages, it is safe to assume that, in the next decade, depopulation will reach a critical level, and the region will have an extremely polarized space.

The Republic of Karelia. Although the Republic of Karelia has a milder climate than the Murmansk region, some of its districts lie in the Far North. The population of the republic decreased by 17% from 1998 to 2018. The only city in Karelia is the capital of the republic, Petrozavodsk. Its population was relatively stable over the 20 years. The population of all twelve republican towns declined: in 58% of them, it did by more than a quarter (Fig. 3). The largest decrease in the population was observed in Belomorsk (by 40.3%), Kem (by 35%) and Olonets (by 30.3%). The Karelian town with the most stable population is Kostomuksha, probably thanks to the Karelsky Okatysh mining and processing plant, which is part of Severstal. Almost half of the Karelian urban settlements are classified as single-industry towns.

There are eleven urbanized villages in Karelia. The population of all of them, except Vyartsilya, reduced significantly over the 20 years (in nine of them, by more than a quarter). The most deplorable situation is in the ancient settlement of Chupa in the Loukhsky district. Since the closure of the mining and processing plant in 1998, its population has decreased by half (53.2%).

Most of Karelians live in urban settlements. Rural dwellers account for only 19% of the regional population (2020). There are seventeen districts in the republic, the average population density of which ranges from 0.5 people per km^2 (Kalevala and Loukhsky districts) to 14.1 people per km^2 (Sortavala district). The average rural

Fig. 3 20 years of change in the population of Karelian towns (stable: 0 ± 10%; declining: 10–25%; significantly declining: >more than 25%)

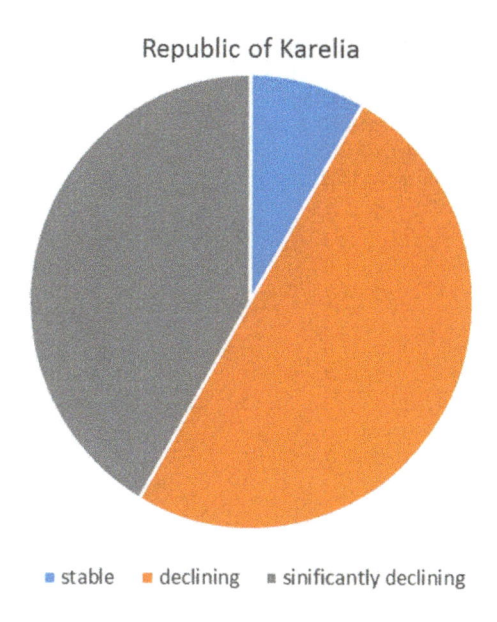

population size in sparsely populated areas is 108.6 people. For example, in the Medvezhyegorsk district (a population density of 2 people per km^2), there are few larger villages, such as Velikaya Guba and Tolvuya, and many tiny ones where only several people live.

The settlement pattern of the republic is uneven and belt-type. Settlements are concentrated along highways and railways. An important factor affecting the pattern is the White Sea-Baltic Canal. Some single-industry villages and towns have only one organization—a nursing home or a mental health treatment facility (Velikaya Guba, Padany, Vidlitsa), an airport (Chalna-1) or a handicrafts center (Shunga). A promising factor is support for ethnic districts (Kalevala, Olonets and Pryazhinsky), the Karelian language and tourism.

The Leningrad region. The Leningrad region, despite its relatively high average population density (22.4 people per km^2) and migration-driven population growth, has many sparsely populated corners. Although most regional towns either growth or are stable, there are several depressed towns (Fig. 4). There are most depopulated districts of Podporozhsky (an average population density 3.6 people per km^2), Lodeynopolsky (5.7 people per km^2), Boksitogorsky (6.7 people per km^2) km) and Tikhvin (9.9 people per km^2). These areas are home to towns that sustained the most substantial population loss: Boksitogorsk (a 24.2%-decrease over the 20 years), Pikalevo (20.9%), Lodeinoe Pole (26.3%), Podporozhye (22.1%) and Tikhvin (19%). The population of local urbanized villages also declined by 18–28%. Although there are many rural settlements in the region, a third of them did not have any permanent residents in 2017. The average population density of permanently populated villages is low: from 0.9 people per 100 km^2 in the Podporozhsky district to 3.6 people per 100 km^2 in the Boksitogorsky district. Most settlements are very small; their

Fig. 4 Population change in the Leningrad region in 1998–2018 (growing: 10–25%; stable: 0 ± 10%; declining: 10–25%; significantly declining: >25%)

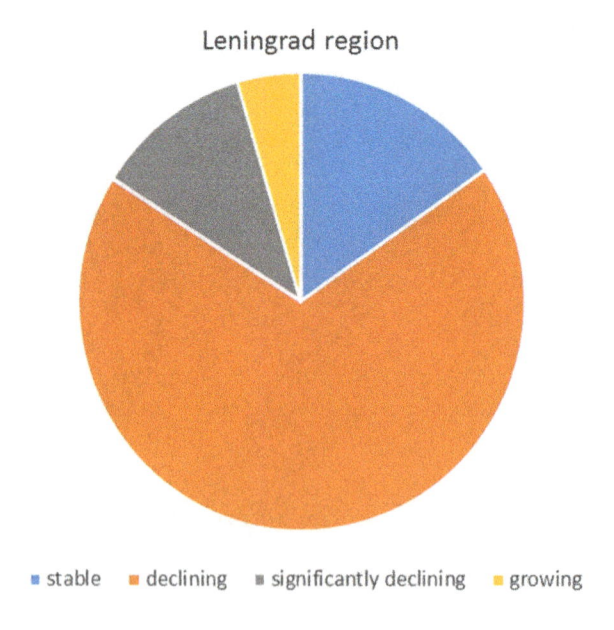

average population size ranged from 48 people (Podporozhsky district) to 96 people (Lodeynopolsky district). Almost 84% of the residents of the Tikhvin district live in its central town. There are 154 villages. Yet if the population of the three largest locations (Bor, Tsvylevo and Shchugozero) is subtracted from the total number, the average population size in the remaining ones will be as little as 50 people.

The settlement pattern of the Leningrad region is centripetal, gravitating towards St. Petersburg. It is rather patchy, and spatial polarization is growing in the region.

The Pskov region. The Pskov region has been experiencing steady depopulation for many decades (Manakov 2016). Over the study period, its population fell by 22%. There is only one city in the region—Pskov, one larger town—Velikiye Luki, and 12 smaller towns. Only Pskov and Velikiye Luki are relatively stable in terms of population change. The population of all Pskov towns decreased by more than a quarter over the 20 years. The most substantial population decline occurred in Sebezh (by 43.2%), Gdov (by 40.4%) and Porkhov (by 36.6%). There are fourteen urbanized villages in the Pskov region. Only three of them have a stable population—Strugi Krasnye, Idritsa and Sosnovy Bor. Prisons are located in the two latter settlements, whereas the backbone of the village of Sosnovy Bor is a nursing home. The rest are depressed settlements. Among the most affected are Krasny Luch (its populated declined by 61.5% over the study period), Bezhanitsy (45.2%) and Zaplyusye (42.6%).

The most depopulated areas of the Pskov region are the Plyussky (an average population density of 2.3 people per km^2), Bezhanitsky (2.7 people per km^2), Kuninsky (3.3 people per km^2) and Gdovsky (3.5 people per km^2) districts. The settlement density is very low, ranging from 3.0 (Plyussky district) to 5.7 units per 100 km^2 (Kuninsky district). The average number of residents varies from 31.2 people (Bezhanitsky district) to 49.7 people (Gdovsky). Among balancing factors are the administrative status (that of the center of a *volost*) and the presence of a military garrison or a railway station.

The current situation in the Pskov region can be described as stable depopulation with settlement pattern irregularities (the network of settlements and transport infrastructure is shrinking).

The Kaliningrad region. The Kaliningrad region is the smallest and southernmost territory of Russia's western borderlands. Its population increased by 4.4% in 1998–2018, and it continues to grow. The only city in the region is its administrative center, the population of which increased by 11.5% over these 20 years. There are 21 towns in the area; eight of them were growing over the study period. Guryevsk is the absolute leader (an 84%-increase over the 20 years), followed by the resort town of Zelenogradsk (28.25%). The population of six towns is stable; of seven, is decreasing (Fig. 5). The most significant population decline occurred over the 20 years in Ozersk (37.5%), Nesterov (20.4%). %) and Krasnoznamensk (20.3%).

As compared to the other western borderlands of Russia, the Kaliningrad region is densely populated. The most sparsely populated district, Krasnoznamensky, has an average population density of 9.4 people per km^2 and a settlement density of 4.1 units per 100 km^2, which is comparable to the remote areas of the Leningrad and Pskov regions, but the average number of residents in rural settlements is much higher—140 people. The most densely populated district of the Kaliningrad region, Guryevsky, has an average population density of 51.9 per km^2. The settlement density in the area is 10.7 units per 100 km^2, whereas the average number of residents reaches 270 people. The settlement system of the region is stable, as well as the settlement pattern. Moreover, the region boasts a well-developed transport infrastructure.

Fig. 5 Population change in the towns of the Kaliningrad region in 1998–2018 (growing: 10–25%; stable: 0 ± 10%; declining: 10–25%; significantly declining: >25%)

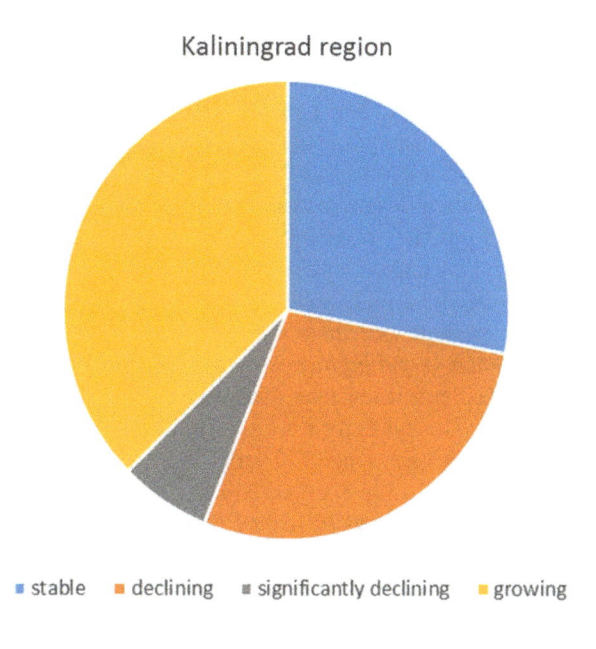

3.2 The Population and Settlement Patterns of Russia's Borderlands and Their Western and Eastern Neighbors: Similarities and Differences

Population change in border regions in 1998–2018. Most border areas, both Russian and international, are experiencing depopulation. The process is most pronounced in the Murmansk and Pskov region, which sustained an above 25%-decrease over the 20 years. The population of Karelia, Latvia, Lithuania and northern Finland fell by 11–21%. Over the study period, the Kaliningrad and Leningrad regions witnessed an increase in the population. The number of residents remained practically unchanged over the years in Estonia, the Norwegian province of Finnmark and the northern regions of Poland. The eastern and southern neighbors of Russia's western borderlands had similar trends. In the Arkhangelsk and Tver regions, the population decreased by 19%; in the Novgorod region; by 17%; in Russian Smolensk and Belarusian Vitebsk, by 16%; in the Vologda region, by 11%.

The current population density in the border regions (2018). The population density of Russia's western borderlands varies from north to south: from 5.2 people per km^2 in the Murmansk region, 3.4 per km^2 in Karelia, 11.1 in Pskov region, 21.6 in the Leningrad region to 65.7 in the Kaliningrad region. The population density trends is the same in the international border areas. It varies from 2.5 people per km^2 in the Sør-Varanger commune (Finnmark, Norway), 2 in the Lappi region (Finland) to 24.8 per km^2 in South Karelia (Finland), 29.3 in Estonia, 43 in Lithuania, 59.6 in Poland's Warmian-Masurian voivodeship. The north–south gradient is also observed in the eastern neighbors of the border regions. The population density changes from 2.8 people per km^2 in the Arkhangelsk region, 8.1 per km^2 in the Vologda region to 15.1 per km^2 in the Tver region, 18.9 per km^2 in the Smolensk region, 29.5 per km^2 in the Vitebsk region of Belarus. The figures are about the same on either side of the border: 24.8 people per km^2 in South Karelia (Finland) and 21.6 in the neighboring Leningrad region; 65.7 per km^2 in the Kaliningrad region and 59.6 in the neighboring Warmian-Masurian voivodeship (Poland). There are contrasts in the population density between Russia's Pskov region and Belarus's Vitebsk: 11.5 and 29.5 per km^2 respectively. These trends will persist over time and are likely to intensify: the population density continues to fall depopulated regions and increases in the rest of the study area.

The current settlement pattern of border regions expressed as the settlement (per 100 km^2) and highway density (per 1000 km^2).

The settlement density is the lowest in the Far North, gradually increasing towards the south. This holds for regions on either side of the border. The average settlement density is 0.1 units per 100 km^2 in the Murmansk region and northern Norway and Finland, 0.45 in Karelia (3–3.4 in the neighboring Finnish regions), 4.3 in the Leningrad region, and 7.2 in the Pskov region. In the Kaliningrad region and Estonia, it is 7.2 per 100 km^2 (just as in Pskov region). The settlement density reaches 6.4 units per 100 km^2 in Latvia, 27.2 in Lithuania, and 33 per 100 km^2 in Poland's Warmian-Masurian voivodeship. The eastern neighbors demonstrate a similar gradient. There are 0.7 settlements per 100 km² in the Arkhangelsk region, 8.7 in the Tver region, and 7.8 per 100 km² in Smolensk. In Belarus's Vitebsk, this figure is as high as 16.2 units per 100 km².

The settlement density cannot be viewed in isolation from the population density. A high settlement density in combination with a low population density (which is the case in the Pskov region) points to a patchy settlement pattern and a small population size: most settlements have just a few residents. When both settlement and population density figures are high (as is the case Lithuania Warmian-Masurian voivodeship), there is a stable settlement structure comprising all types of populated areas—cities, towns, and villages.

Central to the settlement pattern is the road network. In Russian borderlands, the highway density is highest in the Kaliningrad region (604 km per 1000 km², of which 86.3% paved) and the lowest in the Murmansk region (24.5 km per 1000 km², 95.3% paved). On the other side of the border, the maximum highway density is registered in Poland (1350 km per 1000 km²). As to the eastern neighborhood, the highway density ranges from 48 km per 1000 km² in the north (Arkhangelsk region) to 486 km per 1000 km² in the south (Smolensk region). In the Vitebsk region of Belarus, it reaches 500 km per 1000 km².

Population change in cities and towns in 1998—2018. Western neighbors. In Poland, Norway and Finland, the urban population was relatively stable during the study period. There are 47 urban settlements in Estonia. Population growth was observed in the capital city of Tallinn and Saue, which was granted the town rights in 1993. The population of Tartu, Pärnu, Maardu and some other cities stable. Most towns are experiencing depopulation—the losses reach up to 20%. A substantial proportion of towns, 38.3%, witnessed a population decline of above 25%. There are 67 cities and towns in Latvia, but only Riga has a stable population. Depopulation is observed in most Latvian towns. It reaches the maximum rate in Ainaži (its 2018 population was only 57% that 1998). Lithuania has 103 cities and towns. In most of them, the population is stable, especially in those with a population below 10 thousand people. The population of only six towns fell by more than 25%.

Eastern and southern neighbors. In the Arhangelsk region, the population of the regional capital is relatively stable. Twelve regional towns (the only exception is Mirny) were losing the population over the study period. More than half of them saw a population decrease of above 25% (Solvychegodsk, by 45%; Mezen, by 32.2%; Shenkursk, by 31.8%). In the Vologda region, only the population of the eponymous city is growing. The number of residents is stable in another regional city, Cherepovets. The population of 13 local towns is decreasing: 23% lost over a quarter of their dwellers. In the Novgorod region, there are only 12 cities and towns. The population of all of them, except for the regional center, decreased significantly over the 20 years; 75% of Novgorod towns lost over 25% of their residents. In the Tver region, the population of the capital was relatively stable, whereas that of its other 25 towns decreased over the study period—in half of them, by more than a quarter (in Bely, by 38.5%; in Bologoye, by 38%; in Nelidovo, by 35.6%). There are 18 cities and towns in the Smolensk region. Only the regional center was not losing residents during the study period, whereas the population of 33% of regional towns decreased by more than a quarter (that of Demidov, by 36.6%; of Dorogobuzh, by 26.4%). Belarus's Vitebsk region has 19 cities and towns. In six of them (including Vitebsk and Novopolotsk), the population increased over the 20 years. In six other, it did not change. The rest witnessed a slight decrease of no more than 10%. The only exception is the towns of Disna (a 34.8% decline) and Dubrovno (a 25% decline).

3.3 The Current State of Residential Development on Either Side of the Border

The Murmansk region—Finnmark (Norway) and Lappi (Finland). There are three border districts in the Murmansk region: Pechengsky, bordered by Norway's Ser-Varanger commune, Kola and Kandalaksha, bordered by Finland's Inari and Salla communities. In terms of the population density, only the industrially developed Pechenga district stands out. With one town and two urbanized villages, it has a density of 4.3 people per km^2. In all other areas, the population density is low on either side of the border, ranging from 0.7 to 3 people per km^2. The settlement density is below 0.2 per 100 km^2. Residential development is sluggish in all the border areas located in the Far North).

The Republic of Karelia—Northern Ostrobothnia, Kainuu and North Karelia (Finland). There are seven border districts in Karelia (Louhsky, Kalevalsky, Kostomukshsky, Muezersky, Suoyarvsky, Sortavalsky and Lahdenpohsky) that border seven municipalities of three Finnish regions. The average population density of border areas is not much higher in Finland (about 3 people per km^2) than in the neighboring districts of Karelia (0.6 per km^2 in the Muezersky district and 1.2 in the Suoyarvsky district). The average settlement density is nearly the same on either side of the border, varying from 0.1–0.2 units per 100 km^2 in the north to 2–3 in the south.

The Republic of Karelia—the Murmansk, Arkhangelsk, Vologda and Leningrad regions. Residential development in neighboring regions is essentially at the same level. The exceptions are the Lakhdenpokhsky district in Karelia and the Vyborgsky district in the Leningrad region, where residential development is much more active. The population density there is four times that in neighboring districts, whereas the settlement density differs only slightly.

The Leningrad region—South Karelia (Finland). Finland borders the Vyborg district of the Leningrad region, which has the same population density as the territory across the border. The settlement density in South Karelia is 3.5 times that in the Vyborg district, which is explained by differences in the settlement patterns (Finland has numerous hamlets).

The Leningrad region—Ida-Virumaa (Estonia). Two districts of the Leningrad region—Kingiseppsky and Slantsevsky—border Estonia. Although the settlement density on either side of the national border is the same (6.5–7 units per 100 km^2), the population density in Estonia is 1.6 times as high. This may be accounted for by the fact that the population of Estonia's Narva (57,842 people in 2019) is five times that of Russia's Ivangorod. Residential development is at an identical level in the two areas.

The Leningrad region—the Vologda, Novgorod and Pskov regions. Most of the neighboring districts of these regions have the same level of residential development. The exceptions are the better performing Slantsevsky district of the Leningrad region and Gdovsky districts of the Pskov region. Remarkably there is a sixfold population density difference between the two districts.

The Pskov region—Põlvamaa (Estonia). The Pechora district of the Pskov region has a land border with Estonia, namely the Värsk parish of Põlvamaa. The settlement density on either side of the border is about 18.4 units per 100 km^2. This is explained by the territories, which were part of the same country in 1920–1945, having the same settlement pattern. The population density is higher on the Russian side of the border (15.5 per km^2).

The Pskov region—five regions* (novadi) *of former Latgale (Latvia). Four districts of the Pskov region (Palkinsky, Pytalovsky, Krasnogorodsky and Sebezhsky) border on five Latvian *novadi* (Viļakas, Baltinavas, Kārsavas, Zilupes and Ludza). The northern border districts on either side of the border have almost the same population density (7–11 people per km^2) and settlement density (5.8–14 units per 100 km^2), whereas the Sebezhsky district lags behind its neighbors in this respect.

The Pskov region—the Novgorod, Tver and Smolensk regions. There is no contrast between the neighboring districts of these regions in the levels of residential development.

The Pskov region—the Vitebsk region (Belarus). The Verkhnedvinsky and Polotsky districts of the Vitebsk region have a higher level of residential development than the neighboring districts of the Pskov region (Sebezhsky, Nevelsky). The other border districts (Rossonsky and Gorodoksky) do not differ from their counterparts in the Pskov region. This is explained by that the Verkhnedvinsk and Polotsk districts of the Vitebsk region have bigger towns, which statistically affect the average population and settlement density (the latter is higher in rural areas).

The Kaliningrad region—the Klaipeda, Taurage and Marijampole counties (Lithuania). Six districts of the Kaliningrad region border Lithuania, three of them have a land border. The difference in the level of residential development between the neighboring districts is the most substantial on the Curonian Spit. The population density on the Lithuanian side is twice that on the Russian. Lithuania's part of the spit constitutes one municipality—Neringa. The settlement density on the Russian side is as little as 5 units per 100 km^2. Water barriers—the rivers Neman, Šešupė, Širvinta and Lake Vištytis—split the rest of the Russian–Lithuanian borderland. The settlement density on the Lithuanian side is 17.4–37.5 units per 100 km^2. There are many historical hamlets, and the density of cities and towns is high. In the Kaliningrad region, the settlement density is 4.1–7.2 settlements per 100 km^2. There are large villages and few towns.

The Kaliningrad region—the Warmian-Masurian voivodeship (Poland). These border areas are of particular interest since they belonged to one country and comprised one region before World War II. In 1945, the new Soviet-Polish border divided the two areas. The isolated regions developed autonomously and integrated into the administrative hierarchies of their respective states. The northern counties of the Warmian-Masurian voivodeship—Braniewski, Bartoszycki, Kentrzyński, Węgorzewski and Goldap—once constituted the southern part of former East Prussian *Kreise.* Five southern districts of the Kaliningrad region—Mamonovo, Bagrationovsk, Pravdinsk, Ozersk and Nesterov—lay in the north of the same East Prussian administrative units. Although sharing a common history and natural landscape, these territories do not look the same because of differences in residential development levels. The southern part of the divided historical districts has a higher population and settlement density than the northern one. The population density on the Polish side is 35–52 people per km^2; on the Russian side, 14.4–32.6 (the one-town district of Mamonovo, where this figure reaches 77.7, is not taken into account). The settlement density on the Polish side is 3–5 times that on the Russian.

3.4 The State Border as a New Landscape Boundary. The Case of the Kaliningrad Region and the Warmian-Masurian Voivodeship

The state border between Poland's Warmian-Masurian voivodeship of Poland and Russia's Kaliningrad region does not coincide with any natural boundary. Genetically, most transboundary areas of the south-eastern Baltic are landscapes of glacial origin (main moraine plains, end-moraine elevations, lacustrine-glacial and fluvioglacial plains), fluvial origin (deltaic lowlands, ancient alluvial plains, modern valley complexes) and marine and lagoon origin (coastal lagoon lowlands, coastal eolian formations, abrasive-accumulative shores). A large part of the Baltic Ridge is situated on the Polish side of the area. It is associated with end-moraine hills, lacustrine-glacial and fluvioglacial plains (Narodowy Atlas Polski, 1978). In the Kaliningrad region, there are only two visible spurs of the ridge—the Warmia and Vishtynetsky hills. Most of the area is the plains of the main moraine and lacustrine-glacial lowlands (Romanova, Vinogradova, 2011). There are spits—large accumulative sedimentary landforms created by eolian processes—on either side of the border. After a border divided the territory, its two parts started to develop differently. This affected the appearance of modern landscapes. Differences in landscape patterns have affected the landscape structure as follows. In the north, there are much fewer residential landscapes. The local transport network shrank as small villages and hamlets disappeared after the war. Road landscapes became relict. Land use has an even greater effect on landscape features. The south of the "divided region" has a large percentage of agricultural lands varying from 62 to 82%. Crops and perennial plants (fruits and berries) occupy most of the land. In the north, agricultural lands account for about half of the area (an average of 55.5%), and arable lands, for less than third (29.7%). Today the border runs between two different landscapes—the southern shaped by human occupation and the northern mostly affected by natural factors.

Another case of dramatic differences in settlement patterns and land use within a genetically homogenous region is the Vistula Spit. The Polish side is better developed than the Russian one: there are four settlements with a total population of over 2300 people. It has a good transport infrastructure and a vibrant tourism industry. The Kaliningrad part of the Vistula Spit is practically unaffected by human occupation. The population is concentrated in the north of the area; most of the territory is uninhabited; there are very few paved roads.

All border areas of the Warmian-Masurian voivodeship and the Kaliningrad region have common features. They have a lower population density than the rest of the territory. The settlement pattern demonstrates irregularities: the transport network becomes sparse and the population density lower towards the border. Many settlements in the border areas have lost their town rights. The settlement pattern of the Russian borderland is mirrored in Poland.

4 Conclusions

The spatial–temporal dynamics of the urban and rural settlement systems Russia's western borderlands changed as follows over the study period.

- The settlement systems were unstable throughout the study area: the number of villages reduced everywhere except for the Kaliningrad region; the population of most towns and urbanized villages decreased;
- Transformations in the settlement system had unique characteristics in different parts of the study area. In the Murmansk region and Karelia, the settlement pattern was becoming increasing patchy; in the Leningrad region, it was centripetal with a center in St. Petersburg. The situation was less unhappy only in the Kaliningrad region.

Russia's western borderlands and their western and eastern neighbors had more similarities than differences.

- Most border areas, both Russian and foreign, were experiencing depopulation;
- A north–south gradient is visible in Russia and the other countries: the population density was the highest in the south;
- In terms of depopulation, Russia's western borderlands constituted something of a buffer zone between the other countries of the Baltic region and Norway in the west and the belt of adjacent Russian regions and Belarus in the east.
- There were pronounced south-north and west–east gradients in population size, population density urban population, highway density and, human pressure on the landscape.
- The worst situation was in the Pskov region, which lagged behind both its western and eastern neighbors in terms of population change.

The landscapes of the Baltic borderlands are of similar genesis. Differences in their modern appearance and structure are explained by the degree of human occupation, primarily, residential pressure. The most marked contrast is between the border areas of Poland and Russia. The same processes were observed in the border areas of the Warmian-Masurian voivodeship and the Kaliningrad region, which had once been part of the same administrative unit: the settlement pattern was becoming distorted and the transport network, sparser.

Russia's north-western borderlands follow the path of traditional development. The leaders are the Leningrad and Kaliningrad regions. Northern territories require innovative strategies for strengthening settlement systems and creating a modern infrastructure.

Sources of Statistical Information https:/ec.europa.eu/Eurostat; https:/ stat.fi; https:/ stat-gov.pl; https:/ osp.stat.gov.lt; https:/ csb.gov.lv; https:/ stat.ee; https://www.gks.ru/dbscripts/munst/, https:/ rosreestr.ru; https:/vitebsk.belstat.gov.by

References

Bürgi M, Hersperger A, Schneeberger N (2004) Driving forces of landscape change—current and new directions. Landscape Ecol 19:857–868, 2004. https://core.ac.uk/download/pdf/159157274.pdf. Last Accessed 23 May 2019

European Landscape Convention (2000) https://www.coe.int/en/web/landscape/about-the-convention. Last Accessed 20 June 2020

Güney T (2017) Population growth and sustainable development in developed-developing countries: an IV (2sls) approach, Suleyman Demirel University. J Fac Econ Adm Sci 22(4):1255–1277. https://www.researchgate.net/publication/321167892. Last Accessed 20 Aug 2020

Gunko MS (2015) The relationship of towns and rural areas in central Russia, AV-TOREFERAT of the thesis for the degree of candidate of geographical sciences, Moscow, 2015, pp 25. http://emsu.ru/face/dissert/avtoreferat_gunkoms.pdf. (in Russian) Last Accessed 28 May 2020

Isachenko AG (1991) Landscape science and physiographic zoning: a textbook for universities. M.: Higher school, pp 366 (in Russian)

Isachenko AG (2001) Ecological geography of Russia/AG Isachenko; Saint Petersburg State University. - St. Petersburg: Publishing House of St. Petersburg State University, pp 328 (in Russian)

Isachenko AG (2008) Landshaftnaya Struktura Zemli, Rasselenie, Prirodopolzovanie [Landscapes of the Earth: Settlement and Land Use]. Petersburg University Press, Russia, St (in Russian)

Lambin EF, Geist HJ (2007) Causes of land use and land cover change. Encyclopedia of Earth, Environmental Information Coalition, National Council for Science and the Environment, Washington, DC. http://www.eoearth.org/article/Land-use_and_land-cover_change

Landscapes in transition (2017) An account of 25 years of land cover change in Europe, EEA report No 10/2017 — 84 pp. https:// eea.europa.eu/enquiries. https://doi.org/10.2800/81075. Last Accessed 12 Feb 2019

Leksin VN (2012) The crisis of the settlement system in the context of a cardinal transformation of the territorial organization of Russian society. Russ Econ J 1:3–44 (in Russian)

Lipský Z, Romportl D (2015) Classification and typology of cultural landscapes: methods and applications, https://www.researchgate.net/publication/267680328. Last Accessed 13 May 2019

Manakov AG (2016) Depopulation processes in the Pskov region against the background of polarization of the population of NorthWestern Russia, Questions of geography/Mosk. Branch of the

USSR GO - M. Published since 1946 Sat. 141: Problems of regional development of Russia. In: Kotlyakov VM, Streletsky VN, Glezer OB, Safronov SG (eds). M .: Publishing house "Kodeks", 2016. pp 317–337 (in Russian)

Metzger MJ (2018) The environmental stratification of Europe. University of Edinburgh. https://doi.org/10.7488/ds/2356

Milkov FN, Gvozdetsky NA (1986) Physical geography of the USSR. General review. European part of the USSR. Caucasus, Textbook for students. - 5th edn., Rev. and add. - M.: Higher school, pp 376 (in Russian)

Mücher CA, Bunce RGH, Jongman, RHG, Klijn JA, Koomen AJM, Metzger MJ, Wascher DM (2003) Identification and characterisation of environments and landscapes in Europe, Alterra, Wageningen, pp 832. https://www.wur.nl/en/Publication-details.htm?publicationId=publication-way-333237313030. Last Accessed 23 Nov 2018

Narodowy Atlas Polski, Wroclaw – Warszawa – Krakow – Gdansk – Zaklad Narodowy imenia ossolinskich wydawnictwo polskiej Akademii Nauk, 1973 – 1978, pp 127. (in Polish)

Olišarová L, Cillis G, Statuto D, Picuno P (2018) Analysis of the impact of settlement patterns on landscape protection in two different European rural areas, https://www.researchgate.net/publication/324991304_Analysis_of_the_impact_of_settlement_patterns_on_landscape_protection_in_two_different_european_rural_areas Last Accessed 28 May 2020

Pedroli B, Pinto Correia T, Primdahl J (2016) Challenges for a shared European countryside of uncertain future. Towards Mod Community-Based Landscape Perspect, Landscape Res 41(4):450–460. https://doi.org/10.1080/01426397.2016.1156072

Roberts BK (1996) Landscapes of settlement: prehistory to the present. Routledge, NY, pp 181. https://doi.org/10.4324/9780203430729

Romanova E (2015) Land use, settlement, and modern landscapes. Int J Econ Financ Issues 5(Special Issue):25–29. http: www.econjournals.com. Last Accessed 23 Jan 2020

Romanova E (2017) Socioeconomic conditionality of the baltic macroregion landscape development and zoning. J Settlements Spatial Plan 8(2):131–137, http://jssp.reviste.ubbcluj.ro. Last Accessed 23 May 2020

Romanova EA Vinogradova OL (2011) Landscapes, map and text, Kaliningrad region. Atlas of the World. Orlenok V; deputy head G. Fedorov. Workshop "Collection", Kaliningrad, pp 26–27 (in Russian)

Smirnov IP (2019) Medium cities of Central Russia. Tver State un-t, Tver, pp 165 (in Russian)

Vaishar A, Zapletalová J, Nováková E (2016) Between urban and rural: sustainability of towns in the Czech republic. Europ Countrys 4:351–372. https://doi.org/10.1515/euco-2016-0025

Van der Zandena EH, Leversb C, Verburga PH, Kuemmerleb T (2016) Representing composition, spatial structure and management intensity of European agricultural landscapes: a new typology. Landscape Urban Plan 150:36–49 https://doi.org/10.1016/j.landurbplan.2016.02.005. Last Accessed 13 May 2020

Waack C (1995) Persistenz und Dynamik der Sied-lungsstruktur im polnisch-russischen Grenzgebiet. In: Europa Regional vol 31. pp 35–46. http://nbn-resolving.de/urn:nbn:de:0168-ssoar-48589-8. Last Accessed 20 Apr 2019

Innovations for Sustainable Production of Traditional and Artisan Unrefined Non-centrifugal Cane Sugar in Mexico

Noé Aguilar-Rivera and Luis Alberto Olvera-Vargas

Abstract Traditional and artisanal unrefined non-centrifugal cane sugar (NCS) is a solid obtained by concentrating clarified sugarcane juice at atmospheric pressure, which is then crystallized and traditionally molded into blocks of different colors, shapes and sizes. Like traditional brown sugar, NCS is made in sugarcane regions worldwide. However, in most NCS-producing countries, the product's agro-industrial value chain faces numerous socio-economic constraints and a lack of technological innovations that hinder its ability to achieve competitiveness and sustainability for this natural product, such as excessive use of agrochemicals, harvesting with burning of sugarcane fields and GHG generation. Other problems include underuse of equipment built with food-grade stainless steel in industrial facilities or mills, inadequate safety measures in these workplaces, lack of standardization of processes and processes and unstable NCS markets. We present a systematic review of technological innovations and fieldwork with producers in this sector and its prospects in Mexico, one of the world's leading NCS producers with a great tradition and history in this area, to analyze its socio-economic and environmental impact and develop strategies for its sustainability. Therefore, to innovate within traditional and artisanal processes, it is necessary to create public policies and agribusiness proposals linked to universities and research centers aimed at improving the productive value chain. Currently, the sector lacks legal certainty and a scientific basis to support sustainability, quality, production and marketing. It also faces diverse and complex problems during the production process, including issues with production capacity, sugarcane varieties, availability of infrastructure and equipment, energy use, environmental impact, organization of producers, uses of by-products, and inconsistency of product quality and marketing. Other aspects discussed include regional economic impact, political aspects surrounding legislation, and international certification.

N. Aguilar-Rivera (✉)
Facultad de Ciencias Biológicas y Agropecuarias, Universidad Veracruzana, Km. 1 Carretera Peñuela Amatlán de los Reyes S/N. C.P. 94945, Córdoba, Veracruz, México
e-mail: naguilar@uv.mx

L. A. Olvera-Vargas
Centro de Investigación y Asistencia en Tecnología y Diseño del Estado de Jalisco, A.C. (CIATEJ), Av. Normalistas 800 Colinas de la Normal 44270, Guadalajara, Jalisco, México
e-mail: lolvera@ciatej.mx

Keywords Unrefined non-centrifugal cane sugar · Trapiche · Constraints · Prospective analysis

1 What is Non-centrifugal Cane Sugar (NCS)?

Unrefined non-centrifugal cane sugar (NCS) is a sweetener obtained by concentrating sugarcane (*Saccharum officinarum*) juice, generally over an open fire. It is presented in solid form and contains amorphous micro crystals not visible to the human eye that maintain its constituent elements such as sucrose, glucose, fructose and various minerals, without chemical additives. It is usually sold in rectangular, hemispherical or trapezoidal blocks in varying weights (between 0.5 and 1.5 kg), or as granulated powder, depending on the production region, technology used and target market.

One example is powdered *panela*, a type of NCS obtained by evaporation and concentration of cane juice. It differs from the block form of NCS in its concentration of total soluble solids, pH, and percentage of sucrose, reducing sugars and moisture in the finished product. It is produced through several methods ranging from the most traditional to highly mechanized and automated systems (Gutiérrez-Mosquera et al. 2018) (Fig. 1).

NCS is an agro-industrial bioproduct. It can be considered a natural sweetener that preserves most of the nutrients present in sugarcane. Its physical properties, such as color, texture, turbidity and hardness, are important attributes for the acceptance of the product by the consumer (Jaffé 2012; Eggleston 2019): products with greater clarity and sweetness are considered better. It must also be free of attacks by fungi, molds, insects and rodents according to the standards of the Codex Alimentarius.

Fig. 1 Types of Non-centrifugal cane sugar in Mexico (Photography taken in Huatusco Veracruz México)

2 Methodology

A systematic multidisciplinary review of scientific literature was carried out in academic databases worldwide. At the regional level, the statistical information generated by the federal and state governments was analyzed and field trips were made to the producing areas of Veracruz and San Luis Potosí with piloncillo producers to interview them and their technological, socioeconomic and environmental situation. The objective of the fieldwork was to exchange knowledge and experiences in relation to the variables that affect the profitability of the pylon production process, as well as to help determine the sector's technology transfer and research needs. Likewise, the interaction with the producers made it possible to verify, through techniques such as interviews and participatory mapping, the information reported in federal government sources, to take photographs of the real situation of the sugarcane fields and agroindustrial facilities using a digital camera, and subsequently to generate mapping with the use of geographic information system software.

3 Production of NCS

Recently, the economic importance of NCS as a sweetener has increased after a long process of displacement by refined sugar from cane or beet. This renewed interest stems from its nutritionally and functionally significant quantities of minerals, vitamins and phenolics, among other constituents, as well as antioxidant capacities that have ranked it as safe and nutritious high-quality food (Barrera et al. 2020; Verma et al. 2019; Meerod et al. 2019; Lee et al. 2018; Orlandi et al. 2017; García et al. 2017; Jaffe 2015; Seguí et al. 2015; De Maria 2013). According to Galvis et al. (2020), NCS has a high calorific value ($14,684.9 \, Jg^{-1}$) and could be used as a natural sweetener in energy drinks. It is also a source of minerals and has other health benefits. Nagarajan et al. (2019) carried out an extensive review of nutraceutical properties, nutritional benefits and medicinal values of traditionally-processed sugar products, including NCS, and refined sugars.

Extensive research has been conducted on existing NCS processing technologies at various stages (Galvis et al. 2020), from sugarcane cultivation (Vera-Gutiérrez et al. 2019) to product certification. Asikin et al. (2014) and Guerra and Mujica (2010) evaluated the physical and chemical properties of NCS. Velásquez et al. (2019) reviewed recent advances in terms of process variables and their effects on the final product and food safety. Espitia et al. (2020) differentiated three thermal stages during the production process: (1) clarification: juice impurities are removed; (2) evaporation: the concentration of total soluble solids increases from 22 ± 2 to $60 \pm 10°Brix$; and (3) concentration: water evaporates to a concentration of $92 \pm 4°Brix$ to obtain NCS. Khattak et al. (2018) carried out an energy efficiency analysis by modeling process flows in terms of exergy. Alarcón et al. (2020) performed an evaluation and correlation of major thermal and rheological properties of sugarcane juices as intermediaries in the production of NCS. Asikin et al. (2014) studied the changes in physicochemical characteristics of NCS during storage, while Kumar and Kumar (2018) reviewed the packaging material in terms of preventing ingress of atmospheric moisture and maintaining the quality of NCS.

Despite the fact that there are quantitative data regarding the nutritional composition of NCS, mainly for India and Colombia, these are highly relative given the global and regional variations in factors such as different varieties of sugarcane, crop management practices and production processes.

NCS is obtained by concentrating sugarcane juice without centrifuging, generally over an open fire, using a handcrafted sugar mill which in Mexico is called a *trapiche*. The trapiche or mill extracts the juice from the sugarcane, a product known in Latin America as *Guarapo*. Ouerfelli (2009), Signorello (2006) and Silberman (1998) have reported on its origins as grinding equipment. Most are made up of cast iron parts, and those made of wood or stone that still exist are used in the production of cane syrup, spirits or juices for direct consumption by producers and their family members and neighbors. These mills are often inefficient and depend on firewood for fuel, which is costly and contributes to deforestation.

In the producing regions, sugarcane is more than just an agricultural activity; it is the main or only commercial crop for farmers, and the main source of income for almost entire populations who transform it into various products, including NCS in different shapes, sizes and qualities, as well as cane syrup, desserts and spirits. In recent years, stakeholders have focused on training, technology transfer and organizational processes, in order to move from industrial NCS (low quality) to innocuous white and powdered or granular forms, as well as towards organic production. They propose that its characteristics as a sweetener and food supplement, as well as its natural origin, offer possibilities for new competitive markets through a comprehensive study of the production chain and the application of technological, administrative and market innovations.

4 The Market for NCS: Traditional and Modern Uses

NCS has been traditionally consumed as a sweetener in most of the sugarcane growing regions of the world, where it is known by different local names: *chancaca* (Chile, Bolivia, Ecuador and Peru); *kokuto*, black sugar or *kuro sato* (Japan); *gur*, *jaggery* or *khandsari* (India, Pakistan, Nigeria, Kenya, South Africa); *gula java* or *gula merah* (Indonesia); *hakuru* or *vellam* (Sri Lanka); *htanyet* (Myanmar); *gula melaka* (Malaysia); *namtan tanode* (Thailand); *panela* (Colombia, Guatemala, Honduras, Nicaragua, Panama); *papelón* (Venezuela, Central America); *rapadou* (Haiti); *piloncillo* (Mexico); *rapadura* (Brazil and Cuba); *palisade* (Bolivia); *tapa de dulce* or *dulce granulado* (Costa Rica); *moscavado*, *panocha* or *panutsa* (Philippines); *nam oy* (Laos); unrefined muscovado or black sugar (United Kingdom); *cassonade* (France); *Vollrohrzucker* (Germany); raw sugar or evaporated cane juice

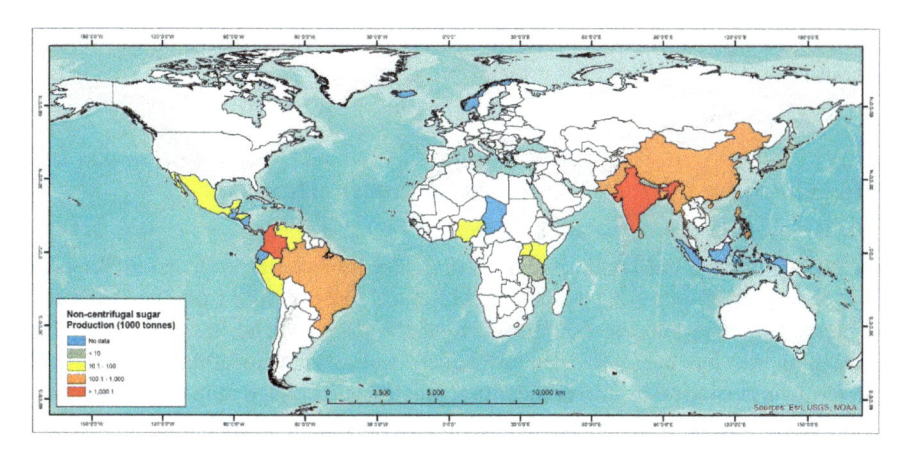

Fig. 2 Non-centrifugal cane sugar producing countries (Data from FAOSTAT 2020)

(United States). Colombia is the highest per-capita consumer of NCS (37.4 kg/Inhab) and the second largest producer in the world (after India) (Fig. 2).

Its use as food derives from a cultural context based on traditional knowledge and experience that families have passed down from generation to generation (Martínez Zarate et al. 2019). This makes it an extremely important gastronomical input in Latin America, the Caribbean, Asia and Africa. NCS is the result of cultural exchange based on the miscegenation of ancestral European and indigenous techniques, and the regional variations in names and production methods give it a sense of belonging to each culture where it occurs. In other words, it is produced under various local conditions, with handcrafted methods prevailing over industrial processes, and through heterogeneous relationships between various social and economic stakeholders. However, the particularities of the territory have not been considered as elements in proposals for strengthening the NCS production chain (Requier-Desjardins and Borray 2004; Martínez and Aguilar-Rivera 2019).

Today, NCS is a marginal food, basically consumed in producing countries. It is still known as "medicinal sugar" in India (Rao et al. 2007) and is widely used in traditional medicine in South Asia, Latin America and the Caribbean, where its potential health benefits are highlighted (Chinnadurai 2017; Weerawatanakorn et al. 2016). Given the controversial nature of sugars in the diet, it may be advantageous that the sweetening capacity of NCS is lower than muscovado and refined sugar.

In contrast to sugar production in Latin America, which is carried out in sugar mills or biorefineries with large-scale industrial structures, NCS production is carried out in small trapiches, usually in mountain areas with low soil fertility and limited mechanization in the sugarcane fields, where peasant families try to diversify production with livestock, agricultural and processing activities, mainly using their family labor (Reyes et al. 2017). Therefore, the NCS agribusiness is a subsistence activity that is rural, traditional and artisanal in all sugarcane regions, but especially in Mexico. It is characterized by little organized production, very low technification and quality control, and low sustainability due to inexistent control of the technical process operations.

4.1 The Case of 'piloncillo' Production in Mexico

The NCS produced in Mexico is called *piloncillo*. It has been produced in Veracruz since the introduction of sugarcane cultivation in 1519 (brought from the Caribbean by Hernan Cortes) and the establishment of the first trapiche in 1524. Since that time, piloncillo has been the main sweetener used by rural inhabitants in the area, and trapiche operation is among the oldest small-scale cottage industries (Fig. 3).

In Mexico, 60% of the established trapiches have an intermediate level of technology and capacity, 38% have low technology and capacity, and only 2% have high technology and capacity. Small sugarcane producers supply both trapiches and sugar mills, using small-scale family labor and in some cases continuing to use traditional production and transportation techniques. Each mill is estimated to support 15

Fig. 3 Ancient and traditional trapiche (Photography taken from the Museum of Mexican History of Monterrey, Nuevo Leon, Mexico, http://www.3museos.com/)

families on average. Piloncillo production is carried out in 50% of the Mexican territory, that is, in 16 states. Veracruz and San Luis Potosi are the main producers at the domestic level, but it is also present, at various technological levels, in 370 municipalities in the states of Tamaulipas, Campeche, Chiapas, Colima, Jalisco, Michoacán, Nayarit, Oaxaca, Puebla, Quintana Roo, Sinaloa, Tabasco, Sonora, Guerrero and Hidalgo with a planted area of 14,678.35 ha and a production volume of 439,715.19 t of sugar cane with an average yield of 30.5 tha^{-1} (Figs. 4 and 5).

Three types of piloncillo are produced: (1) white or cone; (2) black or industrial; and (3) granulated or powdered. Variations during the production process give rise to each different type. Black or industrial piloncillo is the lowest quality product and the most abundantly produced since it has the least complex and cheapest process with low technological and labor requirements. It is presented in traditional 1 kg cone blocks. Production of granulated piloncillo is very low because it requires specialized equipment, so that only trapiches with a high technological level and some with medium-level technology can produce it.

Within the piloncillo value chain at the local, regional and national level, a number of direct actors can be identified. Firstly, there are the growers who sell and/or process the sugarcane used for piloncillo production, who generally have not increased the area planted with sugarcane or the productivity and yield per hectare because this is a residual cane that could not be sold to the sugar mills during the harvest season. Secondly, there are the personnel in charge, owners or administrators of the trapiche mills. Lastly, there are the intermediaries who carry out various actions; for example,

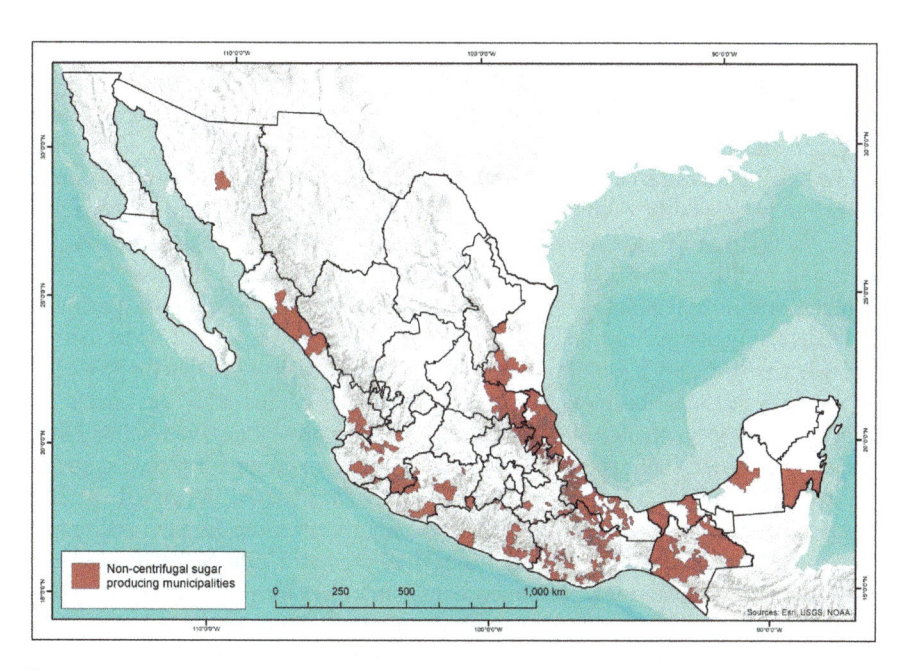

Fig. 4 Piloncillo producing municipalities in Mexico (Data from SIAP 2020)

Yield (t/ha)

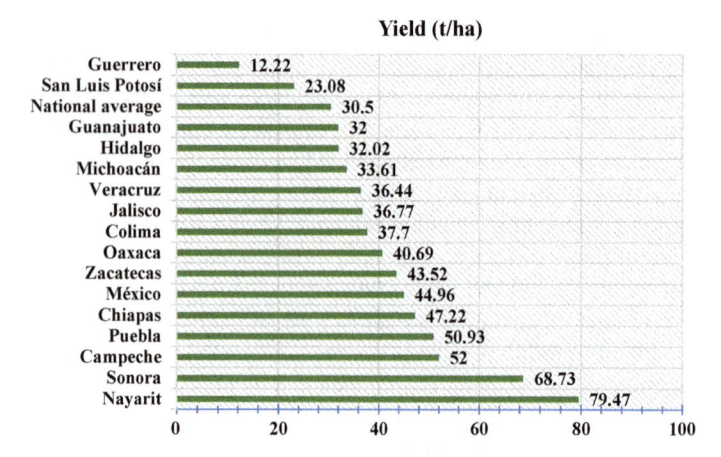

State	Yield (t/ha)
Guerrero	12.22
San Luis Potosí	23.08
National average	30.5
Guanajuato	32
Hidalgo	32.02
Michoacán	33.61
Veracruz	36.44
Jalisco	36.77
Colima	37.7
Oaxaca	40.69
Zacatecas	43.52
México	44.96
Chiapas	47.22
Puebla	50.93
Campeche	52
Sonora	68.73
Nayarit	79.47

Fig. 5 Sugarcane productivity of state producers for piloncillo production in Mexico (Data from SIAP 2020)

they buy piloncillo from the trapiches to sell it to retail consumers, local commercial chains or central supply markets, and they also sell raw material previously bought from sugarcane producers to the trapiches, since the intermediaries have trucks to transport cane and piloncillo.

Processing in a trapiche consists of a series of phases. In general, the first is the reception and weighing of the sugarcane, followed by three technical phases: the extraction of juices (milling, treatment and clarification of the juice), then the concentration and evaporation of the juice (with firewood, bagasse or tires), and finally the molding of the piloncillo according to the destination market. The piloncillo is then packaged and stored and finally marketed. To obtain a ton of piloncillo, 10–12 tons of raw sugarcane material are required, that is, the processing yield of one ton of sugarcane corresponds to 98.6 kg of piloncillo. In addition, the trapiche generates by-products (bagasse and filter mud), which, for the most part, are marketed for use as organic fertilizers or livestock feed or are used within the same farms.

Incorporation of technology in small-scale production is still very low; it could be said that the biggest technical innovation has been the introduction of motors to drive the mills, to replace horses, mules or donkeys. In this sense, it seems that for producers with a diversified agribusiness, having, for example, cattle, sugarcane and other crops, technological change alone is not enough to improve their well-being. Other aspects, such as the natural resources of the productive environment as well as social, cultural and economic aspects of such communities, must be integrated in order to seek comprehensive alternatives that favor a higher income for the rural family (Reyes et al. 2017); however, most agro-industrial trapiche facilities still have serious technological and operational problems; for example, they have little or no food safety precautions in place as the areas for storage of sugarcane or raw material, milling and production are not properly separated. The staff has only empirical knowledge about the operation and capacity of the production process. Most mills do not have

Fig. 6 Trapiche of type of medium technological level in Huatusco, Veracruz, Mexico: **a** view of the agro-industrial facility, **b**, **c** electric mill, **d** piloncillo production area

adequate sanitary facilities, sanitation of the establishment, quality assurance and food safety, use of protective equipment and HACCP certification (Fig. 6).

The commercialization of piloncillo depends on the reception and processing capacity of the trapiches, as well as market demand. Generally, each mill owner indicates to the farmers the weekly quantity of cane that they will require, and they are in charge of supplying it. The demand for piloncillo is variable during the year, increasing markedly from October to March mainly due to the religious festivities around Christmas. The shelf life of the piloncillo, as a final product, is seven to eight months depending on the environmental conditions at storage.

The critical factors in determining the productivity of a trapiche as a rural socioeconomic system include the following (Martinez and Aguilar 2019): sugarcane area cultivated per productive unit (farm or land); sugarcane and sucrose yield; frequency of milling in trapiches; the seasonality of production and competition with sugar, substitute or adulterated products; type, shape and quality of the piloncillo produced, technology used, sanitary regulations and handling of by-products; quality of brown sugar; good manufacturing practice and innocuousness for the food industry; cultural identity and organization of producers.

A lack of precise field data has hindered the evaluation of the productive potential for piloncillo production from sugarcane. Agricultural surveys and field observations are often carried out by the farmers themselves and may lack technical rigor. There is a need for more precise information on the spatial distribution of the planted area and the variations in its productive potential as a function of edaphoclimatic factors.

Current Mexican regulation (Law for the Sustainable Development of Sugarcane 2005) guarantees a minimum price for sugarcane for sale in sugar mills, while in the case of piloncillo, prices remain subject to supply and demand and speculation by intermediaries. When the price per ton of sugarcane increases, it encourages those producers who delivered their cane to the trapiches to stop doing so because the price is better at the sugar mill, and producers who have their own trapiche stop grinding and sell to the closest sugar mill. This constraint of the sugarcane market has created a situation where most trapiches do not operate continuously and are largely technologically obsolete or abandoned (Fig. 7).

Better dissemination of the nutritional and health values of piloncillo, and perhaps replacing its traditional geometric shapes with new designs, might give the product greater market value and increase competitiveness against sucrose from sugar mills.

Fig. 7 Trapiche in Huasteca Potosina, Mexico of low tech (Photographs taken from San Miguel El Naranjo San Luis Potosí, Mexico) **a** View of the old manual or animal traction mill, **b** piloncillo molding instruments, **c** View of the agro-industrial facility, **d** mechanical traction mill, **e** piloncillo molds

4.2 Uses of Piloncillo in Mexico

Piloncillo is an important component of the Mexican diet, partially satisfying nutritional requirements in terms of carbohydrates, minerals and vitamins. It is used in the preparation of cold drinks with lemon and orange, or hot drinks such as coffee, chocolate, herbal infusions and tea. It is also broadly used for cooking meat, sauces, canned fruits and vegetables, typical artisanal breads, cookies, desserts, jams, crystallized fruits and confectionery, and various typical dishes such as the classical Christmas *ponche*. Its rich aromas and flavors give depth and character to meals, as well as a certain warmth due to its robust sweetness and woody and aniseed aromas. It is the ideal complement for strong and aggressive flavors, such as coffee and chili, and can be considered the sweetener par excellence of Mexican cuisine.

The main consumer sectors are the coffee industry, distilled agave beverages, soft drinks, fruit packaging and baking. In addition, the trapiches have made an effort to diversify its use, including making candies with brown sugar, peanuts, sesame seeds, lemon rind and pumpkin seeds. However, the main factor that governs consumer preference and marketing is its external appearance, that is, shape, color and texture, and storage capacity. There is greater demand for granulated or powdered piloncillo due to its good quality, presentation and instant use like conventional sugar.

5 Barriers to the Commercialization of NCS

Recent studies have identified the biggest challenges to NCS production as: incorporating food-grade stainless steel equipment, automating the production process, and adopting good manufacturing practices, certifications, and energy self-sufficiency with bagasse to reach more differentiated markets (Alarcón et al. 2020; Espitia et al. 2020; Gil 2017).

NCS also suffers from lack of consistency in its production and use. Jaffe (2015) states that each producer country has a different research approach regarding the formulation of differentiated products, as well as in the improvement of the stages of the production process, focusing on higher quality and innocuousness. But there is also variation in the components of NCS within the same product in one country or region as well as among products from different countries. It is essential for the causes of these variations to be better evaluated in order to allow for international standardization and optimization of the agro-industrial process (Galicia-Romero et al. 2017; Ramírez and Salazar 2017).

As an agribusiness, it currently has problems related to a dominant productive scheme of land ownership, with a peasant economy involving high use of labor and a weak organizational and associative culture that makes it difficult to obtain economies of scale. It also suffers from low levels of investment in technological improvements and quality standards, low availability of economic resources and difficult access to credit, lack of updated and efficient public policies, and precarious access to markets for the commercialization of NCS at the domestic level and for export.

No progress has been made in the generation of new presentations to make NCS attractive and practical for the consumer. The commercial links of the incipient value chain are made up of local, municipal and regional wholesale markets and, in the case of powdered panela, wholesalers who distribute to the supply centers of nearby cities or large centers in capitals, supermarkets and other consumer industries. The retail market is covered by rural and urban stores. Store owners are major direct agents, getting the majority of the product to the end consumer locally and regionally.

Other important factors that have limited the commercialization of NCS are increasing urbanization and, more importantly, the expansion of fast foods. Companies selling these foods invest heavily in marketing and their products are considered more convenient and easier to access, but the substitution towards consumption of products and carbonated drinks that are high in fats, flours and refined sugars, with high caloric intake but low nutritional value, has been characterized as a "nutritional transition" since the first half of the twentieth century and is linked to the increase in diseases such as diabetes, cardiovascular conditions and others.

This nutritional transition derives from the worldview that industrial is better (Mann and Fleck 2014; Eggleston 2019). The challenge for NCS commercialization is to establish it as a healthier, legitimate alternative to refined and centrifugal sugars, which implies overcoming the negative assumption that conventional NCS is a low-quality, primitive sugar. In some countries, its consumption (in particular the granulated form) is increasing due to its greater ease of use and also due to a tendency towards eating natural and organic products (Quintero-Lesmes and Herran 2019; Kumar and Singh 2020).

5.1 Issues and Prospects for Piloncillo Commercialization in Mexico

Through participatory diagnoses carried out in Mexico with groups of sugarcane producers, trapiche owners and technicians in Veracruz and San Luis Potosí, a number of varied and complex impediments to the production of high-quality piloncillo have been identified (Table 1):

Despite all its potential benefits, there is no strategy for innovation or differentiation of piloncillo as a product. It is sold in rustic forms or packed in cardboard boxes, which do not provide safety and hygiene. Only intermediaries selling to supermarkets invest in giving the product a better presentation in order to charge higher prices.

There is no official norm in Mexico that regulates the milling process or the quality of raw materials and piloncillo. This generates conflicts between producer states, as each mill produces its product according to the particularities and tolerance of their consumers. This has made it difficult to create a consistent market for the product. It has also prevented producers from benefitting from the types of institutional programs

Table 1 Barriers to the production of quality piloncillo identified through participatory diagnoses at fieldwork with producers

Problems with sugarcane production	Problems with piloncillo production	Market forces and policy issues
Labor shortage due to the high emigration of workers to northern Mexico or the United States	Lack of high-technology equipment for making granulated or powdered piloncillo	High price of inputs
Unsustainable management of sugarcane cultivation for piloncillo	Low levels of juice extraction in the mill	Producer uncertainty due to the high fluctuation in piloncillo prices
Existence of very old varieties; lack of renewal of sugarcane cultivars	Poor cleaning and clarification practices for juices	Difficulty accessing loans and high interest rates resulting from the country's economic crisis
Lack of knowledge on how to manage different sugarcane varieties	Use of firewood and tires as burner fuel, which causes problems of deforestation, erosion and pollution	Absence / deficiency of technical assistance to the producer by the local and federal government or universities to increase sugarcane yield and piloncillo production
Low soil fertility	Impact by respiration compounds due to the use of bagasse as fuel in open combustion process	
Phytosanitary problems	Underutilization of piloncillo mills	No use of bromatological, toxicological and food safety analyses of final products
Inadequate weed control practices	Use of undesirable additives in the production of piloncillo as required by intermediaries	Low level of quality, diversification of forms and presentations of final products (Pilon and powdered or granulated)
High costs of cutting, transporting and handling the sugarcane	Low use of by-products due to lack of familiarity with the required technology	Low level of complementary activities to income such as agrotourism and bioproducts (spirits, cane juice and syrup)
Excessive use of agrochemicals without technical advice	No adoption of safety occupational and innocuousness, hygiene and good manufacturing standards	
Sugarcane burning and GHG generation		

and actions available to the sugar industry, such as financing, the adoption of new technologies, acquisition of modern machinery, and the search for related markets such as the artisanal food and liquor industries (Morales-Ramos et al. 2017).

In this sense, it corresponds to the State as a regulatory system to analyze and promote socioeconomic, environmental and technological policies that contribute to the improvement of the production system, mainly its competitiveness and sustainability. All participants in the value chain must be considered through direct lines of action. For example, sugarcane growers should have access to training and workshops on the sustainable use of territory through spatial analysis tools such as agroecological zoning and management practices to obtain higher quality and lower cost raw materials; moreover, trapiche owners require such instruction on engineering and technological innovations, as well as on good manufacturing and safety practices in the food sector, such as those provided for in HACCP standards (Fig. 8).

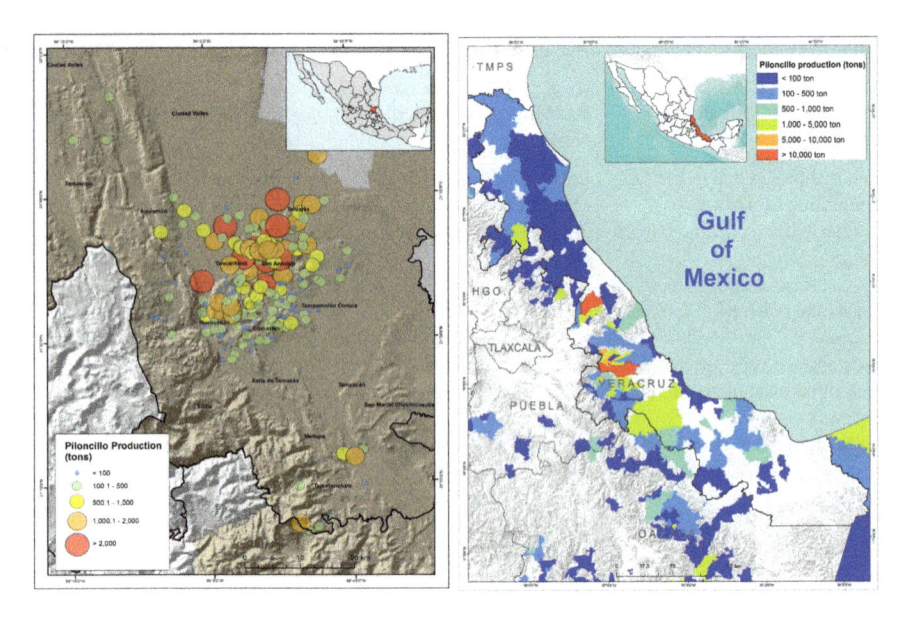

Fig. 8 Spatial allocation of the piloncillo producers in Huasteca (San Luis Potosí) and Veracruz (Data from SIAP 2020; SEDARH 2020 and fieldwork)

It is also important to promote green harvesting practices and the transition to organic production. The use of renewable energies is another key area of opportunity, entailing such strategies as using bagasse for steam generation and energy cogeneration for trapiche processing, taking advantage of solar energy, transitioning to electric mills and, in general, decreasing the use of fossil fuels such as diesel, gas, and gasoline in current mills. The burning of tires and firewood must be curved for health and environmental reasons, and more generally, the adoption of good manufacturing practices must be promoted (Morales-Ramos et al. 2017).

An improvement in the market appeal of piloncillo must be accompanied by the application of technologies and innovations that make sugarcane production more efficient and effective. There is an imminent need to adapt to new plant varieties and implement sustainable agronomic management practices such as organic and agroecological production. Ultimately, success will depend on obtaining high-quality raw materials with greater quantities of sucrose, optimizing production processes (i.e. reducing grinding times), and switching to the manufacture of products with high demand (white piloncillo, granulated or pulverized, spirits, cane syrup). This requires the development and implementation of spatial and agronomic models that allow estimating the productive potential of the existing cane varieties and evaluating new genotypes of sugarcane cultivars (Fig. 9).

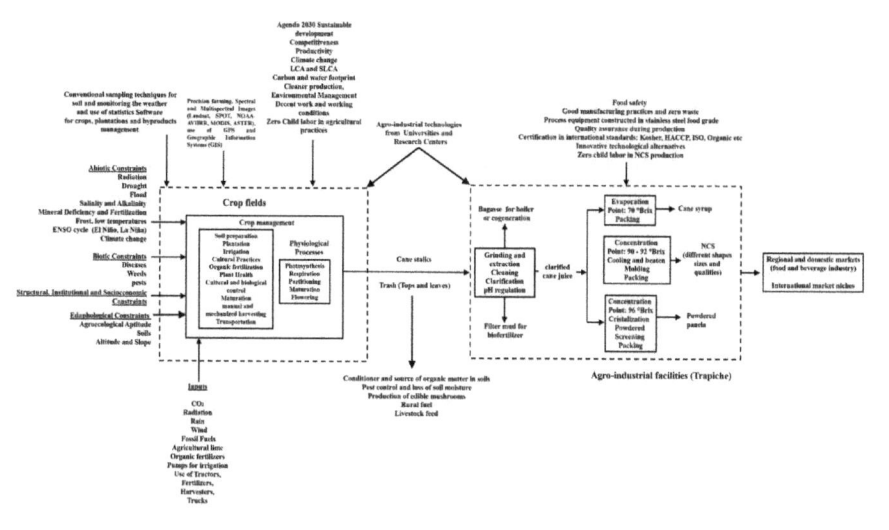

Fig. 9 Innovations for NCS value chain

6 Conclusion

Production of non-centrifugal cane sugar (NCS) derives from one of the rural agro-industries with the longest tradition in Mexico. In contrast to the production of refined sugar, it is mostly carried out in small peasant farms with high intensity of family labor and low rates of introduction of mechanized technologies through artisanal and rudimentary practices in the manufacturing process which have limited its development. This agro-industry is also characterized by a negative environmental impact and high energy consumption from fossil fuel or external electricity associated with the process. While its transformation, technological modernization and access to domestic and international markets would require a high capital influx and much-improved safety standards, NCS production has remained virtually unchanged over recent decades, retaining a rural, artisanal and traditional character, with little project diversification (white and powdered piloncillo, sugarcane juice, trapiche honey, liquor, spirits and rums, traditional sweets, etc.) and a low level of technological innovation. However, actions can be taken to increase competitiveness, income and job opportunities in the producing regions, including: transitioning to sustainable and organic sugarcane cultivation, processing sugar cane with food-grade stainless steel equipment, instituting standardization and certification of processes and products, establishing direct marketing channels that eliminate the need for intermediaries to distribute and sell their final products, using by-products such as bagasse and cane straw in the production of energy and filtering sludge for biofertilizer.

Moreover, it is necessary to create initiatives and proposals aimed at improving NCS production and its value chain. Several countries, such as Mexico, currently lack the necessary legislation, public policies and regulations to make it a competitive agroindustry. There is also a lack of precise spatial and data production statistics for optimizing agricultural output. Future multidisciplinary studies should focus on the potential for strategies such as agroecological zoning to identify local resources, capacities and territorial dynamics. The involvement of researchers and technicians from universities can help increase raw material quantity and quality, introduce adequate innovative technological equipment, from a technical and engineering perspective, throughout the production process, and incentivize the adoption of good management practices and certifications for competitiveness. These interventions would improve sustainability and standardize the quality of NCS products, taking them from a seasonal survival strategy to an established market position.

References

Alarcón ÁL, Orjuela A, Narváez PC, Camacho EC (2020) Thermal and rheological properties of juices and syrups during non-centrifugal sugar cane (jaggery) production. Food Bioprod Process 121:76–90

Asikin Y, Kamiya A, Mizu M, Takara K, Tamaki H, Wada K (2014) Changes in the physicochemical characteristics, including flavour components and Maillard reaction products, of non-centrifugal cane brown sugar during storage. Food Chem 149:170–177

Barrera C, Betoret N, Seguí L (2020) Phenolic profile of cane sugar derivatives exhibiting antioxidant and antibacterial properties. Sugar Tech 1–14

Chinnadurai C (2017) Potential health benefits of sugarcane. In Sugarcane biotechnology: challenges and prospects. Springer, Cham, pp 1–12

Codex Alimentarius (2019) FAO-Roma Italy. http://www.fao.org/fao-who-codexalimentarius/sh-proxy/en/?lnk=1&url=https%253A%252F%252Fworkspace.fao.org%252Fsites%252Fcodex%252FCircular%252520Letters%252FCL%2525202019-34%252Fcl19_34s.pdf

De Maria G (2013) Panela: the natural nutritional sweetener. Agro FOOD Industry Hi Tech 24(6):44–48

Eggleston G (2019) History of sugar and sweeteners. In: Chemistry's role in food production and sustainability: past and present. American Chemical Society, pp 63–74

Espitia J, Velásquez F, López R, Escobar S, Rodríguez J (2020) An engineering approach to design a non-centrifugal cane sugar production module: a heat transfer study to improve the energy use. J Food Eng 274: 109843

FAOSTAT (2020) Sugar non-centrifugal http://www.fao.org/faostat/en/#data/FBS

Galicia-Romero M, Hernández-Cázares AS, Debernardi de la Vequia H, Velasco-Velasco J, Hidalgo-Contreras JV (2017) Evaluation of the quality and innocuousness of raw cane sugar in Veracruz Mexico. Agroproductividad 10(11):35–40

Galvis KN, Hidrobo LD, García MC, Mendieta Menjura OA, Tarazona-Díaz MP (2020) Effect of processing technology on the physicochemical properties of non-centrifugal cane sugar (NCS). Revista Facultad De Ingeniería Universidad De Antioquia 95:64–72

García JM, Narváez PC, Heredia FJ, Orjuela Á, Osorio C (2017) Physicochemical and sensory (aroma and colour) characterisation of a non-centrifugal cane sugar ("panela") beverage. Food Chem 228:7–13

Gil JGR (2017) Characterization of traditional production systems of sugarcane for panela and some prospects for improving their sustainability. Revista Facultad Nacional De Agronomía Medellín 70(1):8045–8055

Guerra MJ, Mujica MV (2010) Physical and chemical properties of granulated cane sugar "panelas." Food Sci Technol 30(1):250–257

Gutiérrez-Mosquera LF, Arias-Giraldo S, Ceballos-Peñaloza AM (2018) Advances in traditional production of panela in Colombia: analysis of technological improvements and alternatives. Ingeniería y Competitividad 20(1):107–123

Jaffe W (2015) Nutritional and functional components of non centrifugal cane sugar: a compilation of the data from the analytical literature. Food Anal Compos 43(1):194–202

Jaffé W (2012) Panela Monitor: azúcar no centrifugada (panela), producción mundial y comercio. Caracas (Venezuela): Panela Monitor. http://www.panelamonitor.org/media/docrepo/document/files/azucar-no-centrifugada-(panela)-produccion-mundial-y-comercio.pdf

Khattak S, Greenough R, Sardeshpande V, Brown N (2018) Exergy analysis of a four pan jaggery making process. Energy Rep 4:470–477

Kumar A, Singh S (2020) The benefit of Indian jaggery over sugar on human health. In: Dietary sugar, salt and fat in human health. Academic Press, pp 347–359

Kumar R, Kumar M (2018) Upgradation of jaggery production and preservation technologies. Renew Sustain Energy Rev 96:167–180

Lee JS, Ramalingam S, Jo IG, Kwon YS, Bahuguna A, Oh YS, Kim M (2018) Comparative study of the physicochemical, nutritional, and antioxidant properties of some commercial refined and non-centrifugal sugars. Food Res Int 109:614–625

Martínez Zarate N, Bokelmann W, Pachón Ariza FA (2019) Value chain analysis of panela production in Utica, Colombia and alternatives for improving its practices. Agronomía Colombiana 37(3):297–310

Mann J, Fleck F (2014) The science behind the sweetness in our diets. Bull World Health Organ 92(11):780

Martínez HC, Aguilar-Rivera N (2019) Competitiveness of the piloncillo agribusiness in the central region of Veracruz. Revista Textual 73:297–330

Meerod K, Weerawatanakorn M, Pansak W (2019) Impact of sugarcane juice clarification on physicochemical properties, some nutraceuticals and antioxidant activities of non-centrifugal sugar. Sugar Tech 21(3):471–480

Morales-Ramos V, Osorio-Mirón A, Rodríguez-Campos J (2017) Innovations in the raw cane sugar mill: production of granulated raw cane sugar. Agroproductividad 10(11):41–47

Nagarajan K, Balkrishna A, Gowda P (2019) Alternative and supplementary health model on traditional sugars. J Nutr Weight Loss 4(115):1–10

Orlandi RDM, Verruma-Bernardi MR, Sartorio SD, Borges MTMR (2017) Physicochemical and sensory quality of brown sugar: variables of processing study. J Agric Sci 9(2):115–121

Ouerfelli M (2009) L'impact de la production du sucre sur les campagnes méditerranéennes à la fin du Moyen Âge. Revue des mondes musulmans et de la Méditerranée, (126). http://journals.openedition.org/remmm/6363. https://doi.org/10.4000/remmm.6363

Quintero-Lesmes DC, Herran OF (2019) Food changes and geography: dietary transition in Colombia. Ann Global Health 85(1): 28, 1–10. https://doi.org/10.5334/aogh.1643

Ramírez JA, Salazar JAV (2017) La cadena de valor de la panela y el fortalecimiento de la agricultura familiar en Costa Rica. Revista Abra 37(55):1–29

Rao J, Das M, Das SK (2007) Jaggery: a traditional Indian sweetener. Indian J Traditional Knowl 6(1):95–102

Requier-Desjardins D, Borray GR (2004) Environmental impact of panela food-processing industry: sustainable agriculture and local agri-food production systems. Int J Sustain Dev 7(3):237–256

Reyes VC, del Moral JB, Bravo MB, Ramírez JFG, Martínez GR (2017) Family agriculture and technology for the preparation of granulated piloncillo in the community of Aldzulup Poytzén San Luis Potosí. Nova Scientia 9(19):481–501

Seguí L, Calabuig-Jiménez L, Betoret N, Fito P (2015) Physicochemical and antioxidant properties of non-refined sugarcane alternatives to white sugar. Int J Food Sci Technol 50(12):2579–2588

SEDARH (2020) Caracterización de productores de caña de azúcar de la Huasteca Potosina, dedicados a la producción de piloncillo (2013/2014) http://www.campopotosino.gob.mx/index.php/biblioteca-digital/category/11-cana-de-azucar%3Fdownload%3D4036:padron-piloncill

ero&sa=U&ved=2ahUKEwir9YWxtarrAhUC8hQKHVJHCjUQFjAAegQIAhAB&usg=AOvVaw0uJSNJS66IKcWlbIBGpe-H
SIAP (2020) Anuario Estadístico de la Producción Agrícola https://nube.siap.gob.mx/cierreagricola/
Signorello M (2006) Canna da zucchero e trappeti a Marsala. Mediterranea Ricerche storiche, 3. http://www.storiamediterranea.it/public/md1_dir/r479
Silberman HC (1998) Sugar in the middle ages. Pharmaceutical Historian 28(4):59–65
Sugarcane Sustainable Development Law (2005) http://www.diputados.gob.mx/LeyesBiblio/pdf/LDSCA.pdf
Velásquez F, Espitia J, Mendieta O, Escobar S, Rodríguez J (2019) Non-centrifugal cane sugar processing: a review on recent advances and the influence of process variables on qualities attributes of final products. J Food Eng 255:32–40
Vera-Gutiérrez T, García-Munoz MC, Otálvaro-Alvarez AM, Mendieta-Menjura O (2019) Effect of processing technology and sugarcane varieties on the quality properties of unrefined non-centrifugal sugar. Heliyon 5(10):e02667
Verma P, Shah NG, Mahajani SM (2019) Why jaggery powder is more stable than solid jaggery blocks. LWT 110:299–306
Weerawatanakorn M, Asikin Y, Takahashi M, Tamaki H, Wada K, Ho CT, Chuekittisak R (2016) Physico-chemical properties, wax composition, aroma profiles, and antioxidant activity of granulated non-centrifugal sugars from sugarcane cultivars of Thailand. J Food Sci Technol 53(11):4084–4092

Traditions and Innovations in the North Caucasus Nature Management

Khava Zaburaeva and Evgeny Krasnov

Abstract The article is devoted to the analysis of the theoretic and methodological bases of the categories of "tradition" and "innovation" in the mountain environment. Differences between traditional and innovative forms of environmental management and points of contact have been defined. Traditional environmental management does not mean opposition to binary antinomy "innovation—tradition" but rather a linear diada of "tradition-innovation" as a balanced (sustainable) developing system. Optimization of environmental management does not involve the loss of traditions, but their reasonable transformation. The essence of traditional environmental management is not the exact reproduction of economic and cultural stereotypes from generation to generation, but to the preservation of harmony in the triad "nature—population—economy." The work proposes the conjugated development of traditional (forest and agro-industrial complex, agriculture, etc.) and innovative (agro- and ecotourism, development of IT technologies in management, etc.) areas of environmental management in the North Caucasus, provided the inexhaustible use of geoecological potential and sustainable functioning of geo-systems.

Keywords Mining · North Caucasus regions · Traditions and innovations · Principles and models of optimization

---

K. Zaburaeva (✉)
Kh.I. Ibragimov Complex Institute of the Russian Academy of Sciences, Staropromyslovskoe shosse, 21A, Grozny 364906, Russia

E. Krasnov
Department of Geoecology, Immanuel Kant Baltic Federal University, st. Alexander Nevsky, 14, Kaliningrad 236041, Russia
e-mail: ecogeography@rambler.ru

© The Author(s), under exclusive license to Springer Nature Switzerland AG 2021
W. Leal Filho et al. (eds.), *Innovations and Traditions for Sustainable Development*,
World Sustainability Series, https://doi.org/10.1007/978-3-030-78825-4_20

1 Introduction

The theoretical and applied problems of optimizing the interaction between nature and society have been discussed in the scientific literature for decades (Sabel'nikova and Karavaeva 2015; Bocharnikov and Egidarev 2017; Verburg et al. 2015). Improvements in environmental efficiency can be achieved by managing elements of activities, products and services that have a significant impact on the environment (Shhukina 2012).

The purpose of this study is to identify features of traditional environmental management and assess opportunities for innovation in the North Caucasus environmental management. The following objectives have been consistently addressed: (1) identify the natural and ethnogenetic features of the research region; (2) explore the modern structure of traditional nature management; (3) analyse emerging new areas of nature management; (4) assess the prospects of innovative directions in nature management.

Traditional environmental management of ethnic groups is considered the most conservative category, which is emphasized already in the name, indicating the preservation of the process in time and space (Abalakov 2013; Novikov and Gil'fanova 2018; Kochurov et al. 2018). The main element of the territorial structures of traditional environmental management is the ethnic local group (Gavrilov 2010). For mining, which is often associated with traditional and archaic ways of life (which in many cases is not true), the topic of innovation is very relevant (Savchenko 2012). At the same time, some authors recommend to allocate buffer zones, the environmental use of which would initially be formed consistent, tolerant to the traditional (Dmitrieva and Naprasnikov 2013).

The North Caucasus is a special region that stands out sharply on the map of Russia among other many peoples and ethnic groups living in a fairly limited space, historical and cultural diversity, originality and breadth of the spectrum of natural and climatic characteristics, acuity and abundance of problems and conflict situations. This is a constant factor of politics, economy and culture in the history of the whole of Russia. It occupies an important geopolitical place. Until now, the "Caucasian" factor of Russia's development has not been transformed into a long-term policy of sustainable and balanced development.

2 Methodology

The article is based on the concept of balanced nature management, which includes on equivalent terms the categories of "use," "protection" and "reproduction" of natural resource potential. At the heart of the concept of nature management and optimization, the authors used a system of methodological principles: interaction, optimality, complementarity, relativity, conservation and development. The work is characterized by the combined use of geosystem and geosituational approaches, principles and models to solve the problem of optimizing mountain area management. Also the article is based on the development of a conceptual and terminological framework that made it possible to analyze traditions and innovations in environmental management of the peoples of the multinational North Caucasus. The main methods are: comparative-geographical, empirical-statistical, ethnohistorical, SWOT and Content-analysis of published sources. The factual basis of the work was made up of publications in leading peer-reviewed publications.

3 Results and Discussion

Nature, population, traditions of economy and culture are closely intertwined in the mountainous regions of the Caucasus. The investigated regions of the North Caucasus on the map of Eurasia are hatched (Fig. 1). Regional specificity is reflected in the traditions of environmental management, landscape planning, language and other features. Environmental management systems in the North Caucasus have been formed for many millennia since the Paleolithic (Zaburaeva and Krasnov

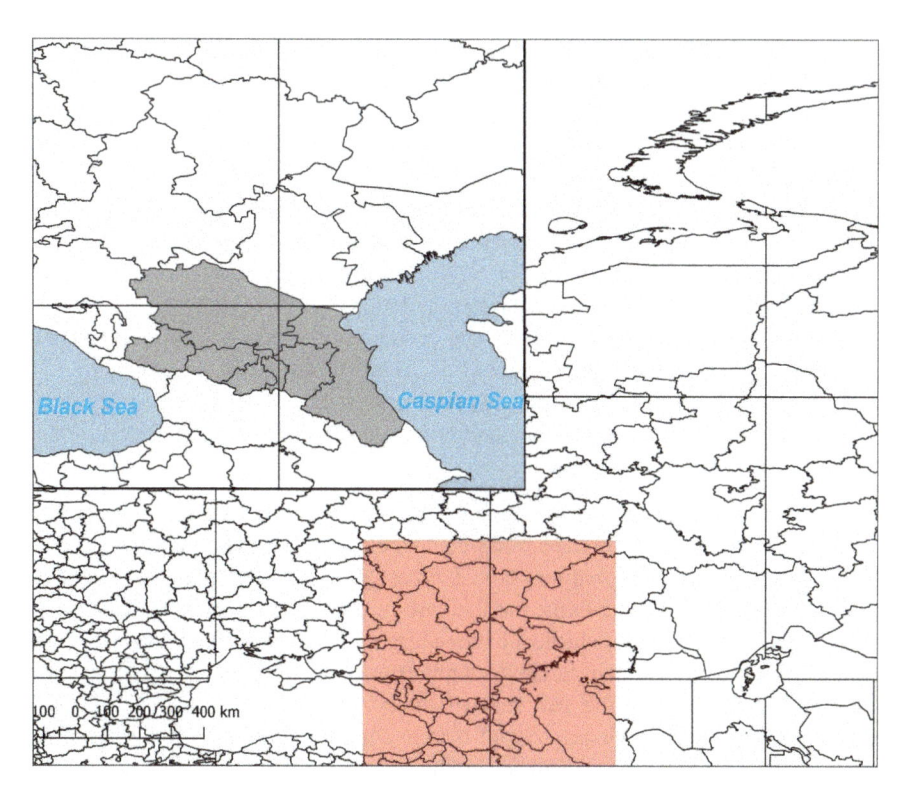

Fig. 1 The regions of the North Caucasus on the map of Eurasia

2018a, b). Traditional environmental management here has historically been based on agricultural production, although oil, mining, engineering, etc. have developed.

Traditional environmental management does not mean opposition to binary antinomy "innovation—tradition" but rather a linear diada of "tradition-innovation" as a balanced (sustainable) developing system. Optimization of environmental management does not involve the loss of traditions, but their reasonable transformation (Zaburaeva and Daukaev 2019). The essence of traditional environmental management is not the exact reproduction of economic and cultural stereotypes from generation to generation, but to the preservation of harmony in the triad "nature—population—economy."

4 Specifics of Nature Management in the Research Region

At the current stage of development, the following features of nature management in the North Caucasus (Gunja 2016) should be highlighted:

1. High potential of bioclimatic resources, allowing the development of productive agriculture, which is of all-Russian importance (the breadbasket of the country).
2. A unique combination of recreational resources (from the use of resources of high-altitude landscapes to seaside).
3. The border situation, which determines the specialization of certain industries, the transit function of roads, pipelines and energy systems, and affects migration flows.
4. Population growth is due to the positive balance in the reproduction of the population in most regions, as well as the influx of migrants. At the same time there is a concentration of the population in the most favorable areas, while, for example, in the highlands there is depopulation.
5. High landscape and biodiversity, the presence of a network of conservation areas, including world importance (Caucasian Biosphere Reserve).
6. The high diversity of natural, ethnocultural and socio-economic conditions determines the large disparities and contrasts in the development of different territories and types of resources.

The North Caucasus is distinguished by the presence of mountains, which form water resources necessary for the development of the foothill and flat steppes, concentrated most of the recreational and bioresources (Barilenko 2018; Zaburaeva et al. 2018a, b). Typical for mountains, the increase in precipitation and the decrease in temperature with height is disturbed in the intermountain basins. They form a special climate: the amount of precipitation does not exceed 200–300 mm per year, and the number of sunny days in the year reaches in some hollows up to 270. Low-snow winters and relatively large amounts of heat in summer allow to combine in intermountain basins water farming and pasture farming. The diversity of climate, terrain, rocks, as well as long-term economic activity led to the formation of a modern structure of natural and cultural landscapes. From the north to the mountainous and foothill areas adjoin the plain steppes on black soil. Now they are almost all plowed.

The modern development of the North Caucasian regions varies greatly. Differences in resettlement, population density, transport development of the region are amplified by the heterogeneity of the landscape structure, ethnic differentiation, the consequences of military actions. All this in general creates a very heterogeneous picture of the natural use of nature from the western resort and agricultural-industrial North Caucasus to the eastern, agrarian, with a wide spread of traditional forms of agriculture.

The situation of the region at the crossroads between Europe and Asia, on the border of the steppes, on which for thousands of years moved nomadic peoples from east to west and from north to south, played an important role in the process of natural resource development. For a long time, steppe landscapes were used mainly as grazing resources, while agriculture was of a focal nature. Ancient agricultural areas penetrate high into the mountains. With the active development of agriculture on the plains, mountain farming gradually loses its importance. Terrace complexes on the slopes begin to collapse, and the mountain population gradually migrates to the foothills and plains (Gunja 2016). Positive characteristics of the North Caucasus on the National background, such as the high degree of development of the territory, the availability of labor resources, the high potential of the agricultural sector, combined with favorable conditions for agriculture face slow institutional transformations. For sustainable development and optimization of environmental management in the North Caucasus, the positive and negative aspects and characteristics of the entire region must be taken into account.

5 The Role of Innovation in Nature Management

Innovation acts as catalysts for development processes taking place in territorial systems, increasing their organization, ensuring self-development, self-organization and acceleration. It is innovation that accelerates the evolution of systems, while evolving themselves, generating new innovations (Kaplina and Pelevina 2018). At the same time, natural and economic systems react either by diversifying functions, which are associated with technological, industrial and other socio-economic revolutions, or through spatial expansion, diffusion of innovations. These processes may be working at the same time. The innovation mechanism connects the past, present and future environmental management through procedures to select from possible possible innovations of environmental management the most acceptable, based on ethnoconstant (Zaburaeva et al. 2018a, b).

The "penetration" of innovation into traditional environmental management has two shells that reflect the nature and intensity of the process. However, the core remains unchanged, which allows the process to be called traditional. Flexibility of traditional environmental management is manifested only outside the nucleus—in the shells (Kochurov et al. 2018).

The essence of the "innovation" category for environmental management is based on the principles of ensuring the optimal balance between the processes of meeting the needs of the population, public production and the preservation of natural resource potential and the quality of the environment. The effectiveness of innovations in optimizing regional environmental management consists of direct (preservation of natural resource potential, minimization/prevention of environmental and social, economic damage from the degradation of natural ecosystems, environmental safety) and indirect (receiving products and services in adjacent areas) effects (Vlasova 2008).

Since ancient times, the population of the North Caucasus has been engaged in various crafts (making household utensils, tools, burkas, palas, carpets, pottery and jewelry, etc.) and at the same time has mastered modern technologies of agricultural and industrial production. Most Caucasian cultures are difficult to modifi, they tend to preserve the traditional way of life for indigenous ethnic groups.

In mountainous areas, agricultural, recreational and environmental areas are among the top priorities. Taking into account its natural-climatic and soil-geomorphological diversity in the agro-industrial complex, the development of crop production, livestock, crafts and crafts is the most significant.

The culture of terrace farming is one of the features of traditional natural use. Dagestan is the northwestern boundary of the great terrace culture. The culture of terrace farming originated in the Middle East and spread over the golden belt of ancient agricultural civilizations of the globe, intersecting with the hotbeds of plant domicile, including in Dagestan (Nabieva 2009). Terrace farming in conditions of shortage of arable land (gardening, vegetable production) can be profitable using biocenotic methods of intensification of production without disrupting the balance of agrobiogeocenosis. Bioenergy in livestock and crop production will simultaneously solve the problems of increasing the productivity of farms, the use of waste, revitalization of disturbed land, etc.

In mountainous areas, terraced gardening and livestock farming are promising. The largest republic of the North Caucasus (Dagestan), for example, in the number of cattle is the 3rd place in Russia, derived from the waste of livestock methane and carbon dioxide, can find a variety of applications (for welding, refueling fire extinguishers, etc.). With a scientifically sound approach to the use, protection and reproduction of agro-potential, mountain regions can not only cover domestic needs, but also act as exporters of diverse fruit and vegetable products. The main condition for this is the balanced use, reproduction and protection of fertile agricultural land.

The authors propose the conjugated development of traditional (forest and agro-industrial complex, agriculture, etc.) and innovative (agro- and eco-tourism, development of IT technologies in management, etc.) areas of environmental management in the North Caucasus, provided the non-attractive use of geoecological potential and sustainable functioning of geo-ecosystems (Fig. 2).

The targeted approach to environmental management requires the author to follow a certain sequence:

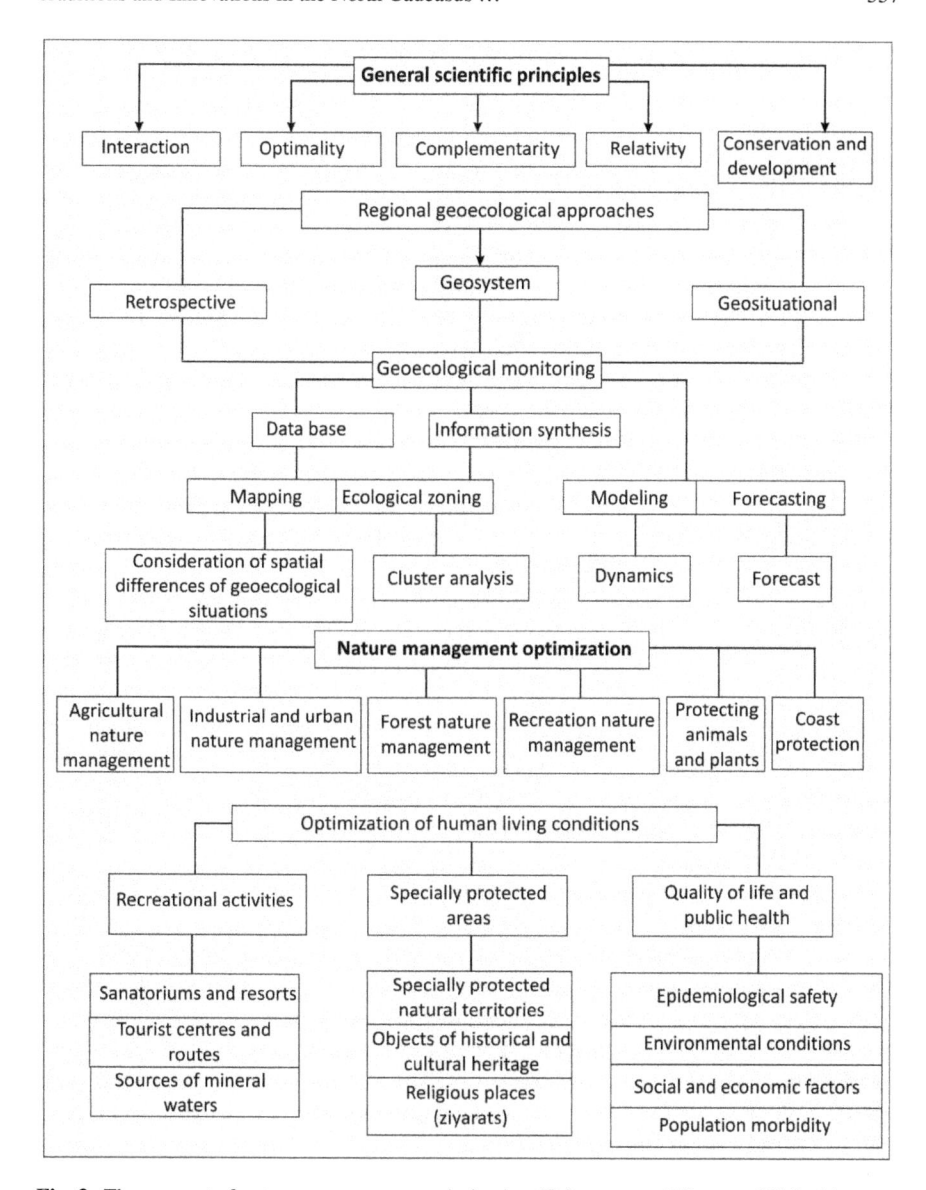

Fig. 2 The concept of nature management optimization (Zaburaeva and Krasnov 2018a, b)

Before making a decision, identify and rank goals (setting a hierarchy of goals);
Identify existing alternatives to meet targets;
Select criteria to identify from existing alternatives to achieve the best possible targets that will measure upcoming costs or resources;
Develop a logical (or different) model that adequately displays the links between goals, alternative means of achieving them, external conditions and resource needs;

Compare costs and results according to the chosen scenario.

6 Optimization of the North Caucasus Nature Management

Achieving the goals of balanced environmental management in the regions of the North Caucasus is possible only on the basis of a systematic approach to solving geoecological and socio-economic problems of land use, combating soil degradation and depletion of their fertility. To this end, a consistent algorithm for assessing and optimizing land use, including three stages—appraisal, analytical and strategic. At the initial stage, "end-to-end" monitoring of the state of the land—a set of observations and assessments of their quality of humus content, acidity, soil erosion, etc., a critical analysis of the legal status of the land and the effectiveness of environmental measures. Particular attention should be paid to the targeted use of land and the validity of the transfer of land to other categories in connection with the highest priority of agricultural land (Ivanova and Ivanov 2019).

In the second stage, geoecological potential (land-resource, biodiversity, human capital) should be assessed, taking into account the goals of balanced environmental management. The final stage is the development of priority and interconnected strategies for the further development of each region, ensuring the preservation and reproduction of their natural resource potential, bio- and ethnocultural diversity. An important role in the proposed algorithm is given to a systematic assessment of the impact of the implementation of previously developed strategies and recommendations (ecoaudit) with further correction of land use forecasts based on the interim results achieved.

The history of mental forms of human relations with natural objects goes back centuries Zaburaeva and Krasnov 2018 (reverence of mountains, groves, springs, etc.) and continues to this day (pilgrimage). Due to its special importance for the Muslims of the Caucasus, it is possible to distinguish mental nature use in a special type. Pilgrimage is a priority for indigenous peoples (ethnic groups) making the walk to Mecca (hajj) and other holy places (springs, burial sites, etc.) (Zaburaeva and Krasnov 2018a, b). In contrast to purely consumer tourism, pilgrimage is a special kind of recreation associated with the improvement of a person's spirituality and moral and ethical standards of conduct, which is of paramount importance in today's conditions of increasing inter-ethnic conflicts and uncertainty for each of us.

To optimize the nature the mental factors to the present time practically were not considered, although the seriousness of the lack of moral and ethical evaluations of economic and other activities and their relationship to ecosystem sustainability are obvious to most sane people. In the future should consider the inclusion of mental activity as a fundamental mental concept (design) of nature.

Tourism has long been an important source of development for the North Caucasus (Zaburaeva et al. 2019a, b). In this area, the state and local private entrepreneurs are both allies (the development of tourism infrastructure benefits everyone) and competitors (especially with large and relatively fast-moving public investment).

The introduction of a tourist cluster, supported by huge public funds, is, in fact, a political project aimed at strengthening the image of the North Caucasus and creating growth points (Shardan et al. 2019). Land prices in prestigious mountainous areas have increased tenfold.

The structure of land use can be improved by introducing the best crop rotation systems, increasing the area of forest strips in semi-desert and deforested mountain areas at the expense of "downed" pastures, around residential areas, in river valleys, etc. (Sokolova and Potapova 2019). The multiplier effect of land use optimization allows to strengthen the processes of clustering of economic entities. These are, for example, "anchor" clusters (grain, oil refining, etc.) in Stavropol, Krasnodar region.

Cluster structure in the field of environmental management is formed at the expense of productive forces, industrial relations, enterprises and organizations, various organizational and economic forms of activity, using natural and resource potential, having a territorial link to certain economic structures and structured by types of economic and environmental activities (Zemcov et al. 2018).

Hypothetically, the geographical boundaries of the cluster can be defined on a fairly large scale—from the city, the area to the international space. They are often formed at the regional level. However, in our view, inter-regional cooperation and the creation of clusters will contribute to more sustainable socio-economic and environmentally friendly development of regions, particularly in the Caucasus. After all, the capacity of tourist flow is almost always higher in companies that are part of the cluster structure than in those working independently of it (Segarra-Oña et al. 2012).

Interregional fruit and vegetable clusters of Chechnya and Ingushetia on the basis of the existing agricultural holdings "Gardens of Chechnya" and "Sad-Giant Ingushetia", tourist-recreation complex on the high-altitude lake "Kezenoy-Am" and the international mountain-ski resort "Veduchi" in the zone of alpine meadows—the first examples of the implementation of the cluster approach in the Northeast Caucasus (Zaburaeva et al. 2019a, b). Traditional and innovative activities are very important for the development of competitive enterprises in the agricultural and recreational areas of natural use for the mountain regions of Russia as a whole.

Thus, clusters (agricultural, tourist-recreational, environmental, etc.) in the North Caucasus suggest a form of voluntary association of enterprises located in territorial proximity and functionally dependent in the production and sale of goods and services. Since the cluster is only a form of economic cooperation based on long-term contractual relations, in fact clusters will earn only if projects are attractive and enterprises will benefit from the introduction of innovative technologies. It is obvious that the clustering of environmental management, optimization of the socio-economic development of the regions under study involve, first of all, the modernization of infrastructure and the training of highly qualified personnel (managers, land builders, marketers, etc.).

An important role is played in the system of maintaining ecological equilibrium in this macro-region, the use of environmentally friendly (safe) technologies and industries, the integration of livestock and crop production, which will reduce waste, obtain organic fertilizers and thus increase the natural fertility of soils. The development and implementation of regional agricultural environmental management programs, with the integration of crop and livestock production with strict control of the parameters of the use of pesticides will have a positive impact on the economic indicators of production and the quality of conditions—the health of the population and ecosystems, their biodiversity, etc. According to a study conducted on the lands of several regions of Germany (Gilhaus et al. 2017), the composition of pasture vegetation depends on many factors: management specifics, soil characteristics, grazing period, pasture loads, etc. The highest variety of vegetation cover was distinguished by seasonal pastures.

In order to achieve the goals of balanced environmental management, the topic of "ecosystem services" is of considerable interest. Ecosystem functions contribute to the preservation of the biodiversity of wild animals and plants, soil fertility, the quality of surface and groundwater, natural landscapes (Deng et al. 2016; Jin et al. 2017).

The model is implemented only if there are able-bodied highly qualified scientific personnel—land builders, managers, agronomists, etc. In addition, environmental management and auditing (enterprises and municipal-level territories) play an important role (Sysuev 2017).

Environmental management and audit, as a system using international environmental standards ISO 14 000, is designed to ensure the environmental safety of production, gradually becoming part of the management activities of regional socio-economic systems of the North Caucasus. Thus, the components of environmental education are included in educational standards. In the North Caucasus, the norms of traditional religions are involved.

The associated directions of the development of balanced environmental management in the north-east of the Caucasus (recreational and environmental areas) at the municipal level are mapped by rice (Fig. 3). Taking into account the natural-climatic and soil-geomorphological diversity of municipal conditions, the development of crop, livestock, crafts, pilgrimage and conservation activities is a priority. The zone of agriculture includes areas (mostly steppe), in which agricultural land is mainly arable land (Malgobek, Sunzhensky, Grozny, Khasavyurtovsky, Babayurtovsky, etc.). There are concentrated black earth, meadow-chestnut, chestnut and other varieties of soils. The increase in productivity and gross harvest of agricultural products will also contribute to the involvement in the turnover of unused arable land, including after clearance, sufficient application of organic and mineral fertilizers. The development of animal husbandry is most promising in lowland and mountainous areas.

The recommendation of priority orientation of agricultural nature management does not exclude the possibility of using land for other purposes. In a number of mountainous areas, conditions are favorable for the cultivation of organic potatoes and legumes. Among the factors limiting balanced land use are land degradation processes (water and wind erosion, salinization, dehumification, etc.) and the concentration of more than half of all arable land in an acrid zone.

Traditional folk crafts in mountain villages and cities are not only witnesses to the ancient and long history of the peoples of the North-Eastern Caucasus, part of their cultural heritage, but also a huge potential for the development of recreational nature management (Zav'jalova 2005). The study of the territorial organization of ancient

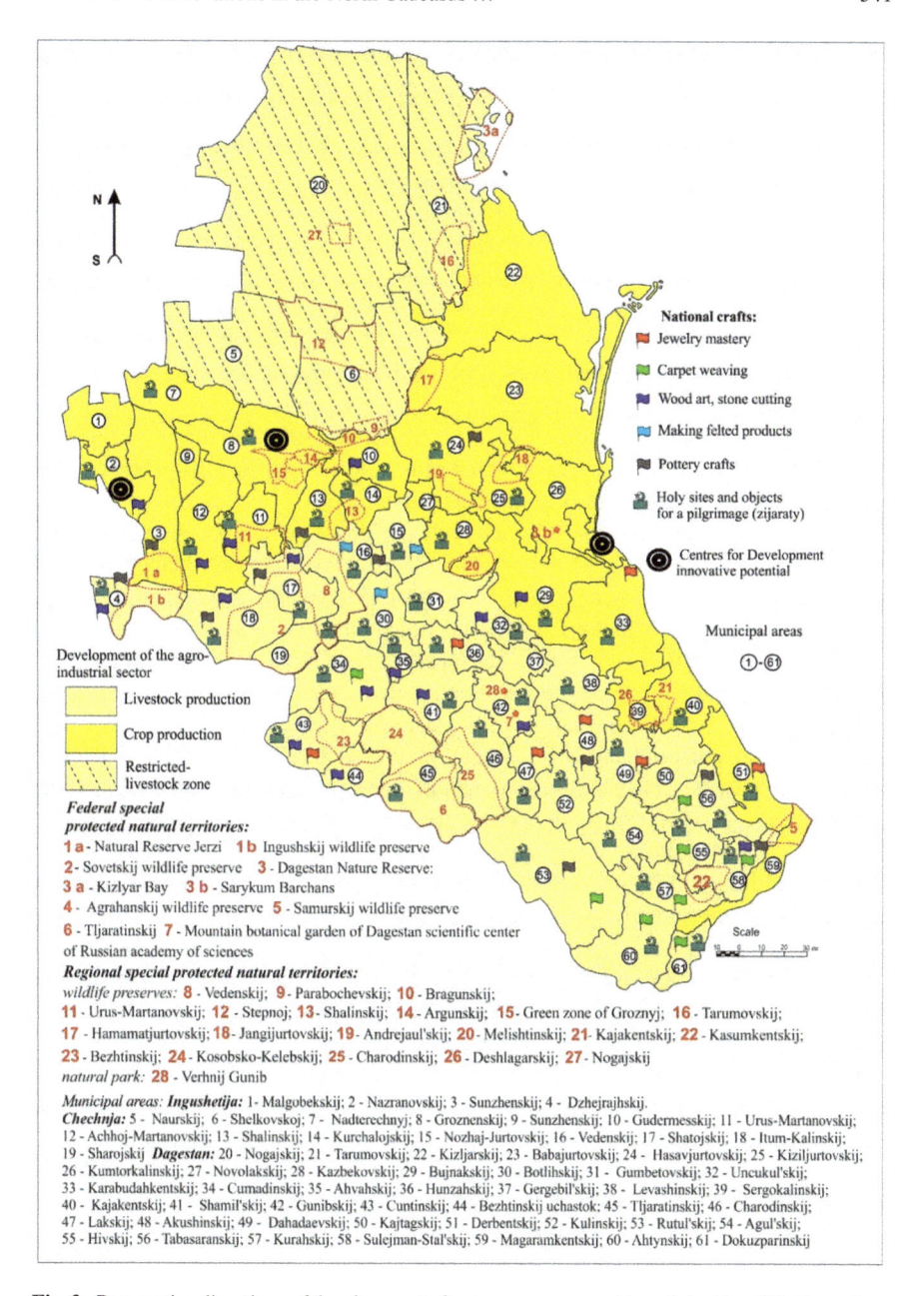

Fig. 3 Perspective directions of development of nature management (municipal level) in the regions of the North-Eastern Caucasus

crafts and crafts will help to identify ways to preserve them during the formation of tourist and recreational areas of nature management. Many types of artistic creativity of the peoples of the Caucasus (carpet weaving, jewelry processing of metals, metal carving on wood, pottery, etc.), objects of pilgrimage (ziyarats, mosques) and cultural and historical heritage (crypts, tower buildings, etc.) have been preserved to this day.

A significant share of the creative potential of the peoples of the North-Eastern Caucasus is concentrated in mountainous and high-altitude areas. Family secrets of masters of jewelry, pottery, etc. have been preserved and continue to be passed down from generation to generation, largely due to the intra-community integrity. The highest art of execution is embodied in women's jewelry, table setting items, writing instruments, various types of weapons (swords, bows, knives, daggers, helmets). Of course, the range of products, in particular, jewelry has undergone significant changes since its inception. If in the middle ages the craft of blacksmiths in many ways meant chain mail, swords and sabres, by now the most relevant jewelry (rings, bracelets), decorative weapons, embroidery in gold and silver, etc.

Jewelry was preserved in the Dakhadayevsky (Kubachi village), hunzakh (Gotsatl village), Lak (Kumukh village) and other districts of Dagestan. There are also carpet weaving masters of Akhtyn (village of Akhty), dokuzparin (village of Mikrah), Tabasaran (village of Khuchni, Arkit) and other districts. Masters of pottery are easily found in the flat part—in the Khasavyurt (s. Sulevkent), Suleiman-Stalsky (s. Ispik) districts, in the mountainous Chechnya—in the Shatoysky district (s. Shatoy), and in the mountainous Ingushetia—in the Dzheyrakhsky (s. Lyazhgi) district. The manufacture of felt products is less common and according to our data has been preserved only in the Botlikh (village of rahata) district of Dagestan, Vedensk (village of Vedeno) and Nozhay-Yurtovsky (village of Simsir) districts of Chechnya.

Historical and cultural monuments are widespread in the North-East of the Caucasus, although differentiated by type of terrain. On the plain, they are mainly represented by monuments to soldiers of the great Patriotic war and other wars, houses where celebrities lived, and in the mountainous part-watchtowers, defensive, signal towers, bas-reliefs, etc. The system of specially protected natural and historical and cultural territories is designed to ensure "balance" of anthropogenic impact and promote sustainable development of regions (Zaburaeva et al. 2019a, b). However, currently the territories and objects of the North-Eastern Caucasus that are particularly valuable in natural or historical and cultural relations are still not a system. They are scattered and almost unguarded. If they are combined into a system (geo-ecological framework), they can become the basis for sustainable regional development. The study (Leibenath et al. 2010) justifies the need to enhance the flow of knowledge and information among practitioners in this field, the synthesis of spatial planning and the formation of environmental networks in a transboundary context.

As the world practice (Hammer and Siegrist 2008) of natural (scientific and cognitive) tourism within specially protected natural areas shows, make a significant contribution to the development of the regions. However, given the fragility and vulnerability of mountain ecosystems, no contradictions can be allowed between the interests of the tourism industry and conservation (Borges de Lima and Green 2017).

7 Conclusion

Based on the our analysis, the following conclusions can be drawn.

Firstly, mountain nature management is a complex system, multi-factor, temporary and multi-aspect process, mutually conditioned by natural, social, ethnic and other development institutions. The conditions and factors of the natural resource management process are always territorial. The formation of natural resource management systems in the North Caucasus has been going on for many millennia since the Paleolithic.

Secondly, the traditional systems of nature management include agro-industrial, industrial, residential, craft, etc., and the emerging ones—innovative (landscape design, geoecological projects, etc.), mental (pilgrimage), etc. Traditional nature management here has historically been based on agricultural production, although oil, mining, mechanical engineering, and others have developed.

Thirdly, in mountainous areas, terraced gardening and animal husbandry are promising. In conditions of shortage of arable land, terraced agriculture (horticulture, vegetable growing) can. Bioenergy in livestock and crop production will simultaneously solve the problems of increasing farm productivity, using waste, revitalizing disturbed land, etc.

Fourthly, in the conditions of the North Caucasus, taking into account its original ethnocultural heritage, an important role should be given to the preservation of national crafts.

References

Abalakov AD (2013) Traditional nature management: challenges and development prospects. Geogr Nat Resour 1:182–188

Barilenko VI (2018) Problems of the development of the north Caucasus of Russia recreational. Mod J Language Teach Methods 8(6):244–250

Bocharnikov VN, Egidarev EG (2017) Wildlife in landscapes and ecoregions of Russia. Geogr Nat Resour 4:38–49

Borges de Lima I, Green RJ (2017) Wildlife tourism, environmental learning and ethical encounters: ecological and conservation aspects. Springer International Publishing, Australia, p 292

Deng X, Li Z, Gibson J (2016) A review on trade-off analysis of ecosystem services for sustainable land-use management. J Geog Sci 26:953–968

Dmitrieva VT, Naprasnikov AT (2013) A geo-ecological experience of educing territories of traditional nature management in Russia. Bull Mos Municipal Pedagog Univ 1:59–69

Gavrilov EV (2010) Territories of traditional managing of natural resources: concept, features, history. Agrarian Land Law 4:55–57

Gilhaus K, Boch S, Fischer M, Hölzel N, Kleinebecker T, Prati D, Rupprecht D, Schmitt B, Klaus VH (2017) Grassland management in Germany: effects on plant diversity and vegetation composition. Tuexenia 37:379–397

Gunja AN (2016) Mountain nature management and processes of modernization on North Caucasus. Effective development of the mountain territories of Russia. Mountain Forum: materials of the intern. Sci Pract Conf 33–50

Hammer T, Siegrist D (2008) Protected areas in the Alps. The success factors of nature-based tourism and the challenge for regional policy. Gaia 17/S1:152–160

Ivanova ZSh, Ivanov ZZ (2019) Conflicts in the land use of mountainous areas. News Kabardino-Balkarian Sci Center Russ Acad Sci 4:40–45

Jin G, Deng X, Chu X, Li Zh, Wang Y et al (2017) Optimization of land-use management for ecosystem service improvement: a review. Phys Chem Earth 101:70–77

Kaplina MS, Pelevina AB (2018) Environmental and economic innovation in nature management as a factor of environmentally sustainable agricultural development. Econ Enterpr 9:160–166

Kochurov BI, Sh ZH, Kerimov IA et al (2018) Modern problems of environmental management in the North Caucasus and the way of their decision. Grozny Nat Sci Bull 3:29–33

Leibenath M, Blum A, Stutzriemer S (2010) Transboundary cooperation in establishing ecological networks: the case of Germany's external borders. Landsc Urban Plan 94:84–93

Nabieva UN (2009) Cultural-geographic aspects of traditional nature management of Republic Dagestan. South of Russia: Ecol Dev 3:9–13

Novikov AN, Gil'fanova VI (2018) Traditional natural resources management: innovations—institutions—traditions. Bull Eurasian Sci 5:39–50

Sabel'nikova EI, Karavaeva NM (2015) The role of innovation in environmental management. Forecasting the innovative development of the national economy within the framework of environmental management: materials of the intern. Sci Pract Conf 218–225

Savchenko IM (2012) Traditional environmental management as the basis for the formation and functioning of the tourist and recreational area. Prob Reg Ecol 4:95–99

Segarra-Oña M et al (2012) The effects of localization on economic performance: analysis of Spanish tourism clusters. Eur Plan Stud 20:1319–1334

Shardan SK, Yandarbayeva LA, Karaeva FE, Misakov AV, Misakov VS (2019) Social infrastructure development issues of mountain area rural territories in depressive republics of Northern Caucasus. Int Trans J Eng Manag Appl Sci Technol 2:275–282

Shhukina VN (2012) Sustainable development in traditional lands. Interexpo GEO-Siberia 1:165–168

Sokolova OE, Potapova EV (2019) The structural features of green areas in settlements. Bull Voronezh State Univ 1:19–24

Sysuev VV (2017) Optimization of environmental management: theoretical approach. A healthy environment is the basis for regional security: collection of proc. of the intern. Ecological Forum 362–375

Verburg PH, Crossman N, Ellis EC, Heinimann A, Hostert P, Mertz O, Nagendra H, Sikor T, Erb K, Golubiewski N, Grau R, Grove M et al (2015) Land system science and sustainable development of the earth system: a global land project perspective. Anthropocene 12:29–41

Vlasova EJ (2008) Efficiency of innovations is in nature management. J New Econ 1:163–170

Zaburaeva KhSh, Krasnov EV (2018a) The balanced mountainous nature management concept in the Regions of the North-East Caucasus. Reg Aspects Soc Pol Yearbook 20:77–88

Zaburaeva KhSh, Krasnov EV (2018b) Formation history of North-Eastern Caucasus regional systems of nature management (the comparative analysis experience). Bull Acad Sci Chechen Repub 3:27–32

Zaburaeva KhSh, Daukaev AsA, Zaburaev ChSh, Sedieva MB (2018a) Biodiversity as the basis for sustainable development of mountain areas on the example of North-East Caucasus. Adv Eng Res 151:975–980

Zaburaeva KhSh, Tajmashanov HJe, Zaurbekov ShSh (2018b) Land use optimization in mountainous regions of Northeast Caucasus. Sustain Dev Mt Territ 1:35–45

Zaburaeva KhSh, Daukaev AsA (2019) Strategy of Forming the Geoecological framework of the territory: on the example of the Chechen Republic. Atlantis highlights in material sciences and technology (AHMST), vol 1. International Symposium "Engineering and Earth Sciences: Applied and Fundamental Research" (ISEES) pp 137–143

Zaburaeva KhSh, Gatsaeva LS, Sarsakov MS, Daukaev AslA, Abumuslimov AA, Abumuslimova IA (2019a) Hydro potential assessment on the territory of the Chechen Republic for recreational purposes. In: International scientific and practical conference 'AgroSMART—smart solutions for agriculture', KnE Life Sciences pp 497–506

Zaburaeva KhSh, Krasnov EV, Zaburaev ChSh (2019b) Cluster approach in environmental management: experience, problems and prospects for further development. Adv Eng Res 182:343–348

Zav'jalova OG (2005) Zones of ethnocontact nature management: mechanism of evolution. Geogr Bull 1:23–35

Zemcov SP, Baburin VL, Kidjaeva VM (2018) Innovation clusters and prospects for environmental management in Russia. Geogr Nat Resour 1:15–21

Khava Shahidovna Zaburaeva Doctor of Geographical Sciences, Chief Researcher of the Department of problems of the fuel and energy complex, Complex Institute named after Kh.I. Ibragimov of the Russian Academy of Sciences, Staropromyslovskoe shosse, 21A, Grozny, 364051, Russia Author of more than 120 scientific publications. Research interests: Geoecology, Nature management, medical Ecology.

Yevgeny Vasilyevich Krasnov Doctor of Geological and Mineralogical Sciences, Professor of the Department of Geoecology, Immanuel Kant Baltic Federal University, st. Alexander Nevsky, 14, Kaliningrad, 236041, Russia. Author of more than 300 scientific publications. Research interests: problems of regional geoecology.

Revitalization of Local Traditional Culture for Sustainable Development of National Character Building in Indonesia

Cahyono Agus, Sri Ratna Saktimulya, Priyo Dwiarso, Bambang Widodo, Siti Rochmiyati, and Mulyanto Darmowiyono

Abstract Education is a cultural and civilized effort to advance human life and to improve human dignity. Ki Hadjar Dewantara (KHD) established Tamansiswa Indonesia in 1922 and proposed the Tri-centra Education concept as harmonious cooperation among formal, non-formal, and informal-education at family, school, and community. An education system with an influential culture and values educate the head, heart, and hand, respectively, based on love, care, and dedication. The Among tutoring system is one with a familial spirit that is based on nature and independence. The cultural concept of "Three Excellent Souls" consists of *cipta* (create), *rasa* (feel), and *karsa* (intend), describe the reason, emotion, and intention, respectively. The national character education is not merely a schooling process, but also in family and social-cultural environment. The development of a superior new civilization could produce prominent Indonesian people based on God, humanity, nationality, family, and justice; by relying on the essential capital of culture and education. Tri-centra education is fully aware of extraordinary character, ethics, noble character, and individual responsibility that must be contributed in real terms, respecting

---

C. Agus (✉)
Faculty of Forestry, UGM, Yogyakarta 55281, Indonesia
e-mail: cahyonoagus@gadjahmada.edu; acahyono@ugm.ac.id

The Association of Tamansiswa Society (PKBTS), Yogyakarta 55151, Indonesia

C. Agus · P. Dwiarso
Majelis Luhur Persatuan Tamansiswa (MLPTS), Yogyakarta 55151, Indonesia

S. R. Saktimulya
Faculty of Culture, UGM, Yogyakarta 55281, Indonesia
e-mail: ratna.saktimulya@ugm.ac.id

B. Widodo
Museum Colloquy Board (Barahmus) DIY, Yogyakarta, Indonesia

S. Rochmiyati
Faculty of Teacher Training and Education, Universitas Sarjanawiyata Tamansiswa, Yogyakarta 55151, Indonesia

M. Darmowiyono
Universitas Sarjanawiyata Tamansiswa, Yogyakarta 55151, Indonesia
e-mail: mulyanto@ustjogja.ac.id

© The Author(s), under exclusive license to Springer Nature Switzerland AG 2021
W. Leal Filho et al. (eds.), *Innovations and Traditions for Sustainable Development*, World Sustainability Series, https://doi.org/10.1007/978-3-030-78825-4_21

others' rights, nature, and diversity and determining responsible choices/decisions. According to the times, TRI-CON (convergent, concentric, and continuous) concept with the Among system in managing culture evolved dynamically. Local culture, folklore, fairy tales, ancient puppets, both orally or in writing, have been used as learning material to instill and preserve moral stories, leadership, local wisdom, and culture. The development of digital media as educational media enables millennial students to be directly involved in stories, take moral wisdom, motivate, be creative, innovate as character education rooted firmly in the ancestral culture. Revitalization of traditional regional culture is essential to support the sustainable development of national character building in Indonesia.

Keywords Character building · Digital learning · Edutainment · Golden generation · Tamansiswa · Tutorial

1 Introduction

According to several international survey institutions, Indonesian education's quality and index are among the worst at the international level (Agus et al. 2020). Indonesia launched five revolutionary education policy packages, "Freedom of Learning," as part of national education restoration. The policy packages consist of primary and secondary education policy packages, Independent Campus, Student Operational Assistance, Movement Organization Program, and Motivating Teacher (Kemendikbud 2020). There are still many misconceptions about the meaning of freedom because they are still multi-perceptual and do not have essential equality. The method of freedom developed by the international system tends to be absolute. Meanwhile, Ki Hadjar Dewantara (KHD) independent mental education to support the national character building must also be supported by responsible self-discipline, disturbing the orderliness of a peaceful, greeting, and happy life (Agus 2017).

The education system is to become the primary human being who develops personal talent potential, contributing significantly to an environment and life that is dignified and sustainable (Agus et al. 2020). Furthermore, it can also contribute significantly to creating blue earth by increasing the added value of the economy, environment, and socio-culture in harmony (Agus 2013, 2018; Cahyanti and Agus 2017). The system needs to be developed through the system, independent, able to determine destiny, independent on orders, strength, and competence in an orderly manner.

Ki Hadjar Dewantara developed national education through Tamansiswa Institution during the colonial era in 1922 (Agus et al. 2020). The Tamansiswa education system is developed based on its own local culture with a peaceful acculturation process worldwide by developing schools such as parks without walls. The Tri-Central Education system includes families, schools, and communities through formal, informal, and non-formal education in a synergistic and integrated manner, not just one party (Agus 2017). *Muhammadiyah*, founded by KH Ahmad

Dahlan and *Nahdlatul Ulama* by KH Wahid Hasyim, also developed the Tri-Centra Education concept. Integrated universal education aims to form team cohesiveness, cooperation, information communication, leadership, focus, the concentration of thought, innovation, creativity, strategic management, analytical power, and increased self-confidence (Cahyanti et al. 2019).

No one knows for sure what will happen in the future. Meanwhile, as time goes on, everything that unfolds in the world moves according to the times. In the face of the times' effects, two controversial camps were found, namely the group which still held fast to the old custom without regard to the development of the times, and the group that tried to keep going by adjusting to the demands of the times. Each of them defended his principle under the pretext of saving culture. Ki Hadjar Dewantara stated that everyone hopes to live safely and happily; there is no need for disputes due to customs differences (Dewantara 2013a). They must dare to move forward, not merely bound by custom. Thus, they are expected to sort out the relevance of the adat in the present. Cultural ideas consisting of cultural values, norms, laws, and related rules become one system or so-called tradition at any time, and the conditions are continually changing. Customs that are no longer by the era, if maintained, will be painful and less useful (Dewantara 2013a).

To deal with cultural development and the times, the "Tri-Con Theory," namely: continuity, convergence, and concentric, could be considered. Supriyoko (2012) explained that the meaning of continuity is always to maintain the predecessors' cultural values and continue their implementation in everyday life. Convergence: providing a meeting room between our culture and foreign cultures for mutual dialogue to create new cultures. Concentricity: can ensure that the new culture created by the convergence of our culture with foreign cultures is constructive and is more beneficial to people's lives (Supriyoko 2012).

Nurhayati (2019) stated that innovation is a change that has the nature of updating the old to the new, increasing the quality and quantity, and may even change differently from the old to something completely new. Based on these changes' nature, innovation can positively and negatively impact individual human life and/or society. Therefore, social innovation will also have an impact on cultural change (Nurhayati 2019). The phenomenon of cultural innovation in Java's academic literacy is exciting and needs to be explored to know the impact and solution. Sari (2019) stated that literacy is a literary skill in reading and reading literary works. Literacy is related to understanding content and interpreting meaning in a complex reading (Sari 2019). Thus, Javanese literacy can be interpreted as an activity of reading and writing ideas based on Javanese manuscripts' material objects and understanding and interpreting their meanings. How to respond to today's readers who might not even know Javanese Manuscripts containing messages from our ancestors?

The COVID-19 pandemic has caused humanitarian, economic, social, cultural, civilization, and environmental tragedies that are changing lives worldwide. The lockdown policy due to COVID-19 also significantly impacts the learning process, research, community service, culture, community, and traditions (Kemendikbud 2020). Innovation and creativity in using the latest information technology online are needed to overcome Covid-19 and the future. The concept of Three-Centra Education based on the culture in families, schools, and communities must be developed in millennial students' style. Adjustment of the millennial approach can be made by developing a tutoring system through student-centered cultural learning. The concept and edutainment process will trigger the learning system to be more enjoyable, but the quality is maintained (Cahyanti et al. 2019). The co-learning and working from a home system in cyber school 4.0 for millennial students in the era of lockdown and destructive innovation with the support of sophisticated information technology and significant data access seem to be the most appropriate learning media. The online empowerment culture learning system must apply win–win solutions, co-creation, co-finance, flexibility, and program sustainability.

This chapter will discuss the revitalization of local traditional Javanese culture based on ancient manuscripts and current realities for national character building and sustainable development in Indonesia. This paper covers cultural innovation, limited to academic literacy concerning information technology.

2 Material and Methods

This chapter was conducted using primary and secondary research data. The primary data collection uses two techniques: (1) field observation: data collection by directly observing the object, including specimens, communities, museums; (2) interviews, questionnaires, communication, WhatsApp group, and focus group discussion. The secondary data were obtained from reports, books, journals, and the internet and included theories, precedents, and standards used in the field. The thematic study method was used to describe current and future educational issues and concepts that become an analytical tool to see the conception of traditional and local culture on sustainable development of character building in Indonesia. Qualitative and quantitative analyses of the collected site data are continuing to identify solutions for similar problems.

3 The Innovation of Local Culture

Ki Hadjar Dewantara stated that culture is a human gift, whereas character is a ripe soul capable and able to create. Human nature encompasses all movements of mind, taste, and will. Thus culture is the fruit of thought (in the form of science, education, teaching), the fruit of feeling (in the form of beauty, nobility, purity, customs), and the fruit of will, among others, tangible buildings, industry, agriculture, and forth (Dewantara 2013a; 2013b).

Information about culture in the past can be known, among others, through the reading of a text (manuscript). Old manuscripts are likened to a window to the past world because these manuscripts are stored information about the values prevailing at that time, thoughts, feelings, beliefs, customs, and others (Baroroh-Baried 1994). Some manuscripts are loaded with manuscripts about law, religion, history, health, arts (dance, musicians, puppets), architecture, agriculture, customs, *primbon* (horoscope, almanac), *pawukon* (a character based on the *Wuku* cycle, every 30 weeks), literature, languages, and *piwulang* (lesson, learning). The following learning manuscript will be shown at first, *Batara Bayu's* manuscript on perseverance, written about two centuries ago (Baroroh-Baried 1994). Second, *Sotya Rinonce*'s manuscript about the importance of fostering a love for the work being done was copyrighted a century ago. Third, *Wasita Rini's* manuscript on the principle of decency for women, in a century, and fourth, *Maskumambang* manuscript about Covid-19 *Sirep Pageblug*, was written Coronavirus hit the world in 2020. The four manuscripts here will be described (1) manuscript as psychotherapy, and (2) Javanese culture strengthens national culture.

3.1 Manuscript as Psychotherapy

The manuscript consists of words, phrases, and sentences that certainly have specific meanings and intentions that the author of the manuscript wishes to convey to his readers. A series of manuscripts can be used to convey teachings, consolation, and triggers of courage. Manuscripts can also function like psychotherapy, that is, as medicine, by using psychic powers' influence through the methods of suggestion, advice, entertainment, and the like (Baroroh-Baried 1994). The following four manuscripts are examined based on their function as psychotherapy.

3.1.1 Manuscript of *Batara Bayu*

The manuscript is quoted from a manuscript titled *Sestra Ageng Adidarma* and *Sestradisuhul*, from the Pakualaman Temple Library collection. This manuscript is part of the *Astabrata* manuscript, as a learning manuscript that contains the teachings of the eight gods guarding the universe, namely *Batara Indra, Yama, Surya, Candra, Bayu, Vishnu, Brama,* and *Baruna* (Baroroh-Baried 1994). Digest Manuscript of *Batara Bayu* according to KGPAA Paku Alam X as follows:

"An ideal leader is required to have the courage and not be easily provoked. The persistence of his attitude causes the people he leads will not to act recklessly. They will submit and obey their leader because he applies stringent sanctions. No one can change his mind because his actions are based on a firm belief and always stand on the truth. Those who still do not follow their example will be allowed to improve their performance with various facilities intended as a 'subtle allusion' parable for their inadequacy. The facility is continuously being disbursed until the person concerned realizes his mistake. If they still cannot understand the parable, stringent sanctions will be imposed. Nevertheless, if they can immediately read the parable and realize their mistake, they will be forgiven".

The Batara Bayu figure with *Pancanaka* nails in the *Sestra Ageng Adidarma* manuscript wears *Poleng Bangbintulu* patterned cloth and carries a *Rujakpolo* mace in its flagship heirloom. Three identities are also found in the manuscript space about the *Batara Bayu* in the *Sestradisuhul* manuscript. Current readers must be equipped with knowledge about the characters of puppets, clothing, and weapons they wear to understand the manuscript's illumination (Fig. 1).

In 2014 the meaningful ancestral message was enshrined by GBRAA Paku Alam, consort KGPAA Paku Alam X in *Asthabrata* series batik cloth. It has hoped that present and future generations can still enjoy the work of ancestors originally inscribed in ancient scripts that not just anyone can open or flip through pages of manuscripts because the paper is fragile. The batik cloth with aesthetic *Batara Bayu* motif from Pakualaman was given the name "*Bayu Krastala*," whereas the *Asthabrata Jangkep* motif contains the eight gods' elements (Fig. 2). Through the message written in the manuscript and the interpretation of the illumination, it is understood that by imitating *Batara Bayu*, who has a strong personality and is not easily provoked, it is hoped that the reader or user of the *Bayu Krastala* patterned cloth always strives for resilience.

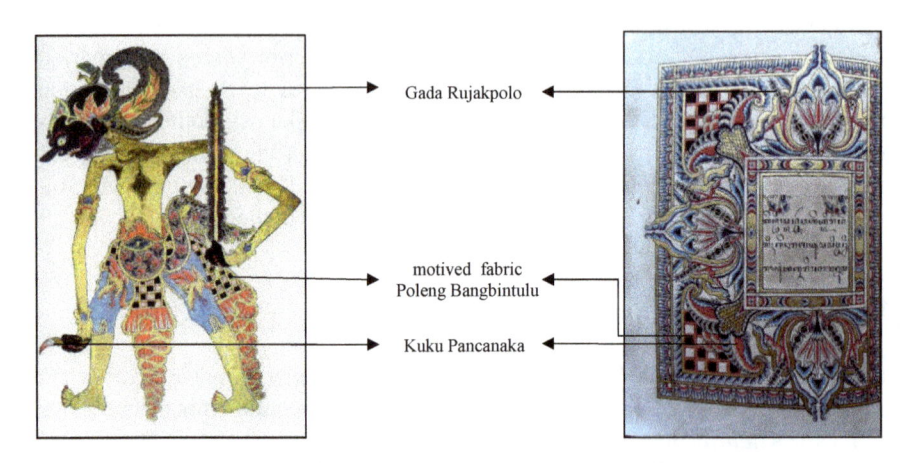

Fig. 1 Batara Bayu Illumination in *Sestra Agĕng Adidarma* (left) and *Sestradisuhul* (right)

Fig. 2 The batik motif of *Bayu Krastala* (left) and *Asthabrata Jangkep* (right)

3.1.2 Sotya Rinonce's Manuscript

The learning's manuscript compiled by Suryapranata (1917) contains 44 messages that should be implemented daily. The contents of the *Sotya Rinonce* manuscript, among others, contain suggestions for getting used to being disciplined, thinking positive and optimistic, introspective, diligent, determined, always seeking, tolerant, and others. One manuscript is stated as follows (Dewantara 1961).

"If you do something, you should be with a happy and sincere heart".

- Strive to understand and be able to, because there is a possibility that the work will occur only briefly or for a long time and is not well-liked. Therefore, it must force it to like it to not complicate its implementation because it is already relieved thanks to the urge to like it. The work done with a sense of displeasure will certainly feel heavy and complicated, sometimes even causing failure. Although it can be done, the results are not acceptable.
- If carrying out tasks accompanied by pleasure will give strength and peace of mind. Even if a job has not been carried out for a long time, face it with a happy heart. Do not turn to others, and do not get bored with thoughts because they have done it before.
- If they know that it turns out the job choice is wrong, then stop then look for an excellent job. However, even if it intends to quit the job, it should do a job well.

The manuscript's essence is in doing a job should be based on a happy and sincere heart. Something that is done with pleasure and sincerity will produce optimal results. Therefore, always think positive and optimistic.

3.1.3 Wasita Rini's Manuscript

Wasita Rini 'Advice for women', an *Asmaradana* melody manuscript with the enjoyment of Ki Hadjar Dewantara, was created because of the concern of witnessing girls' freedom of association in that era. Learning for these women contains an invitation for women to be able to maintain their dignity. At that time, the women wanted to achieve independence from the confines of tradition feared to be out of control. Therefore, *Wasita Rini* song (Dewantara 1961) was created as follows.

- In the past and present, the principle of feminine teachings is no different; both want to protect the women so that they are holy and safe, avoiding distress.
- The difference at this time, which is called the era of independence, is that all people oppose other parties' power and control. That is how women do not like to be treated as others wish.
- Remember that women being independent means being free from other people's commands and being healthy and capable of controlling oneself. Therefore, do not forget that rights and obligations are not separated.

- What is called mandatory is all the readiness and willingness, physically and mentally, then following the right to carry out one's own will; because you already have a balance sheet to weigh what is right and what is not good.
- Women who nurture a spirit of independence must know their interests: intelligence and a sense of freedom give life independence, as for decency is a safety fence.

A century ago, when Ki Hadjar Dewantara delivered a message through *Wasita Rini's* song, there was already a feeling of turmoil among women who wanted to be free. Such turmoil continues today. Ki Hadjar Dewantara's message is very relevant to the current situation and conditions where many women are demanding equal rights with men without regard to the obligations that must be done. The core of KHD lesson to women is that since ancient times the things that need to be maintained by women are physical and spiritual purity. His whole attitude was based on decency. Suppose they want to be free from oppression. In that case, they must equip themselves with an understanding of "independence" that is strong and able to be independent and in harmony between seeking rights and obligations.

3.1.4 Maskumambang Teks of Sirep Pageblug

Maskumambang is the name of one of the Javanese *macapat* melodies/songs. Sastrowiryono (1978) stated that each song of *macapat* and having specific rules is also adhered to by the song's character. *Maskumambang* consists of four lines with the rules (12i, 6a, 8i, 8a), contains 'worry and worry,' and miserable 'sad'. The difference between the character of songs in *macapat* is influenced by the composition of the notation and rhythm of each song to develop a certain sense for the singer and listener of the chanting.

An interesting phenomenon when the Covid-19 pandemic swept across the world, which began in the early 2020s until sometime, has caused all the earth inhabitants to feel and witness the enormity of a viral epidemic that is rapidly causing casualties and devastating impacts. During the Covid-19 pandemic, *the Rengeng-Rengeng* community in Java developed the "Maskumambang-Sirep Pageblug" song through online facilities. There is an accent of *Ngayogyan, Banyumasan, Osing, Bali, Sundane*se, and others. The sense of togetherness in this suffering can increase one's immunity because singing and listening to the chanting delivered wholeheartedly causes a smooth flow of blood. The vibrational energy of sincerity voicing poetry with specific tones and rhythms attached to the *Masumambang* song is self-medication of suffering by crystallizing problems as psychotherapy. Increased immunity causes the body and soul to become primed, so they avoid the Covid-19. The expression expressed by the song of *Maskumambang* tells us that the corona creature that caused the pandemic is a creature sent by God to carry out the destiny provisions to remind all people on earth not to be complacent and should recognize God's power. We must realize that this pandemic event has become His will, and we can only ask for His mercy.

Table 1 Manuscript functions

No	Title	Figure	Behavior	Useful as
1	*Asthabrata*	Batara Bayu	Be firm in a heart for the truth	Heart booster when hesitating
2	*Sotya Rinonce*	General	Doing something sincerely and happy	Self-encouragement so that they can work optimally without being burdened with strings attached
3	*Wasita Rini*	Woman	Striving for rights and obligations in a balanced and ethical manner	Self-control so as not to be a selfish/demanding woman
4	*Maskumambang "Sirep Pageblug"*	General	*Total surrender* 'surrendered to God'	Peace of mind from anxiety, fear, and helplessness in facing the Covid-19 pandemic, as a pandemic spell

If singing with all the heart, the four verses of the *Maskumambang* song will foster a sense of resignation to the Creator. Like a spell, the whining chants can erase the pain, anxiety, and despair into an understanding of optimism. The Corona-virus will stay away from it because the singer believes that he is in the protection of God (Table 1).

About 200 packets of the *Maskumambang* manuscript have been created until the end of May 2020. This manuscript proves that Javanese literary literacy, especially *macapat* song today, is still alive and cared for, even though only a small percentage care about it. Based on the four manuscripts' observations, the following table shows the manuscript function, as seen from the narrated character and its behavior, up to its usefulness for today's readers.

The ancestral messages outlined in the old manuscripts/books will not be known to the broader community because they are only stored in the library cupboard. Readers are also constrained by Javanese letters and languages used in ancient books. The following table shows the hopes and efforts to disseminate *piwulang* (lesson) by considering the year of writing and the vehicle used at that time, and its existence in the present (Table 2).

3.2 Javanese Culture Strengthens National Culture

Ki Hadjar Dewantara stated that national culture is the essence of regional culture throughout the Indonesian archipelago, both old and new. As for development, it must go through the "TriCon" way, which is continuous with the past, converging

Table 2 Vehicle Innovations and Efforts to Spread the Lesson

No	Title manuscript, production year	Mode		Hopefulness	Effort
		In the past	At present		
1	*Asthabrata* (1841 dan 1847)	Imported paper; Javanese letters	On cloth in the form of batik	The messages can be conveyed to the public without having to open old manuscripts that are easily damaged	Messages are conveyed through exhibitions and seminars, and batik cloth laden with messages can be bought and sold
2	*Sotya Rinonce* (1917)	Local paper mills; Javanese letters	Limited to being delivered in the classroom	So that 44 lesson Manuscripts can be implemented	The manuscript must be transcribed and translated, then presented in the form of an e-book
3	*Wasita Rini* (±1925)	Local paper mills; Latin letters	The document is just a recording on a cassette for personal listening	Lessons about decency can be understood by women and implemented	distributed online with attractive packaging
4	Maskumambang "*Sirep Pageblug*"	During Covid-19 pandemic	online	Able to strengthen self-immunity through battens	distributed online with attractive packaging

with other cultures, and concentric in a broad or united unity but still has personality traits (Dewantara 2013a).

The elements of Javanese culture are closely embedded, include (1) efforts to harmonize thought-sense-intention and (2) inheritance in the form of literary and Javanese literary works and works of art (batik, puppets, and Javanese songs). Through an interpretation of *Wasita Rini*'s Manuscript (see Table 3) and *Masku-*

Table 3 Interpretation of *Wasita Rini's* Manuscript

Case	Complaint	Impact	Solution	Acquisition
Women want to live independently, without restraint	Feeling confined by obligations as a woman and feeling oppressed by male authority	Leaving morals because of the demands of realizing rights and forgetting obligations	Align creativity-feeling-initiative with wise considerations	Happy, prosperous, happy birth and mind

Table 4 Interpretation of the Masumambang "Sirep Pageblug" Manuscript

Case	Complaint	Impact	Solution	Acquisition
Pandemic Covid-19	Anxious, worried, afraid of being exposed and affected by a deadly virus	Experiencing paranoia and feeling helpless	Calming down with the nuances of warmth towards God's will	• Increase immunity • Have many friends in the same fate • Happy physically and mentally

mambang Sirep Pageblug's manuscript (Table 4), efforts to harmonize thought-sense-intention can be seen as follows.

Table 3 shows that to defuse a case to end happily requires synchronization of copyright-initiative. Starting from the creativity (thought) that women want independence because they feel oppressed, the intention (will) is driven to rebellion without decency. Because the impulses of thought will have been overflowing, resulting in a thinning policy and then disappearing, causing anger. Javanese culture denies wrath. An attitude of wrath will not occur if the offender has a 'humble' attitude and is sincere in doing an action. With humility and a sincere heart, he can harmonize his creativity and taste so that his rights and obligations can be done with joy, which is what is called an independent life.

Likewise, with Table 4, to eliminate anxiety and powerlessness in facing the Covid-19 pandemic, the *Sirep Pageblug* community group jointly conducts self-therapy through the whining by creating and chanting the *Maskumambang* song, which is nuanced to the nostalgic nuances, containing the sense of 'surrender to God'. At this point, humans are led to negate the ego's attitude so that in the end, they grow sincere recognition that it is on God's power and love that all events occur. Thus, the corona pandemic is no longer a worrying and frightening threat. God indeed created him for a purpose that would certainly bear fruit. Therefore, they should be mindful and alert 'always remember and be vigilant'. Remember God and past events, which bring wisdom, so be alert. Humility, humble, *total surrender,* consciousness, and vigilance are part of Javanese culture, strengthening national culture and is expected to be a heritage facing the times' challenges.

4 Tri-Con Culture Promoting Young Generation Souls

Ki Hadjar Dewantara wrote that the nature of education is a cultural process, both of which cannot be separated (Dewantara 2013a). In the context of cultural management, we must understand the concept of Tri-Con (Convergent, Concentric, and Continuous).

4.1 Tri-Con to Manage Nation's Culture

A pre-World War I in Europe, there was a restoration of education; first, the teaching was top-down from the teaching teacher changed bottom-up considering students' talents and abilities. KHD convergently absorbs knowledge from Europe and concentrates on the archipelago's local wisdom for the previous hundred years. This culture must be managed continuously and born, among other things, *"Among"* tutorial method a blend of a western classical education system with the institution cottage in Java (Dewantara 2013a).

- Convergent: Do not close with the association of world culture, selectively sort out and choose a universal culture that is beneficial to enrich, enhance the development of the nation's own culture, towards the point of convergence of humanity that is just and civilized.
- Concentric: In wading and blending into the universal culture, hold fast to own cultural character to strengthen its personality. A healthy nation certainly has the character of the nation's cultural character.
- Continuous: Management and development of the nation's culture must be carried out continuously, from period to period, from system to system, and from generation to generation without interruption.

The three elements of the Tri-Con are interrelated and inseparable in the culture of the nation. Only the importance of convergence will tend to print intellectuals that lack character. Concentricity will build a person who tends to be utopian away from the progress of nature and age, and continuity feels empty because there is no underlying ideology (Dewantara 2013b). The order of virtue of the Tri-Con elements may differ depending on the carrier's duties and responsibilities. Students will undoubtedly focus more on if convergence absorbs knowledge and knowledge to provide the future. For tutors, advisers, and traditional stakeholders, there is much focus on the process of concentricity so that students and students remain national. The management of primary education must focus more on continuity so that the results of culture can be sustainable. The three elements of Tri-Con should be synergistically harmonious, not unbalanced to any one element. The Tri-Con culture manages the present culture in harmony with the progress of nature and age, based on the nobility of the past culture to develop the future culture (Dewantara 2013b; Agus et al. 2020).

4.2 Adoption of Foreign Culture with Adaptation in Japan

In Japan, the Meiji Restoration of 1866 adopted convergence to absorb western cultural progress in a convergent manner adapted concentrically with the Samurai, Bushido, and Kamikaze's local wisdom. The allocation of Japanese scholarships to study abroad was substantial at 43% of the state budget. The development of Japanese culture was very rapid in the fields of education, technology, industry, economy, trade, and military, so that in 1904 Japan was able to subdue the Russian military. This event inspired the eastern people that a colored nation could defeat a white nation. Then there was a revival of the eastern movements including in India and Southeast Asia, among others marked by the establishment of *Budi Utomo* 1908 (Agus et al. 2020).

Aside from being convergent and concentric, Japan conducts culture continuously. After the atomic bomb defeat, Japan can still rise in economic culture, industry, trade, and technology. The Tri-Con culture has also been successfully carried out in a foreign country, even Japan, which has adopted a foreign culture (convergent) followed by its own cultural (concentric) adaptation to improve the nation's cultural tradition (continuous).

4.3 Tutorial "Among" System as an Implementation of Tri-Con Culture

Among means carrying/fostering with a sincere heart without strings attached. Education at Tamansiswa College is carried out according to the *Among* System, which is a character education system with a family spirit and is based on:

4.3.1 Nature-Based

Nature-based is an acknowledgment of God's power and a condition for achieving the fastest and fastest possible progress by the natural progress of nature and the times (Dewantara 2013a; Agus 2018)).

(a) Children's nature is always peaceful and comfortable in the family realm; the among systems bring an atmosphere of family warmth in the teaching and learning activities in the dormitory (home-schooling) with teaching 24 h a day. The tutor acted and functioned as a father/mother to the school students with all the care, and vice versa. The family's nature causes students to feel at ease, not uprooted from the warmth of their natural compassion. Now, the home-schooling prototype Tamansiswa is still held at *Taruna Nusantara* High School Magelang.

(b) The nature of the child who likes to play (game/kinder spellen) is facilitated in the teaching and learning activities with a load of child games and simulation. For example, memorizing teaching material with song, making a map on the page by playing in the sand, the content of character in the song, fieldwork practice, Student Community Services (SCS), out bond, scout, game to hone a child's character.

(c) *Tut wuri handayani* develops the nature of the child's talents. *Tut wuri* gives creative freedom to the child, *handayani* guides the child's natural talents. Students are encouraged to self-learning, actively seeking out passively waiting to be told, as an independent learning essence.

(d) Nature is always advancing to adjust to nature and times. Nature is demanding that there is always progress in change over time. Nature as a struggling agency must not change, but the form-content-rhythm could adjust nature and time.
(e) The natural demands of children always want independence from the womb. The system's primary purpose is to free the soul of the child physically and mentally, and energy. In the colonial era, the soul of independence was used to prepare for independence, then the soul of independence to fill and strengthen its independence and state.
(f) The nature of the homeland with various races, tribes, and religions in the Unity in Diversity with more than 17,000 islands must be grateful to build a multi-cultural nature rather than monoculture. The nature of Indonesia's motherland with more than 17,000 islands, 740 tribes, and various religions, must be grateful as a complementary blessing of God. There must be no majority domination over the minority because it is not under the fairy and fairy of humanity (1945 Constitution).

4.3.2 Independence

Independence carries the meaning of God's gift to humans by giving them the right to self-regulate by observing self-discipline requirements to maintain order and peace in social life (Dewantara 2013a).

a. The independence of the child's soul is born, his mind and energy are the tutorial system's primary goals of "among". The freedom of the soul and soul is born, and the strength of the Indonesian people is needed in the pre-independence era, in the realm of independence so that the influence of other countries will not sway us. Sovereign in politics, self-reliant in economics, personality in culture is a hallmark of a nation's true independence (the Trilogy of Independence).
b. Independence must be followed by self-discipline if it does not interfere with other people/groups' independence to form an orderly and peaceful, greeting, and happy society. Self-discipline means prioritizing the Obligations of Human Rights (defending the State, deliberation, consensus, filial piety) rather than the Human Rights of individuals or groups. As the anticipation of "continuity" in the euphoria of democracy.
c. The "among" system prohibits punishment and coercion in the learning activities because it will hinder the child's independent soul's growth. The punishment of coercion in a disciplined cadaver will inhibit an independent soul's growth and inhibit student talents development.
d. Kinder spellen's attitude as an "independent soul embryo" students launched synergized *wiraga, wirama, and wirasa* (sport, art, and taste) processes in harmony with creation, taste, and will. *"Tut wuri handayani"* facilitates students to play and simulate nature so that their independent talents develop according to their abilities.

4.3.3 Character Building Education

As a complete system among students is given character-building education, including nationality. So, the students have a distinctive personality and spirit of national defense and the Indonesian homeland (Dewantara 2013a; 2013b). The school proactively coordinates with students' parents and families and extra-curricular activities in scouting and other skills, as a synergism in the Tri-Centra Education. Character education subjects are included in local content and national character building. All the only nations in the world begin with national character characteristics (Japan, Korea, USA, United Kingdom, China).

4.3.4 Tri-Centra Education

The educational environment that affects the mental development of students there are three environments (Dewantara 2013a), all of which must be conducive to each other, namely:

a. A sincere family environment, loving education from unconditional parents, is a student's primary character education.
b. Students' parents formally entrust the school environment to the school where they study science and manners.
c. Community Environment that educates the child by learning by doing in the community. Positive community activities help shape children's character and must be kept away from damaging social contamination.

If one factor is not conducive, Tri-Centra Education for the child (broken home family, student brawl at school, drugs in the community) will inhibit the growth of children's character. KHD plans broader education in Universal Education, namely, "Every place is a school, and everyone is a teacher".

4.3.5 Trilogy of Education

a. *Tut wuri handayani*: freeing students to develop their creativity, while the teacher/tutor fosters from behind must not merely dictate, at the elementary school level.
b. *Ing Madya Mangun Karso*: encouraging students to be proactive in mingling and motivating the learning environment to improve the quality of education (peer, competition, creativity, innovation, analysis) at the level of Basic Education to Higher Education.
c. *Ing Ngarso Sung Tulodho*: When becoming an official/leader, one must become a role model for others and their juniors. Community service with the motto of science and scientific charity will benefit the wider community, not just for the group or personal.

This concept was also developed as a Trilogy of Leadership Management for clean governance or conflict management and in family-based economics in preparing the 1945 Constitution. Principle Tamansiswa 1922 stated that by not being bound physically and spiritually, and with a pure heart intend tutor/teacher close to the child (Dewantara 2013a; 2013b). We do not ask for any rights, but we will surrender ourselves to the child. KHD puts the function of civil/teacher as a fighter in the cultural and community development agency through educational means in the broadest sense. We prioritize the obligation to foster the child rather than his rights, fostering a generation of national and independent spirits.

5 The Role of Museums in Education of Nation Character

According to the International Council Of Museum (ICOM 2017), a museum is a permanent institution that serves the community's interests and its progress, is open to the public, does not aim to seek profit. Museum collects, maintains, examines, exhibits, and communicates material evidence of human material and its environment for study, education, and pleasure. Meanwhile, according to Government Regulation No. 66 of 2015, the museum is an institution that functions to protect, develop, utilize collections, and communicate to the public for study, education, and pleasure. Based on the two definitions, the museum task is no longer focused on managing collection objects but needs to adapt to its role in the family, school, and community. In this millennial era, museums plan a target number of visits solely but must consider the opinions and perspectives of families, schools, and surrounding communities towards the museum.

The magnitude of the interest of families, schools, and communities cannot be measured only by the number of museum visits and the museum's role in the activities carried out by schools and communities (Kemendikbud 2020). A museum is a valuable place in a nation's life journey and stores various valuable cultural objects and historical objects of the nation's struggle, essential for learning. There are various historical values, heroism, struggle, and culture of the museum's Indonesian people. The museum's existence as an institution that stores, maintains, and exhibits cultural heritage objects of historical value is very relevant as a vehicle to build national character, so the millennial generation always adheres to the Indonesian people's identity. In contrast, national identity can be known and lived a thorough understanding of collecting objects and history stored in museums. Essentially, the museum represents the past of a worthy nation and deserves to be known, understood, and understood by millennial generations so that it is not interrupted by the historical chain of struggle and national culture.

Museums in Indonesia must improve themselves in facing the shift in values due to globalization and information technology (Kemendikbud 2020). The museum's role as a multidimensional medium to build national character and be a source of inspiration and innovation for millennial generations can be an optimal nation's civilization. A museum is a valuable place in a nation's life journey and stores various sublime works of ancestors, reflecting the richness and diversity of national culture necessary for learning.

Nation character education is a conscious and planned effort to develop the nation's character through education (Dewantara 2013b; Agus et al. 2020). Indonesian National Education System is a conscious and planned effort to create an atmosphere of learning and learning. So that students actively develop their potential to have religious, spiritual strength, self-control, personality, intelligence, noble character, and skills, it needs itself, the community, the nation, and the country".

In the context of national character education, education is not just a schooling process but a process in the movement format. The schooling system is an integral part of the character education movement (Dewantara 2013b). Simultaneously, the nation's character education's direction is to give birth to Indonesian humans who are intelligent, noble, and have Indonesian personality. Intelligence is a spirit of continuous learning to improve the quality of self. Cleverness, accompanied by moral morals, will give birth to humans who have dedication and love for Indonesia's homeland. Ki Hadjar Dewantara stated that "to get a teaching system that will be beneficial to the collective effort, it must be adapted to the people's life and livelihood. Therefore, we must investigate all the deficiencies and disappointments in our lives related to the nature of society as we wish…".

Ki Hadjar Dewantara's stated that an education (system) should not be separated from the developing problems and cannot be separated from the community's ideals (Dewantara 2013b). In the context of nationality, the education system must itself be able to answer the nation's present and future challenges. The relationship between education and national issues, history, and national ideals is needed so that every citizen of the nation can know and understand the Indonesian nation's existence early on. Besides, education must improve the quality of human resources to overcome the nation's various problems and utilize its full potential to become the power to build and create a better, fair, and prosperous national life based on Pancasila.

A sense of love for the homeland, nation, and state can be grown with a deep understanding of the history of the nation's struggle, why the Indonesian nation should be independent, what are the direction and future journey of the nation (Agus et al. 2020). Therefore, education is a learning process and a liberation process, making an independent, active, and creative soul develops independently and with dignity. The generation of millennial must be able to understand and learn from historical experience. By understanding the importance of learning from historical experience, it is hoped that the footing to build the present and future is focused. One of the learning media in strengthening character education is the museum. Through the museum, it is expected that the strengthening of character education in the framework of planting historical awareness, the spirit of nationalism, the love of the homeland, can be achieved for the younger generation. The museum serves as a cultural bridge between generations, a window of culture, a means of presenting a new world, seeing the past, develop the nation's culture and civilization in the future. Through the millennial generation museum, we can find out how the Indonesian people's long journey in seizing, filling, and maintaining Indonesian independence is based on Pancasila and the 1945 Constitution.

Our education world has given a large portion of knowledge, but forgetting the development of attitudes, values, and behaviors in the learning process (Dewantara 2013b). One of the weaknesses of our society is the underdeveloped culture of visiting museums. Cultivating visiting museums among the public, especially students and students, is not easy. Even in the current era of information technology, they prioritize buying cell phone pulses for communication rather than buying a museum entry ticket. Other factors that cause a lack of interest in visiting museums include the culture of accessing information via the internet, radio, television and visiting the mall more dominating than visiting a museum. Also, there is still a lack of understanding and appreciation of the values of heroism, history, and cultural arts stored in the museum.

With the flood of information and social media invasion, the museum is now increasingly challenged to become a trusted source of historical information on the nation's civilization, a friend of change, and a vehicle for strengthening character education based on Pancasila. According to Cicero, the world historian states: History is the witness of time, the torch of truth, the life of memory, the teacher of life, and the bearer of news from time to time.

Multidimensional crises have occurred in the nation's life and state, including the failure to socialize Pancasila's values as the nation's character. For this reason, museums must play a role in helping the government strengthen character education by fostering and equipping students. The golden generation of Indonesia in 2045 must have a Pancasila spirit and good character to face the dynamics of change in the future (Agus et al. 2020). The national education foundation development must place character education as the leading soul in state administration with community involvement.

Museums can play a role in helping schools implement programs to strengthen character education by implementing Pancasila values. These values are religious, honest, tolerant, disciplined, hardworking, independent, democratic, curious, national spirit, motherly love, respect for achievement, communicative, peace-loving, fond of reading, caring for the environment, caring socially, and being responsible.

Character education is a deliberate attempt to help people understand, care, act on core ethical values, judge what is right. Also, care about what is right, and do what is believed to be right, even when facing external pressures and internal temptations (Dewantara 2013a). Therefore, character education requires a connection between the role of the school and the museum role. The museum must help schools understand essential values and then support their programs, including strengthening the government's character education. One such effort is to introduce the museum to schools through joint exhibitions. The exhibition is recognized or not; it has become a vehicle that can educate. Because with the exhibition, the community, especially students and students, can know and obtain a variety of knowledge about the history and the past and the nation's valuable culture. As for the museum manager, the museum exhibition is one way to socialize the existence of collections and allow museum managers to interact with visitors regarding the collection presented, whether it is according to the community's wishes. The museum manager can also promote and see first-hand the behavior, interests, and absorptive capacity of visitors to collect items on display.

Besides, the Museum Manager proactively cooperates with the family, school, and community environment. First, through the role of family/parents, the children of their sons and daughters are reawakened to love and love the museum because, in the museum, there are various sources of historical values, heroism, struggle, and culture of the Indonesian people. The family is the beginning of the formation of a museum visiting culture. The habit of visiting museums needs to be instilled from an early age and starting from the family environment. Parents have a very dominant role in directing their sons and daughters to love visiting museums.

Secondly, through the role of the school, it can be explained that the information contained in the museum collections is presented as a vehicle: strengthening educational character, fostering imagination and appreciation for creativity, as well as being a source of educational inspiration and innovation. Third, through the community, friends, and lovers of the museum and the local community, making the museum a place of exciting and comfortable activities for education, research and recreation, discussion, self-performing arts, and cultural, social activities. Of the three roles, the museum collections presented can provide many changes informing one's character for good or bad. The museum collections' constructive role and function can direct a person into a good teaching and learning process by presenting festive events, information, and role models.

Presumably through the role of the museum can strengthen character education through the process of teaching and learning early in the family, school, and community to give birth to new people of Indonesia with a strong character, reflecting personal qualities of noble character, love of the motherland, the spirit of nationality, honesty, tolerance, discipline, caring, independent and responsible.

6 Cultural Aspect on Sustainable Development Goals

The new Sustainable Development Goals (SDGs) 2030 include 17 agendas and 169 specific targets, which should become a reference for the achievement of SDGs worldwide, countries, and cities. The 2030 Agenda is a small step forward in considering the cultural aspects of sustainable development. Even though none of the 17 SDGs agendas mention the cultural sector exclusively, their inputs, processes, outputs, and outputs require a reference from cultural aspects (UCLG 2017). Petti et al. (2020) reported that the cultural aspect plays an essential role as a legacy and an explicit reference in achieving SDGs Agenda 11.4 and an indirect reference to other SDGs agendas.

Cultural aspects have an essential role in achieving the SDGs 2030 Agenda (UCLG 2017). Cultural aspects have been shown to play a significant role in empowering active participation in cultural life, developing individual and collective cultural freedom, protecting tangible and intangible cultural heritage, and protecting and promoting various cultural expressions. All of that is civilization as a core component of human beings and sustainable development. Cultural rights, heritage, diversity, and creativity are the core components of human and sustainable development (UCLG 2017).

The achievement of international targets will occur through triumph at the local and national levels; therefore, it is crucial to understand local communities' empowerment through cultural aspects. It is necessary to understand the intervention strategy on local heritage managed nationally to contribute significantly to sustainable development (Petti et al. 2020). The results of cross-comparisons from samples of national cultural heritage databases confirm the SDGs achievement's general harmonization. Petti et al. (2020) concluded that there is broad agreement on cultural heritage conceptualization with an international framework. There is consistency in the classification and valuation of asset categories. However, cultural integration in sustainability always faces various challenges to achieve a consistent and coherent approach. To that end, efforts to develop a big database based on various indicators can reflect the overall level of achievement of the 2030 Agenda targets. Further research and analysis on the strong correlation between national data sets and international targets are needed.

Traditional and local knowledge is a peak formulation accumulation of knowledge, innovation, and practice of indigenous peoples and local communities worldwide (Subramanian and Pisupati, 2010). Kohsaka and Rogel (2019) state that traditional and local wisdom is a crucial component in human–environment relations in local ecosystems. Indigenous peoples' survival is very dependent on nature from generation to generation. Local communities have developed their tenure in the environment that has shaped and sustained their intergenerational livelihoods based on observations and experiences (Kohsaka and Rogel 2019). Traditional and local knowledge is dynamic, ingrained in cultural and social changes, always evolving, and is not static (Kohsaka and Rogel 2019). Cultural elements are manifested as wisdom and transformed by the social structure, cultural norms, political systems, spiritual beliefs, and social group's biophysical environment. Traditional and local knowledge is highly dependent on the nuances of its formation's context and local conditions.

UNESCO (2020) ensures that culture has an essential role in most of the Sustainable Development Goals. The SDGs' agenda related to quality education, sustainable cities, the environment, economic growth, sustainable consumption and production patterns, peaceful and inclusive societies, gender equality, and food security. Culture supports and promotes sustainable development from economic, social, and environmental dimensions through cultural heritage, cultural industries, and creative industries. UNESCO's task in promoting cultural diversity and UNESCO's Cultural Conventions is central to implementing the 2030 Agenda for Sustainable Development.

The Indonesian government places culture as a high priority scale and an integral part of sustainable development, a significant essential capital as one of the leading resources for development. The Indonesian nation has a very diverse culture because it has more than 1000 ethnic groups and approximately 726 languages (Kemendikbud 2020). The various national culture reflects the national wealth in the form of wisdom, science, technology, and specific and unique expertise. Therefore, all must protect, develop, and promote the diversity of Indonesian culture. Revitalization of traditional and local knowledge, wisdom, and culture is essential to support a dignified and sustainable national character in Indonesia.

7 Conclusion

The local Javanese traditional culture is related to the values and norms of humility, humility, total submission, awareness, and vigilance to remember God. Traditions from various archipelago regions have contributed significantly to enriching the culture and shaping national character building. Traditional cultural literacy will remain alive and useful if its processing uses the Tri-Con theory: continuity, convergence, and concentricity. Old Javanese conventional manuscripts need to be transcribed, translated, and modified through online manuscript digitization. The innovative use of information technology 4.0 can convey ancestral messages written in ancient texts for present and future generations. The development of the latest Maskumambang-Sirep *Pageblug* repertoire through the solidarity action is an example of the psychotherapy tradition in the current Covid-19 pandemic face. Revitalization of traditional regional culture is essential to support a dignified and sustainable national character in Indonesia.

References

Agus C (2013) Management of tropical bio-geo-resources through integrated bio-cycle farming system for healthy food and renewable energy sovereignty: sustainable food, feed, fiber, fertilizer, energy, pharmacy for marginalized communities in Indonesia. In: Proceedings of the 3rd IEEE global humanitarian technology conference, GHTC 2013. 6713695, pp 275–278

Agus C (2017) Revitalisasi Ajaran Luhur Ki Hadjar Dewantara untuk Pendidikan Karakter Bagi Generasi Emas Sebagai Cucuk Lampah Kebangkitan Nasional II Indonesia. Jurnal ABAD 1(1):51–66 (In Indonesian)

Agus C (2018) Development of blue revolution through integrated bio-cycles system on tropical natural resources management. In: Leal Filho W, Pociovălișteanu D, Borges de Brito P, Borges de Lima I (eds) World sustainability series: towards a sustainable bioeconomy: principles, challenges and perspectives. Springer, Cham. pp 155–172

Agus C, Cahyanti PAB, Widodo B, Yulia Y, Rochmiyati S (2020) Cultural-based education of Tamansiswa as a locomotive of indonesian education system. In: Leal Filho W et al (eds) Universities as living labs for sustainable development. World sustainability series. Springer, pp 471–486

Baroroh-Baried (1994) Pengantar Teori Filologi. Yogyakarta: Badan Penelitian dan Publikasi Fakultas (BPPF) Seksi Filologi Fakultas Sastra, Universitas Gadjah Mada. Yogyakarta. 126 hal (In Indonesian)

Cahyanti PAB, Agus C (2017) Development of landscape architecture through geo-eco-tourism in tropical karst area to avoid extractive cement industry for dignified and sustainable environment and life. IOP Conf Ser Earth Environ Sci 83:012028

Cahyanti PAB, Widiastuti K, Agus C, Noviyani P, Kurniawan KR (2019) Development of an edutainment shaft garden for integrated waste management in the UGM green campus. IOP Conf Series Earth Environ Sci 398:012001

Dewantara K (1961) Soal Wanita. Jogjakarta: Madjelis Luhur Taman Siswa. Yogyakarta. 68 hal (In Indonesian)

Dewantara KH (2013a) Prinsip, Konsepsi, Keteladanan, Sikap Merdeka, Bagian I—Pendidikan. Universitas Sarjanawiyata Tamansiswa dan Majelis Luhur Persatuan Tamansiswa. Cetakan kelima. Yogyakarta. 148 hal (In Indonesian)

Dewantara KH (2013b) Prinsip, Konsepsi, Keteladanan, Sikap Merdeka, Bagian II—Kebudayaan. Universitas Sarjanawiyata Tamansiswa dan Majelis Luhur Persatuan Tamansiswa. Cetakan kelima. Yogyakarta. 136 hal (In Indonesian)

ICOM (2017) International Council of Museums (ICOM) Statutes. https://icom.museum/wp-content/uploads/2018/07/2017_ICOM_Statutes_EN.pdf#:~:text=International%20Council%20of%20Museums%20%28ICOM%29%20%E2%80%93%20Statutes%20%282017%29,ICOM%20and%20the%20ICOM%20Code%20of%20Ethics%20for. Accessed 20 Aug 2020

Kemendikbud (2020) Kementerian Pendidikan dan Kebudayaan RI. https://www.kemdikbud.go.id/. Accessed 18 Aug 2020

Kohsaka R, Rogel M (2019) Traditional and local knowledge for sustainable development: empowering the indigenous and local communities of the world. In: Leal Filho W, Azul A, Brandli L, Özuyar P, Wall T (eds) Partnerships for the goals. Encyclopedia of the UN sustainable development goals. Springer. https://doi.org/10.1007/978-3-319-71067-9_17-1

Nurhayati E (2019) Bahasa, Sastra, Seni Pembelajarannya di Era Disrupsi, dalam Suryaman M, Budiyanto D, Wakidi (eds) Jalan Menuju Inovasi Kebudayaan, Perspektif Bahasa, Sastra, Seni, dan Pembelajarannya. Interlude bekerja sama dengan Fakultas Bahasa dan Seni Universitas Negeri Yogyakarta. Yogyakarta. Hal pp 44–56 (In Indonesian)

Petti L, Trillo C, Makore BN (2020) Cultural heritage and sustainable development targets: a possible harmonisation? Insights from the European Perspective. Sustainability 12:926

Sari ES (2019) Literasi Sastra dan Teknologi Informasi, dalam Suryaman M, Budiyanto D, Wakidi (eds) Jalan Menuju Inovasi Kebudayaan, Perspektif Bahasa, Sastra, Seni, dan Pembelajarannya. Interlude bekerja sama dengan Fakultas Bahasa dan Seni Universitas Negeri Yogyakarta. Yogyakarta. Hal 76–84 (In Indonesian)

Sastrowiryono W (1978) Sekar Macapat. Bimbingan Kesenian Majelis Luhur Persatuan Taman Siswa. Yogyakarta. 64 hal (In Indonesian)

Subramanian SM, Pisupati B (2010) Traditional knowledge in policy and practice: approaches to development and human well-being, pp 577, United Nations University Press, Hong Kong

Supriyoko K (2012) Tamansiswa sebagai Pilar Pendidikan, dalam Swasono SE, Macaryus S (eds) Kebudayaan Mendesain Masa Depan. UST-Press bekerja sama dengan Majelis Luhur Persatuan Tamansiswa. Yogyakarta. Hal 36–45 (In Indonesian)

Suryapranata RM (1917) Serat Sotya Rinonce. Weltevreden: Albrecht & Co. Yogyakarta. 82 hal

UCLG (2017) Culture in the sustainable development goals: a guide for local action. UCLG Committee on Culture, Barcelona, p 40

Unesco (2020) Culture for Sustainable Development. https://en.unesco.org/themes/culture-sustainable-development. Accessed 5 Oct 2020

Ki Prof. Dr. Cahyono Agus DK is a Professor at Universitas Gadjah Mada Yogyakarta Indonesia, was born in Yogyakarta, March 10, 1965. The Doctorate was obtained from Tokyo University of Agriculture and Technology, Tokyo, Japan, in 2003. He was head of UGM University Farm 2008–2015. He currently serves as Chairman of the Association of Tamansiswa Society (PP PKBTS) 2016–2021, a *Majelis Luhur Persatuan Tamansiswa* (MLPTS) member 2016–2021, and member of Education Board Daerah Istimewa Indonesia. Active as a reviewer in research, community development, scientific publications, and institutional development in Higher Education, Indonesia. He published many scientific works in international seminars and journals and has several awards and copyrights from various agencies.

Nyi Dr. Sri Ratna Saktimulya is a lecturer at Universitas Gadjah Mada Faculty of Cultural Sciences—Javanese Literature Study Program; Head of Pura Pakualaman Yogyakarta Library (2011—present); Member of the Cultural Council of Special Region of Yogyakarta (2020–2022); Head of UGM Center for Cultural Studies (2021–2023). She was born in Yogyakarta on September 18, 1960. Area of interest: philology. Published scholarly works include *Gusti Kanjeng, Perempuan Perkasa dari Pura Pakualaman* (2019), *Transformasi Naskah dalam Ekspresi Seni*—a book chapter in *Naskah Nusantara antara Kekunoan dan Kekinian (2018), Pesan Leluhur dalam Naskah Kuna Pura Pakualaman: Pelestariannya pada Motif Batik Modern* (2018), Following the Footsteps of Sunan Kalijaga through Illumination in the Pakualaman Scriptorium (2018), *Naskah-naskah Skriptorium Pakualaman Periode Paku Alam II* (1830–1858) (2016), Javanese Script and Manuscript in Inscribing Identity—The Developments of Indonesia Writing System (2015), and *Katalog Naskah-naskah Perpustakaan Pura Pakualaman* (2005).

Ki Priyo Dwiarso is Ki Hadisukatno's son, one of the first teachers at the Tamansiswa Yogyakarta school, and maestro of *dolanan anak* songs (Javanese children's games). Since the 1960s, Ki Priyo has taught arts and music education to young learners while also composing many children's songs and operettas. Ki Priyo's writing on Ki Hadjar Dewantara and Tamansiswa education has been published in several books, journal articles, and newspapers. Ki Priyo is the former Secretary-General of the Tamansiswa Executive Council (Majelis Luhur Persatuan Tamansiswa (MLPTS)) and is currently a Special Advisor to the MLPTS Yogyakarta.

Ki R. Bambang Widodo, S.Pd., M.Pd born in Yogyakarta, July 5, 1958. The last education is Magister University Sarjanawiyata Tamansiswa (UST) in 2014. Journalist. He currently serves as Chairman of the Barahmus Museum Association Chapter Daerah Istimewa Yogyakarta, Chairman I of the National Executive Board of the Indonesian Museum Association, Management of the Indonesian Journalists Association (PWI) Daerah Istimewa Yogyakarta, and Management Board of Regional Main Cooperative Board (Dekopinda) Yogyakarta.

Nyi Dr. Siti Rochmiyati has been a lecturer in the Indonesian Language and Literature Education Department of Sarjanawiyata Tamansiswa University (UST) Yogyakarta since 1992. She earned her master's degree in Educational Research and Evaluation from Yogyakarta State University. Currently, she is working on her Doctorate in Language Education. Her research activities and expertise relate to curriculum, lesson plan, model and instructional media, character education, nationalism, and language education policy. She also possesses a certificate of assessor for elementary, junior, and senior high school teachers in Central Java and Yogyakarta.

Ki Dr. Mulyanto Darmowiyono is an assistant Professor at Educational Management Study Program, Postgraduate of Educational Management of Sarjanawiyata Tamansiswa University, teaches MIS and Management Strategy. He received his doctor degree from Yogyakarta State University, with research focused on implementing quality management at Vocational High Schools. He was a former English lecturer at Bina Karya Vocational High School Kebumen–Central Java and became a headmaster for seven years.

Modern Technologies in Tourism as a Tool to Increase International Tourism Attractiveness and Sustainable Development of the Kaliningrad Region

Anna V. Belova, Nikolay Belov, and Ivan Gumeniuk

Abstract The development strategy of the tourism industry is based on the need to attract international tourist flows. Border regions often act as a springboard for promoting international tourism within a country. The more tourist opportunities a region has, the more stable and diversified external tourist flows are. The Kaliningrad region, as a seaside region, has historically specialized in coastal and health tourism. It has a rich cultural, historical and natural tourist potential, which is used only by the local population and is practically not known to international tourists. Modern technologies make it possible to expand tourism opportunities of the region, to offer new tourism products and services, and involve new groups of the local population in tourist activities. People with disabilities is one of the groups benefiting from technological advancement in the tourism industry. New technologies in tourism (first of all, virtual excursion tours) can ensure the implementation of environmental approaches, in which unique natural, cultural and historical objects of the region can be explored without any increase in the anthropogenic load on them. New technologies in tourism create conditions for the transformation of tourist services and products. The authors describe stages of the development of technology-based tourist products—excursion routes designed for different target groups of tourists. The article shows that only larger actors can afford designing virtual tours using augmented reality (AR) since it requires significant financial resources. The Kaliningrad region, benefiting from the funds of the Poland-Russia 2014–2020 Cross-Border Cooperation Programme, is now creating a technology-based tourism product. This new type of product will allow the region to preserve natural sites, attract investments and restore its rich historical and cultural heritage. The project will facilitate the further improvement of the social and economic situation in the region and will contribute to the stability of border areas.

A. V. Belova (✉)
Immanuel Kant Baltic Federal University, Nevskogo str 14, Kaliningrad, Russian Federation 236000

N. Belov · I. Gumeniuk
Institute of Environmental Management, Urban Development and Spatial Planning, Immanuel Kant Baltic Federal University, Nevskogo str 14, Kaliningrad, Russian Federation 236000

Keywords Tourism · Baltic region · Virtual reality · Sustainable development · Historical and cultural heritage · Kaliningrad region

1 Introduction

Historical and cultural tourism in Europe attracts a large number of international and domestic tourists. European macroregions have their own historical and cultural peculiarities that form specific tourist points of attraction and popular destinations. These are the Mediterranean region, the Balkan countries, the Baltic region, where the Scandinavian region can be distinguished as a separate tourist destination. The development of historical and cultural tourism in the European macroregions contributes to the cultural and socio-economic development of the areas, thus creating opportunities for the sustainable development of these regions.

The Kaliningrad Region is part of the Baltic macroregion. Integrated into a single network of historical and cultural sites of the Baltic Sea region, Kaliningrad has an enormous potential for the development of historical and cultural tourism. It boasts many tourist routes related to the historical and cultural heritage of the region, including cross-border ones with the neighboring states—Poland and Lithuania. However, due to a lack of resources and economic perturbations, a significant part of the historical and cultural heritage of the region is in a deplorable state, being unsuitable for tourism purposes (Belova and Fedina-Zhurbina 2020).

The use of virtual (VR) and augmented reality (AR) in excursion routes can be considered an effective tool for the development of tourism and sustainable development of the Kaliningrad Region. These technologies help to visually present the lost sites of the historical and cultural heritage of the region since even the lost sites are integral parts of these virtual routes. In addition, the use of augmented reality in tourism products can attract attention to the historical and cultural heritage objects that need to be restored and can also serve as a tool of attracting investment to territories with underdeveloped tourist infrastructure.

In the Kaliningrad region, modern trends of the tourism industry reflect the need for sustainable development and are mainly associated with ecological tourism (Halme 2001). Several areas of the Kaliningrad region, primarily border areas, participate in the development of tourist products in cooperation with local population and small businesses (Wall 1997; Berno 2001; Dunets 2019).

2 Theory and Methodology

The use of information technology for expanding conventional tourism and achieving sustainability is reflected in the studies dating back to the 2000s (Milne et al. 2004; Sheldon et al. 2001; Brown and Chalmers 2003). An exponential growth of publications has been observed since the 2010s (Guttentag 2010), and this was primarily due to the beginning of the era of the smartphone. Since then, high technologies have become an indispensable part of everyday life. At the same time, it is necessary to note inconsistencies in the interpretation of the terms "virtual tours", "virtual reality" and "augmented reality". There are also opposing points of view on the need and the degree of integration of the above-mentioned technologies in tourist activities. Yet, recent publications have described these technologies not only as an addition to existing tourist products, but also as an independent tourism product (Thomas et al. 2017; Skamantzari and Georgopoulos 2016).

The situation of 2020 showed that the tourism sector is very unstable. A more active development and integration of augmented (AR) and mixed reality (MR) in tourist routes will make it more sustainable and attractive (Hoque and Ashikul 2020). These technologies will not replace a 'real' visit to a spot, but will give people an opportunity to have a look at heritage sites in advance (Ferimar 2017). At the same time, the use of mixed reality makes it possible to expand and improve existing tourism products (routes), customizing them for different target groups and creating new experiences (Mesáro et al. 2016).

In this article, the following research methods were used: an overview and analysis of research literature on the topic and an anonymous sociological survey "Tourism virtualization' held among researchers, guides, representatives of municipal authorities and decision-makers. The survey included three open-ended questions:

- How do you assess the prospects for tourism after Covid-19?
- What do you think about the virtualization of tourism and museums?
- Would you be interested in a virtual tour?

In total, 35 people were interviewed. Based on their responses flowcharts were built and included in this article.

3 Research Results

3.1 Current Trends in the Tourism Industry Development

The tourism industry is one of the smooth-running economic activities in the world, generating about 10% of global GDP and 10% of global jobs (EIR 2020 Importance of Travel and Tourism Infographic 2020). The development of tourism can be based on a favorable geographical location or historical and cultural potential, which is unique for each region. The example of the European Union demonstrates the cost-effectiveness of historical and cultural tourism (Research For TRAN Committee 2015), which on the one hand, can counter-balance the seasonality of tourism activities (Łonyszyn and Terefenko 2014), and, on the other, can ensure a wider involvement of the population and local business in tourism (Aref 2011). Many EU countries, including such countries of the Baltic region as Sweden, Denmark, Germany, starkly demonstrate how historical and cultural tourism can be developed rather effectively. These countries have set a good example for countries of Eastern Europe and neighboring regions.

The Kaliningrad Region, due to its geographical position and historical development, has a rich potential for promoting historical and cultural tourism. The tourism industry in the Russian exclave has demonstrated an upward trend in recent decades. However, this progress is mainly based on the active use of environmentally sensitive coastal landscapes. The only exception is the development of historical, cultural and event tourism in the administrative center of the region—the city of Kaliningrad). Until recently, the majority of tourists visiting the region came from other parts of Russia. So far, international tourist flows have been minimal. The constraints included a whole variety of factors ranging from institutional (the need to apply for a visa), socio-economic (limited and poor quality of tourism infrastructure and related services, a low quality of transport infrastructure) to the information ones (weak promotion of tourism opportunities of the region on the world market). The preparation for and hosting the games of the World Football Cup in 2018 significantly improved the situation, primarily in terms of institutional procedures and the regional infrastructure (the reconstruction of the airport, the construction of new hotels and restaurants that are part of the world leading hotel chains). The launch of the electronic visa procedure for foreign citizens visiting the Kaliningrad Region created favorable institutional conditions for the development of international tourism in the Russian exclave. Along with the development of coastal tourism, the region plans to further develop historical and cultural forms of tourism, including transboundary ones. It is done in cooperation with the border regions of the neighboring republics of Lithuania and Poland, which have a similar historical and cultural tourism potential.

To be competitive on the world market of tourist services the region, on the one hand, should build on the best practices of promoting regional tourism, particularly those based on modern technology and innovation. On the other hand, it should take into account global trends in tourism development, which reflect the need for sustainability, accessibility and a more active involvement of local communities in the planning and implementation of tourism activities. The development of virtual sightseeing routes provides a good opportunity for combining these principles, marketing the region, reaching new tourist markets, involving local communities in the development of such tours and providing equal opportunities for different age and social population groups, including people with disabilities.

International research reveals of marketing efficiency of virtual sightseeing routes as a way to promote travel services. For instance, Jacob et al. (2010) have proved that tourists who take a virtual tour can understand their destination better. The information presented in virtual tours forms certain expectations in tourists (Cho et al. 2002). Consequently, they tend to be more satisfied when visiting the site since their impressions are based on the previously acquired knowledge of the place (Cheong 1995).

The development of virtual tourist routes promotes closer cooperation between local actors, companies, builds effective economic ties (Hopeniene et al. 2009), as well as facilitates the involvement of the local population in decision-making (Garau and Ilardi 2014). The advantages of virtual reality technologies in tourism are analyzed in detail in (Guttentag 2010). As the author notes, virtual reality in tourism is used in planning and management, marketing, entertainment, education, accessibility improvement and the preservation of cultural and historical heritage.

3.2 Stages and Limitations of the Visualization Information Technology in Tourism

It should be noted that virtual tours of historical and cultural heritage sites or unique natural sites are not a new invention (Cho 2002). However, there has been an exponential growth in the development of visualization technologies and augmented reality lately. In most cases, these technologies are represented by so-called 3600-panoramas with an option to move between pre-set points. A mobile audio guide can be considered as the beginning of the era of virtual tours and digital technologies. It contained only audio content, without any visualization of the sites of historical and cultural heritage. Later, the idea of using virtual reality appeared (Guttentag 2010). However, relatively low capacity did not allow this technology to develop in the tourism sector, and even currently available tours developed with this technology leave visitors with a feeling of "artificiality" (Tussyadiah et al. 2018). In addition, a number of factors hamper a complete integration of this technology into everyday practices. The first one is extremely high costs of preparatory work. Museums and small municipalities cannot afford the development of full-fledged virtual tours. The second obstacle is a conservative bias that the number of tourists will decrease dramatically since there will be no need to come and see the sites (Kask 2018). In a way, augmented reality may become a game-changer since its popularity has been increasing: the end user does not need any special equipment, a mobile phone is enough to enjoy the benefits of this technology (Saxena 2020; Cranmer 2020). Augmented reality (AR) has become almost ubiquitous in the following areas: city tours, museum tours and travel transport. The technology is particularly popular among millennials as the main tourist flow generator. Figure 1 presents a comprehensive picture of existing tourist information technologies:

As it can be seen from the diagram, augmented reality (AR) has incorporated the best features of all previous technologies. Increasing capacity of portable devices made it possible to use both virtual models and photogrammetric models. In addition, other gadgets that were not originally intended for virtual tours appeared on the market of tourist technologies. Ground-based laser scanners, which are high-precision geodetic equipment operationally generating huge data arrays, are becoming increasingly popular (De Paolis 2017).

At the very beginning, the AR technology was perceived exclusively as a game. Gradually, its application proved to be effective in many business sectors:

- It is used by hotels to provide their information online to increase the demand for reservations;

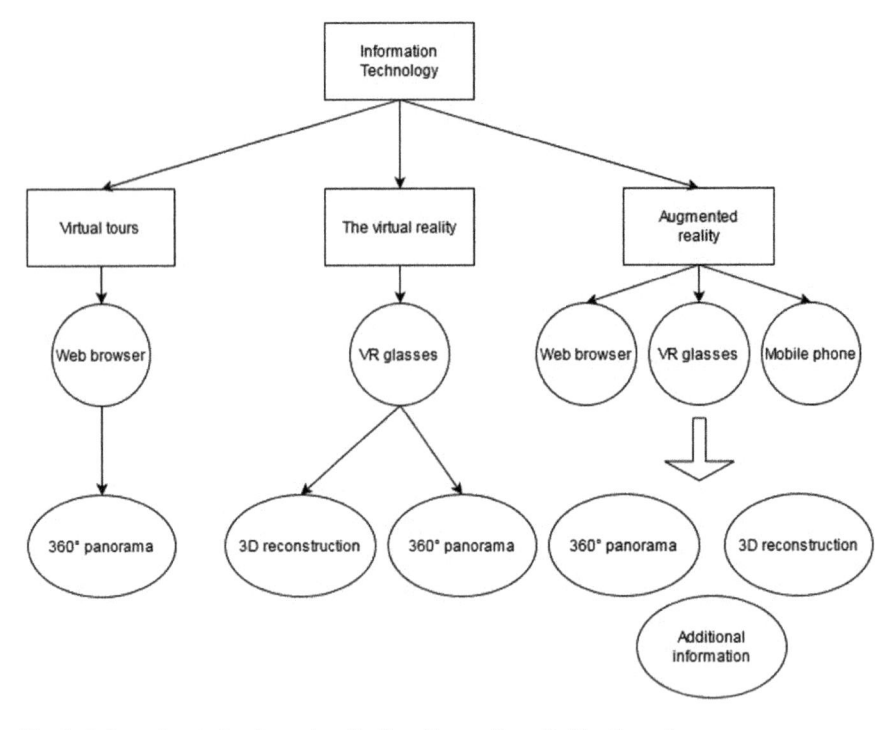

Fig. 1 Information technology visualization. *Source* Compiled by the authors

- Increasing tourist flows require more efficient communication of information. Audio guides and information messages appearing on the screen of mobile phones when pointing to a tour object, thus expanding the customer's experience (Rahimi 2020).
- Spatial orientation involves the integration of individual AR elements into a familiar map of services. Google was the first company to integrate elements of AR in its navigation application in 2018. The introduction of interactive elements, such as arrows or pointers, removes communication barriers (Chen 2020).
- Local urban transportation. Only the initial steps have been taken in this direction in major tourist centers. In the future, AR elements will be integrated into public transport networks.
- Introduction of the elements of AR entertainment. So far, it has been used as a form of entertainment for children (for instance, incorporating popular characters, objects, etc.). There have been several attempts to customize certain elements for adult users (Xu 2016).
- The use of AR in specialized city tours. The integration of this technology makes it possible to evaluate the guide's performance as well as to experience new emotions, recreating sites of historical and cultural heritage.

The target audience → Specialized equipment → Technological platform → Scenario → Filling content → Testing

Fig. 2 Implementation stages of AR content. *Source* Compiled by the authors

- AR for museums. It gives an opportunity to obtain additional information about sites and to demonstrate their principles of functioning. Visitors receive more information in a very unobtrusive way and feel like real researchers.

The advantages of AR seem truly endless, but there are certain difficulties with AR implementation. To design a tour based on AR, one should go through a number of stages (Ross 2020; Pranz 2020; Kerr 2020; Argyriou 2020) (Fig. 2).

During the first stage, the main target audience is determined, since the requirements for tourist routes are different for different age groups. In addition, the needs of disabled people with hearing or vision problems must be taken into account. Wrong identification of the target group may significantly reduce tourist flows.

The next step is choosing and purchasing specialized equipment. At this stage, it is also important to decide how the content of the AR product will be selected. Currently, the tourist technology is moving further away from the classic 360 panoramas towards a more detailed 3D modeling. Accordingly, the range of equipment used ranges from professional cameras and drones to high-precision laser scanners and specialized 3D cameras.

The next step is choosing software and a specific platform for the AR tour. A good example of this type of platform is the Matterport service (Ravikumar 2017). The choice of software packages for developers is quite extensive—from enterprise-level leadership solutions to open-source software. Additionally, there are web companies specializing in providing this kind of service.

The following step is the development of a tour script. An improperly chosen script may drastically reduce the attractiveness of the route. This is a creative stage, since it is here that not only the degree of filling the routes with sites, but also the amount of information, ways and timing of its presenting are determined (Han 2018).

The most important and the most time-consuming stage is filling in the content. This stage consists of the integration of graphic, text and audio information in the would-be tour. Here, well-coordinated work of specialists in the field of photogrammetry, 3D modeling, video operators and sound engineers is required.

The final step is testing the route. It is carried out first individually for various age categories and then in mixed groups.

It is clear that a full-fledged implementation of the created tourist product requires well-coordinated work of a sufficiently large team. Another issue to be considered is high production costs. Therefore, this kind of technology is difficult to launch for small companies, museums or small municipalities.

3.3 Introduction of Mixed/augmented Reality Technologies in the Tourism Industry of the Kaliningrad Region

The notions of AR (augmented reality) and MR (mixed reality) tend to be confused. MR is the next step in the development of tourist technologies. The essence of MR is to bring virtual images into the user's space–time, to visualize and fix their location according to the objects of real space so that the user perceives them as real. In this case, there is no division in the perception of the world into real and virtual (Fig. 3).

It has become clear that in the changing world, such technologies are a necessary condition for sustainable tourism since they help integrate not only the established and popular destinations, but also small and medium-sized cities, small municipal museums, historical and cultural heritage sites into tourist packages, which otherwise would interest researchers only. At the same time, high costs of implementation mean that designing a technology-based tour is impossible or extremely difficult without a grant or state support.

So far, there have been no projects of using MR in the Kaliningrad Region. Nevertheless, there are two projects, which are noteworthy: "Personal Guide to the National Park" implemented by the National Park of the Curonian Spit (available only as a mobile app for Android / iOS) and the "VR Tour for the Royal Castle Kaliningrad" from Future Technologies LLC (a simulated site). Other developments either are the results of the work of individual enthusiasts or are based on the 3600 panoramas that have already become the classic of the tourist industry even though this technology does not provide full interaction.

Municipalities of the Kaliningrad Region, especially those having limited budgets, are aware of the cost of designing a technology-based tourist product. They cannot afford covering the cost of production. So many municipalities try to attract external

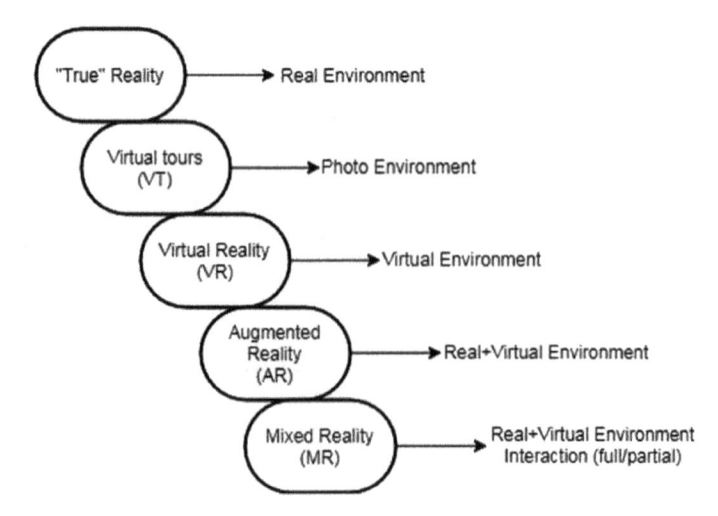

Fig. 3 Transition from the "real" reality to the mixed one

investment or apply for state grants. Cross-border cooperation programs are one of the most effective financial tools of supporting the development of border territories. Border regions of Russia and neighboring countries can take part and receive grant funds for the development new projects and tourist products (Belova and Korshuk 2017). Cross-border cooperation programs, in which the Kaliningrad Region takes an active part, help to develop tourism in border areas with the neighboring countries—Lithuania and Poland. Special attention is given to designing new tourist products based on AR.

One of such projects is the project "Baltic Odyssey—Creating a Common Historical and Cultural Space" of Poland-Russia 2014–2020 Cross-Border Cooperation Program. Within the framework of this project, virtual tours of the historical and cultural heritage of the Kaliningrad Region and border regions of Poland are being made. The idea is to create a cross-border route, a quest for tourists. Using an electronic gadget (for example, a smartphone), the tourist can travel and explore a territory, learn about its history, explore its heritage sites and traditions, etc. A virtual quest tour developed within the framework of the Baltic Odyssey project is a very special tourist product. It allows the user to immerse in the history of the region, explore little-known natural and architectural monuments, visit villages and small towns without physically getting there since many remote areas lack the necessary tourist infrastructure but are extremely attractive. The implementation of the above-mentioned project will contribute to the conservation and restoration of heritage sites and will become a driver for sustainable development of the territory, its economy and social sphere.

4 Conclusions

The results of this study have led to the following conclusions. At the beginning, the introduction of virtual technologies in tourism was considered as a marketing technique that made it possible to increase the attractiveness of a tourist product and to expand the potential target audience. Gradually, virtual tours have become a separate area of tourist activity, having their own target audience. Although they still perform an important function of promoting tourism in general, modern technologies allow specialists to develop a new type of tourist products aimed at completely different target groups. Virtual tours are now becoming tools for the promotion of cultural or historical sites. This conclusion clearly illustrates the broad prospects of virtual tourist routes as a separate direction in the development of tourism in the region, which requires an integrated interdisciplinary approach.

At the same time, visualization technologies (primarily augmented reality) has a wide range of applications in tourism, while still encountering objective implementation difficulties. First of all, this is due to the need to go through a number of stages when designing a technology-based product. Incorporating augmented reality in tours is resource-intensive since it requires the participation of many specialists and considerable investment. At the same time, a consistent trend of technological transition in tourism is obvious, i.e. it is a shift from augmented reality to mixed reality-based tourist products.

References

Aref F (2011) Sense of community and participation for tourism development. Life Sci J 8(1):20–25

Argyriou AS (2020) The situation in the tourism industry in Greece before and after the outbreak of the Covid-19 pandemic

Belova AV, Fedina-Zhurbina IV (2020) The potential of the Kaliningrad Region in the development of health Tourism. In: Baltic Region—the region of cooperation. Springer, Cham, pp 285–296

Belova AV, Korshuk EV (2017) Rol' mezhdunarodnyh proektov v povyshenii potencialaistoriko-kul'turnogo naslediya dlya razvitiya transgranichnogo turizma (The role of international projects in enhancing the opportunities of historical and cultural heritage for the development of cross-border tourism) //Etnosocial'nye i konfessional'nye processy v sovremennom obshchestve: sb. nauch. st. /GrGU im. YA. Kupaly; redkol.: M. A. Mozhejko (otv. red.) [i dr.]. – Grodno : «YUrSaPrint», 2017.c.337–342

Berno T, Bricker K (2001) Sustainable tourism development: the long road from theory to practice. Int J Econ Dev 3(3):1–18

Brown B, Chalmers M (2003) Tourism and mobile technology. In: ECSCW 2003. Springer, Dordrecht, pp 335–354

Chen L, Xie X, Lin L, Wang B, Lin W (2020) Research on smart navigation system based on AR technology. In: Fifth international workshop on pattern recognition, vol. 11526. International Society for Optics and Photonics, p 115260J

Cheong R (1995) The virtual threat to travel and tourism. Tour Manage 16(6):417–422

Cho YH, Wang Y, Fesenmaier DR (2002) Searching for experiences: the web-based virtual tour in tourism marketing. J Travel Tour Mark 12(4):1–17

Cranmer EE, tom Dieck MC, Fountoulaki P (2020) Exploring the value of augmented reality for tourism. Tourism Manag Persp 35:100672

De Paolis LT, Bourdot P, Mongelli A (2017) Augmented reality, virtual reality, and computer graphics. In: 4th International conference, AVR 2017, Ugento, June 12–15, 2017, proceedings Springer

Dunets AN, Ivanova VN, Poltarykhin AL (2019) Cross-border tourism cooperation as a basis for sustainable development: a case study. Entrepreneurship Sustain Issues 6(4):2207–2215

EIR (2020) Importance of Travel and Tourism Infographic, The World Travel & Tourism Council (WTTC) https://wttc.org/Research/Economic-Impact/moduleId/1226/itemId/67/controller/DownloadRequest/action/QuickDownload. Accept 09 July 2020

Garau C, Ilardi E (2014) The "Non-Places" meet the "Places:" Virtual tours on smartphones for the enhancement of cultural heritage. J Urban Technol 21(1):79–91

Guttentag DA (2010) Virtual reality: applications and implications for tourism. Tour Manage 31(5):637–651

Halme M (2001) Learning for sustainable development in tourism networks. Bus Strateg Environ 10(2):100–114

Han DI, tom Dieck MC, Jung T (2018) User experience model for augmented reality applications in urban heritage tourism. J Heritage Tourism 13(1):46–61

Hopeniene R, Railiene G, Kazlauskiene E (2009) Potential of virtual organizing of tourism business system actors. Eng Econ 63(4).

Hoque A, Shikha F, Hasanat MW, Arif I, Hamid ABA (2020) The effect of Coronavirus (COVID-19) in the tourism industry in China. Asian J Multi Stud 3(1):52–58

Jacob C, Guéguen N, Petr C (2010) Media richness and internet exploration. Int J Tour Res 12(3):303–305

Kask S (2018) Virtual reality in support of sustainable tourism. Experiences from Eastern Europe (Doctoral dissertation, Eesti Maaülikool)

Kerr L (2020) Sour grapes: mitigating the risk of overtourism in British Columbia's eno-tourism

Łonyszyn P, Terefenko O (2014) Creation of an alternative season based on sustainable tourism as an opportunity for Baltic Sea Region. J Coastal Res 70:454–460

Mesáro P, Mandičák T, Hernandez MF, Sido C, Molokáč M, Hvizdák L, Delina R (2016) Use of augmented reality and gamification techniques in tourism. E-review Tourism Res 2

Milne S, Mason D, Hasse J (2004) Tourism, information technology, and development: revolution or reinforcement? Companion Tourism 12(2):184–194

Peeters PM, Eijgelaar E, Dubois G, Strasdas W, Lootvoet M, Zeppenfeld R, Weston R (2015) Research for TRAN committee-from responsible best practices to sustainable tourism development. European Parliament, Directorate General for Internal Policies, Policy Department B: Structural and Cohesion Policies, Transport and Tourism

Pranz S, Nestler S, Neuburg K (2020) Digital topographies. Using AR to represent archival material in Urban space. In: Augmented Reality and Virtual Reality. Springer, Cham, pp. 139–149

Priolo FR, Mediavillo, RJB, Austria AND, Angeles AGSD, Fernando MCG, Cheng RS (2017) Virtual heritage tour: a 3D interactive virtual tour musealisation application. In: Proceedings of the 3rd International Conference on Communication and Information Processing (pp. 190-195)

Rahimi R, Hassan A, Tekin O (2020) Augmented reality apps for tourism destination promotion. In: Destination management and marketing: breakthroughs in research and practice. IGI Global, pp 1066–1077

Ravikumar P, Gobinath K, Bhaskaran G, Chandrasekar V (2017) Eco conservation and development of tourism along Yelagiri hills using view shed analysis of geographical information system

Ross D (2020) Towards meaningful co-creation: a study of creative heritage tourism in Alentejo, Portugal. Curr Issues Tourism 23(22):2811–2824

Saxena U, Kumar V (2020) Augmented reality: new era of technological tourism. UGC CARE J 19(8):187–199. Tathapi with ISSN 2320–0693

Sheldon PJ, Wober KW, Fesenmaier DR (2001) Information and communication technologies in tourism 2001. In: Proceedings of the international conference in Montreal, Canada. Springer, Wien

Skamantzari M, Georgopoulos A (2016) 3D Visualization for virtual museum development. Int Arch Photogrammetry, Remote Sens Spatial Inf Sci 41:961

Thomas et al (2017) Mii-vitaliSe: a pilot randomised controlled trial of a home gaming system (Nintendo Wii) to increase activity levels, vitality and well-being in people with multiple sclerosis. BMJ open 7(9):e016966

Tussyadiah IP, Wang D, Jung TH, tom Dieck MC (2018) Virtual reality, presence, and attitude change: empirical evidence from tourism. Tourism Manag 66:140–154

Wall G (1997) Sustainable tourism–unsustainable development. In: Tourism, development and growth: the challenge of sustainability, pp 33–49

Xu F, Tian F, Buhalis D, Weber J, Zhang H (2016) Tourists as mobile gamers: Gamification for tourism marketing. J Travel Tour Mark 33(8):1124–1142

Cross-Border Cooperation Programmes as a Sustainable Tool for Tourism Development: The Case of the Kaliningrad Region

Anna V. Belova and Irina V. Fedina-Zhurbina

Abstract This research analyses cross-border cooperation programmes as a sustainable tourism development tool used in Russian and EU border regions. Special attention is paid to the Kaliningrad region and its role in the development of cross-border tourist routes. The natural and cultural heritage of the region contributes to its tourism potential. In recent years, Kaliningrad has established itself as a popular destination for domestic and international tourists. Tourism has become a decisive factor in achieving regional sustainability and in increasing the attractiveness of the territory. This paper analyses three types of tourism that are central to cross-border cooperation programmes: nature tourism, health tourism, and cultural tourism. The emphasis is placed on the recent advancement in these areas and their effect on regional social and economic development and the well-being of Kaliningraders. Nature tourism projects launched within cross-border cooperation programmes have increased the number of natural sites open to visitors and contributed to their better protection. Health tourism projects have built on healthcare technologies and stimulated the purchasing of new equipment, which is now used by healthcare organisations. Cultural tourism projects have been important for the revitalisation of cultural and historical sites, the introduction of new technologies, the development of new tourist routes and the organisation of new events for tourists.

Keywords Nature tourism · Health tourism · Cultural heritage · Cross-border cooperation · Sustainable development · Kaliningrad region

1 State of Knowledge and Methodology

Sustainable tourism development in the European border regions was extensively studied in the late 1990s–early 2000s (Blatter 2000; Zaucha 1998; Hall 2000). In their works, many Russian and international researchers analyse the role of the Kaliningrad region, which is part of the European macro-region, and, more precisely, the Baltic Sea region, in the development of sustainable cross-border tourism (Pacuk

---

A. V. Belova (✉) · I. V. Fedina-Zhurbina
Immanuel Kant Baltic Federal University, 14 A. Nevskogo St, Kaliningrad, Russia

© The Author(s), under exclusive license to Springer Nature Switzerland AG 2021
W. Leal Filho et al. (eds.), *Innovations and Traditions for Sustainable Development*,
World Sustainability Series, https://doi.org/10.1007/978-3-030-78825-4_23

and Palmowski 1998; Anisiewicz and Palmowski 2014; Studzieniecki et al. 2016; Batyk and Semenova 2013; Spiriajevas 2019; Korneevets et al. 2019; Palmowski and Fedorov 2019; Belova and Fedina-Zhurbina 2020; Soldatenko and Backer 2020).

From the tourism point of view, the Kaliningrad region is a cross-border region of Russia in Europe. There are many tourism development tools that the region have benefited from—federal target programmes, regional and national strategic instruments, and programmes for cross-border and transboundary cooperation. One of their common priorities is to increase tourism competitiveness of border regions (*see* Lithuania-Poland-Russia Cross-border Cooperation Programme 2007–2013, the final version of Operational Programme with comments of the European Commission, 2008; Triple Jump, 2008; Lithuania-Russia Cross-border Cooperation Programme 2014–2020, 2016) (Fedorov et al. 2015).

This study aims to explore the efficiency of cross-border cooperation programmes regarded as a tool for sustainable development of tourism in the Kaliningrad region. To achieve this aim, we employed methods of statistical data analysis and analysed international projects with the participation of the Kaliningrad region. Diagram and tables were built to visualise the research findings. Special emphasis was laid on the analysis of literature on cross-border cooperation, the Kaliningrad region as a partner in European cooperation projects, and the development of tourist destinations in the region.

2 Introduction

Trends in tourism development have shifted to tourism products designed to meet the needs of visitors and to preserve the natural, historical and cultural heritage of a territory. Consequently, it is important to include seldom visited sites in new tourist routes and tourism products, thus ensuring their socio-economic development. Practice shows that such natural, historical and cultural sites tend to be unpopular among tourists because of poor infrastructure.

The Kaliningrad region is part of the Baltic region. On the one hand, its geographical location is an advantage for the development of tourism. On the other hand, it creates a number of barriers and limitations resulting from the exclave position of the region: borders with neighbouring states, no common borders with mainland Russia and the states, with which the country has open borders, for instance, Belarus. The spatial configuration of tourist flows in the Kaliningrad region differs significantly from that in the European states and mainland Russia. Inbound travel was simplified only in July 2019 when electronic visas were introduced. Most international visitors come to the region by car. Russian tourists prefer travelling by air since, when travelling to the region by land, they have to cross two foreign states, and a Schengen visa is required.

In recent years, the tourism industry has been actively developing in the region: new tourist routes have been created, upgraded destinations promoted, multi-sight tourism products comprising historical, cultural, and active recreational elements developed. It is noteworthy that investment in the tourism industry has been gradually increasing. Investors get more interested in using cross-border cooperation tools—cross-border cooperation programmes, which facilitate the development of tourism in border areas, preserve and promote local cultural and historical heritage to make regions more attractive, and create new infrastructure that stimulates regional development, particularly that in in the field of tourism.

3 Research Results

The Kaliningrad region is a Russian border region located between two European states—Poland and Lithuania. It is a dynamic actor in cross-border cooperation, particularly in programmes for cross-border and transboundary cooperation. Since 2005, the region has successfully participated in joint projects with its neighbouring states, including in cross-border cooperation programmes aimed at tourism development. Tourism is one of the main priorities of joint EU—Kaliningrad projects as well as of the regional and federal authorities. Located on the Baltic Sea coast and boasting historical, cultural, and natural riches, the region has all the necessary prerequisites for making the tourism industry more dynamic (Kropinova 2020). Regional and federal authorities have introduced measures to increase tourist flows, for instance, electronic visas have been issued without a visa fee.

In 2005–2016, a total of 41 international cross-border cooperation projects centred on tourism were implemented within the Lithuania-Poland-the Kaliningrad region Neighbourhood Programme (24 projects) and the Lithuania-Poland-Russia 2007–2013 Cross-border Cooperation Programme (17 projects) (Triple Jump. Projects of Lithuania, Poland and Kaliningrad Region Neighbourhood Programme, 2008). Currently, thirty projects are being implemented in the framework of the ongoing cross-border cooperation programmes—Poland-Russia 2014–2020 and Lithuania-Russia 2014–2020. The overall tourism development budget of the Lithuania-Poland-Russia 2007–2013 Cross-border Cooperation Programme is 40.4 million euros; 29.4 million euros were allocated within the Poland-Russia 2014–2020 programme; the Lithuania-Russia 2014–2020 programme totals approximately 10.0 million euros.

Tourism development projects implemented within cross-border cooperation programmes involving the Kaliningrad region can be divided into several categories aimed at:

- preservation and sustainable use of infrastructure, including the one intended for tourists;
- promotion of health tourism;
- preservation of regional natural heritage.

Figure 1 shows the distribution of tourism development projects run within the Lithuania-Poland-the Kaliningrad region Neighbourhood Programme.

Seventeen projects supported tourism development in border areas within the Lithuania-Poland-Russia 2007–2013 Cross-border Cooperation Programme.

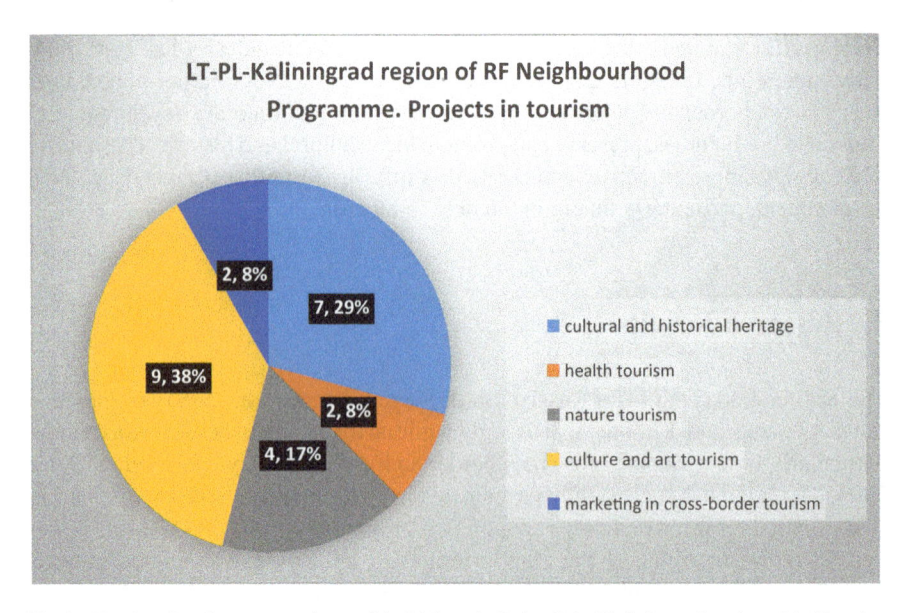

Fig. 1 Tourism development projects of the Lithuania-Poland-the Kaliningrad region of the Russian Federation Neighbourhood Programme (number of projects and its percentage). *Source* Prepared by the authors

Figure 2 is a pie chart visualising the distribution of project category.

Twelve tourism development projects launched in cross-border areas involving the Kaliningrad region aimed at the development of infrastructure. Several tourism infrastructure objects were built, restored, or renovated. This increased the attractiveness of the region for domestic and international tourists as well as provided access to some earlier isolated historical and cultural heritage sites. Tourism infrastructure projects completed in the Kaliningrad region included:

- the restoration and improvement of parks and their infrastructure in Svetly, Sovetsk and Gusev;
- the renovation and construction of sports facilities in the towns of Svetlogorsk, Bagrationovsk, and Mamonovo;
- the opening of the Ancient Sambia Viking-age open-air museum on the Russian part of the Curonian Spit;
- the construction of a high ropes park in Ozersk.

Other projects concentrated on health tourism and natural heritage preservation. During the period of the two programmes, the Kaliningrad region received about 5 million euros within the framework of 13 projects. Table 1 shows health and nature tourism projects implemented in 2005–2015 within the two cross-border cooperation programmes.

Currently, thirty projects are being implemented in the Kaliningrad region, with their budget totalling over 20.2 million euros. They are run within two bilateral

Cross-border cooperation Programme LT-PL-RU 2007-2013. Projects in tourism

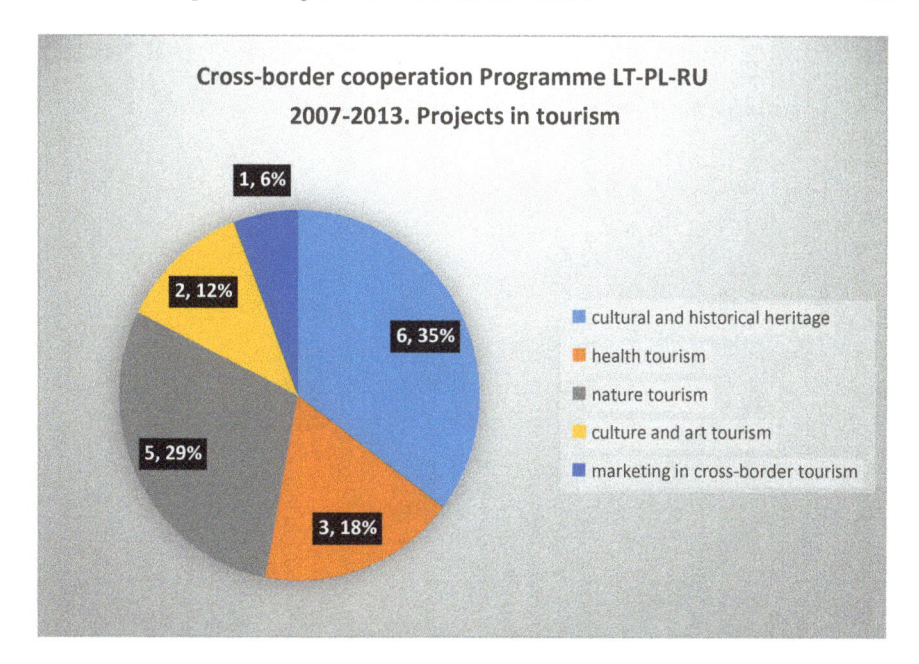

- 1, 6% — marketing in cross-border tourism
- 2, 12% — culture and art tourism
- 3, 18% — health tourism
- 5, 29% — nature tourism
- 6, 35% — cultural and historical heritage

Fig. 2 The distribution of tourism development projects by category. The Lithuania-Poland-Russia 2007–2013 Cross-border Cooperation Programme. *Source* Prepared by the authors

cross-border cooperation programmes of the European Neighbourhood Instrument: Poland-Russia 2014–2020 and Lithuania-Russia 2014–2020 (PL-RU 2014–2020 CBC Programme, LT-RU 2014–2020 CBC Programme). There are 14 Russian-Polish tourism projects (total budget—16.8 million euros) and 16 Russian-Lithuanian ones (total budget exceeding 3.4 million euros). Almost all these projects include an infrastructure component. The expected results of the projects involving the Kaliningrad region are as follows:

- a new cycling route along the Baltic Sea coast from Zelenogradsk to Svetlogorsk via Pionersky, 34 km long;
- the beatification of a park in Svetlogorsk, a new footpath in Sovetsk; the reconstruction of the fence and an activities area in the Kaliningrad Zoo;
- the partial reconstruction and renovation of the Friedland Gate in Kaliningrad; the renovation of historical barracks in Ozersk and the roof of the Tilsit Theatre building in Sovetsk; the renovation of the Kristionas Donelaitis Museum;
- the overhaul of the Vityaz and Irbenskiy museum ships; the purchase of a Viking ship for the Kaup open-air museum; the construction of a mini-shipyard in the Ancient Sambia open-air museum;
- an ecological exposition about the natural heritage of the region and birds at the Curonian Spit museum.
- Improving conditions for more affordable medical rehabilitation tourism for children.

Table 1 International projects launched within cross-border cooperation programmes aimed to develop nature and health tourism in the Kaliningrad region and the bordering countries

No.	Project title	Project objective	Project budget
Lithuania-Poland-the Kaliningrad region of Russian Federation neighbourhood programme			
1.	2006/338 Creation of Cross-border Bicycle Trail Along Old Post Road on Curonian Spit: EUROVELO-BALTICA	The main result was a feasibility study for the cycling route on the Russian part of the Curonian Spit. The route is to be integrated into the Eurovelo network. The Curonian Spit is a unique natural reserve of amazing beauty, boasting clean and fresh air. The cycling route on the Curonian Spit will be a perfect site for open-air exercise	224,880 euros (including 194,380 euros from TACIS) Website: www.europroekt-kosa.ru
2.	2006/383 Support to the Development of Common Recreation Space in the Macro-region	The project aimed to establish a tourist centre in the Kaliningrad region, which was to be part of the tourism centre network created in border areas. Publicity and promotion campaigns were conducted to attract tourists to the Kaliningrad region and to acquaint Russian tourists with the attractions of Palanga. The new tourist centre will provide visitors with necessary information and aid them in choosing the most suitable options	209,774 euros (including 178,937.22 euros from TACIS)
3.	2005/155 Opening of Water Tourism Route Connection in the Curonian Lagoon: Klaipėda-Kaliningrad	A water route from Klaipėda to Nida (Lithuania) was created within the project. As soon as the border-crossing problems are solved, the route will be extended to Rybachy in the Kaliningrad region. The project partners ran a strong promotion campaign to revive water tourism in the Curonian Lagoon between Klaipėda and Kaliningrad. Water tourism may include a wellness component in the future	143,500 euros (including 107,625 euros from the ERDF) Website: www.mariuturas.lt

(continued)

Table 1 (continued)

No.	Project title	Project objective	Project budget
4.	2005/196 Increasing Accessibility to Tourist Objects in the Baltic Coast Border	Partners of the project shared their expertise, experience and knowledge and elaborated a common tourism development strategy for the Baltic Sea coastline region. A series of publications promoted the region as a common space for tourism. Natural parks, cycling routes, and water sports became important elements of health tourism programmes	257,282 euros (including 192,961.25 euros from the ERDF)
5.	2006/391 Lagoons as Cultural and Historical Crossroads of Peoples in South-Eastern Baltic Area	Seven partners from Russia, Lithuania, and Poland took part in expeditions aimed to collect and register heritage sites of the territory, which historically was East Prussia (Lithuania Minor). The data were digitalised and presented to the public in printed and electronic media. A strategy and an action plan for the development of maritime cultural and historical heritage were prepared for the Russian part of the Curonian and Vistula Lagoons. A feasibility study was carried out	299,990 euros (including 269,990 euros from TACIS)
6.	2006/313 Tourist Infrastructure for Berlin-Kaliningrad-Klaipėda Waterway in Tczew and Klaipėda	The objective of the project was to increase the attractiveness of the Berlin-Kaliningrad-Klaipėda international waterway route	1,333,333.20 euros (including 999,999.90 euros from the ERDF)

(continued)

Table 1 (continued)

No.	Project title	Project objective	Project budget
Lithuania-Poland-Russia 2007–2013 cross-border cooperation programme			
1.	LPR1/010/001 Lagoons as crossroads for tourism and interactions of peoples of south-east Baltic: from the history to present	The project centred on the coast of the Curonian and Vistula Lagoons. Three states took part in the project: Lithuania, Poland, and Russia. There were 13 project partners. The activities were aimed at the development of infrastructure, education, cultural events, and active tourism in the cross-border region. The majority of activities incorporated elements of 'live history' (folk festivals, handicrafts master-classes, etc.)	1,840,847.80 euros
2.	LPR1/010/038 Development of tourist and recreation infrastructure on the basis of the restoration and preservation of historical and cultural heritage of urban parks	The objective of the project was to stimulate the economic and social development of the Jurbarkas district in Lithuania and the city of Kaliningrad in Russia by solving common tourism infrastructure problems. The Jurbarkas mansion park was completely renovated: dead trees and bushes were uprooted; walking and cycling paths were renovated; new flower beds were made; new greenery was planted Tourist infrastructure was a fundamental component in the development of any type of tourism. This was particularly true for the health and recreation component of the project	2,863,199.51 euros
3.	LPR1/010/044 Baltic amber coast. Development of the cross-border area through building up and modernization of tourism infrastructure. Part II	The project was aimed at improving tourism infrastructure. A 2-km long wooden promenade was built in Yantarny (Kaliningrad region). Litterbins and benches were installed on the promenade Amber Beach has always attracted tourists who come to unwind. The wooden promenade is perfect for long walks along the sandy coast	1,363,066.00 euros

(continued)

Table 1 (continued)

No.	Project title	Project objective	Project budget
4.	LPR1/010/056 Cross-border tourism dimension	Within the project, tourism organisations from Bialystok in Poland, Druskininkai in Lithuania, and Kaliningrad in Russia forged cross-border ties. The aim was to facilitate the exchange of knowledge and experience in the field of health tourism in the border areas of Lithuania, Poland, and Russia	233,514.70 euros
5.	LPR1/010/093 The development of active tourism as a common ground for the Polish-Russian cooperation	The result of the project was two high ropes parks for children, teenagers, and adults. One was built in Elk (Poland), and the other in Ozersk (Russia). The project stimulated the development of active tourism, increased the effectiveness of promotion campaigns, and inspired an international programme for the development of active tourism in both partner towns. Considered good for health, high ropes parks are extremely popular with locals	458,268.84 euro
6.	LPR1/010/181 Baltic touristic games—know-how for the development of tourism potential of the Baltic region	The project sought to develop sports and active tourism in the Baltic Sea region by organising joint sports events and competitions	2,853,500.00 euros
7.	LPR1/010/143 Opportunities and benefits of joint use of the Vistula Lagoon	The main aim of the project was to strengthen socio-economic cooperation between the neighbouring regions of the Vistula Lagoon and to assess the natural and social resources for the development of the area and tourism	1,078,270.00 euros

The participation of the Kaliningrad region in cross-border cooperation projects has had a positive effect on the development of the regional tourism infrastructure. The attractiveness of the region for tourists has increased. Opportunities for unlocking the tourism potential of the area and including new objects in the range of tourist attractions have arisen.

In an exclave region, which is the case of Kaliningrad, cross-border and transboundary cooperation programmes for socio-economic and sustainable development are currently acquiring profound importance. Although some of these programmes were launched 15 years ago, they have proved their effectiveness and a long-term positive impact. Many of them put emphasis on tourism, and the budgets of tourism-related programmes have been growing.

4 Conclusions

This study leads to the following conclusions.

1. Over 34.0 million euros have been allocated for tourism development within cross-border and transboundary cooperation projects involving the Kaliningrad region. New partnerships have been established to connect 85 organisations in Poland, Lithuania, and the Kaliningrad region.
2. The exclave Kaliningrad region, which borders on the EU member-states (Lithuania and Poland), is a dynamic actor in cross-border and transboundary cooperation programmes. This translates into increased investment in the sustainable development of the region.
3. The geographical location of the Kaliningrad region and its historical and cultural heritage provide a solid basis for the development of tourism as a component of sustainable development. The region's 15 years of experience in cross-border and transboundary cooperation programmes as well as substantial funds allocated for tourism within them make these programmes a sustainable tool for tourism development in the Kaliningrad region.
4. Some of the tourism infrastructure objects created within cross-border cooperation projects are unique to the region (the Ancient Sambia Viking-age open-air museum). These projects had the most positive effect on the development of active tourism and the preservation of the cultural and historical heritage of the region. Active tourism infrastructure (sports facilities, pedestrian paths, cycling routes along the coastline of the Baltic Sea) required fewer resources. The greatest challenge for heritage preservation is that some of the objects belong to the Russian Orthodox Church and thus cannot be included in the cooperation projects.

Some projects have engaged earlier unknown heritage sites, which now have the necessary infrastructure for visitors; they introduce environmentally friendly technologies and make a significant contribution to the sustainable development of the Kaliningrad region. All this has contributed to a positive image of the region among locals and abroad.

References

Anisiewicz R, Palmowski T (2014) Small border traffic and cross-border tourism between Poland and the Kaliningrad oblast of the Russian Federation. Quaestiones Geographicae 33(2):79–86

Belova AV, Fedina-Zhurbina IV (2020) The potential of the Kaliningrad Region in the development of health tourism. In: Baltic Region—the region of cooperation. Springer, Cham, pp 285–296

Blatter J (2000) Emerging cross-border regions as a step towards sustainable development? Experiences and considerations from examples in Europe and North America. Int J Econ Dev 2(3):402–440

Fedorov G, Belova AV, Osmolovskaya LG (2015) On the future role of Kaliningrad oblast of Russia as an international development corridor. Eurolimes 19: 57–68

Hall D (2000) Sustainable tourism development and transformation in Central and Eastern Europe. J Sustain Tour 8(6):441–457

Iwona B, Lyudmila S (2013) Cross-border cooperation in tourism between the Warmian-masurian voivodeship and the Kaliningrad region. Baltic Region 5(3)

Joint Technical Secretariat (2008) Triple Jump: projects of Lithuania, Poland and Kaliningrad region of Russian Federation Neighbourhood Programme. Vilnius

Korneevets VS et al (2019) Influence of border regions relations on the tourist choices of the population. GeoJ Tourism Geosites 25(2):569–579

Kropinova EG (2020) The role of tourism in cross-border region formation in the Baltic Region. In: Baltic Region—the region of cooperation. Springer, Cham, 2020, pp 83–97

Lithuania-Poland-Russia Cross-Border Co-Operation Programme 2007–2013 Final Version of Operational Programme after EC Comments. Document adopted by The European Commission on 17 December 2008 (2008). [online] Available at www.ec.europa.eu/regional_policy/sources/docoffic/official/communic/wider/wider_en.pdf. Accessed 30 June 2020

Lithuania-Russia Cross-Border Cooperation Programme. Approved projects. Available at http://www.eni-cbc.eu/lr/en/projects/650. Date of access 02.07.2020

Lithuania-Russia Cross-Border Cooperation Programme 2014–2020. Approved by European Commission on 19 December 2016 (2016). [online] Available at www.ec.europa.eu. Accessed 30 June. 2020

Maciunaite J, Petrovich Y, Simkunaite V (eds) (2016) Lithuania-Poland-Russia ENPI cross-border cooperation programme 2007–2013. Center of European Project, Warsaw

Pacuk M, Palmowski T (1998) The development of Kaliningrad in the light of Baltic co-operation. In: The Nebi yearbook 1998. Springer, Berlin, Heidelberg, pp 267–282

Palmowski T, Fedorov GM (2019) The development of a Russian-Polish cross-border region: the role of the Kaliningrad agglomeration and the Tri-City (Gdansk—Gdynia—Sopot). Baltic Region/Baltijskij Region 11.4 (2019)

Poland-Russia cross-border cooperation programme. Heritage. Available at https://plru.eu/files/uploads/Projekty/pl-ru_publikacja_heritage.pdf. Date of access 02.07.2020

Soldatenko D, Backer E (2020) Trans-border territories in tourism in 2020: new perspectives. In: CAUTHE 2020: 20: 20 Vision: New Perspectives on the Diversity of Hospitality, Tourism and Events, p 241

Spiriajevas E (2019) Borderlands of Lithuania and Kaliningrad region of Russia: preconditions for comparative geographic approach and spatial interaction. In: Borderology: cross-disciplinary Insights from the Border Zone. Springer, Cham, pp 15–29

Studzieniecki T, Palmowski T, Korneevets V (2016) The system of cross-border tourism in the Polish-Russian borderland. Procedia Econ Finan 39:545–552

Zaucha J (1998) VASAB 2010 transnational cooperation in the spatial development of the Baltic Sea Region. In: Sustainable development for Central and Eastern Europe. Springer, Berlin, Heidelberg, pp 163–179

New Approaches to Sustainable Management of Wetland and Forest Ecosystems as a Response to Changing Socio-Economic Development Contexts

M. G. Napreenko, O. A. Antsiferova, A. V. Aldushin, A. K. Samerkhanova,
Y. K. Aldushina, P. N. Baranovskiy, T. V. Napreenko-Dorokhova,
V. V. Panov, and E. V. Konshu

Abstract Mires and forested wetlands are terrestrial ecosystems that perform many vital biosphere functions. In the Kaliningrad region, large wetland and wet forest blocks have been considered since the early twentieth century as core habitats that have great significance for sustainable development goals. Nevertheless, high-priority protected areas have not been designated there. Most wetland ecosystems have been transformed to farmlands, forestry sites and peat fields. The remaining habitats are still under threat of extinction. In a market-driven environment of capitalist society, a nature conservation system has to adopt new approaches and seize opportunities stemming from interactions between the academic and business

M. G. Napreenko (✉)
Natural Heritage Centre, Kaliningrad, Russia

M. G. Napreenko · T. V. Napreenko-Dorokhova
Shirshov Institute of Oceanology, Russian Academy of Sciences, Moscow, Russia

O. A. Antsiferova · A. V. Aldushin · Y. K. Aldushina · P. N. Baranovskiy
Kaliningrad State Technical University, Kaliningrad, Russia
e-mail: anciferova@inbox.ru

A. V. Aldushin
e-mail: aldushin@klgtu.ru

Y. K. Aldushina
e-mail: yuliya.aldushina@klgtu.ru

P. N. Baranovskiy
e-mail: baranovskiy@klgtu.ru

A. K. Samerkhanova
Vishtynetsky Nature Park, Kaliningrad, Russia

T. V. Napreenko-Dorokhova
Immanuel Kant Baltic Federal University, Kaliningrad, Russia

V. V. Panov
Tver State Technical University, Tver, Russia

E. V. Konshu
Baltic Forest Company, Guryevsk, Kaliningrad Region, Russia

communities. Our experience of successful nature conservation projects suggests the following avenues to explore in the Kaliningrad region:

(1) environmental rehabilitation projects for drained peatland to ensure fire prevention, biodiversity restoration and reduction in greenhouse gas emission;
(2) sustainable peatland management for paludiculture, nutrient retention and other ecosystem services;
(3) FSC forest management certification for biodiversity preservation and the identification of high conservation value forests (HCVF).

Keywords Ecosystem conservation · Peatland rehabilitation · Paludiculture · Forest certification · Sustainable wetlands management · High conservation value forests

1 Introduction

The protection and reconstruction of terrestrial ecosystems are among the 17 Sustainable Development Goals (2020) adopted by the UN. Mires and forested wetlands perform a variety of vital functions that benefit humanity (Katz 1971; Welsch et al. 1995; Joosten et al. 2017). One of them is closely linked to the atmospheric composition since these lands are part of the global biogeochemical carbon cycle and perform CO_2 sequestration by storing carbon in peat (Joosten et al. 2012). They also accumulate freshwater, regulate the water flow and affect natural water composition, which serves as a biogeochemical barrier capable of excess nutrient retention. These ecosystems contribute to biodiversity and habitat conservation, preserve natural aesthetics and meet human recreational needs (Minayeva and Sirin 2012; Bonn et al. 2014).

In the early twentieth century, Carl Weber, one of the pioneers of scientific telmathology, expressed his concerns about wetland conservation in East Prussia (Weber 1902). After the war, most large mires of the Kaliningrad region were designated as valuable ecosystems that need to be protected by nature conservation programmes like the UNESCO Telma Project, the Ramsar Convention's Wetlands of International Importance, the Important Bird Areas, Red Data Lists, and various regional conservation initiatives (Napreenko 2002).

Nevertheless, high-priority protected areas have not been designated there, and most wetland ecosystems have been transformed to farmlands, forestry sites and peat fields. Drained peatland was abandoned after occasional fires. Remaining habitats are still under threat of extinction (Napreenko 2015).

In a market economy, the state rarely allocates sufficient funds to regional environmental projects. Large areas of wetland and forest habitats end up being exploited by private natural resource users, whose only objective is to profit from selling raw materials.

The nature conservation system has to adopt new approaches and seize opportunities stemming from interactions between the academic and business communities.

Our experience of successful nature conservation projects suggests several avenues to ensure sustainable wetland and forest management.

2 Methods

In this article, our findings are placed in the context of several result-focused project investigations. Each survey was carried out using different approaches, which are described in the specific section of the paper and which correspond to the purposes and scope of the project. Generally accepted ecological survey methods and procedures were applied at project sites.

Analysis of documentation and cartographic data. For each project, the authors thoroughly analysed available open-access data sets such as archival papers, technical documentation, and cartographic information. When developing the rewetting concept, a comprehensive study of the land fund and concomitant legal issues was conducted.

Surveying the peatland drainage system. The investigation of watercourses consisted of assessing their basic morphometric characteristics: the length, the mean width and depth of ditches, and hydraulic connection with other watercourses. The study of drainage ditches was accompanied by mapping artificial barriers across watercourses and describing their condition, origin, and spatial features.

Biodiversity study. The species composition of local plants, animals, and fungi was identified by observation. Observations were carried out along both circular routes skirting the study area and transverse routes crossing different sectors of the study sites. The geographic coordinates of rare species locations were recorded.

Phytosociological survey on sample plots. Vegetation was described using relevé and plant community data. The precise coordinates of the sites were recorded. The sample plot size was $100m^2$ for open-land habitats and $400m^2$ for sites with a dense canopy closure.

Vegetation and ecosystem mapping. Mapping was performed by analysing open-access satellite imagery; the distribution of vegetation units across the study area was described. UAV images and other cartographic materials were used in the process. The data were further adjusted and verified on site during the field inspection of the study area.

Soil survey. Soils were examined along geomorphologic profiles running across the main elements of the peatland. The peat bed was probed and samples retrieved using either a Russian peat corer, or an Edelman auger or an Ismailsky auger, depending on the properties and structure of soils. Soil description and diagnosis were performed for control sections. Samples were taken from each 10 cm of a soil section or core retrieved. The field water capacity, hygroscopic water content, and pH_{H2O}, pH_{KCl}, and ash content in soil and peat samples were analysed in the laboratory.

Analysis of the botanical composition of peat. The laboratory treatment of peat samples included cleaning, elutriation and estimation of the peat decomposition degree. Specimens were studied under the microscope to identify plant macrofossils and estimate their percentage in each taxonomic group.

3 Environmental Rehabilitation Projects for Human-Disturbed Peatland

The principal aims of rehabilitation initiatives are fire prevention, reduction in greenhouse gas emission and biodiversity restoration. The general idea is to bring back the water regime of the disturbed peatland in order to create conditions for vegetation development and peat formation.

When implemented, this will aid in solving nature-oriented and practical problems.

It will also be possible to carry out various types of economic activity in the area, such as biomass production, berry cultivation, and ecotourism.

The pilot project entitled PeatRus: 'Restoring Peatlands in Russia—for Fire Prevention and Climate Change Mitigation' (2020) has been bringing the above idea to life (Report 2018b). This project is run at two abandoned mined-out peat deposits—Vittgirrenskoye and Vyshnyovoye (Fig. 1). Now it is funded by Russian and international environmental programmes but later it will be taken over later by regional authorities and the business community. The project is also carried out in some other Russian regions—Moscow, Tver, Vladimir, Kaluga, Ryazan, and Nizhny Novgorod.

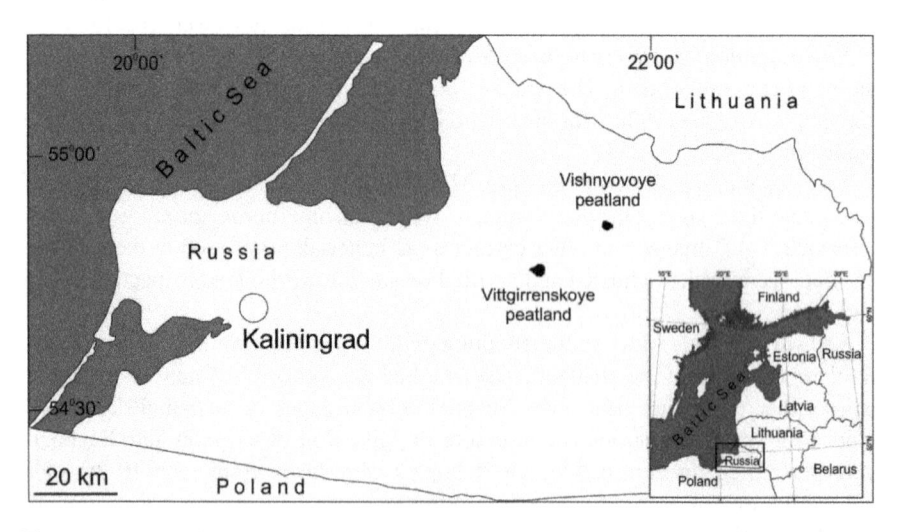

Fig. 1 PeatRus project pilot areas in the Kaliningrad region (Vittgirrenskoye and Vyshnyovoye)

Methodologically, the peatland restoration project consists of three main stages: (1) a comprehensive survey of the disturbed peatland; (2) developing a rewetting/mire restoration concept; (3) restoration works on site.

During the first survey of the area, three aspects should be taken into account: hydrography and the hydrological regime, habitat diversity and the conservation status, and the characteristics of peat deposits and their place in the landscape. A separate study has addressed current land use on the project site as well as legal matter management.

3.1 A Study of Hydrography and the State of the Drainage Network as a Preliminary to a Rewetting Concept for Disturbed Peatland

The hydrological features of the drainage network should be taken into account when rehabilitating wetlands ecosystem functions and creating conditions for restoring natural wetland ecosystems.

The main objective of a hydrological survey of human-disturbed peatland in the Kaliningrad region is developing project design solutions for the restoration of mire ecosystems. It includes measuring the water level and groundwater depletion after pre-extraction peatland drainage as well as analysing the current properties and morphometric indicators of the drainage network.

The hydrological study revealed the main elements of the water regulation process. These are drainage and transporting channels, artificial and natural dams, culverts and drainage pipes under vehicular passages.

The drainage system is shaped by the size and function of its elements. The largest channels (*first category*) run along the perimeter of peat fields. They fall into the category of interception drains, whose main tasks are the catchment of surface and drainage runoff from the surrounding agricultural topography and water discharge to the nearest trunk drainage channel.

Second-category channels pass through the central part of peatland and receive runoff from smaller, *third-category channels*. The smallest, *fourth-category* channels and underground drainage pipes laid along the routes of peat extraction vehicles catch water directly from peatland fields.

Using the knowledge of the drainage system, optimal project design solutions for a rewetting concept were developed. The main goal of rewetting is the restoration of the hydrological regime in a disturbed peatland. It does not mean, however, the creation of a permanent free water surface; it is aimed at a gradual increase in the moisture content in the upper layer of the peat bed across the entire project area. This way an environment for the natural development of mire vegetation will be formed.

Rewetting measures will not affect first-category channels, which will prevent possible extra moistening of the soil cover of adjacent agricultural lands and thus safeguard the interests of land users.

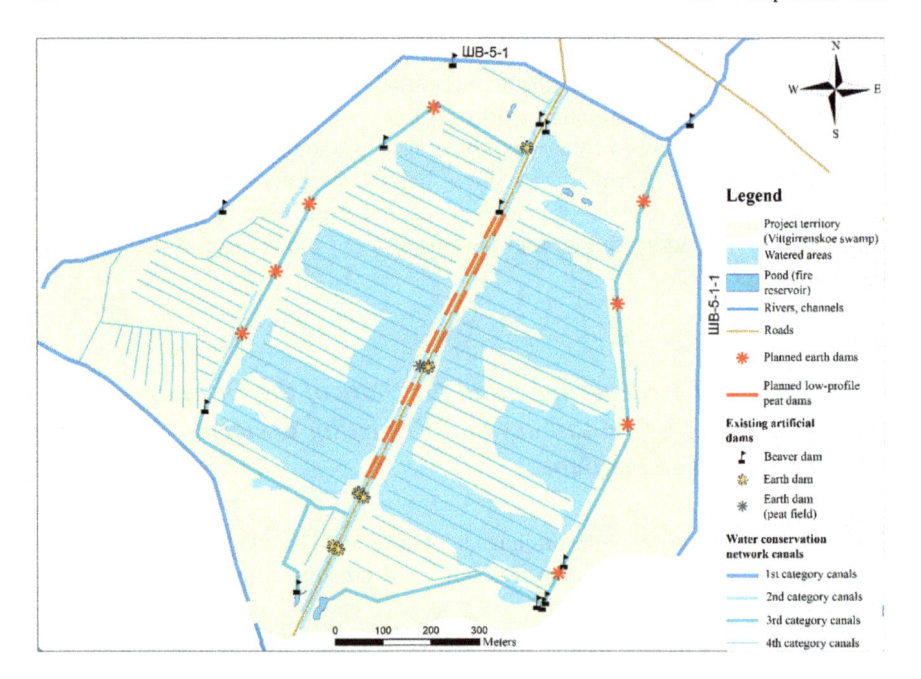

Fig. 2 The layout of the artificial dams in the rewetting concept (Report 2018b) for the Vittgirrenskoye peatland (PeatRus project area)

Because the slope of second-category channels ensures water backup in channels of the third and fourth categories, it is necessary to place *cascade coverage* on peat dams.

In the areas where internal runoff is absent, it is recommended to block channels by a series of *low-profile* (about 0.5 m high) peat dams; 40–50 m long, they should be built every 50–100 m (Figs. 2 and 3) to reduce the evaporation rate. These dams must be 0.2 m above underground drainage pipe outlets to block runoff. This is especially important for the areas where this system is functioning today.

The implementation of project design solutions will lead to the retention of incoming water in the peatland. Water backup in the channels and moisture accumulation in peat will contribute to extra moistening of the upper peat layers. This way is ecologically optimal for the restoration of a wetland ecosystem.

Our hydrological survey showed the effectiveness of existing beaver lodges for runoff regulation in large drainage ditches. The rewetting concept includes a clause that says that the morphometric indicators of the beaver dam should be taken into account at the stage of artificial dam design and spatial arrangement.

The hydrological investigation helped calculate the rate of increase in the water table and thus predict water regime dynamics during mire restoration.

It can be safely assumed that the rewetting period will last from two to five years in the major parts of both peat deposits. In the course of rewetting, the hydrological

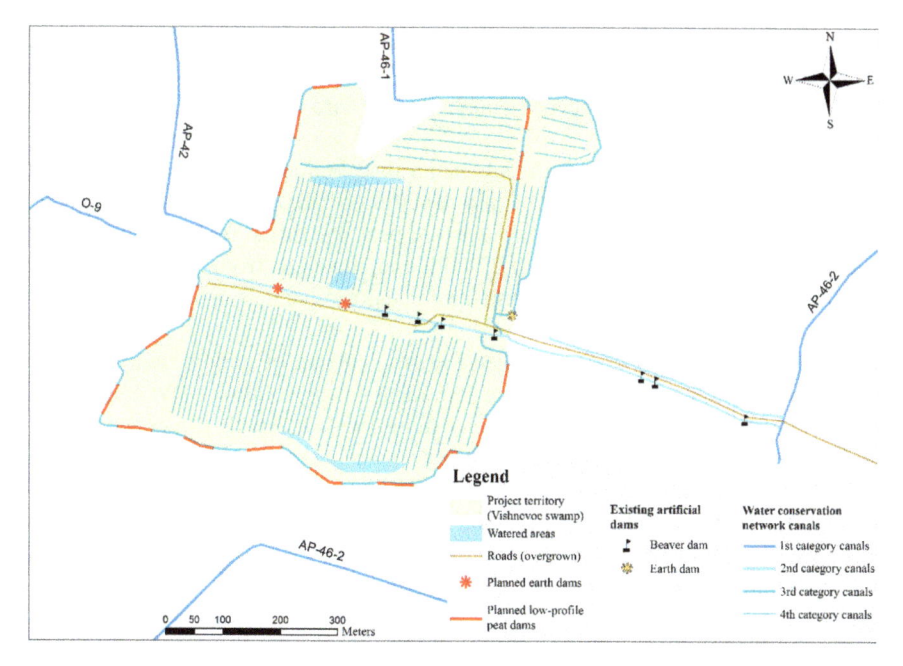

Fig. 3 The layout of artificial dams in the rewetting concept (Report 2018b) for the Vishnyovoye peatland (PeatRus project area)

regime is expected to stabilise. Spring and autumn flood phases will gradually grow longer, and the mean evaporation rate will drop.

3.2 Soil Investigation as a Contribution to Project Solutions for the Rehabilitation of Human-Disturbed Peatland

Just as hydrological and biological studies, the examination of the peat bed is crucial for working out a peatland rewetting concept. Since soils keep information about the entire landscape history, a detailed study of the soil profile helps reconstruct past events (Targulyan and Goryachkin 2008). Analysis of soil morphology and the composition of peat deposits makes it possible to reconstruct the evolution of plant communities and the features of mire formation before starting drainage.

The soil study explored the quantitative and qualitative indicators of the peat bed to reveal the current state of peatland ecosystems and to construct various scenarios for them.

The study showed that the disturbed peatland had distinct spatial zonality as regards the water regime and soil conditions. This spatial distribution pattern is the product of the natural mire development as well as human occupation (land reclamation, peat extraction etc.). The boundaries of various zones were identified and the permissible human pressure calculated. The findings contributed to the background knowledge of the landscape as well as to the ecological zonation of the territory outlined in the rewetting concept (it includes the core area, protection zones and demonstration fields).

A detailed soil survey showed the spatial distribution of residual peat layers, which should be taken into account in a rewetting concept as a characteristic critical to project design solutions. The existing peat resources are sufficient for the construction of low-profile peat dams. To this end, materials from peat storage areas and lanes located along the second- and third-category channels, which are now used by transport, should be exploited.

The study of the soil cover revealed the arrangement of underground drainage pipes in the upper layer of the peat. An additional design solution—blocking runoff using an underground plastic drainage system—was added to the concept. This decision should increase rewetting efficiency.

In the central part of the peatland, the residual layer of the raised-bog peat has retained its physical and chemical characteristics—the density, ash content, degree of decomposition, and pH. Thus, the restoration of the raised bog ecosystems is likely to occur rather rapidly if achieved by increasing the natural moisture of the raised-bog peat rather than by the formation of a water body with a free water table. This idea—a *gradual increase in the moisture content in the upper peat layer*—became the core of the peatland rehabilitation mechanism in the rewetting concept. These measures should provide necessary materials and environmental conditions for the development of bog vegetation.

The soil survey helped identify typical peatland sites where long-term monitoring can be performed, which involves tracking changes in soil (peat) properties and the condition of the entire mire ecosystem.

Burnt trunks and charcoal residues in the surface layers of the peat (Fig. 4a) indicate that the area is subjected to occasional fires. Analysis of soil humidity and peat

Fig. 4 Present-day peat bed in the Vittgirrenskoye peatland (PeatRus project area): **a** a fragment of a burnt pine trunk; **b** hydrothermal degradation on the surface of raised-bog peat (photo: O. Antsiferova)

hydrophobicity confirmed the threat of spring peat fires in the peatland. Rewetting will ensure that an important expected result—a reduced probability of fires—is achieved.

Low soil humidity and hydrophobicity are indicators of the intensive hydrothermal degradation of surface peat. This leads to higher levels of CO_2 emissions (Fig. 4b). The implementation of the rewetting concept will lead to the sequestration of greenhouse gas emissions in the peatland.

Analysis of the botanical composition of peat showed that, before the reclamation and drainage for peat extraction, there had been typical large raised bogs in the area. However, none of the former plant communities that had formed the top peat layers survived in the Vittgirrenskoye and Vishnyovoye peatland deposits or their environs.

Current vegetation characteristics and the properties of the residual peat bed suggest that mesotrophic-transition mire communities (a poor fen) will prevail during rewetting, i.e. over the next 100 years, until peat accumulation accelerates and changes the trophic status of the top peat layer. This conclusion is supported by studies into the self-restoration of peatland (Panov 2013). At the same time, natural and artificial heterogeneity of the mire will speed up oligotrophication and the formation of *Sphagnum* hummocks and raised-bog communities.

3.3 Analysis of Biodiversity for Zonation in Human-Disturbed Peatland and Rehabilitation Forecasts

The examination of biodiversity in the human-disturbed peatland and the mapping of its vegetation cover lay the groundwork for the zonation of the project area as well as for the monitoring of changes in the ecosystems during ecological rehabilitation. This makes it possible to predict how the rewetted territory will develop and how it may be used in the future. If parts of the peatland have dissimilar ecological conditions, their future functions may differ as well.

A necessary condition for rewetting is conserving as many rare species locations as possible. It is important to revise biodiversity findings at a special and habitat level before developing a rewetting concept. This revision will help identify the biodiversity-related natural functions of the future peatland ecosystem as well as its possible ecosystem services.

The Kaliningrad peatland has several vegetation types and corresponding plant formations (Fig. 5), which constitute various biotopes. Their future will depend on the management strategy chosen for the territory (rewetting, drainage, spontaneous development).

Based on GIS vegetation mapping, the project was divided into two main parts. The ***central part*** is covered with heathland regrowth vegetation of different types at different succession stages. It has developed over the past 20–40 years after primary

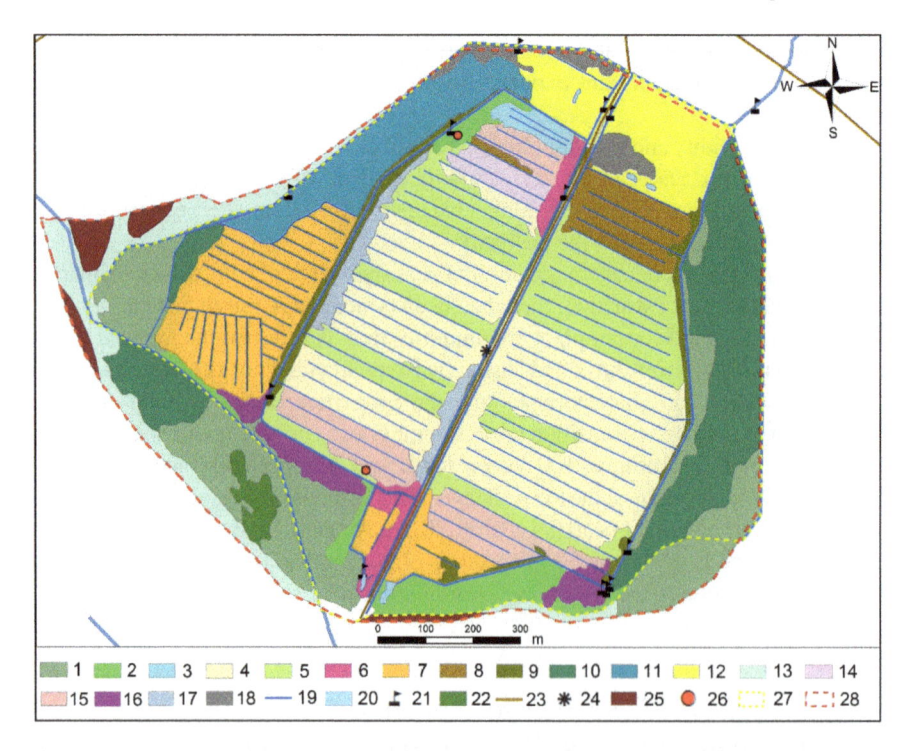

Fig. 5 A Map of current vegetation in the Vittgirrenskoye peatland (Report 2018b). **Forest arboreal vegetation**. 1. Birch and aspen wet forests. 2. Dry birch forest. 3. Birch forest with tall reed. Ligneous vegetation on peaty heathlands. 4. Sparse birch regrowth with Eriophorum. 5. Dense and high birch stand. 6. Closed-canopy birch stand. 7. Closed-canopy birch stand with scattered pine. 8. Post-fire birch stand. 9. High birch stand along large ditches. Shrub vegetation. 10. Inundated willow shrublands with Phragmites. 11. Inundated willow shrublands with Phragmites and dying aspen stand. 12. Willow shrubland at mined-out sites (without peat). **Herbaceous and dwarf shrub vegetation**. 13. Woodland edge. 14. Open areas with bare peat along large ditches. 15. Juncus-dominated poor-fen vegetation. 16. Poor-fen vegetation with birch regrowth. 17. Phragmites-dominated reed beds at inundated sites. 18. Tall-herb vegetation with Phragmites at mined-out sites (without peat). **The vegetation of aquatic habitats**. 19. Hydrophilic vegetation in drainage ditches. Adjacent areas. 20. Deforested area. 21. Agricultural lands. 22. Firewater ponds. 23. Beaver lodges. 24. Peat collection point. **Rare species locations**. 25. Lycopodium clavatum (Common Clubmoss). 26. Gravel roads and field tracks. 27. PeatRus project area. 28. Boundaries of the cadastral land parcel

vegetation was destroyed. The ***peripheral part*** is covered with inundated forests and shrubland, which have been forming for 60–70 years.

All identified plant formations were subsumed under the categories comprising a special classification of land covers of human-disturbed peatland (Sirin et al. 2018). For classification purposes, the vegetation map (Fig. 5) was converted into a map of land-cover classes (Fig. 6). All detected vegetation types were divided into six classes (Fig. 7): bare peat patches (class 1 'bare peat'); herb and reed communities (class 2 'grass'); pine communities (class 3 'pine'); willow and birch-dominated communities (class 4 'willow-birch'); wet-area hydrophilic communities with cattail

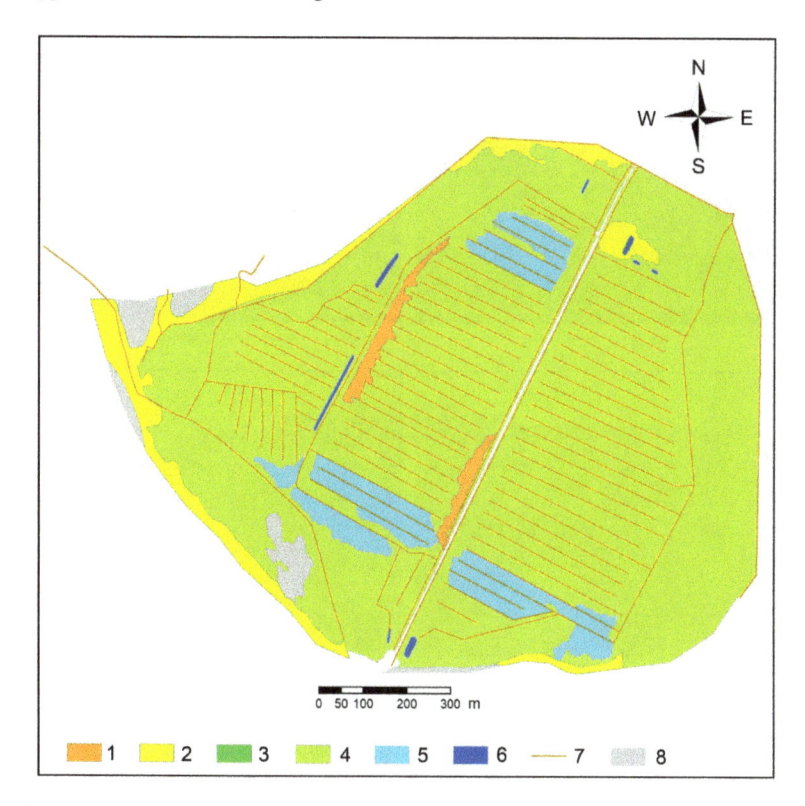

Fig. 6 A map of the present-day land-cover classes (Sirin et al. 2018) for the Vittgirrenskoye peatland (Report 2018b). 1. Bare-peat areas (land-cover class 1). 2. Reed communities (land-cover class 2). 3. Pine communities (land-cover class 3). 4. Willow and birch-dominated communities (land-cover class 4). 5. Hydrophilic communities formed by tall herbs (land-cover class 5). 6. Water bodies (land-cover class 6). 7. Drainage ditches (land-cover class 7). 8. Adjacent areas

and reed (class 5 'hydrophilic communities'); sparsely overgrown water bodies (class 6 'water bodies'). Class 7 'drainage ditches' differs from the other size both in fire-hazard level and environmental properties, including greenhouse gas emissions.

These areas constitute a single system. Thus, vegetation covers of human-disturbed peatland from different regions can be accurately compared. The testing of this system has shown (Sirin et al. 2011, 2020) that, when grouped into the above land-cover classes, patches of vegetation can be properly detected in satellite images of disturbed peatland. This method can also facilitate long-term monitoring of the environmental condition of restored peatland.

The mapping of the Kaliningrad project area helped to predict a gradual change from forest-shrub vegetation to sparse thicket and herb communities (patterned willow-stands, sedge meadows and reeds). The initial rehabilitation of the peatland ecosystem is going to take 10–25 years in the central area. A full functional restoration of the peatland can be expected in 80–150 years.

Fig. 7 Sites (habitats) with different vegetation at the PeatRus project peatland deposits: Vittgirrenskoye and Vyshnyovoye (photo: M. Napreenko): **a** bare-peat area (*land-cover class 1*); **b** dry birch stand (*land-cover class 4*) along the drainage ditch (*land-cover class 7*); **c** hydrophilic poor-fen community with *Eriophorum* and birch regrowth (*land-cover class 5*); **d** *Phragmites*-dominated reed bed (*land-cover class 2*) in the site inundated by beavers with the free water table (*land-cover class 6*)

The concept for the environmental rehabilitation (rewetting) of human-disturbed peatland in the Kaliningrad region, which was explored within the PeatRus project, suggests that plant communities of the study area have retained a core of peatland flora. This core will serve as the starting point for natural ecosystem restoration if suitable conditions are created by the rewetting project.

Various animal species live in the peatland, many of them either regularly breed there, or use it as a feeding biotope, or visit it during seasonal migration (e.g. birds). The restoration of the natural ecosystem will encourage the re-migration of stenotopic species inhabiting raised bogs and transition mires.

4 Peatland Management Projects for Nutrient-Retaining Paludiculture

The main ecological benefit of the projects is the proper functioning of natural wetland ecosystems, which can hold the main nutrients—nitrogen and phosphorus.

Another important result will be the mineralization and eutrophication of wetland habitats and thus greater fertility of rewetted peatland. These measures are meant to develop a new form of agriculture—paludiculture—in areas where agriculture and forestry have poor prospects because of the high cost of drainage reclamation.

Such projects may also contribute to the restoration of degraded peatland so that the wetland can be involved in the controlled use of natural resources in river runoff areas.

Paludiculture means that biogenic substances in the mire plant phytomass are removed through mowing and land use. This significantly reduces biogenic pressure on river systems. Rational management will help preserve and even increase the biodiversity of wetland ecosystems.

The above measures were taken in the Kaliningrad region as part of the international project entitled 'DESIRE—Development of sustainable peatland management by restoration and paludiculture for nutrient retention in the Neman river catchment'. The project focused on the basin of the Neman—the region's largest transboundary river, which carries a significant amount of nutrients causing eutrophication and other problems in the Curonian Lagoon. This waterbody is recognised by HELCOM as one of the Baltic 'hot spots' (HELCOM 2020). Most Kaliningrad wetlands, as well as estuarine sections of many smaller rivers, are located in the Neman delta.

4.1 Peatland Mapping

An important stage of the project was the inventory and detailed mapping of wetland and peatland ecosystems in the Russian sector of the Neman River basin. Remote sensing materials were analysed and on-site research conducted to determine the exact location, type and main characteristics of peatland in the study area.

A database and GIS were created for peatland on the Russian side of the Neman basin, using the collected materials. Cartographic data and an array of attributive information concerning the peat type, drainage degree, environmental status, current use, etc. were obtained. The open-source GIS will make it possible to rank peatland ecosystems according to their conservation significance and involve them in paludiculture.

4.2 Pilot Site Selection

The DESIRE project team led by specialists of the Vishtynetsky nature park examined 15 areas and identified six of them as potential sites (Fig. 8) for testing of paludiculture techniques. The results achieved by German, Poland, and Belarusian partners were taken into account.

Melioration systems in the Neman delta have been deteriorating for 30 years; many plots have been flooded. The restoration of the previous hydrological regime

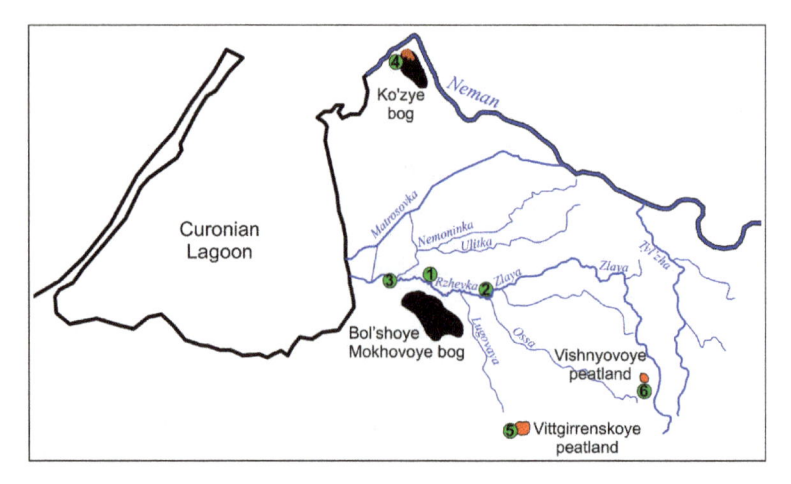

Fig. 8 A map of potential sites for the DESIRE project (natural mires are marked in black; human-disturbed peat deposits with peat extraction fields, in red; DESIRE project pilot sites, in green). Site 1. The Rzhevka floodplain near the village of Gromovo. Site 2. The upper reach of the Rzhevka river. Site 3. The Rzhevka-Nemonin confluence site. Site 4. The Kozye wetland (the section with the peat fields). Site 5. The Vittgirrenskoye peat deposit. Site 6. The Vishnevoye peat deposit.

requires substantial financing. This necessitates a search for new farming methods. One of them is paludiculture, which is the focus of the project. The biomass grown on wetlands following that method can be used in construction, energy, and gardening. Moreover, paludiculture contributes to the removal of nutrients held by wetlands ecosystems.

4.3 On-Site Research

Hydrological studies were carried out at model sites. Their findings will be used in the preparation of feasibility studies for territories intended for wetland crop production (paludiculture). The Bol'shoye Mokhovoye bog in the Slavsk district was chosen as a model site. It is the largest and best-preserved peatland ecosystem in the Kaliningrad Region.

The main objective of the research is to analyse spatial and hydrometric data. To this end, it is necessary to measure the electrical conductivity of groundwater, take piezometric measurements of the water level, and determine the topographic characteristics of ecologically dissimilar plots. These data will be used to create a model of groundwater flow in the peatland.

4.4 Information Campaign

The goal of this project is to select the best sites for paludiculture as well as to demonstrate the possibility of restoring mined-out peat deposits and making sustainable use of them for the benefit of the economy. It is important to raise awareness of new, environmentally friendly approaches to nature management and encourage interdisciplinary dialogue on peatland rewetting and biomass growing in the Neman basin as ways to reduce nutrient discharge into the river waters and the Baltic Sea.

Although paludiculture and its development in the Kaliningrad region is still a subject of discussion within the DESIRE pilot project, its introduction will be the first step towards the integrated and effective management of local bog ecosystems.

5 Forest and Wetland Conservation Through Voluntary Forest Certification in the Kaliningrad Region

This environmental initiative can largely contribute to nature conservation, if environmental scientists, experts, and responsible leasehold forestry businesses collaborate to implement the Forest Stewardship Council (FSC) scheme in line with the standards of the Russian National Voluntary Forest Certification System (2012). For example, one of the authors has experience of cooperation with the Baltic Forest Company, a major forest leaseholder in the central part of Kaliningrad Region.

The forest certification system encompasses a wide range of activities relating to environmental, technological, social and legislative issues. Its key objective is conservation planning, which requires identifying, managing, and monitoring ***high conservation value forests*** (HCVF). The HCVF approach is aimed at ensuring particular care when performing forestry operations and avoiding harm to biodiversity at valuable sites. It helps preserve species and habitat sustainability across the forest area. HCVF sites can include both forest and wetland ecosystems located within the areas of the National Forestry Reserve (*Goslesfond*).

During our HCVF survey, we followed the guidelines from forest certification handbooks by prominent Russian and foreign experts (Jennings et al. 2004; Andersson et al. 2007; Yanitskaya 2008; Kobyakov 2011). Each site defined as HCVF was subsumed under one or more High Conservation Value (HCV) categories as outlined in the FSC classification (Jennings et al. 2004). Our HCVFs comprise sites of the following HCV types: significant concentration of biodiversity, incl. both Protected Areas (HCV 1.1) and threatened/endangered species (HCV 1.2); regionally or nationally significant large landscape level forests (HCV 2); areas that contain rare ecosystems (HCV 3); areas that provide basic services of nature such as critical to water catchment (HVC 4.1) or barriers to destructive fire (HCV 4.3).

The study identified nine natural sites that meet FSC criteria and have been suggested for designation as HCVF sites in the territory leased by the Baltic Forest Company (Table 1; Figs. 9 and 10). Half of them are large raised bogs occupying an

Table 1 Natural sites designated as High Conservation Value Forests (HCVF) in the territory leased by the Baltic Forest Company (Report 2018a)

Site No	HCVF site	Basic criteria	HCV type according to the FSC classification (Jennings et al. 2004)
1	Beech-oak forest on a topography elevation near Klenovoye (Fig. 10a)	– A fragment of natural zonal vegetation of regional significance – A rare type of forest community of regional significance – High aesthetic value	HCV 1.2 HCV 2 HCV 3
2	Tall-herb alder forest in the valley of the Yasenka river	– A nearly intact natural forest site of regional significance	HCV 2 HCV 4.1 HCV 4.3
3	Bol'shoye raised bog (Fig. 10d)	– A reference ecosystem for raised bogs (a remnant habitat) – A nearly intact site natural vegetation of regional significance – Rare plant communities of regional significance – Rare plant species of regional significance – Rare soil types – High aesthetic value – Habitat-formation value	HCV 1.1 (potential) HCV 1.2 HCV 2 HCV 3 HCV 4.1
4	Papushinenskoye raised bog (Fig. 10b)	– A reference ecosystem for raised bogs (a remnant habitat) – A nearly intact site with natural vegetation of regional significance – Rare plant communities of regional significance – Rare plant species of regional significance – Rare soil types – High aesthetic value – Habitat-formation value	HCV 1.1 (potential) HCV 1.2 HCV 2 HCV 4.1
5	Lazheningskoye raised bog	– A typical raised bog ecosystem – A site with natural vegetation of regional significance – Rare plant species of regional significance – High aesthetic value – Habitat-formation value	HCV 1.1 (potential) HCV 1.2 HCV 2 HCV 4.1

(continued)

Table 1 (continued)

Site No	HCVF site	Basic criteria	HCV type according to the FSC classification (Jennings et al. 2004)
6	Rich fen with *Wolffia arrhiza* near Malinovka (Fig. 10c)	– A reference ecosystem for rich fens (remnant habitat) – A nearly intact site with natural vegetation of regional significance – Rare plant species of regional significance – Habitat-formation value	HCV 1.1 (potential) HCV 1.2 HCV 4.1 HCV 4.3
7	Transition mire ecosystem on terrestrialising lakes—Lake Maloye Olen'ye and Lake Bol'shoye Olen'ye	– A reference ecosystem for a transition mire on a terrestrialising lake (remnant habitat) – A nearly intact site with natural vegetation of regional significance – Rare plant communities of regional significance – Rare plant species of regional significance – Rare soil types – High aesthetic value – Habitat-formation value	HCV 1.1 (potential) HCV 1.2 HCV 3 HCV 4.1 HCV 4.3
8	Old-growth wayside trees (hornbeam, oak and lime)	– Unique natural object – High aesthetic value	HCV 1.1 (potential)
9	Swamp and inundation forest communities across the bank of Lake Russkoye	– A potential site with natural vegetation of regional significance – A rare plant community of regional significance	HCV 1.1 (potential) HCV 3 HCV 4.1 HCV 4.3

area of 1100 ha.

Most HCVFs contain habitats of rare plant species that either are of regional conservation significance (Sokolov 2003) or are included in Red Data lists in neighbouring countries (Napreenko 2002).

HCVF areas, identified as large landscape-level ecosystems (HCV 3) of regional significance correspond to units of the Ecological Land Classification of Europe, ELCE (Digital 2018). These are *Baltic-Byelorussian-Ukrainian lime-pedunculate oak-hornbeam forests*, incl. spruce-broadleaved communities (ELCE-index *F40*); *Alder carrs, often in combination with alder-ash forests, tall reed vegetation and sedge swamps* (ELCE-index *T1*); *Baltic Sphagnum magellanicum-raised bogs* (ELCE-index *S9*).

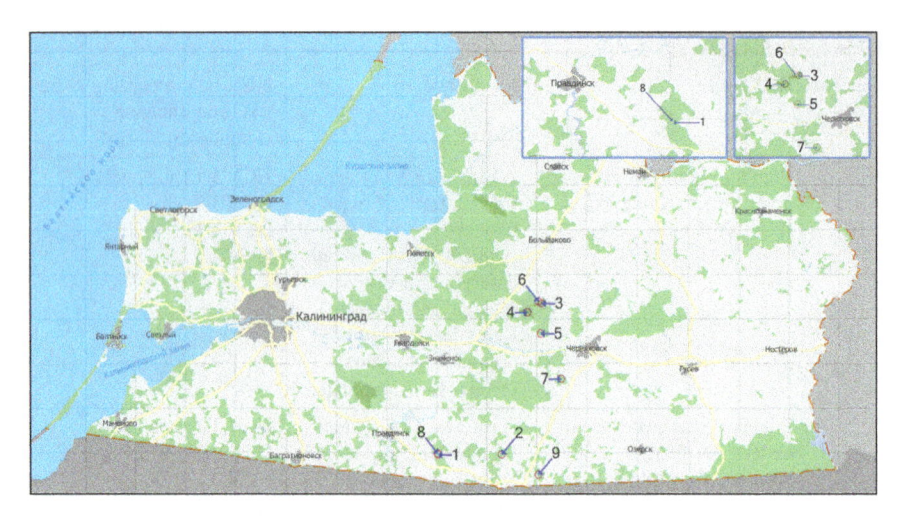

Fig. 9 Areas selected for HCVF designation on a map of the Kaliningrad region (Report 2018a). The indicators on the map match HCVF numbers in Table 1

Most HCVFs have been recommended for inclusion in the Regional List of Protected Natural Areas as natural monuments. Some sites are meant to be involved in ecological tourism (eco-trails).

The HCVFs and rare species habitats were mapped using GIS technology (Fig. 9). The maps could be used for long-term monitoring.

The leaseholder will be given management recommendations concerning each HCVF. The guidelines focus on fire prevention, natural state preservation, further biodiversity research and monitoring. There is a good chance that these areas will be maintained as sites of high conservation value.

6 Conclusion

Strict protection of intact habitats and responsible ecosystem management will contribute to attaining Sustainable Development Goals. It is important that new resource-management approaches continue to be developed and tested in the Kaliningrad region, including:

(1) environmental rehabilitation projects for drained peatland to ensure fire prevention, biodiversity restoration and reduction in greenhouse gas emission;
(2) sustainable peatland management for paludiculture, nutrient retention and other ecosystem services;
(3) FSC forest management certification for biodiversity preservation and the identification of high conservation value forests (HCVF).

Fig. 10 Natural sites designated as High Conservation Value Forests (HCVF) in the territory leased by the Baltic Forest Company (photo: M. Napreenko): **a** a beech-oak forest near Klenovoye (site 1); **b** bog pines—a *Sphagnum* community in the Papushinenskoye raised bog (site 4); **c** a rich fen with *Wolffia arrhiza* near Malinovka (site 6); **d** a hummock-hollow complex in the Bol'shoye raised bog (site 3)

These innovative environmental protection initiatives do not necessarily require that traditional methods be abandoned. In particular, it is important to create a network of protected natural areas. This measure is urgent because large wetland and wet forest blocks constitute the core habitats of the regional environmental framework. Thus, they should be preserved or restored if necessary (Napreenko 2002, 2015). This step will facilitate the achievement of regional Sustainable Development Goals, namely, sustainable use of water and biological resources and soil productivity.

It is crucial to encourage the involvement of natural-resource users in environmental activities by adopting new approaches and methods, for instance, by creating new financial mechanisms for estimating the value of ecosystem services of wetland and forest habitats.

Acknowledgements The study and conservation initiatives were supported by grant WWF951/RU005648/GLO from the World Wildlife Fund, WWF Russia; by project DESIRE No.

R091 of the Interreg Baltic Sea Region Programme and EU-Russia Cross-Border Cooperation Programme; by the Baltic Forest Company (Agreement of 09-01-2017). The work was co-financed by the Natural Heritage Centre and the Russian Academic Excellence Project run at the Immanuel Kant Baltic Federal University.

References

Andersson L, Mariev A, Kutepov D, Neshataev V, Alekseyeva N (2007) Identification and investigation of the biologically valuable forests. Saint-Petersburg, p 171 (in Russian)

Bonn A, Reed MS, Evans CD, Joosten H, Bain C, Farmer J, Emmer I, Couwenberg J, Moxey A, Artz R, Tannenberger F, von Unger M, Smyth MA, Birnie D (2014) Investing in nature: developing ecosystem service markets for peatland restoration. Ecosyst Serv 9:54–65

Digital map of European ecological regions (2018) [online]. Available at: https://www.eea.europa.eu/data-and-maps/data/digital-map-of-european-ecological-regions. Accessed 01 Nov 2017

FSC Forest Stewardship Standard for the Russian Federation (FSC-STD-RUS-V6–1–2012) (2012) [online]. Available at: https://fsc.org/en/document-centre/documents/resource/183. Accessed 08 Oct 2012)

HELCOM hot spots (2020) [online]. Available at: https://helcom.fi/action-areas/industrial-municipal-releases/helcom-hot-spots/. Accessed 01 June 2019

Jennings S, Nussbaum R, Judd N, Evans T (2004) The high conservation value forest toolkit. ProForest, Oxford, p 184

Joosten H, Tanneberger F, Moen A (eds) (2017) Mires and peatlands of Europe: status, distribution and conservation. Schweizerbart Science, Stuttgart, p 780

Joosten H, Tapio-Biström M-L, Tol S (eds) (2012) Peatland—guidance for climate change mitigation through conservation, rehabilitation and sustainable use. FAO and Wetlands International, p 100

Katz NY (1971) Mires of the earth. Nauka, Moscow, p 295

Kobyakov KN (ed) (2011) Conserving the natural areas of special value in North-West Russia. Saint-Petersburg, p 506 (in Russian)

Minayeva TY, Sirin AA (2012) Peatland biodiversity and climate change. Biology Bulletin Review 2(2):164–175

Napreenko MG (2002) The flora and vegetation of the raised bogs in Kaliningrad Region. PhD. Kaliningrad University, p 291 (in Russian)

Napreenko MG (2015) The mire ecosystems. In: Medvedev VA (ed) The nature in Kaliningrad Region. Water objects. Istok, Kaliningrad, pp 56–76 (in Russian)

Panov VV (ed) (2013) Potential use of the mined-out peatlands. Triada, Tver, p 279 (in Russian)

Report on the outcomes of an interaction between the Natural Heritage Centre and Baltic Forest Company LLC in 2017 according to the standards of the Russian National Voluntary Forest Certification System under the Forest Stewardship Council scheme (FSC-STD-RUS-V6–1–2012) (2018a) Natural Heritage, Kaliningrad, p 27 (in Russian)

Report on the scientific research project (2018b) Investigations in the human-disturbed peatlands in Kaliningrad Region for the purposes of mire ecosystem rehabilitation Natural Heritage, Kaliningrad, p 111 (in Russian)

Restoring Peatlands in Russia—for fire prevention and climate change mitigation (2020) (online). Available at: https://russia.wetlands.org/our-approach/peatland-treasures/. Accessed 01 July 2020

Sirin AA, Medvedeva MA, Makarov DA, Maslov AA, Joosten H (2020) Monitoring the vegetation cover of the rewetted peatlands in Moscow Oblast. Vestnik of Saint Petersburg University. Earth Sci 65(2):312–335

Sirin AA, Medvedeva MA, Maslov AA, Vozbrannaya A (2018) Assessing the land and vegetation cover of abandoned fire hazardous and rewetted peatlands: comparing different multispectral satellite data. Land 7(71):1–22

Sirin AA, Minayeva TY, Vozbrannaya A, Bartalev S (2011) How to avoid peat fires? Sci Russ 2:13–21 (in Russian)

Sokolov AA (2003) A conspectus of the rare vascular plants in Kaliningrad Region. In: Proceedings of the Russian geographical society (Kaliningrad Branch). Kaliningrad University, Russian Geographical Society, Kaliningrad, pp 5–116 (in Russian)

Sustainable development goals (2020) [online]. Available at: https://www.un.org/sustainabledevelopment/sustainable-development-goals/. Accessed 01 July 2020

Targulyan VO, Goryachkin SV (eds) (2008) Soil memory: soil as a memory of biosphere-geosphere-anthrosphere interactions. Publishing LKI, Moscow, p 692 (in Russian)

Weber CA (1902) Über die Vegetation und Entstehung des Hochmoors von Augstumal im Memeldelta, mit vergleichenden Ausblicken auf andere Hochmoore der Erde. Verlagsbuchhandlung Paul Parey, Berlin, p 252

Welsch DJ, Smart DL, Boyer JN, Minkin P, Smith HC, McCandless TL (1995) Forested wetlands: functions, benefits and the use of best management practices. U.S. Dept. of Agriculture, Forest Service, National Resources Conservation Service, p 62

Yanitskaya T (2008) Practical guide in defining high conservation value forests in Russia. WWW-Russia, Moscow, p 136 (in Russian)

Dr Maxim Gennadievich Napreenko is graduated in Biology and Chemistry (1994). He has completed a post-graduate programme (1997) and received a doctorate's degree (Ph.D.) in Botany (2002). He has long-term experience working in wetland ecosystems focusing on their biodiversity, vegetation and formation patterns. Currently, he is a researcher at Shirshov Institute of Oceanology, Russian Academy of Sciences (Atlantic Branch). His research interests are in palaeobotany and palaeogeography, wetland and forest ecology, vegetation science and bryology. He is also a head of the Natural Heritage Centre (since 2008) dealing with environmental education and nature conservation.

Dr Olga Alekseevna Antsiferova is graduated in Biology and Chemistry (1997). She has completed a post-graduate programme (1998) and received a doctorate's degree (Ph.D.) in Agrophysics and Soil Science (2001). She's an expert in soil ecology having experience in longstanding investigations of soil cover in agroecosystems and soils water regime monitoring. Currently, she is an associate professor at Kaliningrad State Technical University, Department of Soil Science and Agroecology. Her research interests are in evolution of soils and landscapes, soil hydrology, and agroecological assessment of soils. She is also a head of the Kaliningrad branch of the V.V. Dokuchaev Society of Soil Scientists (since 2015) dealing with regional soil research and soil ecological education.

Andrey Victorovich Aldushin is graduated Kaliningrad State Technical University (2007) with a major in Engineering of Automated Information Processing and Control Systems. In the same year, he entered the full-time postgraduate study at KSTU in Ichthyology (graduated in 2010). Currently, he completes a dissertation. Since 2008, he has been actively involved in the scientific and pedagogical work at the Department of Ichthyology and Ecology of KSTU. He developed and revised a number of academic training courses for students in the areas of "Ecology and Nature Management" and "Aquatic Biological Resources and Aquaculture." His research interests are in fishery and ecology, anthropogenic influences on natural environment, application of hydroacoustic methods in fishery research.

Amalj Karimovna Samerkhanova is graduated in Ichthyology and Ichtyopathology in 2001. She received her Master Degree in Environmental Management and Policy, International Institute for Industrial Environmental Economics at Lund University, Sweden in 2006. Currently she is a deputy director of Nature Park Vishtynetskiy coordinating the strategies for the development of specially protected areas covering wetlands.

Dr Yuliya Kazimirovna Aldushina is graduated Kaliningrad State Technical University (2000) in Fishery Management. She has completed a post-graduate programme (2003) and received a doctorate's degree (Ph.D.) in Ichthyology (2009). Since 2000, she is actively involved in scientific and educational work in Kaliningrad State Technical University. As an assistant professor, she deals with environmental management, environment protection courses. She is also interested in assessing the impact of economic activities on natural environment, ecosystems preservation and implementation of Environmental Management Systems standards at enterprises.

Pavel Nikolaevich Baranovskiy is graduated in Fishery. Currently, he works as a lecturer at the Department of Ichthyology and Ecology in the Kaliningrad State Technical University. He has experience of investigation of freshwater ecosystem. He also participates in regular monitoring research of water bodies. His sphere of research interests is ichthyology and ecology of freshwater ecosystems.

Dr Tatiana Vladimirovna Napreenko-Dorokhova is graduated in Biology (2008). She has completed a post-graduate programme (2011) and received a doctorate's degree (Ph.D.) in Geography (2002). She was responsible for managing a range of palaeogeographical research projects in the Kaliningrad Region. Currently, she is a researcher at Shirshov Institute of Oceanology, Russian Academy of Sciences (Atlantic Branch). Since 2018, she is also a research officer at Immanuel Kant Baltic Federal University. Her research interests are in palaeogeography and palaeoclimatology, stratigraphy of the Holocene, nature conservation.

Professor Vladimir Vladimirovich Panov is graduated Tver State Technical University (1983) in Geology of Peat Mires. He supported his Ph.D. thesis on mire monitoring and received a doctorate's degree in 1992 at Moscow Institute of Geodesy and Cartography. In 2003, he received a higher doctorate (Doctor of Sciences) in Geography at the Institute of Geography, Russian Academy of Sciences after supporting his doctoral dissertation (Dr.Sc) on mire restoration. He is currently a full professor at Tver State Technical University. Director of the Peat Institute. His research interests are in mire development and peatland restoration.

Elena Valerievna Konshu is graduated in Marketing (2006). Currently, she is within the top management team of the Baltic Forest Company LLC. She is responsible for public relations activities and cooperation with scientific community focusing on pilot projects aimed at implementation of FSC standards in the Kaliningrad Region. Her professional interests are in Forest Certification System, marketing and wise use of forests.

SkyrosIsland in the Front Line of Sustainable Development Promotion

Valentina Plaka, Chrysoula Sardi, Iliana–Dimitra Psomadaki, Olga Kouleri, and Constantina Skanavis

Abstract In Skyros Island, three public sectors have established an eco-community, which hosts students and volunteers from Greece and other countries, in order to promote a new eco-lifestyle, known as "SKYROS Ecovillage" model. Through its actions,it aims to spread the message of environmental awareness in order toeducate and transform the upcoming generations into environmentally responsible decision makers. Since 2015, a network of supporters has been established and they strongly believe that humans and nature can live in harmony.It is indispensable, to update the cultural software of our society and to cultivate an attitude of responsibility, consciousness and active environmental participation.This paper provides an analysis of a paradigmatic approach of an environmentally successful innovative community, stationed at a Greek port.

Keywords Sustainable development · Responsible environmental promotion · SKYROS project · Maritime tourism · Eco community

1 Introduction

The issues,the worldfaces, are created by humans and we are in an agony trying to solve at least the ones we understand. Being complex problems, when we try to solve them, most of the times we create new ones.People need a sense of responsibilityand conscience (Andersen and Björkman 2017). Globalization, environmental degradation, overpopulation, rise of technology and the new lifestyle we endorse have altered our environmental perceptions and behavior. In search of a sustainable lifestyle, the

---

V. Plaka (✉)
University of the Aegean, University Hill. Mytilene, 81100 Xenia Lesvos, Greece
e-mail: plaka@env.aegean.gr

C. Sardi · I. Psomadaki · O. Kouleri · C. Skanavis
University of West Attica, 196 Alexandras Avenue, Athens Campus, 11521 Athens, Greece
e-mail: csardi@uniwa.gr

C. Skanavis
e-mail: kskanavi@uniwa.gr

Research Center of Environmental Communication and Education of the Department of Environment at the University of the Aegean collaborated with Skyros Port Authority. This partnership was seen as an intriguing opportunity which gave rise theEnvironmental Campaign, named "SKYROS Project" through which, the aim is to spread environmental awareness and educate in a way that humans and nature can live in harmony.Since 2019, another academic institution, the University of West Attica, becomes the third public sector, which participates in SKYROS Project. The Department of Public and Community Health of the University of West Attica will play a vital role in Skyros Project by offering its knowhow on health promotion.

The island of Skyros is located in the archipelago of the Sporades. It is considered to be an island of crucial importance, as it is located in the heart of the Aegean Sea and connects many destinations together. Its port, Linaria, has been characterized as an environmentally sustainable small community that promotes responsible environmental behavior in the local community and among the visitors (Antonopoulos et al. 2015; Antonopoulos et al. 2017). Skyros Port Authority, as the qualified port management sector, has adopted sustainable agendas, where innovations are created with the goal of the completeness, the competitiveness and a strongenvironmental image of the port (Antonopoulos et al. 2017). Up until the touristic period of 2016, this practice has resulted to an increase of touristic arrivals up to 975% since 2010 (Antonopoulos et al. 2017). In the meantime, according to the gathered data of the Skyros Port Authority, in the summer of 2018, the arrivals spiked up to 1300%.

1.1 Building Eco-Philosophy in a Port Community

In Skyros Island, the strong environmental profile of Linaria Port compliments the needs of the research laboratory in order to test environmental awareness potential in real time set up, promoting in innovative ways Responsible Environmental Behavior to residents, tourists and program participants. This project began as an experiment and six years later, Skyros Project has become an environmental brand name. Participants discover their environmental identity through environmental actions based on educational protocols and research. This multi-awarded Skyros Project has enticed the attention around the world. The public sectors have managed to create an ecologically active community that hosts university students, volunteers, local community and maritime tourists in order to promote a new philosophy of an ecological lifestyle, known as the "SKYROS Eco-Community". Participants become active decision makers, and safeguard the right to a healthy environment for all.

1.2 Marine Observatory of Skyros—Sustainable Tourism Observation Database

Skyros Project consists of 5 action pillars: Summer Academy for Environmental Educators, internships for students, participation in global educational events, environmental education actions and Sustainable Tourism Observation Database.

A Marine Observatory of Skyros has also been established since 2016 as a follow upof the Sustainable Tourism Observation Database.

The Sustainable Tourism Observatory is an effort of the World Tourism Organization (UNWTO), with the Global Observatories for Sustainable Program Tourism (GOST), to support the sustainability of tourist areas in order to be more economically, socially and environmentally sustainable (Karavitakis and Chondromatidou 2016). The main goal was the development of tourism sustainability and boosting the perception of these areas, mainly into promoting elements of nature and the environment in general which are essentially the pillars of the tourism destinations identity, but also how they evolved over time (Karavitakis and Chondromatidou 2016). Based on theabove-mentioned concept, the researchers of SKYROS Project established the 1st Sustainable Tourism Observatory of Skyros Island, in 2015.

The results of the above Observatory showed that the island of Skyros is occupied with a low level of tourism development of the infrastructure, the lack of organization by the institutions, the reduced promotion and exhibition of the cultural—natural beauty of the island etc. (Karavitakis and Chondromatidou 2016). In contrast to the rest of the island of Skyros which shows low tourism (Karavitakis and Chondromatidou 2016), Linaria Port shows an increase of tourist arrival of yachts by 975% from 2010 (Antonopoulos et al. 2017). This port should be a standard for other small tourist ports of the Greek islands, whose operation is hampered by the economic crisis and its consequences (Antonopoulos et al. 2017). The results of the study led the Management Authority of Linaria Port to seek an evaluation of maritime tourism, through which they will promote the upgrade of the port while protecting marine and coastal environment. Finally, in collaboration with theUniversity of the Aegean, the research of the First and Second Observatory of Maritime Tourism was carried out in summers of 2016 and 2017.

In this research the **3rd Marine Observatory**results will be demonstrated. In the first part, an overview will be presented, revolving around the themes of the sustainable touristic growth, marine observatories and sustainable port community of Skyros Island, based on SKYROS Project Campaign. Following that, in the second part, the methodology that was used for the 3rd Marine Observatory, the research questions and the research tools of this study will be analyzed. In the third part the results will be assessed. Finally, in the fourth part the conclusions of the research will be presented.

2 SKYROS Port Eco-Village: The Successful Model

Tourism is one of the largest and most dynamic sectors of the world economy noting continuous growth and diversification with the emergence of new countries—destinations on the world's tourist map (UNWTO 2015). According to the World Council for Sustainable Tourism, sustainable development is essential for managing touristic areas, to prevent adverse consequences as well as balancing the relationships between the local community, the environment, the expectations of visitors and tourism businesses. This balance requires the cooperation of both local and regional authorities, citizens and the private sector but also the creation of a development strategy and management of tourism's development (Karavitakis and Chondromatidou 2016). Sustainable tourism, based on the World Tourism Organization (UNWTO 2015) is defined as "tourism that fully takes into account current and future economic, social and environmental impact, meeting the needs of the visitors, the industry, the environment and its host tourism communities". Maritime tourism is one of the largest economies in the world with significant contribution to the touristic economies of those countries that have developed it (Hall 2001).

Maritime tourism refers to the set of activities that are hosted or focused on marina environment and involve travelling away from one's permanent residence (Orams 1999). In Greece, maritime tourism tends to grow and is equally popular with tourists and the Greek population. The existing network that includes marinas, boat shelters even after the necessary changes for development and improvement, provides several marine tourism options, offering the visitor the pleasure of the sea route and the exploration of island and mainland areas of the country (Diakomihalis 2007).

The port of Linaria, Skyros, is considered a small public port with marina facilities.It is a very recent phenomenon for marinas to be considered as tourist attractions which are related to the destination in which they are located (Favro and Kovacic 2015). Arli (2012) describes marinas as tourism companies, which provide accommodation on private and commercial yachts or leisure crafts for their owners and their crew. Just like other travel destinations, marinas need to group together their customers in important departments, to match their own offered services by specific market segments (Hanlan et al. 2006). According to Heron and Juju (2012) marinas should create their own "unique sale proposal" to distinguish them from the rest.

According the following, port of Linaria has managed to stand out as "the Blue Port with the Green Shade", named such by the United Nations. With minimal financial resources and no permanent employees, the small port has been confirmed in recent years as a role model for small touristic ports. Every year it stands out through its facilities and services, leaving its mark as an innovation in the Greek marinas. The main categories of goods and services provided by marinas are fuel, lubricants, water, other supplies, equipment and reparations (Diakomihalis 2007).The facilities of the Linaria port have adopted a pro—environmental behavior. Facilities such as the use of electronic scooters, photovoltaic installations, reduction of energy and water consumption, attract touristic interest and bring economic growth, but also respect the environment and community's wellbeing.

The port emphasizes on sustainable development in tourism.All of its facilities promote an innovative ecotourism. The adoption of responsible environmental behavior of all stakeholders (individuals, organizations etc.) is one of the key solutions for all modern societies and governments that support environmental protection (Andriopoulos et al 2017; Ganiaris et al. 2018). Environmentally responsible behavior of tourists limits or restricts damage to the ecological environment (Yen-Ting et al. 2014). The search for a sustainable lifestyle brought into light this Greek eco-community at Linaria Port.The idea of the Observatories' actions was to collect data that would determine best practices. These data on tourism, reportpotential environmental pressure, which is of enormous value for preventing deterioration in an area of interest. Filling in comments from visitors in a guest book has proven to produce helpful information for environmental preservation and upgrading the port's image. Therefore, the marine observatory, the tourist observatory and the individual comments from tourists create a portfolio based on which environmental protection can be practiced at all times (Skanavis et al. 2018).This paper emphasized on the last research based on the Marine Observation of summer 2019.

3 Methodology

The present paper evaluates and upgrades the touristic product of Skyros Port. So, based on the pre—existing researches of previous years, the Tourism Observatory (2015 and 2018), the Marine Observatory (2016 and 2017), a new research on Maritime Tourism took place in summer of 2019.

The 1st Marine Observatory (Antonopoulos et al. 2017) was based on an existing study, by researchers of the Aegean University, Department of Shipping, which was focused on Cruise Tourism in Chios (Lekakou et al. 2011). After some necessary changes and adjustments, the first research was carried out, during the summer months of July and August 2016, in the Port of Skyros, Linaria. A questionnaire was distributed to the vessels that were docked in the Port. For a second consecutive year (2017), the 2nd Marine Observatory took place in the same methodology.

The Skyros Port Fund in collaboration with the University of West Attica carried out the research for the 3rd Marine Observatory. Also, this research was conducted by the researchers of the Laboratory of Environmental Communication and Education of the Department of Public and Community Health. The research began in the summer of 2019, at Linaria Port of Skyros Island, where questionnaires were distributed to a sample of 121 tourists that have docked during the period July–September 2019.

Even though the questionnaire that was distributed, was in the same format of the one distributed in 2016, necessary updates have been made.The respondents answered questions that were related to both services and sustainable development and innovation that take place in the port. The questionnaires turned out to draw the attention and be characterized as groundbreaking and innovative by the respondents.

The questionnaire was formulated and divided in two parts. It was guided by another standard questionnaire used elsewhere (Lekakou et al. 2011). The administered questionnaire was composed by two parts:

- The first part is about the Tourists' Profile.
- The second part is about the sojourn at the Port.

The main purpose of this research is to compare the results from the 1st and 2nd Marine Tourism Observatory with the current situation of the port as well as to evaluate or point out problems related to the benefits, services, any needed improvements and also new perspectives. Finally the adaptation of a responsible environmental behavior was an issue of investigation.

4 Results

4.1 Tourists' Profile of Linaria Port

In the first part of the questionnaire, there were some questions about the participants' demographic data, so that the average tourist's profile could be portrayed.

As a result, the average tourist who visited the eco-community of Linaria Port was German (42%) and male (65%). The average visitor was at age of the range 56–65 years old (46%) and with higher education (80%). Also, with an annual income up to 20.001–40.000 (17%).The average tourist had an open sea skipper license (54%) and used private vessel (62%) without a skipper(60%).

The visitor went to vacation with ownvesselmore than once per year (66%), for more than 7 days at a time (91%), and visited 4 or more ports per holiday trip (86%). The average visitor considered sailing as a touristic activity and experience and moved between ports in archipelagoes travelling for maximum 8 h per day (42%). The average tourist considered as very important the following services: easy and affordable mooring (41%), water supply (58%), WiFi (27%), showering services (47%), WC (45%), resupplying (39%), protection from weather conditions (78%) and the environmental policy of the port.

The average tourist considered the following as important services: power supply (34%), reception base, help from the port's personnel, help from the customs' authorities, technical services, connectivity with the mainland of Greece, food services and additional touristic services. The visitor believed that the transportation services to the rest of the island and the information services were of moderate importance, and at the same time the average tourist believed that the outdoors' cinema, exercising services, creative activities for children, secretarial support and lodging are not as important.The average visitorstated that sailing satisfaction is basedon the offered activities and general experience (83%).

4.2 Sojourn at Linaria Port

In the second part of the questionnaire regarding the first question "If they visited the island of Skyros for the first time" 62% responded positively, 36% answered they had visited the island before and a 2% didn't answer the question. The 81% stated that during their visit at Skyros, anchored in Linaria, 1% in Atsitsa, 1% in AgioFoka, 1% in Pefko and 1% in Lalari.

Regarding the overnight stays during their stay on the island, 46% had arranged one day, 37% for two days, 6% for three days, 3% for four days, 2% for seven days, 2% for ten days and only 1% for five days. In the end as it seemed, 22% spent the night only one day, 48% spent two days, 10% stayed on the island four days and finally 6% three days. Regarding the question "During your visit in Skyros, did you stay in a boat or accommodation" 95% replied on a boat while 5% left the answer blank.

Tourists were then asked to rate the facilities and amenities of the port. Starting with port security, 78% rated them at a high level, 20% at a satisfactory level and 2% had no answer. The water supply was considered by 95% of the tourists as high level, 3% as satisfactory while 2% did not answer. Subsequently, the sanitary facilities were rated by 77% at a high level, 17% considered them satisfactory, 3% had no answer and the remaining 3% left the answer blank. Then the tourists evaluated the rating facilities as high level by 78%, satisfactory by 14%, moderate by 3% and 5% did not answer.

Anchorage costs for 54% of the tourists were high, for 33% at a satisfactory level, for 3% were mediocre, 2% had no answer, while 8% left the answer blank. Port staff/ services were rated by 87% at a high level, 9% at a satisfactory level, 2% at a moderate level, while the remaining 2% did not answer. Services from the customs authorities based on the 72% of the tourists were considered as high standard, 16% considered them satisfactory, 1% as moderate, 1% as bad/nonexistent while 8% of the tourists had no answer. Subsequently, the cleanliness of the port was judged by 86% to be at a high standard, 8% as moderate, 3% as satisfactory, while 1% did not answer. The bonding services were considered by 86% of the tourists as high level, 9% as satisfactory, 3% left the question blank while 2% had no answer.

Technical support services were rated by 39% of the tourists as high level, 19% did not have an answer, 17% left the question blank, 12% considered them as mediocre, 11% as satisfactory while only 2% assessed them as not satisfactory/ insufficient. Then, the connection to the port/transport services, were considered by 36% of the tourists as high level, 23% as moderate, 22% as satisfactory, 3% rated them as bad/ nonexistent, 2% as unsatisfactory/insufficient, 9% did not have an answer, while 5% left the question blank. Information services were rated as high by 72% of the tourists, 23% as satisfactory, 2% as moderate while 3% left the answer blank. Regarding the environmental situation/port infrastructure, it was considered by 78% of the tourists as high level, 19% as satisfactory while only 3% rated it as moderate. Port facilities and equipment were 72% were considered 72% high, 25% satisfactory, 2% left the

Table 1 Facilities and amenities of the port assessment

Criteria	The evaluation of facilities (%)
Port security	2
Water and electricity supply	2
Sanitary facilities	3
Anchorage cost	2
Port/Staff services	2
Custom authority services	8
Port cleanliness	1
Mooring services	2
Technical support services	19
Connection to the port/transport services	9
Information services	2
Environmental situation/port infrastructure	3
Port facilities and equipment	1

answer blank while 1% did not have any answer. At this point the tourists' comments about the facilities and amenities of the port should be mentioned:

"Everything was perfect", "I did not expect such services in Linaria", "The best port we have anchored", "Best marina I can find in Greece", "Excellent", "Keep it up", "Perfect Baths", "Flawless", "Well maintained port with friendly stuff", "The number of the WC were insufficient" and "There was a bad smell like a sewage tank"

The evaluation of the facilities and the amenities of the port (Table1) was done with a rating scale of high level, satisfactory, moderate, unsatisfactory—inadequate and bad—nonexistent.

To the question "It is known that the port is state-owned?" only 38% answered yes, the remaining 53% answered no, while 9% left the question blank. Their answers to the question "Do you know another port with similar benefits/services?" 72% had a negative answer, only 16% answered positively while 12% left the question blank. From those who answered yes, to continuation of the question "If yes, which one?" 25% mentioned Marina Sani, 13% mentioned the Dodecanese, 13% in the Cyclades, 12% in Hydra, 12% in Paros, 12% in Samos, while the remaining 13% didn't answer the question.

Then, on the following question "Can you classify the port of Skyros, in relation to other Greek ports?" 75% judges it as excellent, 20% as very good, 2% as good while 3% left the question blank. "In relation to other European—International Ports?" 42% of the tourists rated it as excellent, the 36% as very good, 17% as good, 2% as satisfactory while 3% left the question unanswered.

In addition, tourists were asked to express their satisfaction on the products and services of Skyros. 89% of tourists described the behavior quality of the port staff as excellent, 6% as good, 3% as very good while 2% did not answer the question.

The quality of services with others (shops, taverns, etc.) in the port of Linaria were rated 47% as very good, 23% as excellent, 20% as good, 5% satisfactory and 5% left the question blank. Regarding now the quality/price ratio of other services (shops, taverns, etc.) in Linaria Port, 52% rated it as very good, 20% as good, 16% as excellent, 1% as satisfactory while 11% left the question unanswered. The evaluation of services outside Linaria showed that 42% rated it as very good, 16% as excellent, 14% as good and 28% did not answer. The behavior quality of the residents and professionals in the Linaria area were rated by 42% of the tourists as very good, 33% as excellent, 16% left the answer blank, 8% rated it good while only 1% as satisfactory.

The tourists then evaluated the sights of the island and specifically 36% rated them as very good, 27% as excellent, 22% as good, 1% as satisfactory and the remaining 14% did not answer the question. Access to attractions, by 44% of the tourists was considered as very good, 30% as good, 9% as excellent, 2% as satisfactory, 1% as insufficient and the remaining 14% did not answer. They also had to evaluate the advertising information where 33% considered it very good, 17% excellent, 17% good and 33% did not answer to the question. The evaluation of the roads was estimated by 36% as very good, 16% as good, 11% as excellent, 8% as satisfactory, only 1% as insufficient while 28% left the question blank. Tourists also had to judge the beaches where 31% of them considered them excellent, 25% very good, 10% good, 3% insufficient and the remaining 31% left the answer unanswered. The cuisine was judged by 51% of the tourists as very good, 19% as good, 14% as excellent and 16% left a blank answer. Tourists also rated the cleanliness where 47% considered it very good, 42% excellent, 2% good and 9% didn't answer the question.

The total quality of the touristic products of Skyros was evaluated by 56% as very good, 20% as excellent, 5% as good, 2% as satisfactory while 17% preferred not to answer the question.

5 Discussions and Conclusions

When a port incorporates environmentally friendly strategies and promotes a sustainable lifestyle, then tourism development rates exhibit a satisfactory increase. Linaria Port's continuous gathering of pertinent to tourism and environment data, generates useful results.Implementing, collating, interpreting and disseminating environmental information related to impacts from tourism activities, enables the wisely use of scientific information in the decision-making at all levels of port's operation. This small port community has understood and established practices through whicha sustainable tourism is being promoted. Furthermore, circular economy in tourism can affect positively the planning and management of all small eco marina communities all over Greece (Fountas et al. 2018).

Every summer, the data obtained fromthe Marine Observatory of Skyros present an accurate base in the assessment of facilities and amenities of the Linaria port. Another point of interest is the extrapolation of information related to tourists' environmental consciousness intentions. The lifestyle that this port eco-community promotes acts as an interactive tool and promotes a new lifestyle based on environmental sustainability. The port of Linaria operates with high environmental standards, serving as an example for small touristic ports, promoting a unique ecotourism model and enforcing sustainable development.

The adoption of responsible environmental behavior of all stakeholders (society, individuals, organizations, governments, etc.) is one of the major elements if sustainability is the end goal (Andriopoulos et al. 2017; Ganiaris et al. 2018). The Port of Skyros Island is in the front line of this pro-environmental movement. SKYROS Project plays a crucial role in the promotion of a responsible environmental profile, while it gives a boost in buildingan eco-positive life-style in this port community. Creative environmental education projects delivered at the port, can promote environmentally responsible behavior for both visitors and residents (Antonopoulos et al. 2017).

The port community has an environmental identity, which attracts tourists to adapt and increase their own standards of environmental consciousness. When consumers participate in activities using simple materials and imagination, they develop unique experiences known as "creative experience" (Richards and Wilson 2006). This community has as a goal to environmentally empower all whocome in contact with it. It is also a way of assisting people who might feel the urge to live ecologically and to discover their eco-ego.

The Eco-community of Skyros Port is introducing a way for a vacationing site for those interested to actively participate in the environmental protection and sustainable living during their personal time off (Ganiaris et al. 2018). At the same time, this port can be used as a learning and research environment, revealing new methods that can promote sustainable development in the tourism sector (Antonopoulos et al. 2017). In this way, Skyros Island is the unique Greek island, which promotes the Sustainable Development as a lifestyle. These practices are an environmental wake up signal for other small islands around the world. A new sustainable thinking needs to be established all over the tourist-oriented communities. The search for a sustainable lifestyle, focusing on the reduction of environmental threatening issues should be the end goal of the agenda of all governments and societies all over the world.

References

Andersen LR, Björkman T (2017) The nordic secret: a european story of beauty and freedom. FriTankeFörlag, Scandbook, Falun

Andriopoulos C, Avgerinos E, Skanavis C (2017) Is an Ecovillage type of living arrangement a promising pathway to responsible environmental behavior?". In: Innovation Arabia 10: health and environment conference, Dubai, pp 210–219

Antonopoulos K, Skanavis C, Plaka V (2015) Exploting futher potential of linariaport-Skyros: From vision to realization (in Greek). In:1st Hellenic conference on marines, NTUA, Athens, Greece, pp 101–111

Antonopoulos K, Plaka V, Mparmpakonstanti A, Dimitriadou D, Skanavis C (2017) The blue port with a shade of green: the case study of Skyros Island. Conference paper. In: 7th Health and environment conference, Dubai, Unated Arab Emirates, 6th to 8th March, 2017, pp 175–187

Antonopoulos K, Plaka V, Skanavis C (2017) Innovative implements to collect information: effectiveness and safety in linaria port, Skyros (in Greek). In: 7th Hellenic conference in management and improvement on costal zone, Athens, 20–22 November

Arli E (2012) Examining the level of influence from promotion components in Marina management in terms of demographic features. J Commerce Tourism Educ Faculty 1(1):25–52 (In turkish)

Diakomihalis MN (2007) Greek maritime tourism: evolution, structures and prospects. Res Transport Econ 21:419–455

Favro S, Kovacic M (2015) Construction of marinas in the Croatian coastal cities of Split and Rijeka as attractive nautical destinations. In: Rodriguez GR, Brebbia CA (eds) Coastal cities and their sustainable future, Southampton, UK, WIT Press, pp 137e14

Fountas Ch, Plaka V, Skanavis C (2018) Circular tourism through small eco-marinas. In: 10th International conference on Islands Tourism, 7th to 8th September, Palermo, Italy, pp 195–206

Ganiaris M, Zouridaki F, Plaka V, Skanavis C, Antonopoulos K, Aygerinos M (2018) In search for an island to host an ecovillage. In: 14th Protection and restoration of the environment, 3rd to 6thJuly, 2018, Thessaloniki, Greece, pp 209–218

Hall M (2001) Trends in ocean and coastal tourism: the end of the last frontier? Ocean Coast Manag 44(9–10):601–648

Hanlan J, Fuller D, Wilde S (2006) An evaluation of how market segmentation approaches aid destination marketing. J Hospital Leisure Market 15(1):5e26

Heron R, Juju W (2012) The marina: sustainable solutions for a profitable business. CreateSpace Independent Publishing Platform

Karavitakis E, Chondromatidou AM (2016) The tourist observatory as a tool for sustainable tourism development. Thesis, Department of Environment, University of the Aegean, Mytilene, The case of Skyros

Lekakou M, Stefanidaki E, Vaggelas G (2011) The economic impact of cruise to local economies. In: The case of an Island, proceedings of the Athens tourism symposium, Athens

Orams M (1999) Marine tourism: development, impacts and management. London and New York, Routlege

Richards G, Wilson J (2006) Developing creativity in tourist experiences: a solution to the serial reproduction of culture? Tour Manage 27:1209–1223. https://doi.org/10.1016/j.tourman.2005.06.002

Skanavis C, Antonopoulos K, Plaka V, Pollaki SP, Tsagaki-Rekleitou E, Koresi G, Oursouzidou C (2018) Linaria port: an interactive tool for climate change awareness in Greece. In: Leal Filho W, Lackner B, McGhie (eds) Addressing the challenges in communicating climate change across various audiences, Springer, Switzerland, pp 281–295

UNWTO (2015) Observatory on sustainable tourism under the auspices of the world tourism organization: provisional rules for the operation and management of observatories

Yen-Ting HCh, WanI L, Tsung-Hsiung Ch (2014) Environmentally responsible behavior in ecotourism: Antecedents and implications. Tour Manage 40:321–329

Valentina Plaka Valentina Plaka is an environmental scientist, who holds a Master's Degree in Environmental Policy and Biodiversity Conservation from the University of the Aegean. Today, she is a PhD candidate in Environmental Communication and Education in the University of the Aegean. She was awarded with Award of Excellence as a postgraduate student from the University of the Aegean. Also, Valentina is author ofa list of researches andpresentation on international and national conferences andarticles in magazines and book chapters. She participates in environmental articles in an educational journal and in creatingof educational kits about environmental issues. She aims to educate and communicate environmental issues to the society towards a new lifestyle, according to sustainable development

Chrysoula N. Sardis Chrysoula N. Sardis is an Environmental Scientist. Her postgraduate studies, MSc: " Water Resources Science and Technology", specialization: "Port Management" was held at the School of Civil Engineering of the National Technical University of Athens. Today she is a PhD candidate in Department of Public and Community Health at the University of West Attica. She is currently a Senior Researcher at the Epidemiology Laboratory of the Department of Public and Community Health in the field of Environmental Education & Communication.She has contributed in scientific publications and presentations at international scientific conferences.She has received the "ECOCITY" Environmental Sensitivity Award in 2016 and she was distinguished at the University of the Aegean Excellence Awards Ceremony, for her exceptional contribution during her undergraduate studies

Iliana Psomadakiis Iliana Psomadakiis an undergraduate student in the Department of Public and Community Health, University of West Attica. She is Junior Researcher at the Epidemiology Laboratory of the Department of Public and Community Health in the field of Environmental Education & Communication.

Olga Kouleriis Olga Kouleriis an undergraduate student in the Department of Public and Community Health, University of West Attica. She is Junior Researcher at the Epidemiology Laboratory of the Department of Public and Community Health in the field of Environmental Education & Communication.

Constantina Skanavis Dr. Constantina Skanavis is a Professor in Environmental Communication and Education at the Department of Public and Community Health at the University of West Attica. She is also the Head of the Module of Environmental Education and Communication. Prior to this position she served as a Professor of the University of the Aegean for 18 years. Before that she was a Professor at California State University, Los Angeles for 12 years. She has developed several courses on issues of environmental communication and education and health promotion. Professor Skanavis has numerous publications on an international basis and has given presentations all over the world. Professor Skanavis is the co founder of Skyros Project an innovative environmental awareness program with over 7500 followers

Transformations of Trolleybus Transport in Belarus, Russia and Ukraine in 1990–2020

Marcin Połom

Abstract In the past decades, the functioning of urban transport in the countries of Central and Eastern Europe has been based on traditional means of urban transport, primarily trams, trolley buses and diesel buses. The technological development of power sources has caused significant transformations in urban transport. On the one hand, the new possibilities have allowed for increasing the availability of non-emission means of transport in a larger area; on the other one, they allow the introduction of non-durable connections served by electric buses in place of trolleybuses or trams. The sustainability of infrastructure, which until now was the guarantor of public transport functioning, remains key. Lines operated by electric buses do not have such a guarantor. This leads to the conclusion that technological development can contribute to the unsustainable development of urban transport. Post-socialist countries, due to the underdevelopment of individual motorisation, had numerous tram and trolleybus transport networks. Now, fascinated by electric buses, they could willingly replace classic means with new vehicles without, however, conducting appropriate research as to the legitimacy of such actions. The article will outline the activities that are undertaken by exemplary cities in the countries of Central and Eastern Europe that subscribe to or deny the idea of sustainable development of transport in relation to new power technologies in urban transport.

Keywords Public transport · Electromobility · Trolleybus · Tramway · Electric bus · Transport policy

1 Introduction

In June 2020, there were 279 trolleybus transport systems in the world. Among them, 7 in Belarus, 79 in Russia and 43 in Ukraine. Over 46% of all trolleybus systems were located in the discussed area. Therefore, it is legitimate to state that trolleybus transport is an important branch of urban transport in the mentioned countries. It

M. Połom (✉)
Division of Regional Development, Institute of Geography, Faculty of Oceanography and Geography, University of Gdańsk, Gdańsk, Poland
e-mail: marcin.polom@ug.edu.pl

was historically significant and it is today. Despite the significant share of trolleybus transport in Eastern European countries, many urban electric transport systems have been closed in the last 20 years. Among the reasons were economic issues, a poor technical condition caused by years of underinvestment, but also political aspects.

The title of the article refers to the book by S. Tarkhov (2000) entitled *Empire of the trolleybus. Volume 1—Russia*, which outlined the functioning of all trolleybus transport systems in Russia at that time. The information contained in this publication became the basis for comparative research with the current situation in this country. Russia was undeniably a kind of trolleybus empire, so one of the research questions was trying to find the answer whether it still is.

This article attempts to systematise information on the contemporary functioning of trolleybus transport in three countries of Eastern Europe—Belarus, Russia and Ukraine. Against this background, in-depth research on the operation and development of trolleybus systems in selected cities was conducted. In particular, the directions of development or reasons for limiting or even eliminating trolleybus transport were analysed. The research task in this article was to find models of the functioning of trolleybus transport and to identify the factors that affect them and then to classify all trolleybus networks.

Trolleybus transport fits the idea of sustainable development as well as the idea of a smart city. The operation of trolleybuses allows reducing the effects of transport on the environment, especially in the places of operation. In the case of trolleybuses, the lack of pollutant emissions improves the air condition in city centres, which is particularly important due to the increase in the motorisation index in all countries of Central and Eastern Europe. Therefore, maintenance and development of trolleybus transport are important from the point of view of climate protection. The article attempts to find an answer to an additional question whether political decisions may affect the closure of trolleybus transport, and therefore whether they are in conflict with the idea of sustainable development.

2 Scientific Background, Methods and Research Area

Trolleybuses are not as popular as other means of public transport. Therefore, the number of scientific papers devoted to trolleybuses is not large. Although the number of trolleybus systems in Eastern Europe is the largest in the world, there are very few scientific studies on this subject in Belarus, Russia and Ukraine. Individual studies concern the functioning of trolleybus transport in individual cities (Alexandrova et al. 2004; Khairullina et al. 2020; Rudakewych 2017; Sustainable 2018; Tarkhov et al. 2010), competition from marshrutkas, and production of trolleybuses (Dmitriiev and Shevchenko 2017; Hutyria et al. 2020; Olander and Glinsky 2005), or the prospects of using trolleybus transport also in relation to electric buses (Blinkin and Muleev 2016). The identification of the lack of comprehensive studies on trolleybus transport, in particular its state and prospects for operation, shows the need to prepare this article (Bogodistyy 2017; Mjakota 2010; Oh and Gwilliam 2013; Rudakewych et al. 2017;

Stepanov 2019; Voznyanov 2017). It is particularly interesting to be able to compare the situation in the countries of Eastern Europe with the transport policies of the Americas, Asia and Western Europe in the future (Sgibnev 2014). The more so as the global urban electric transport, including trolleybuses, is gaining in popularity due to unfavourable climate changes caused by man (Barbosa 2016; Bezruchonak 2019; Borowik and Cywiński 2016; Połom 2018; Tica et al. 2011).

The article uses published scientific papers and materials from websites of urban transport enthusiasts and transport organisations that accurately describe all existing and decommissioned trolleybus systems. The work was created mainly by the intimate method through a thorough analysis of the functioning of all trolleybus transport systems in Belarus, Russia and Ukraine.

Urban electric transport in the Soviet Union played an important role and its development was always supported by the central authorities. With the collapse of the Soviet Union, the situation in the individual new countries varied. The approach of the Soviet Union to supporting the development of urban trolleybus transport was inherited by Belarus. The collapse of urban transport, including trolleybuses, occurred mainly in Ukraine, which did not implement significant economic reforms. As a result, the cities lost the ability to maintain urban transport at a good technical and operational level. Additionally, both Belarus and Russia had trolleybus factories (Połom 2016), so their purchase on the local market was easier than in Ukraine, where the production of trolleybuses was recreated in the 1990s.

Based on selected features such as the size and regularity of purchases of new trolleybuses, acquiring trolleybuses from the secondary market in the absence of new vehicles, the quality of the existing rolling stock, spatial development of infrastructure, modernisation of the existing infrastructure, use of new technologies (e.g. batteries), spatial development of trolleybus lines and political decisions of local authorities, five models of trolleybus transport development in Eastern Europe were distinguished.

3 The Current State of Trolleybus Transport in Belarus, Russia and Ukraine

As mentioned in the introduction, trolleybus transport in the countries of the former Soviet Union (Belarus, Russia and Ukraine) is very popular. Trolleybus systems are more common than in other countries. In 1990, in the area of the three analysed countries, there were 146 trolleybus transport systems in total (cf. Gorodskoj èlektrotransport; Trolley: motion).

In the last decade of the twentieth century, two systems in Russia were liquidated. In Kamyshin, despite the purchase of trolleybuses and the construction of infrastructure in 1988–1991, it was never completed and no transport was launched. The second city to lose trolleybus transport was Grozny, where trolleybuses were liquidated due to war damage in 1994. At the beginning of the twenty-first century, the cities of

Russia and Ukraine began to feel the effects of the economic collapse and the lack of financial resources for the maintenance of municipal electric transport. Additionally, the emerging competition from private minibuses worsened ticket revenues due to declining passenger numbers. The phenomenon of marshrutkas (minibuses), typical of Eastern European countries, is the main cause of the decline in public transport, in particular trams and trolleybuses (Muleev 2020; Ryzkov and Zyuzin 2016; Ryzhkov 2018; Sanina 2011).

In the first decade of the twenty-first century, five trolleybus networks in Russia (Arkhangelsk, Kurgan, Shakhty, Syzran, Tyumen) and two in Ukraine (Pierevalsk, Toreck) were closed, with one of the dismantled trolleybus networks in Russia (Syzran) being launched in 2002 and only existing until 2009. In that decade, three new networks were also created—two in Russia (Vidnoye in 2000 and Podolsk in 2001) and one in Ukraine (Kerch in 2004). The lack of funds influenced, above all, lower vehicle purchases, the deteriorating condition of the existing rolling stock and infrastructure. Some cities gave up further spatial development of trolleybus transport (cf. Gorodskoj èlektrotransport; Trolley: motion).

In the next decade, trolleybus transport regressed further. No new system was launched, and the existing ones suffered from various existential problems. There were also new conditions that influenced the changing approach of local authorities to trolleybus transport. There were noticeable decisions dictated by political conditions and the attitude of local authorities to trolleybuses, which were not supported by economic or technical analysis. Moreover, the development of the technology of trolleybus autonomy and non-catenary power supply from batteries created new opportunities for spatial development, which some systems took advantage of. In the second decade of the twenty-first century, three trolleybus networks were liquidated in Ukraine (Dobropole, Kadijówka, Vyglegivsk) and six in Russia (Kamensk-Uralsky, Perm, Astrakhan, Lipetsk, Vladikavkaz, Blagoveschensk) (cf. Gorodskoj èlektrotransport; Trolley: motion).

Throughout the analysed period, there was an equal number of trolleybus networks in Belarus. The policy of the Belarusian authorities regarding electric transport is very favourable, both at the national and local level. Seven cities had trolleybus transport and in the years 1990–2020, they regularly developed it and invested in the rolling stock and infrastructure.

In the analysed period, the number of trolleybus networks decreased by seventeen, which meant the elimination of 12% of the existing systems (cf. Table 1). Most networks were closed in Russia (13).

Table 1 Number of trolleybus transport systems and change between 1990 and 2020

State	1990	2000	2010	2020	Change 1990–2020 (number)	Change 1990–2020 (%)
Belarus	7	7	7	7	0	0
Russia	92	90	85	79	−13	−14
Ukraine	47	47	46	43	−4	−9
Sum	146	144	138	129	−17	−12

4 Detailed Analysis of Selected Cities

After a preliminary analysis, cities located in all the discussed countries were selected for detailed analysis. These are cities that stand out with separate transport policies in terms of trolleybus transport. The selection of cities enables distinguishing models of trolleybus transport development in Eastern Europe. From the scientific point of view, it is interesting to see how the local governments approach the development of public transport, including trolleybuses, in different ways, despite the similar historical heritage, socio-economic, cultural, and legal conditions.

4.1 Kaliningrad

Kaliningrad is a large city (almost 490,000 inhabitants in 2020), located on the Baltic Sea. As an exclave of the Russian Federation, it borders Belarus, Lithuania, and Poland. The trolleybus network was launched in 1943 when Kaliningrad was known as the Prussian city of Königsberg. It was a typical phenomenon for the areas of the German Reich, which in the second half of World War II experienced significant shortages of liquid fuels and developed electrified urban transport, both trams, and trolleybuses. As a result of the war, the city was completely destroyed, including the trolleybus network. Trolleybus transport in Russian Kaliningrad was restarted in 1975 for a similar reason as in 1943. At the turn of the 1960s and 1970s, the global fuel crisis caused a renewed interest in electric transport in cities.

Currently, 45 trolleybuses of four types are in use, which were purchased in batches in 2006, 2009, 2012, and 2013. All vehicles were manufactured in Russia or purchased from Belarus, and 40% of them are low-floor. While the trolleybus park is in relatively good condition, the number of lines served by them has been systematically decreasing. The Kaliningrad municipal self-government is characterised by a slow decay of urban electric transport. The operation of tram lines has been significantly limited to just one, and trolleybus lines to three. In addition, the range of their operation was limited by liquidating subsequent sections of the traction network. Back in the 1990s, there were six trolleybus routes, and the local authorities announced the construction of further sections in the first decade of the twenty-first century to replace the liquidated tram lines. In fact, only a small fragment was built along Gorky str. At the same time, four trolleybus routes were closed and two were shortened.

A significant opportunity to improve the functioning of trolleybus transport was seen in the organisation of the Football World Cup in 2018. Both the national, regional, and city authorities announced significant investments in urban transport, in particular pro-ecological. In fact, the announced infrastructure and rolling stock investments have not been implemented. In 2019, the Kaliningrad city authorities organised an online plebiscite to answer the question of how to develop electric transport. The questions were designed in such a way that they were mutually exclusive, one could vote for the development of tram and trolleybus transport or for electric buses. The survey was preceded by a biased opinion that tram transport brings annual losses of 105 million rubles and trolleybus transport 125 million rubles; 1.3 billion rubles will be needed to renew the tram rolling stock, and the renovation of the tram infrastructure will cost 1.25 billion rubles. Instead, it was suggested that electric buses, costing just 30 million rubles, are a worthy alternative. Such an approach did not show the real costs of operation and functioning of transport based on electric buses. Despite this, the existing means of electric transport clearly won in the Internet poll by a vote of 64 to 36%.

The actions of Kaliningrad's municipal authorities can be described as neglecting municipal electric transport, which contributed to the systematic extinction of its operation. The lack of infrastructure and rolling stock investments leads to the liquidation of traditional means of electric transport in Kaliningrad, and the suggestive questions included in the internet survey show the actual attitude of the city authorities to trolleybus transport.

4.2 Lutsk

Lutsk is a quite large city located in the western part of Ukraine. It has over 217,000 inhabitants (2020). Trolleybus transport was launched in this city on April 8, 1972. In the following years, new lines were opened and the fleet of trolleybuses expanded. As in other cities of Ukraine, the political and economic transformation resulted in a serious crisis in financing trolleybus transport. In the years 1993–2009, only 16 new trolleybuses of various types were purchased. Such a number did not allow for proper maintenance of the technical condition of the rolling stock, it led to decapitalisation and reduction of operational possibilities. The variety of new trolleybuses (Bogdan, YuMZ, ZiU) made the functioning of the technical facilities difficult. The lack of sufficient financial resources led to a further reduction in trolleybus transport. A partial solution to this problem was the purchase of used trolleybuses from another city. Due to the limited secondary market for trolleybuses, Lutsk took the opportunity to purchase vehicles in Poland in Lublin. These were high-floor trolleybuses of various types of Jelcz brand, which were not operated in Ukraine at all. Therefore, this solution was subject to risk. In the years 2012–2015, 23 trolleybuses were purchased in Lublin. They were vehicles of seven types, differing from each other in terms of both electrical equipment and bodies. Additionally, in 2015, three used trolleybuses were purchased from another Polish city in Gdynia. There were three different vehicles, including two Jelcz and one Mercedes-Benz. Initially, Lutsk planned to purchase a significant number of Mercedes-Benz trolleybuses from Gdynia, but the change of city authorities and the public transport company resulted in a departure from this idea. Mercedes trolleybuses were converted from diesel buses in Gdynia in 2004–2010. Their main advantage was, above all, the low floor and the adaptation for the disabled people. In the following years, Lutsk did not purchase new trolleybuses, until 2019, when a contract was signed for the supply of 29 factory-new low-floor

Bogdan trolleybuses. The purchase was possible thanks to a loan from the European Bank for Reconstruction and Development.

Trolleybus transport in Lutsk can be classified as a typical example of an important transport subsystem in the city, which was gradually degraded due to the low level of financing. However, the opportunity was used to improve the situation by purchasing a significant number of used vehicles in order to survive the most difficult period. Thanks to the commissioning of the atypical, but a large number of Jelcz brand vehicles, trolleybus transport was maintained at an appropriate level until it was possible to purchase new vehicles.

4.3 Lviv

Trolleybuses in Ukraine are as popular as in Russia. As in other eastern-European countries, their functioning in Ukraine was related to the underdevelopment of individual motorisation. Tramways and trolleybuses were the pillars of public transport in Lviv during the communist period. For Lviv, well-functioning urban transport is important because it is the largest city in western Ukraine, with just under 725,000 inhabitants in 2020.

The first idea for putting trolleybuses into operation in Lviv appeared at the beginning of the twentieth century, and then again just before the outbreak of World War II in 1938–1939. The idea was revived in 1947. It was then decided to replace one of the tram routes with a trolleybus line. The first trolleybuses left the streets of Lviv on November 27, 1952. In the following years, more sections of the trolleybus line were put into operation and the trolleybus traffic increased. Until the end of the communist period, trolleybus transport in Lviv was in relatively good condition. Regular deliveries of new trolleybuses of Ukrainian and then Czechoslovak production made it possible to recreate the rolling stock and keep it in good technical condition. Along with the independence of Ukraine, as a result of the economic downturn, some trolleybus routes were closed in the early 1990s. Despite the difficulties, from 1994, new trolleybuses of the domestic production of the LAZ brand were put into operation. In the years 1994–1998, 39 trolleybuses were purchased. In the following years, the purchase of used trolleybuses in the Czech Republic was a way to improve the condition of the rolling stock. The advantage of this solution was familiarity with the vehicles, as they were similar or identical to those already operated in Lviv. In 2006, new trolleybuses were introduced after a long break. These were the first LAZ low-floor trolleybuses in Lviv. In 2011, one of the trolleybus lines was extended, and the mayor of Lviv announced the acquisition of 60 new trolleybuses in connection with the organisation of the EURO 2012 football championship (Kuczabski and Połom 2019). However, the project was not implemented, although the organisation of a mega-event such as the European championship always gives hope for an impulse for reform and change. In 2016–2017, the trolleybus route was extended to the new terminal at Lviv passenger airport.

Significant changes in the functioning of trolleybus transport began when Lviv obtained a loan from the European Bank for Reconstruction and Development, which allowed the purchase of 50 new low-floor trolleybuses. Locally manufactured Electron vehicles were ordered to further strengthen the local labour market. The local authorities announced the purchase of 100 new trolleybuses with alternative power sources (batteries) in the coming years, which would allow for the spatial development of trolleybus lines based on the existing traction infrastructure. This approach refers to the European trend of sustainable development—the use of existing infrastructure and the advantages of the developed battery technology.

The regression of trolleybus transport in Lviv was related to the economic problems of Ukraine and the central financing of transport. In addition, numerous public transport routes served by minibuses, often stopping at any destination, won an unequal competition with trams and trolleybuses. Along with the transfer of some economic powers to local governments, opportunities for the renewal of urban transport were created. The Lviv authorities took advantage of this opportunity and, having analysed various solutions, chose the most economically and technically effective project for the renovation of the existing trolleybus transport as well as its development with the use of new technologies.

4.4 Minsk

With the liquidation of trolleybus transport in Moscow, the system in Minsk has become one of the two largest in the world (together with the one in Kyiv). The first trolleybuses on the streets of Minsk were launched in September 1952, and in the following years, new routes and trolleybuses were added. In 1956, there were 39 trolleybuses, and five years later, there were already 100 vehicles. The number of trolleybus routes was growing until the 1990s. Significant changes in the functioning of trolleybus connections took place at the beginning of the twenty-first century when parts of the traction infrastructure were liquidated and several lines were shortened. The spatial development of the trolleybus network in Minsk was resumed less than 10 years later, and subsequent sections were put into operation. Regular investments in the following years allowed for an increase in trolleybus transport. The number of vehicles was about 1000 and slightly decreased to less than 900 at the end of the second decade of the twenty-first century. The number of passengers transported by trolleybuses in 2000 was over 340 million, and over the next 20 years, it decreased to the level of 150 million. This is a result of the development of individual motorisation in Belarus.

Regular investments, in particular in the rolling stock, are a characteristic feature of trolleybus transport in Minsk. Currently, over 800 trolleybuses are in operation. All of them have been produced at the local plants of MAZ and Bielkommunmash and are low-floor. For example, in 2004, 42 trolleybuses were purchased, 2005–54, 2006–101, 2007–238, 2008–290 (Połom 2016). In the years 2019–2020, the first trolleybuses were put into operation, also using battery power–70 MAZ 203 T. In the following years, there is a plan to purchase trolleybuses with extended range when powered by batteries, which will allow for the spatial development of connections without the need to build the traction infrastructure in areas with smaller passenger flows.

Minsk also took steps to introduce electric buses. There were voices in the media that trolleybuses are an obsolete means of transport and should be replaced by modern electric buses. However, the first operational experiences showed that electric buses are very sensitive to congestion and the cost of their operation may be higher than that of trolleybuses, even including the maintenance of the catenary. Therefore, the decision was made to introduce limited operation of electric buses only on bus lines.

4.5 Moscow

Moscow is the largest city in Europe that has had trolleybuses in recent years. At the same time, for last decades until 2015, it was the largest trolleybus system in the world, both in terms of the number of lines, the length of the traction network, and the number of vehicles. Currently, the largest system exists in Kyiv in terms of the number of lines, and in Minsk in terms of the number of lines and vehicles. It is also significant that the regression of trolleybus transport in the world in the late 1950s began with the liquidation of the then-largest trolleybus network in London. The case of Moscow is special because trolleybus transport is enjoying a renaissance all over the world. Most of the systems are being developed, modernised, and even new ones are created, e.g. in Italy. Moscow took a different path, in a very short period eliminating the entire trolleybus system and replacing it with electric buses.

Trolleybus transport had been a distinguishing feature of the Russian capital for almost 90 years. The first trolleybuses hit the streets of Moscow in 1933. This system was developing almost all the time. In 2014, over 100 routes with a length of over 600 km were already in operation. There were nine depots, servicing almost 1600 trolleybuses, including 750 low-floor ones. On August 25, 2020, trolleybus traffic was completely stopped, and electric and diesel buses were put on the trolleybus routes. Since September 4, 2020, a special museum line "T" has been operating.

The emergence of trolleybus transport was initially related to the replacement of tramlines. The dynamic development of the network took place in the 1950s. From the 1970s, spatial development was stopped. The extensive network of connections serving the main arteries of the city suffered from particular difficulties in functioning along with the development of individual motorisation since the beginning of the 1990s. Moscow is one of the most crowded and congested cities in the world (Gehl 2013; Trubina 2020). In the mid-1990s, trolleybuses in Moscow transported over 1.5 million passengers a year. This value decreased in the following years. In 2010, only over 300 million passengers were transported in trolleybus transport. The first ideas for the elimination of trolleybuses in Moscow appeared at the end of 2010 and were related to Sergei Sobyanin starting the office of Moscow mayor. In 2014, local authorities announced the reduction of the network along with the liquidation of sections in the city centre. It was then that the reorganisation of public transport in Moscow was planned and the existing trolleybus routes were to be operated by buses, both diesel and electric one. As a result of the actions taken, the purchases of new vehicles were reduced. The project to eliminate trolleybus transport was criticised by experts and residents. In January 2017, a rally in defence of trolleybuses took place (Muleev 2016; Vereŝagina 2016). Since 2016, excess trolleybuses have been transferred to other Russian cities. Since 2016, 440 vehicles have been delivered. In 2020, the delivery of another 366 trolleybuses is planned. In 2018, 200 electric buses were ordered, which meant the end of trolleybus transport. A total of 600 electric buses are planned to be put into service in Moscow by the end of 2019. In fact, only 100 electric buses had been delivered by May 2019. In the summer of 2018, a decision was made to close the trolleybus system by the end of 2020. In the spring of 2019, a massive disassembly of the traction infrastructure began. On August 25, 2020, the last trolleybus serving the M4 line left the depot, thus terminating regular services in Moscow.

The decision of the Moscow authorities must be viewed on three levels. Firstly, the argument about the difficulties in the operation of trolleybuses that used the same streets as other road users, and thus were delayed by congestion, is true. However, it is no different to operate the same routes by buses, which are also delayed by congestion. Secondly, it seems that the decision to eliminate trolleybus transport was political and resulted only from the reluctance of the local authorities to this mean of transport. Electric buses are an alternative that does not require traction infrastructure but requires chargers and adequate facilities, which also entails significant investments. Thirdly, the battery technology is still imperfect, and it is difficult to predict what their service life will be, especially under the heavy working regime in crowded Moscow. A solution commonly used all over the world are trolleybuses with a hybrid power supply, traction, and battery system, which allows to combine the advantages of the trolleybus and the electric bus. The Moscow authorities did not see such arguments, and the decision to dismantle the world's largest trolleybus system may have many years of consequences.

4.6 Vinnytsia

Vinnytsia is an example of a local government that pursues a different policy in the field of trolleybus transport compared to the cities described above. It is a large city with over 370,000 residents, which is located in the central-western part of Ukraine. It has a public transport system, which includes trolleybuses, as well as buses, minibuses, and trams. Trolleybus transport was launched in Vinnytsia as a result of the need to connect two industrial areas with the city centre. It officially started its operation in 1964. In the following years, until the declaration of independence of Ukraine, the trolleybus network was being expanded, and the trolleybus fleet increased to over 180 vehicles. In 1994, a record transport was achieved, amounting to less than 95 million passengers during the year. In the following years, due to the difficult economic situation of the country, the situation of trolleybus transport deteriorated, although until 1998, there were regular investments in infrastructure, and new routes were put into operation. The number of transported passengers decreased by approximately 20 million. However, the several years' collapse was quickly resolved. In 2006, another new trolleybus route was commissioned, and the modernisation of the rolling stock began, first, by modernising the existing trolleybuses and then (2006) by purchasing new low-floor LAZ trolleybuses. In 2012–2017, most of the older ZiU trolleybuses were comprehensively modernised. At the same time, investments were made in new rolling stock. In 2001–2003, two YuMZ trolleybuses and one ZiU trolleybus were purchased. Serious rolling stock investments started a little later. In 2006, 5 low-floor LAZ trolleybuses and 11 ZiU trolleybuses were put into operation, and in 2014–40 Bogdan low-floor trolleybuses. Although no new vehicles were purchased from 1993 to 2001, the condition of the rolling stock, thanks to the conducted renovations, was relatively good compared to the situation in other trolleybus systems in Ukraine.

The next chapter in the history of Vinnytsia trolleybuses began in 2018 when activities related to the production of trolleybuses were undertaken in its own workshops. The Belarusian MAZ body was selected for assembly, on the basis of which the trolleybus was built under the name PTS-12 VinLine. By mid-2020, 6 trolleybuses of this type had been built. Thanks to the assembly of new vehicles with the use of the depot's workshop facilities, costs were reduced and, at the same time, employment was provided. This type of policy distinguishes Vinnytsia from most Ukrainian cities and is an example of good practice that allows for the renewal of the rolling stock while reducing costs. An additional attribute of the new vehicles is the possibility of driving on battery power for a distance of up to 20 km. Thanks to this solution, extension of selected trolleybus lines is planned in areas with smaller passenger flows. Nevertheless, the construction of traction network sections is still planned.

5 Discussion and Conclusions

Trolleybus transport remains an essential part of urban transport in Eastern European countries. The underdevelopment of the automotive industry during the communist period prevented the liquidation of urban electric transport, which was the case in the developed countries of Western Europe and the United States, and today, due to climate change, it is perfectly suitable for use and development as non-emission at the place of operation. In the three analysed countries, Belarus, Russia and Ukraine, there are nearly 50% of all trolleybus systems in the world. Extensive experience in the operation of trolleybuses is now an important heritage and allows planning sustainable development in the field of urban transport.

The article analyses the condition of trolleybus transport in three Eastern European countries in order to define the functioning transport policies in relation to trolleybuses. A similar economic situation, as well as socio-cultural and legal conditions led to the assumption that in most cities the situation is similar. However, thanks to an in-depth analysis, five distinctive patterns of conduct in the functioning of trolleybus transport were identified, which are described in Table 2.

On the basis of the developed functioning models, all trolleybus transport systems in Belarus, Russia and Ukraine existing in the years 1990–2020 were classified (Table 3). Among all the trolleybus systems existing in the analysed period, 6 were classified as model I. These are only Russian cities where the process of making decisions about and eliminating trolleybus transport was very quick. In Model II, 19 Russian and Ukrainian cities were classified, including 4 Russian and

Table 2 Trolleybus transport functioning models in Eastern Europe

	Functioning model	Characteristic	Model city
I	Dynamic liquidation	Closing trolleybus services in the short term. Quick disassembly of traction infrastructure. Replacing trolleybuses with electric buses without appropriate economic and technical analyses. Politically dictated decisions	Moscow
II	Slow liquidation	Slow decommissioning of trolleybus transport by closing subsequent lines and the lack of rolling stock and infrastructural investments	Kaliningrad
III	Save by all means	An attempt to maintain trolleybus transport, despite the lack of adequate financial resources, e.g. rescuing the rolling stock park by purchasing used trolleybuses from other cities	Lutsk
IV	One-time renewal of the entire system	Slow degradation of trolleybus transport. Lack of rolling stock investments at a satisfactory level until the trolleybus transport is renewed thanks to large projects financed from external funds	Lviv
V	Regular investments (I)	Despite economic difficulties, regular spatial and rolling stock development, e.g. replacement of trolleybuses with new ones—built with the use of own facilities to reduce costs. Introduction of new trolleybuses with batteries, etc	Vinnytsia
	Regular investments (II)	Regular spatial development and rolling stock investments. Adaptation of trolleybus transport to the needs of disabled people. Despite the possibility of introducing electric buses, limited to a small number. The perspective of using trolleybuses to an even greater extent	Minsk

Table 3 Classification of trolleybus networks in Belarus, Russia and Ukraine according to functioning models in the years 1990–2020 (bold: liquidated trolleybus systems)

Functioning models	State	Trolleybus networks	Number of trolleybus networks
I	RUS	**Moscow, Perm, Tyumen, Lipetsk, Arkhangelsk, Blagoveschensk**	6
II	RUS	Kamensk-Uralsky, Astrakhan, Belgorod, Tver, Kaliningrad, **Shakhty, Vladikavkaz, Kurgan, Syzran**	9
	UKR	Alchevsk, Antratsyt, **Dobropillia**, Horlivka, **Kadiyivka**, Krasnodon, Lisichansk, **Perevalsk, Toreck, Vuhlehirsk**	10
III	RUS	Chelyabinsk, Dsershink, Engels, Ivanovo, Izhevsk, Leninsk-Kuznetskij, Novosibirsk, Oryol, Penza, Petrozavodks, Miass, Smolensk, Taganrog, Ujanovsk, Velikiy Novgorod, Vladivostok, Vologda, Yakaterinburg, Yoshkar-Ola	20
	UKR	Bakhmut, Bila Tserkva, Cherkasy, Chernivtsi, Donetsk, Khartsyzk, Luhansk, Lutsk, Makiivka, Mykolaiv, Rivne, Severodonetsk, Ternopil	13
IV	RUS	Armavir, Balakovo, Ckerkassk, Vladimir	4
	UKR	Lviv	1
V	BLR	Babruysk, Brest, Hrodna, Gomel, Minsk, Mogilev, Vitebsk	7
	RUS	Abakan, Almetyev, Barnaul, Beresniki, Bratsk, Brjansk, Cheboksary, Chita, Irkutsk, Kaluga, Kazan, Kemerovo, Khabarovsk, Khimki, Kirov, Kostroma, Kovrov, Krasnodar, Krasnoyarsk, Kursk, Makhachkala, Maykop, Murmansk, Nalchik, Nizhni Novigrod, Novocheboksarsk, Novokuybyshevsk, Novokuznetsk, Novorossiysk, Omsk, Orenburg, Podolsk, Rostov-na-Donu, Rubzovsk, Ryazan, Rybinsk, Saint Petersburg, Samara, Saransk, Saratov, Stavropol, Sterlitamak, Tambov, Tolyatti, Tomsk, Tula, Ufa, Vidnoye, Volgodonsk, Volgograd, Voronezh, Yaroslav	52
	UKR	Alushta, Chernihiv, Dnipro, Ivano-Frankivsk, Kherson, Kerch, Kharkiv, Khmelnytskyi, Kryvyi Rih, Kyiv, Mariupol, Odessa, Poltava, Sevastopol, Simferopol, Sloviansk, Sumy, Vinnytsia, Yalta, Zaporizhzhia, Zhtytomyr	21

5 Ukrainian cities that have definitely eliminated trolleybuses. Other cities are in danger of decommissioning their systems. The main factors are insufficient financial resources. Model III, according to which, with insufficient financial resources, local authorities and trolleybus operators perform a number of activities to ensure that the trolleybus transport functions properly; 20 Russian and 13 Ukrainian cities were classified here. To some extent, these systems are at risk of decommissioning, but there is a chance that they will survive, for example in Ukrainian cities, thanks to investment loans from the European Bank for Reconstruction and Development and loans from the World Bank. The cities classified as models IV and V are in the best situation. Accordingly, in model IV there are 4 Russian and 1 Ukrainian. In model V, these are 7 Belarusian cities, 52 Russian cities and 21 Ukrainian cities. It should be emphasised that all systems in Belarus were classified only in model V.

When doing research on trolleybus systems in post-communist countries of the Soviet Union, it should be concluded that trolleybuses are still an important means of urban transport. With ruling out political decisions, the situation is either good or very good in most cities. A smaller part of trolleybus networks is at risk of closure. First of all, the factors influencing such a situation are the bad financial situation of the city and strong competition from private transport (marshrutkas). Trolleybus transport, due to its scale of popularity in Belarus, Russia and Ukraine, can play an important role in environmental and climate protection. If decisions on its operation are dictated by social, environmental, economic and technical issues, trolleybus transport will still be able to function and develop. When decisions are dictated by politicians' subjective opinions, then electric transport, which is more expensive to maintain than buses and minibuses, is at risk.

Acknowledgements The research was conducted and funded by the National Science Center—decision number: UMO-2016/23/D/HS4/03085.

References

Alexandrova A, Hamilton E, Kuznetsova P (2004) Housing and public services in a medium-sized Russian City: case study of Tomsk. Eurasian Geogr Econ 45(2):114–133

Barbosa F (2016) Modern trolleybus systems as a technological option for greening bus corridors—a technical economical assessment. SAE Technical Paper Technical Paper 2016-36-0177

Bezruchonak A (2019) Geographic features of zero-emissions urban mobility: the case of electric buses in Europe and Belarus. Eur Spat Res Policy 26(1):81–99

Верещагина Л (2016) Без «Б»: Обречён ли московский троллейбус. The Village. https://www.the-village.ru/city/ustory/236457-moskovskiy-trolleybus [Vereŝagina L (2016) Bez «B»: Obrečën li moskovskij trollejbus. The Village]. Last accessed 28 June 2020

Blinkin M, Muleev E (2016) Russian cities mobility culture: international comparison. In: Blinkin M, Koncheva E (eds) Transport systems of Russian Cities. Transportation Research Economics and Policy, Springer, Cham, pp 259–272

Богодистый ПА, Збарский ЛВ, Палант АЮ (2017) Троллейбусы Украины. Золотые страницы, Харков:1–480. [Bogodistyy PA, Zbarskiy LV, Palant AYU (2017) Trolleybusy Ukrainy. Zolot'ie stranits'i, KHarkov:1–480.]

Borowik L, Cywiński A (2016) Modernization of a trolleybus line system in Tychy as an example of eco-efficient initiative towards a sustainable transport system. J Clean Prod 117:188–198

Dmitriiev I, Shevchenko I (2017) Problems and prospects of development of the automotive industry in Ukraine. Sci j Pol Univer 20(1):11–23

Gehl J (2013) Moscow on the way to the city for people: Public spaces and public life. Institute of the General Plan of Moscow, Moscow, pp 1–128

Городской электротранспорт. https://transphoto.org/ [Gorodskoj èlektrotransport]. Last accessed 30 June 2020

Hutyria S, Chanchin A, Yaglinskyi V, Khomiak Y, Popov V (2020) Evolution of trolley-bus: directions, indicators, trends. Diagnostyka 21(1):11–26

Khairullina E, Santos Y, Ganges L (2020) Tram, trolleybus and bus services in Eastern-European socialist urban planning: case studies of Magdeburg, Ostrava and Oryol (1950s and 1960s). J Trans Hist 1–32

Kuczabski A, Połom M (2019) Wpływ organizacji mistrzostw piłkarskich EURO 2012 na zmiany w transporcie publicznym Lwowa. Prace Komisji Geografii Komunikacji PTG 22(4):46–58

Mjakota D (2010) The analysis of the urban electric passenger transport condition in republic of belarus. BNTU Minsk, pp 28–30

Мулеев О (2016) Остановка по требованию: зачем убирают троллейбусы в Москве?, РБК. https://www.rbc.ru/opinions/society/13/04/2016/570def3b9a79471cd79d25fc [Muleev O (2016) Ostanovka po trebovaniû: začem ubiraût trollejbusy v Moskve?, RBK.]. Last accessed 28 June 2020

Muleev E (2020) Why do marshrutkas exist in one city and not in others? Toward a political economy of routes in Russian Urban Public Transportation. J Eco Sociol 21(2):99–112

Oh JE, Gwilliam K (2013) Review of the urban transport sector in the Russian Federation. Transition to Long-Term Sustainability. The World Bank, Washington, pp 1–36

Olander A, Glinsky S (2005) Trolleybus production in Russia, Belarus and Ukraine. Trolleybus Magazine 41(260):26–32

Połom M (2016) Międzynarodowe powiązania na rynku producentów trolejbusów w Europie w latach 2000–2014. Prace Komisji Geografii Przemysłu Polskiego Towarzystwa Geograficznego 30(3):75–90

Połom M (2018) Trends in the development of trolleybus transport in Poland at the end of the second decade of the 21st century. Prace Komisji Geografii Komunikacji PTG 19(4):44–59

Rudakewych I (2017) Współczesne tendencje rozwoju komunikacji trolejbusowej w dużych miastach zachodniej Ukrainy. Prace Komisji Geografii Komunikacji PTG 20(2):19–30

Rudakewych I, Sitek S, Soczówka A (2017) Transformations of urban electric transport in Ukraine after 1991 in the view of transport policy. Eur Spat Res Policy 26(1):61–80

Ryzhkov A (2018) Local public transport in Russia: regulation, ownership and competition. Res Transp Econ 69:207–217

Ryzkov A, Zyuzin P (2016) Urban public transport development in Russia: trends and reforms. Higher School of Economics Research Paper No. WP BRP 05/URB/2016

Sanina A (2011) The marshrutkaas a socio-cultural phenomenon of a Russian megacity. City Cult Soc 2:211–218

Sgibnev W (2014) Urban Public Transport and the State in Post-Soviet Central Asia. In: Burrell et al. (eds) Mobilities in Socialist and Post-Socialist States. Palgrave Macmillan, Hampshire, pp 194–216

Stepanov P (2019) Characteristics of construction and operation of trolleybus systems in the world. Prace Komisji Geografii Komunikacji PTG 22(3):64–72

Sustainable Mobility for Odessa (2018) A road map for improving accessibility and Energy efficiency (2018) The World Bank, Washington, pp 1–177

Tarkhov S (2000) Empire of the trolleybus. Volume 1—Russia. Rapid Transit Publications. London.

Тархов С, Козлов К, Оландер А (2010) Електротранспорт України. Енциклопедичний путівник. Варто, Київ:1–864. [Tarkhov S, Kozlov K, Olander A (2010) Elektrotransport Ukrayiny. Entsyklopedychnyy putivnyk. Varto, Kyyiv:1–864.]

Tica S, Filipović S, Živanović P, Bajčetić P (2011) Development of trolleybus passenger transport subsystems in terms of sustainable development and quality of life in cities. Int J Traffic Trans Eng 1(4):196–205

Trolley:motion. Urban e-mobility. https://www.trolleymotion.eu/. Last Accessed 30 June 2020

Trubina E (2020) Sidewalk fix, elite maneuvering and improvement sensibilities: The urban improvement campaign in Moscow. J Trans Geogr 83:102655

Vozyanov A (2017) Urban electric public transport in eastern and southeastern Europe. Toward a historical anthropology of infrastructural crises. Mobility in History 8:63–76

Doctor Marcin Połom is master's degree in Geography (2008) and Ph.D. in Transport Geography (2016). He is currently Assistant Professor at the University of Gdańsk, Poland. His current research interests sustainability in urban transport. He conducts research on the development of electromobility in public transport, in particular regarding trolleybus transport and electric buses.

Legal Instruments for Business Communities

Business Communities as a Tool for Sustainable Development

Rafael Gustavo de Lima and Samara da Silva Neiva

Abstract The Sustainable Development Goals proposed by the United Nations after 2015, contributes to the formation of an international sustainability regime, evidencing the capacity to build internationally shared frames of reference for the theme of sustainability. The existence and sharing of networks, rules and guidelines for monitoring international practices highlights the formation of an international sustainability regime based on the 17 goals of the SDGs. In this sense, the work proposes to evaluate the SDGs as a practical product of a global effort for sustainability, based on the theory of International Regimes. Thus, reveals how the main international regimes are based, especially, in business communities that prioritize sustainable economic and environmental relations. In this same sense, the work explain how international legal regulation is justified with such relations, to keep it cohesive and effective for all members of the regime.

Keywords International regulation · Business communities · Sustainable Development Goals · Environmental and economic relations

1 Introduction

The fulfillment of the objectives that permeate the theme of the development of sustainability is an extremely high priority for the entire business community. The United Nations 2030 Agenda in 2015 confirmed the creation of 17 Sustainable Development Goals that pose new challenges for companies, which need to adjust their operations and strategies to the requirements of the SDGs (Tsalis et al. 2020). Based on this, this chapter aims to answer the research question, on how a business communities could contribute for the sustainable development implementation?, since they

---

R. G. de Lima (✉)
Federal University of Santa Catarina, Des. Vítor Lima Street, Florianopolis, Carvoeira 222, Brazil
e-mail: rafael.lima@ufsc.br

S. da Silva Neiva
University of Southern of Santa Catarina, Adolfo Melo Street, 14, Florianopolis, Centro, Brazil
e-mail: samara.neiva@unisul.br

act as major drivers for the implementation of sustainable development objectives on a scale global.

The method used for the development of the research was a systematic review of the literature, with the aim of finding both modern and contemporary authors who address issues related to international regimes, the implementation of the objectives of sustainable development, the business community, presenting as this relationship can be made, and as business communities can occur in different ways, to stop and find the gap that still exists in the literature so that in future studies they can be addressed.

In this sense, the first section theoretically addresses the fundamentals of international regimes, so that is possible to consider as the existence and sharing of networks, rules and guidelines for monitoring international practices shows the formation of an international sustainability regime based on the 17 axes of the SDGs. After are observed the Sustainable Development Goals (SDGs), since its conception to the pretend results. In sequence the theoretical fundamentals of business communities are considered. Lastly, the chapter list practical examples of how international regulation of environmental and economic relations of business communities contribute to the promotion of sustainability in a global scale.

2 Methods

In this chapter the methods were a literature review organized using the scoping proposed by Arksey and O'Malley (2005), and further corroborated by Levac et al. (2010). This method was selected, because the objective of the paper was to examine the literature to answer a specific question. Considering the array of themes involved in Sustainable Development and that this is an emerging field of research, a scoping study is appropriate. Therefore, this review involved the following five stages of a scoping study (Arksey and O'Malley 2005; Levac et al. 2010).

- **Stage 1: Identifying the research question**
  The research question formulated was: how a business community could contribute for the sustainable development implementation? This research question is further addressed in the next section.
- **Stage 2: Identifying relevant studies**
  In this stage, the search for studies in the literature was based on the Scopus, Science Direct, Web of Science and Google Scholar databases. Due to their similarities, Science Direct and Scopus were analyzed with the same criteria, while Web of Science and Google Scholar were analyzed individually. As in systematic literature reviews, the study adopted inclusion and exclusion criteria, favoring breadth rather than depth in the selection. Thus, the Boolean expression was used in this study (Business Communities AND Sustainable Development AND International Legal Regimes). The Boolean expression was used to map papers that address the relations between the themes.

Regarding the inclusion and exclusion criteria considered for Science Direct and Scopus databases, these were a scientific publication (peer reviewed papers published in journals), written in English, with the Boolean expression appearing in the "Title, Abstract, Keywords". For the inclusion and exclusion criteria, due to the limitations of the Google Scholar database in this regard, "citations" and "patents" were eliminated, and the "any time" option was selected for the period of analysis.

- **Stage 3: Study selection**
  After searching the databases for pre-selecting, the material to be studied, the process of organizing the data obtained was carried out, first excluding those documents that could be duplicated due to the possibility of publication in different databases.
- **Stage 4: Charting the data**
  After selecting the material to be analyzed, Excel software was used to organize the data on the selected papers. These data were organized in columns by year, publication title, authors' names, journal title, number of citations in Google Scholar.
- **Stage 5: Collating, summarizing, and reporting the results (analysis and recommendations)**
  The articles that were selected supported the literature review, providing concepts and definitions associated with the theme, as well as the dimensions it presents. This one step was aligned with a qualitative analysis of the selected literature, identifying elements that support the categorization of recommendations for policies, research and development and practices (as suggested by Arksey and O'Malley 2005; Levac et al. 2010). To further corroborate the theme and present a more detailed view, when dealing with data analysis, some practical examples from business communities were added.

3 Literature Review

3.1 International Regimes

The Sustainable Development objectives proposed by the United Nations build, after 2015, an agenda that contributes to the formation of an international sustainability regime, highlighting the capacity to build internationally shared frames of reference for the theme of sustainability. The existence and sharing of networks, rules, and guidelines for monitoring international practices shows the formation of an international sustainability regime based on the 17 axes of the SDGs. In this sense, the work proposes to evaluate the SDGs as a practical product of a global effort for sustainability, based on the theory of International Regimes. When considering the formation of an international regime geared towards sustainability, the instruments of the traditional theory of international regimes need to be considered. One of the main authors on the subject is Krasner (1982), who points out that international regimes can be defined as principles, norms, rules, both implicit and explicit, and decision-making procedures in certain areas that belong to international relations.

They evaluate the regimes as a set of attributes disseminated by the international system and understand that no pattern of behavior can be sustained for any period of time without creating a compatible regime, so that regimes and behaviors "are inexorably interwoven" (Krasner 1982). On the other hand, the authors Breitmeier et al. (2006), state that international regimes can be considered as social institutions that were created with the purpose of answering specific questions in a social environment that is anarchic and has no centralized public authority. It can also be pointed out that the change from one international regime to another undergoes changes both in its principles and in the norms. It is important to highlight that the participants of an international regime can be more than just States, there is a range of actors that seek to realize their interests in new conformations of behaviors and conducts such as international networks that are related to the protection of the environment.

Figure 1 outlines the three views mentioned above and facilitates the understanding of the feedback movement between regimes and behaviors that the third view allows in relation to the other two views.

While for Stone (2008) the criticism of the theory of regimes, he systematizes the central elements for possible complications for international society in reaching common goals for a group or even for the whole set of actors. The first complication may be present in the difficulty of cooperation needed for a regime that imposes on authors the search for public goods through the coordination of their actions. Another complication involves the delegation of normative tasks to third parties, which reflects the difficulty of States in using their control prerogatives in favor of institutions, non-state actors or international networks that make up a regime.

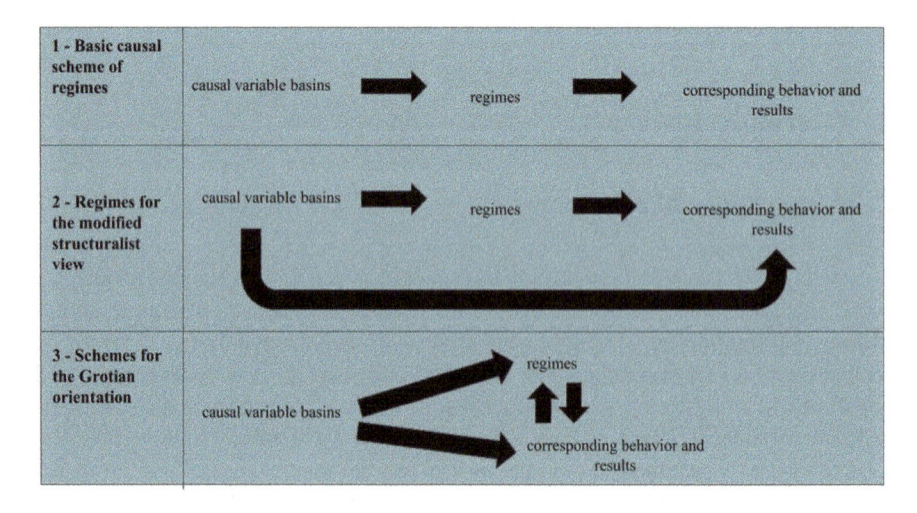

Fig. 1 Schemes of regimes. *Source* Elaborated by the authors based on Krasner 1982, pp 96–99

However, even though susceptible to these complications, an international sustainability regime information shows signs of being able to deal with these difficulties, even if it does not meet the speed expected by environmentalists, civil society and development theorists in general. That is, even with the difficulty of preserving the public good required by all, in this case the collective wellbeing of all the actors involved, States in general conduct discussions in partnership with civil society entities of national or international representation, in a to remove the complication of delegation to some extent. Authors such as Stone (2008) and also Milner (2009) cite the work developed by Keohane (and Nye) several times to develop their analyzes, mainly with regard to the idea of complex interdependence and issues related to the maintenance of international regimes. In this sense, but more concerned with the demand for international regimes and the dynamics that States, and non-governmental actors have in their relations, Keohane (1982) seeks to contextualize the need, strength and extent of international regimes.

Thus, it can be considered that the international regime when it comes to sustainability admits constant revisions and improvements. This permeability Keohane (1982) explores the importance of regimes in view of the simple conformation of international agreements restricted to one or the other actor. Based on approximations with economic theory, the author assesses aspects such as (1) the lack of an established legal framework, (2) the imperfection (cost) of the information; (3) and the positive transaction cost between the actors suggest the need (demand) for international regimes, in order to somehow discipline the behavior of the actors involved and to generate an environment of predictable behaviors. In addition to the influences on the regimes, on the change/substitution of one international regime for another (unlike simple adaptations).

Keohane and Nye (1988) mention that these substitutions can start from typically economic-technological processes in which norms and established procedures are ineffective, outdated or no longer of a regulatory nature. Thus, international regimes must always undergo a careful analysis, in order to identify their real purposes in each specific situation, in order to identify whether it is a regime or the propagation of an ideology induced by one or some actors most powerful in the system. Finally, Keohane and Nye (1988) conclude that international regimes may arise from international agreements and treaties, they may be an expression of the evolution of part of these instruments or they may also be implicit in the relationship between two or more actors (which is commonplace among actors who contribute to the formation of an international sustainability regime). Based on this, it is still important to present a brief historical concept on the emergence of sustainable development objectives that serve as the basis for the construction of this chapter, and to understand how this can be related within a business community.

3.2 UN Sustainable Development Goals

The Sustainable Development Goals (SDGs), created by the United Nations post-2015 agenda, were developed in the wake of events of global relevance. Thus, the SDGs materialize, after 2015, an agenda that contributes to the formation of an international sustainability regime, highlighting the capacity to build internationally shared frames of reference for the theme of sustainable development (Lima et al. 2016). International conferences such as Stockholm 72, Eco 92 and Rio + 20 helped to consolidate an inclusive and transparent intergovernmental process, open to all interested parties and with a view to creating global sustainable development goals to be approved by the UN General Assembly (UN-Sustainable Development Knowledge Platform 2020).

In 1983 the Commission on Environment and Development—Brundtland Commission was created, responsible for preparing a report on the importance of development and how it should be carried out (Duran et al 2015). The Commission published a report entitled "Our Common Future", which defined the concept of sustainable development, characterized as the ability to meet the needs of the present without compromising the ability of future generations to meet their needs without limitations (Sinakou et al. 2018; Buckingham and Kina 2015; Gardner 1989; Yang et al. 2017; Laine 2005). In addition, the report indicates that sustainable development is no longer a goal only for developing countries and becomes a goal common to all countries, shaping the international sustainability regime, as international society starts to act in a favorable way guaranteeing the needs for human life (United Nations 2020; Bond and Morrison-Saunders 2011; Glavič and Lukman 2007; Holden et al. 2014).

In this sense, four primary dimensions to four secondary dimensions on sustainable development were also presented. The primary dimensions, framed as fundamental and non-subjective objectives, cannot be negotiable and aim to safeguard ecological sustainability in the long term, satisfy basic human needs and promote intra-generational solidarity and inter-generational equity. In addition, secondary dimensions include the intrinsic value of nature, the promotion of environmental preservation and the increase of public participation in improving the quality of life (Holden et al. 2014). As the world's first report on sustainable development, the Brundtland Report presented the essential objectives for development and environment policies that highlighted the need for: 1. Resumption of development; 2. Adaptation to sustainable growth; 3. Meeting the basic needs of human beings; 4. Ensuring a sustainable population level; 5. Conservation and improvement of essential resources for maintaining life; 6. Reorientation of technologies for risk management; and 7. Ensuring synergy in decisions that consider the environment and the global economy (Neiva 2019).

This last point is highly related to the theme addressed in this research and allows us to observe that there is no way to consider the synergy of practical actions between sustainable economic and environmental relations without the establishment of an institutional framework for legal regulation that can remain cohesive and effective for all members and in all business communities under the same international sustainability regime (Neiva et al. 2020; Lima et al. 2020). In this sense and counting on the support and broad discussion of organized civil society across the planet, international institutions and permanent global forums, the United Nations on its 70th anniversary, launched, in September 2015, the Sustainable Development Goals.

Structured in 17 axes, the SDGs comprise a universal agenda that seeks, by 2030, 169 specific goals and 303 indicators that demonstrate the scale and ambition of a new Universal Agenda, based on an international regime structured and shared with all the countries that make up the SDGs. United Nations (United Nations Development Programme 2018; Ferranti 2018; Fitchett and Atun 2014; Hák et al. 2016). The SDGs build on the legacy of the Millennium Development Goals and conclude what they have failed to achieve, taking up issues that are sensitive to the majority of the global group of countries such as human rights, gender equality and human empowerment. Thus, integrated and indivisible, the SDGs appear to balance the three dimensions of sustainable development: economic, social and environmental in an institutional arrangement clear and organized global strategy with a central focus on the United Nations (UN 2018; van Vuuren et al. 2015; Persson et al. 2016; Sofeska 2016; Hui 2001).

Thus, the existence and sharing of networks, rules and guidelines for monitoring international practices shows the formation of an international sustainability regime based on the 17 axes of the SDGs, which points to a global effort related to sustainability (Lima 2014). In this sense, the concern of the United Nations in stimulating, in different ways, sustainable development in a comprehensive way is evident. In seeking to understand a universal agenda, there was a movement to contemplate all segments organized in contrast to a policy of choosing a specific area and allocating all possible efforts to it (Fitchett and Atun 2014; Ait-Kadi 2016; Pedersen 2018; Kanter et al. 2016). Thus, while respecting the need for each axis, everyone's effectiveness is sought through goals that unfold in the measurement and materialization of practical, sustainable products and actions, extrapolating the simple theoretical exercise of the mere "objective" (Lima 2014; Ait-Kadi 2016; Wulf et al. 2018).

Thus, by considering the theoretical premises of the institutional establishment of an international sustainability regime, in addition to also considering the globally shared expectations of the SDGs for the global community of countries in addressing the common needs for social, economic and environmental management of resources, if, with the next sessions to the understanding of the business community's as practical promoters of such theoretical reality, especially within the scope of their international legal regulation of environmental and economic relations.

3.3 Business Community

A business community be those that come together and create community businesses to face the challenges they face together. There are many types of community businesses, including stores, farms, pubs and call centers. What they all have in common is that they are accountable to the community and that the profits they generate provide a positive local impact, such as boosting the local economy. Acting in a remarkably similar way to social enterprises, community companies are committed to positively benefiting society through trade in a sustainable manner. One of its main premises is the fact that all the profit from a community business is reinvested in the local area. Unlike social enterprises, community businesses are focused on benefiting a specific local geographic area (Lee et al. 2014).

Another point that needs to be considered is that community businesses also have similarities with location-based charities that manage assets. However, a community business is accountable to the beneficiary community, which can mean that local people are involved in formal participation or even legal ownership. A conceptual definition of 'Business Community', that is, an organization of companies that share a common business area and interact to conduct business transactions, which can be instantiated in many different contexts (from Public Administrations to private organizations). A Business Community is a community of heterogeneous and autonomous users, Business Operators, connected by a common business (Bhattacharyya et al. 2012).

A Business Community can be described as a set of companies, that present as main points: Share a common market sector (such as cargo transportation, in our case); Maintain existing mutual business relationships; They are heterogeneous in relation to their functions, that is, they are involved in different activities within the same market (different part of the product/service chain); They are heterogeneous in terms of their size and level of information technology, that is, they vary from small professional offices to large companies; They are coordinated and supported by a Business Authority (Baglietto 2005).

Some authors such as Ramchandran et al. (2016) present some examples of business communities, when addressing in their study the concept of "micro-networks" as self-sustaining energy systems is driven largely by business models, which must be developed considering the perspectives of all stakeholders, where telecommunications towers, flour mills and rural communities were considered as anchor, commercial and community customers, respectively. To approach the research, the authors used primary surveys to evaluate socioeconomic characteristics, availability of renewable energy resources and energy demand.

Based on this, you can point out that the business community has several characteristics such as: Pervasively: all operators in the Business Community must be able to access the services regardless of the characteristics of their infrastructures (for example, network numbering, firewall policies etc.); Interoperability: services must allow communication not only between human operators, but also between corporate information systems, allowing the integration of applications for interactions between companies; this interaction must occur regardless of the basic software, middleware and application adopted; Openness: operators in the business community must be able to interact with each other in the presence of different technological gaps and different types and organizations of internal systems (Baglietto 2005).

Extensibility: Business Community partners must be able to update data exchange formats and software interoperability protocols with quick, low-cost updates, even in the presence of radical innovations in application domains and technological scenarios; Security: business community partners must be able to exchange messages or transact with guarantees of authenticity, integrity and confidentiality; Service provider independence (dismemberment): operators in the business community should not be forced or obliged to use a unique or privileged service provider to access network services and applications; in other words, support for operators in the business community should be allowed to more than a single actor (Baglietto 2005).

The concept of community business is causally linked to the notion of community ownership and to more widely cooperative ownership models. That in his History of Ownership of Community Assets, Steve Wyler argued that community ownership represents a series of English socio-political thoughts and activism that can be traced back to the progressive removal of commons from the Norman conquest and later, where he periodically revolts, based on this, it can be said that access to common resources was canceled by the state, usually with punitive measures against the demonstrators and the districts from which they had come. At the same time, the notion of recreating common access to wealth has fueled the concept of "community" that animated the thinking of Robert Kett, Gerard Winstanley, the Levellers, the Diggers and the Ranters (Ramchandran et al. 2016).

Having failed to prevent the legal privatization of resources concluded by the Acts of Inclusion, subsequent efforts have focused on the development of intentional communities defined by common control or ownership of assets, mainly land, such as Owenite models, the Chartist National Land Company of Feargus O'Connor and more successfully, Ebenezer Howard's cooperative retail movement and garden cities. While in Scotland, communities that were being threatened by slum deforestation, came together and organized reforms and property protests, the community led actions against absentee landlords, encouraging activism to acquire highlands and islands to belong to the community.

These efforts were driven by the approval of the 2003 Land Reform Act in Scotland, which gave community organizations the right of first refusal on land for sale and the right to buy when it reached an independently assessed purchase price for certain areas of cultivation. Another example that can be analyzed for building the business community as we know it today was the development of railway preservation. It was an early example of community business in the post-war period, mixing volunteers with paid employees and negotiating with the income of passengers (Adler 2002).

In an increasingly interconnected, complex, and turbulent world, businesses are sailing in unknown waters. During this uncertainty, the global community came together in a call for global action to guide all stakeholders—including businesses—in building a more sustainable, equitable and inclusive society. Although the Sustainable Development Goals (SDGs) have been designed and approved by governments, they also constitute a global framework for measuring business contributions to society (Birch 2003).

Authors such as Berry (2020), present in their studies that due to the sustained interest of the business community, it can occur in several spheres such as, for example, commercial agriculture in the global South, it was welcomed for its potential to bring capital to rural areas neglected, but also raised concerns about the implications for customary land rights and the terms of integrating local land and labor in rural areas. global supply chains. In global development policy and discourse, the concept of "inclusive business" has become central to efforts to resolve these tensions, with the idea that integrating small farmers and other disadvantaged actors in partnerships with agribusiness companies can generate benefits for economies nationals, private investors and local livelihoods. The academic treatment of the topic tends to be polarized in profit/loss narratives or points to the contingency and social differentiation of localized experiences. The conclusions point to the need to rethink the political choices behind these trends and how we implement the fiscal, legislative, and state guard functions to shape agrarian trajectories.

In the current environment of growing social and economic problems in society, it is particularly important to increase the level of social responsibility in the business community and strengthen its institutional structure. The subject of the study is the organizational and economic relations that arise when companies and economic entities interact to implement social responsibility (Milchakova and Reshetnikov 2020). A new paradigm for innovative companies, business community, state and local governments and other organizations to design and implement clean energy programs is taking shape. Design can be characterized as a process to break with standard practice, as it is evident that standard practice results in significant conflicts with economic, environmental, social or other values (German et al. 2020).

4 The Business Community Practice

When considering a business community as a group of companies that share a common business area and interact to carry out public and private commercial transactions, it is important to assess that building networks for sustainability relies on the practice of changing the ways in which what is already known is produced and processed. Thus, SDGs 17 promote the creation of partnerships and bring stakeholders together to reinvent their methods that reinforce the global partnership for sustainable development. In this way, multisectoral partnerships shape business communities that mobilize and share knowledge, advance joint expertise, foster new technologies and optimize financial resources to support the achievement of sustainable development goals in all countries, particularly in developing countries (UN 2020; SDSN—UN 2020).

In this sense, it is interesting to observe how economic relations are regulated in favor of environmental issues in business communities with special emphasis on the global promotion of sustainable development, in its economic, social, and environmental dimensions. Thus, for the company Trend Hunter, which identifies consumer ideas and investment opportunities for companies that promote innovation in the world, business communities are largely based on the ideas they promote. With 20,000,000 monthly views, TrendHunter.com consolidates itself as a business community oriented to new global trends, using big data, human researchers and artificial intelligence in its analysis processes that evaluate the content of a global network of 200,000 more than 350,000 innovative ideas, in a universe of more than 3 billion options, captured from 150 million people. Thus, the company lists over 23 thousand articles related to innovative projects and ideas in the economic-environmental area, which involve society in segments of ecological architecture, sustainable transport, consumer insights, in addition to trend reports and targeted research (Trend Hunter 2020).

Among the virtual business communities, companies that defend social welfare stand out by developing engaging and collaborative business communities, bringing together work capacities and urban facilities, such as working in the home office. These communities are specifically designed to promote entrepreneurship and interaction, internally and externally. In this sense, new technologies can replace typical offices physically based to encourage creativity, innovation and team building. As these business communities are not limited to the physical domain, their members access online benefits, including outsourcing tasks with the quick and simple exchange of digital jobs and services. Thus, the development of business communities of the future will greatly affect the relationship between economy and society, impacting the use of the environment around them (199Jobs 2020; Fiverr 2020).

Another important contribution considers the theme of markets in collaborative networks for the purposes of solving human resources. Collaboration is the human force that defines the present and the digital transformation promotes that cooperative and creative work can solve business challenges, with positive externalities in the economic, environmental, and social dimensions. Thus, collaborative networked markets, in which companies lean towards opportunities with a community mindset, are increasing (Firebrand 2020). An example of this is the Berlin-based company Jovoto, which suggests the possibility of a collaborative market model, bringing together tens of thousands of creative professionals from global brands and companies to solve the design and innovation challenges of other organizations. Its collaborators win, Jovoto wins and whoever receives wins, with the value defined first, then produced directly by the community members. While target audiences and interconnected groups are a characteristic of networked markets, communities are the scaffolding of a collaborative market (Jovoto 2020).

Another practical experience that exemplifies how economic and environmental relations are altered by business communities is Permaculture, often defined as sustainable agriculture and which seeks to encourage better practices in the use of natural resources, avoiding wasting and stimulating continuous innovation for world development. By seeking to satisfy the human need for nutrients and improving environmental preservation, permaculture reduces the use of non-renewable resources and encourages the integration of different forms of production. Such practice becomes relevant for the achievement of global goals for sustainable development, as is the case of the 2nd United Nations Sustainable Development Goal, since it seeks to end hunger, food insecurity and malnutrition, offering development opportunities for several families (Deggau et al. 2019).

Permaculture is a concept mentioned and developed for the first time by Mollison and Holmgren (1978), two Australian scientists. They wanted to create a philosophy of life, which could be applied to all branches of science, to support sustainable development. The word comes from the combination of "permanent", "agriculture" and, later, "culture", to mean that it is not limited to agriculture, but is a worldview, which concerns all aspects of human settlements (Ferguson and Lovell 2013; Ulut 2008). Previous American and British literature on agricultural practices suggests that the word "permanent" was used in this context with a value equivalent to what "sustainable" would currently mean (Ferguson and Lovell 2013).

The movement's initial focus was to create a sustainable food production system (Haluza-DeLay and Berezan, 2013), having as principles the coherent management of the land and nature, the possession and community of the land with a view to the collective production of food, shared notions of finance and economy, health and spiritual well-being, culture and education, as well as tools, technologies and constructions that observe not only market precepts, but profit, but essentially that respect the environment and promote social well-being (Deggau et al. 2019).

Thus, sustainable farming business communities can be considered an extremely relevant point for any project aimed at reducing poverty or guaranteeing food security, especially in developing countries. According to World Bank data, 86% of the world's poorest population are in developing countries and survive on subsistence agriculture (Adenle et al. 2012). Some authors as Gafsi et al. (2006) state that sustainable agriculture will solve problems and obstacles brought by standard agriculture to the environment, the availability of resources for cultivation, the deterioration of human health resulting from the growing use of pesticides in food production, the desertification of certain areas and the difficulties of small agricultural production to survive.

The concept of food security comprises the harmony of the environmental relations of production with the economic relations of consumption, food supply, a basic factor in the composition of business communities related to sustainable agriculture. Thus, food security translates into the ideal situation when all people have physical, social and economic access to sufficient, safe and nutritious food, with the capacity to meet their needs and food preferences, aiming at a healthy life (FAO 1983). To guarantee food security, there are four main dimensions to be considered: availability, or physical existence and how food is provided from production or cultivation to commercialization; access, or the resources to obtain sufficient quality and quantity for a nutritious diet, always considering economic, cultural and social aspects of food; consumption, which is reflected in health and nutrition conditions; and stability, related to how the food supply remains constant, including factors such as economic resources, income and the food itself (FAO, FIDA, UNICEF, PMA and OMS 2019).

Furthermore, sustainable agriculture is considered important to promote development in all countries, not just in developing countries, since it guides the creation of business communities focused on food production (Xu et al. 2006). Proof of this is that the approach proposed by Permaculture in relation to energy can be seen as an alternative to the current use of oil, seen as harmful to the environment and with limited time, which means that solutions for a post-oil world must be found. For the realization of sustainable agriculture to be possible, governments must participate by promoting education on this subject and extension projects, which can result in opportunities to increase skills and knowledge. When addressing product quality, it becomes possible to assess that the nutritional value reaches levels consistent with food security, while food prices can be maintained at a standard accessible to the public (Veisi et al. 2016).

Thus, the international legal regulation of environmental and economic relations in business communities linked to sustainable agriculture and permaculture point to the importance of supporting organizations and institutions of a local character, so that they become stronger, provide better access to education, help to better use of important resources such as water and energy and enable food producers to know local needs as ways to combat malnutrition (Veisi et al. 2016).

Another practical example of how business communities affect and are affected by the interrelationship between environment and economy lies in the fact that food production has been following the population and, for that, it is largely based on the use of pesticides in crops. If the estimates mentioned are confirmed, it will be more difficult to achieve food security, unless food production increases or population growth is limited (Carvalho 2006). Campos et al. (2018) and Chen et al. (2018) claim that without the use of pesticides, about 70% of food production would be lost since many microorganisms affect production. Although pesticides have increased production capacity, their overuse has caused problems such as contamination of water, air and soil, creation of toxic microorganisms, human poisoning caused by chemicals in food and better resistance of microorganisms to chemicals.

Another example of the transformation of environmental and economic relations by business communities focuses on the construction sector with the World Green Building Council (WorldGBC), a global network that leads the transformation of the "built environment" to make it more healthy and sustainable. Collectively, with the so-called Green Construction Councils (GBCs) in about 70 countries, the entity promotes actions to fulfill the ambition of the Paris Agreement and the UN Global Goals for Sustainable Development. Through a system change approach and with work that comprises several sectors, the objective is to transform the construction industry into a zero-carbon environment and a sustainable built environment. This is because the entity observed that buildings and constructions represent 39% of CO_2 emissions related to energy and the relevance of this work accelerates the reality that all buildings have zero net emissions by 2050 (World Green Building Council 2020).

In addition, there are academic initiatives that aim to accelerate and expand the impact of business communities around the globe through international partnerships, promoting the sustainability regime in the corporate environment. The UN Sustainable Development Solutions Network (SDSN), created in 2012, acts as a global network, mobilizes global scientific and technological knowledge to promote practical solutions for sustainable development, including the implementation of the Sustainable Development Goals (SDGs) and the Climate Agreement from Paris. The SDSN works closely with United Nations agencies, multilateral financial institutions, the private sector, and civil society. In addition, the SDSN is guided by a Leadership Council, which brings together global leaders and much of the work of the SDSN is led by smaller scale networks, at national or regional levels, which mobilize networks of universities, research centers and other institutions knowledge around the SDGs in their local realities, so that theoretical sustainability efforts meet the practical reality of effective actions (SDSN—UN 2020).

Finally, but no less important for business communities is the performance of Networked platforms, sets of networks that connect multiple stakeholders simultaneously and can, by association, produce effects on a geometric scale, in addition to the different branches of activity that it can reach. In this sense, such clusters of partnerships expand the capacity of the networks, their potential and practical actions when they, when mature and well structured, begin to relate to other equally powerful networks. An example of this is the Connecting Business Initiative (CBI), linked to the United Nations Office for the Coordination of Humanitarian Affairs, which provides a model of national networked platforms that involve the private sector in disaster preparedness, response and recovery.

CBI has eleven networks at the national level, which it supports with access to resources, networks, mechanisms and tools that allow companies to contribute effectively, but each platform is executed and managed locally, without direct CBI management, which reinforces the condition of sustainable development in making use of the capabilities acquired by agents in their oriented business communities, in this specific case to the theme of natural environment disasters, with clear repercussions on the economic initiatives with which they are related (UN 2020; Partnership Platforms Report 2020; SDSN—UN 2020).

5 Conclusion

The construction of this chapter and the development of the research aims to reflect how the Sustainable Development objectives proposed by the United Nations build, from 2015, an agenda that shapes the formation of an international sustainability regime, specifically from the construction of business communities. To demonstrate the ability to build internationally shared frames of reference for the topic of sustainability. Through the Theory of International Regimes, the existence and sharing of networks, rules, and guidelines for monitoring international practices that show the formation of an international sustainability regime supported by the 17 axes of the SDGs.

In this approach, international regimes were understood as defined by principles, norms, implicit or explicit rules and decision-making procedures in a certain area of international relations around which the expectations of the actors converge. These factors are defined by the international actors themselves (state or not) as they establish consensus, formalize treaties and conventions, and share common values. Then, steps were taken to consolidate the 17 United Nations Sustainable Development Goals, materialized by the international community when considering globally shared expectations regarding the treatment of common needs for social, economic, and environmental resource management.

Then, the work presented some premises about the objectives of sustainable development, and about the business communities, seeking to understand entrepreneurs as practical promoters of this theoretical reality, especially within the scope of their international legal regulation of environmental and economic relations. In this sense, when considering that one of the main points for the implantation of sustainability is the participation of the population and their civil engagement, a business community was evaluated as one that gathers and creates community businesses to face the challenges they face together, emerging as community businesses, stores, farms, pubs and call centers. What they all have in common is that they are accountable to the community and that the profits generated provide a positive local impact, such as boosting the local economy. Acting in a remarkably similar way to social enterprises, community-based companies are committed to positively benefiting society through trade in a sustainable manner. One of its main premises is the fact that all profits from a community business are reinvested in the local area. Unlike social companies, community-based companies are focused on benefiting a specific local geographic area or even a professional segment.

Finally, practical examples of business communities related to products, processes and themes of global relevance, such as permaculture, sustainable agriculture and food security, were presented, so that the theoretical concept can be analyzed in reality. Thus, when considering a business community as a group of companies that share a common business area and interact to carry out public and private commercial transactions, it was observed how the construction of networks for sustainability changes the ways they are produced and what is. already known is processed.

In this way, this chapter observed how economic relations are regulated in favor of the environmental issue in business communities with special emphasis on the global promotion of sustainable development, in its economic, social and environmental dimensions, seeking to illustrate with practical examples the international partnerships that promote and reinforce the sustainability regime in the business environment. In order to further contribute to the theme of sustainability and the importance of creating innovations on the topic, it can be suggested as future studies, the construction of multicriteria studies of cases, involving countries in different realities, to analyze how business communities are being implemented by them, and what benefits they are providing.

References

Adenle AA, Sowe SK, Paryil G, Aginan O (2012) Analysis of open source biotechnology in developing countries: an emerging framework for sustainable agriculture. Technol Soc 34:256–269. https://doi.org/10.1016/j.techsoc.2012.07.004

Adler P, Kwon S (2002) Social capital: prospects for a new concept. Acad Manag Rev 27(1):17–40. https://doi.org/10.2307/4134367

Ait-Kadi M (2016) Water for development and development for water: realizing the sustainable development goals (SDGs) vision. Aquatic Procedia [s.l.], 6:106–110, Aug. Elsevier BV. https://doi.org/10.1016/j.aqpro.2016.06.013

Arksey H, O'Malley L (2005) Scoping studies: towards a methodological framework. Int J Soc Res Methodol 8(1):19–32. https://doi.org/10.1080/1364557032000119616

Baglietto P, Maresca M, Parodi A, Zingirian N (2005) Stepwise deployment methodology of a service-oriented architecture for business communities. Inf Softw Technol 47(6):427–436. https://doi.org/10.1016/j.infsof.2004.09.011

Bhattacharyya S (2012) Energy access programmes and sustainable development: a critical review and analysis. 16, 1e28. https://doi.org/10.1016/j.esd.2012.05.002

Birch D (2003) Doing business in new ways: the theory and practice of strategic corporate citizenship with specific reference to Rio Tinto's community partnerships. Deakin University, Melbourne, Corporate Citizenship Research Unit

Bond AJ, Morrison-Saunders A (2011) Re-evaluating sustainability assessment: aligning the vision and the practice. Environ impact assess rev [s.l.], 31(1):1–7, Elsevier BV. https://doi.org/10.1016/j.eiar.2010.01.007

Breitmeier H, Young OR, Zürn M (2006) Analyzing international environmental regimes: from case study to database. The MIT Press, Cambridge, MA

Buckingham S, Kina VJ (2015). Sustainable development and social work. International encyclopedia of the social & behavioral sciences, [s.l.], Elsevier, pp 817–822. https://doi.org/10.1016/b978-0-08-097086-8.28095-1

Campos EVR, Proença PLF, Oliveira JL, Bakshi M, Abhilash PC, Fraceto LF (2018) Use of botanical insecticides for sustainable agriculture: future perspectives. Ecol Ind. https://doi.org/10.1016/j.ecolind.2018.04.038

Carvalho FP (2006) Agriculture, pesticides, food security and food safety. Environ Sci Policy 9(7–8):685–692. https://doi.org/10.1016/j.envsci.2006.08.002

Chen H, Huo Z, Dai X et al (2018) Impact of agricultural water-saving practices on regional evapotranspiration: the role of groundwater in sustainable agriculture in arid and semi-arid areas. Agric for Meteorol 263:156–168. https://doi.org/10.1016/j.agrformet.2018.08.013

Neiva SS da, MacQueen JH, de Andrade Guerra, JBOS (2020) Social business in the context of sustainable development. Decent Work Econ Growth 1–12. https://doi.org/10.1007/978-3-319-71058-7_32-1

Deggau AB, Greuel L, Neiva SS, de Andrade Guerra JBSO (2019) Permaculture, clean production, and food security. In: Leal Filho W, Azul A, Brandli L, Özuyar P, Wall T (eds) Zero hunger. Encyclopedia of the UN Sustainable Development Goals. Springer, Cham. https://doi.org/10.1007/978-3-319-69626-3_37-1

Duran, Dan Cristian et al (2015) The objectives of sustainable development - ways to achieve welfare. Procedia Econ Finance [s.l.], 26:812–817. Elsevier BV. https://doi.org/10.1016/s2212-5671(15)00852-7

FAO (1983) World food security: a reappraisal of the concepts and approaches. Director General's Report. FAO, Rome

FAO, IFAD, UNICEF, WFP, and WHO (2019) The state of food security and nutrition in the world 2019. Safeguarding against economic slowdowns and downturns. Rome, FAO

Ferguson RS, Lovell ST (2013) Permaculture for agroecology: design, movement, practice, and worldview. A Review Agron Sustain Dev 34(2):251–274. https://doi.org/10.1007/s13593-013-0181-6

Ferranti, Pasquale (2018) The United Nations Sustainable Development Goals. Reference Module in Food Science, [s.l.], Elsevier, pp 1–3. https://doi.org/10.1016/b978-0-08-100596-5.22063-5

Firebrand (2020) About firebrand. Retrieved from https://firebrandtalent.com/blog/2017/09/communities-future-of-business/. Accessed 28 June 2020

Fitchett JR, Atun R (2014) Sustainable development goals and country-specific targets. Lancet Global Health [s.l.], 2(9):503–503, Sept. Elsevier BV. https://doi.org/10.1016/s2214-109x(14)70282-7

Fiverr (2020) About fiverr. Retrieved from https://www.fiverr.com/. Accessed 29 June 2020

Gafsi M, Legagneux B, Nguyen G, Robin P (2006) Towards sustainable farming systems: effectiveness and deficiency of the French procedure of sustainable agriculture. Agric Syst 90:226–242. https://doi.org/10.1016/j.agsy.2006.01.002

Gardner JE (1989) Decision making for sustainable development: selected approaches to environmental assessment and management. Environ Impact Assess Rev [s.l.], 9(4):337–366, Dec. Elsevier BV. https://doi.org/10.1016/0195-9255(89)90028-0

German LA, Bonanno AM, Foster LC, Cotula L (2020) Inclusive business in agriculture: evidence from the evolution of agricultural value chains. World Dev 134, 105018. https://doi.org/10.1016/j.worlddev.2020.105018

Glavic P, Lukman R (2007) Review of sustainability terms and their definitions. J Cleaner Prod [s.l.], 15(18):1875–1885, Dec. 2007. Elsevier BV. https://doi.org/10.1016/j.jclepro.2006.12.006

Hák T, Janoułková S, Moldan B (2016) Sustainable development goals: a need for relevant indicators. Ecol Indic [s.l.], 60:565–573, Jan. 2016. Elsevier BV. https://doi.org/10.1016/j.ecolind.2015.08.003

Haluza-DeLay R, Berezan R (2013). Permaculture in the city: ecological habitus and the distributed ecovillage. In: Environmental anthropology engaging ecotopia, bioregionalism, permaculture, and ecovillages, pp 130–145. https://doi.org/10.1017/CBO9781107415324.004

Holden E, Linnerud K, Banister D (2014) Sustainable development: our common future revisited. Glob Environ Change [s.l.], 26:130–139, May 2014. Elsevier BV. https://doi.org/10.1016/j.gloenvcha.2014.04.006

Hui SCM (2001) Low energy building design in high density urban cities. Renew Energy [s.l.], 24(3–4):627–640, Nov. 2001. Elsevier BV. https://doi.org/10.1016/s0960-1481(01)00049-0

199 Jobs (2020) About 199 jobs. Retrieved from https://199jobs.com/. Accessed 27 June 2020

Jovoto (2020) Get inspired. Retrieved from https://www.jovoto.com/. Accessed 28 June 2020

Kanter DR. et al. (2016) Translating the sustainable development goals into action: a participatory backcasting approach for developing national agricultural transformation pathways. Glob Food Secur [s.l.], 10:71–79, Sept. 2016. Elsevier BV. https://doi.org/10.1016/j.gfs.2016.08.002

Keohane RO (1982) The demand for international regimes. International Organization. vol 36. no. 2, 1982. The MIT Press, pp 325–351

Keohane RO, NYE J (1988) Poder e Interdependência. La política mundial en transición. Grupo Editor Latinoamericano. Buenos Aires, 1988

Krasner SD (1982) Structural causes and regime consequences: regimes as intervening variables. International Organization. Cambridge (MA), vol 36, no. 2, pp 185–205, Spring. In: Causas estruturais e consequências dos regimes internacionais: regimes como variáveis intervenientes. Rev. Sociol. Polit. [online]. 2012, vol. 20, no. 42, pp. 93–110. ISSN 0104-4478

Laine M (2005) Meanings of the term sustainable development in Finnish corporate disclosures. Accounting Forum [s.l.], 29(4):395–413, Dec. 2005. Elsevier BV. https://doi.org/10.1016/j.accfor.2005.04.001.

Lee M, Soto D, Modi V (2014) Cost versus reliability sizing strategy for isolated photovoltaic micro-grids in the developing world. Renew Energy 69:16–24. Elsevier BV. https://doi.org/10.1016/j.renene.2014.03.019

Levac C, Colquhoun H, O'Brien KK (2010) Scoping studies: advancing the methodology. Implement Sci 5(69):1–9. https://doi.org/10.1186/1748-5908-5-69

Lima RG, Lins HN, Pfitscher ED, Garcia J, Suni A, de Andrade JBSO, Delle FCR (2016) A sustainability evaluation framework for science and technology institutes: an international comparative analysis. J Cleaner Prod 125:145–158. Elsevier BV. https://doi.org/10.1016/j.jclepro.2016.03.028

Lima RG, de Oliveira Veras M, da Silva Neiva S, de Andrade Guerra JBSO (2020) Sustainability assessment using governance indicators. In: Leal Filho W, Marisa Azul A, Brandli L, Gökçin Özuyar P, Wall T (eds) Sustainable cities and communities. Encyclopedia of the UN Sustainable Development Goals. Springer, Cham. https://doi.org/10.1007/978-3-319-95717-3

LIMA RG de (2014) A contribuição dos institutos de ciência e tecnologia para a formação de um regime internacional de sustentabilidade (2014). 248 p. Dissertação (Mestrado) - Universidade

Federal de Santa Catarina, Centro Sócio Econômico, Programa de Pós-graduação em Relações Internacionais, Florianópolis.

Milchakova N, Reshetnikov L (2020) Model of business interaction in corporate social responsibility strategy. E3S Web of Conferences, 164, 10017. https://doi.org/10.1051/e3sconf/202016410017

MILNER HV (2009) Power, interdependence, and nonstate actors in world politics. In: Milner HV, Moravcsik A (eds) Power, interdependence, and nonstate actors in world politics, Research Frontiers, Princeton University Press, Princeton, pp 3–27

Mollison B, Holmgren D (1978) Permaculture one: a perennial agricultural system for human settlements. Tagari Publications, Tyalgum, International Tree Crop Institute USA

Neiva SS (2019) A percepção do legislativo municipal sobre a implantação de cidades sustentáveis. 262p. Dissertação (Mestrado) - Universidade do Sul de Santa Catarina, Programa de Pós-graduação em Administração, Florianópolis

Partnership Platforms Report (2020) Platforms for Partnership Report. Retrieved from https://sustainabledevelopment.un.org/content/documents/2699Platforms_for_Partnership_Report_v0.92.pdf. Accessed 22 May 2020

Pedersen CS (2018) The UN Sustainable Development Goals (SDGs) are a great gift to business! Procedia Cirp [s.l.], 69:21–24, 2018. Elsevier BV. https://doi.org/10.1016/j.procir.2018.01.003

Persson Å, Weitz N, Nilsson M (2016) Follow-up and review of the Sustainable Development Goals: alignment vs. internalization. Rev Eur, Comp Int Environ Law [s.l.], 25(1):59–68, Abr. 2016. Wiley. https://doi.org/10.1111/reel.12150

Ramchandran N, Pai R, Parihar AKS (2016) Feasibility assessment of anchor-business-community model for off-grid rural electrification in India. Renew Energy 97:197–209. Elsevier BV. https://doi.org/10.1016/j.renene.2016.05.036

SDSN - UN (2020) About us. Retrieved from https://www.unsdsn.org/about-us. Accessed 02 June 2020

Sinakou E et al (2018) Academics in the field of education for sustainable development: their conceptions of sustainable development. J Cleaner Prod [s.l.], 184:321–332. Elsevier BV. https://doi.org/10.1016/j.jclepro.2018.02.279

Sofeska E (2016) Relevant factors in sustainable urban development of urban planning methodology and implementation of concepts for sustainable planning (Planning documentation for the master plan Skopje 2001–2020). Procedia Environ Sci [s.l.], 34:140–151, 2016. Elsevier BV. https://doi.org/10.1016/j.proenv.2016.04.014

Stone RW (2008) Institutions, power, and interdependence. In: Milner HV. Moravcsik A (eds), Princeton University Press, Princeton, pp 31–49

Trend Hunter (2020) About trend hunter. Retrieved from https://www.trendhunter.com/about-trend-hunter. Last Accessed 03 May 2020

Ulut ZB (2008) Permaculture playgrounds as a new design approach for sustainable society. Int J Nat Eng Sci 2(2):35–40

United Nations (2018). Report of the world commission on environment and development: our common future. Retrieved from http://www.un-documents.net/our-common-future.pdf. Last Accessed 10 July 2020

United Nations (2020) Brazil. ODS-17. Retrieved from https://nacoesunidas.org/pos2015/ods17/. Accessed 30 Apr 2020

United Nations Development Programme (2018). Sustainable development goals. Retrieved from http://www.undp.org/content/undp/en/home/sustainable-development-goals.html. Last Accessed 03 May 2020

United Nations-UN (2020). Sustainable Development Goals. Retrieved from https://www.un.org/sustainabledevelopment/. Last Accessed 03 May 2020

UN-Sustainable Development Knowledge Platform (2020) Sustainable development. Retrieved from http://sustainabledevelopment.un.org/owg.html. Last Accessed 03 May 2020

Van Vuuren, DP et al. (2015) Pathways to achieve a set of ambitious global sustainability objectives by 2050: explorations using the IMAGE integrated assessment model. Technol Forecast Soc Change [s.l.], 98:303–323. Elsevier BV. https://doi.org/10.1016/j.techfore.2015.03.005.

Veisi H, Liaghati H, Alipour A (2016) Developing an ethics-based approach to indicators of sustainable agriculture using analytic hierarchy process (AHP). Ecol Ind 60:644–654. Elsevier BV. https://doi.org/10.1016/j.ecolind.2015.08.012

World Green Building Council (2018) About green buildings. Retrieved from https://www.worldgbc.org/. Accessed 15 June 2020

Wulf C et al. (2018) Sustainable development goals as a guideline for indicator selection in life cycle sustainability assessment. Procedia Cirp [s.l.], 69:59–65. Elsevier BV. https://doi.org/10.1016/j.procir.2017.11.144

Xu X, Hou L, Lin H, Liu W (2006) Zoning of sustainable agricultural development in China. Agric Syst 87:38–62. Elsevier BV. https://doi.org/10.1016/j.agsy.2004.11.003

Yang BXU, Tong SL (2017). Analysis on sustainable urban development levels and trends in China's cities. J Cleaner Prod [s.l.], 141:868–880. Elsevier BV. https://doi.org/10.1016/j.jclepro.2016.09.121

Rafael Gustavo de Lima has a degree in international relations, economics and foreign trade and a master's degree in international relations. Develops studies on the theme of sustainable development for more than 5 years, with a main focus on Higher Education Institutions and on transforming sustainability into something tangible for all people.

Samara da Silva Neiva has a degree in International Relations, and a master's degree in Business Management. is a member of the Center for Sustainable Development/Research Group on Energy Efficiency and Sustainability (GREENS) and works mainly on themes aimed at creating sustainable cities and communities.

Sustainability Practices Among Russian Business Communities: Drivers and Barriers Towards Change (The Cases of Moscow and Kazan)

Polina Ermolaeva and Ksenia Agapeeva

Abstract The study provides holistic insights on the Russian business communities' corporate sustainability practices based on a representative survey (n = 400) and semi-structured interviews (n = 30) with Russian senior executives across various sectors, scales and cities. The given research added to the scarce academic debates on Russian corporate sustainability practices by providing managers' perspectives on these practices in retrospective and prospective terms. The study revealed their reflections on the key rationales and drivers for the implementation and enhancement of sustainability practices, opportunities and barriers associated with adopting sustainability-led initiatives and their self-perceived role towards securing a better environment. Another unique contribution is the authors' typology of the Russian enterprises into 'Green' and 'Brown' clusters based on companies' environmental performance. The findings show that the most popular sustainability initiatives include water metering devices, paper saving, duplex printing, separate waste collection, heat and lighting regulation, and the collection of batteries, light bulbs and wastepaper. The key rationale behind implementing these initiatives is the environmental care factor. The main benefits are seen in contributions to environmental improvement and the creation of a positive corporate image. The key barriers include the lack of governmental supportive measures, economic and technical difficulties. Therefore, tax incentives and other state administrative support could become the most important measures for promoting environmental initiatives in Russian firms.

Keywords Corporate environmental responsibility · Corporate sustainability practices · Business sustainability transitions · Russia

P. Ermolaeva (✉) · K. Agapeeva
Kazan Federal University, Kremlevskaya 18, Kazan, Russia

1 Introduction

The concept of sustainable development has recently begun to be formalized in Russia by means of appropriate changes in legislative, consumer, educational, and communication areas (Bobylev and Solov'eva 2017). Practical aspects of the concept's formalization have resulted in updating business strategies that include social and environmental dimensions of profit, reviewing the fundamentals of corporate governance and expanding the scope of business responsibility. Thus, corporate sustainability initiatives have recently started receiving deliberate attention from the Russian government authorities. Since 2019, new government regulations have been introduced. They include business transition to the principles of best available technologies, differentiated measures of state regulation depending on the environmental hazard degree of an enterprise's activities, strengthening of environmental expertise and others. Some large-scale industrial Russian companies working abroad progressed in implementing cutting-edge environmental programmes and act as 'green champions' for the national business communities (Salmi 2008; Tynkkynen 2005).

Despite these growing 'green' initiatives, in general, Russian enterprises face numerous institutional, legislative, administrative and financial barriers against adopting environmentally friendly practices and norms; thus, modest progress has been made in meeting the environmental standards (Fedosova 2020). During the past 30 years, Russia has experienced several major social and economic reforms which precipitated a dramatic redistribution and privatization of public goods. There was a change of legal, social and political institutes, including privatization of large governmental enterprises, with decentralization of the political power to the regional elites and oligarchs that create crony capitalism coupled with a resource-intensive economy. The Soviet legacy of the monopolization of the state companies put into question the transparency and legitimacy of the environmental tests and performance of large industrial companies. All these processes in turn produce apathy and cynicism among Russian business communities towards their companies' sustainable transitions.

Taking into account Russia's enormous environmental impact as a major world emitter of greenhouse gases and a leading global supplier of fossil fuels, and its challenging socio-political and business contexts, a better understanding of current Russian corporate sustainability practices is needed.

The literature analysis shows that the number of publications dealing with sustainability commitments by Russian business communities has grown over the past 10 years (Orazalin et al. 2019; Orazalin and Mahmood 2018; Andreassen 2016; Finley et al. 2019; Furrer et al. 2010). The majority of studies are analyzed via the framework of social corporate responsibility (e.g. Ponomarenko et al. 2016; Tret'yak 2015; Visser et al. 2019; Belyaeva 2013). For example, Ponomarenko et al. (2016) offer an assessment of the present-day state of corporate social responsibility policies (further referred to as CSR) in the coal companies of Russia and Poland. The authors came to conclusion that the majority of the largest Russian companies are in the initial stage of CSR development and do not pay enough attention to the protection of the environment. Tretyak (2015) demonstrated that the Russian companies have been following the concept of sustainable development while minor firms are associated CSR with social policies or follow the concept of corporate citizenship with emphasis on social and environmental issues.

Modest research (e.g. Crotty and Rodgers 2012a; Graves et al 2019; Molchanova et al 2020) has been undertaken on the factors that drive corporate environmental initiatives in Russia, including management's environmental behaviour, concerns, and governmental support (e.g. Crotty and Hall 2014; Crotty and Rodgers 2012b). For example, the most recent study by Graves et al. (2019) focuses on the pro-environmental behaviour of Russian employees via the lenses of environmental transformational leadership theory which describes managers as role models in the communication of environmental values to their employers. Molchanova et al. (2020) examine the environmental concerns reflected by the mission statements of the top Russian companies.

While a significant corpus of literature has been undertaken on corporate environmental behaviour from CSR and managerial perspectives, there are limited studies on the Russian business communities that aim at analysis of the contemporary Russian corporate sustainability practices holistically, including the drivers for their adoption by the business executives and key barriers that prevent environmental changes.

This article begins to fill this gap, and to contribute to global discussions noted above, by providing insights into the corporate sustainability practices of Russian senior executives through a case study of the cities of Moscow and Kazan. More specifically, the authors critically analyze the dynamics of Russian corporate sustainability practices (five years ago, currently and plans for the next two years), key rationales for the implementation of sustainability-led practices, opportunities and barriers associated with adopting them, main drivers towards the increasement of sustainability-led initiatives, and the role of business in contributing to the environment via the lenses of the senior executives. Furthermore, based on the different environmental performance of the enterprises, they would be clustered into distinguished types.

Following these parameters, this research seeks to generate a better understanding of the corporate sustainability practices implemented by Russian enterprises of various sectors, scales and across different Russian cities. To the authors' best knowledge, this is the first study of this kind that is focused on the Russian context.

2 Theoretical and Methodological Grounds of the Study

In the literature sustainability practices are understood under a triple-bottom line approach integrating economic, environmental and social aspects that aims to consider future generations and intergenerational justice (Schaltegger and Burritt 2010). Businesses can implement principles of sustainability in the preparation and review of corporate strategies, supporting new agreements and negotiations that promote sustainable practices, and by developing new projects driven by sustainability principles and, finally, by broadening their vision of sustainability beyond the limits of the company (Schaltegger and Burritt 2006).

Conceptually, in this study we understand corporate sustainability practices as environmentally friendly activities that help to minimize the company's ecological footprint. These practices aim at developing and applying ideas for reducing the company's environmental impact such as recycling and reusing, pollution reduction, energy efficiency, decrease in consumption of hazardous, harmful and toxic materials, implementing sustainability programmes, rethinking supply chains, etc.

The research, in the form of a case study of Moscow and Kazan, used a mixed methods approach. The rationale behind choosing the cities of Moscow and Kazan is based on their similarities and differences. They are the largest urban centres in Russia with a rapid innovative development and higher than average levels of education and income in comparison to other similar Russian cities. However, there are also differences in terms of geographic locations, the different structure of industrial and employment consolidations, and the different levels of environmental pollution and eco-conflicts. In both cities a large number of innovative local initiatives led by municipalities and/or civil society aimed to improve the environment by reducing air and water pollution and providing efficient waste management and sustainable transport. Both cities have undergone significant recent urbanization and have higher than average levels of infrastructural advancements in comparison to other large Russian cities (Ermolaeva et al. 2020). In both cities civil society stands actively for the environment, including independent air pollution monitoring, protection of green and blue zones, protests against incineration plants, and initiatives around sustainable food consumption, cycling etc.

The quantitative component of the study employed semi-structured online interviews with heads of enterprises in Moscow and Kazan quota-controlled by scale and industry, based on statistics for each city (total n = 400). The quantitative strategy was supplemented by in-depth interviews (n = 30) with heads of enterprises in the two cities to address the in-depth assessments of the key features, challenges and tendencies of the sustainability transitions of their enterprises. The survey and interviews were carried out at the end of 2019.

The mixed methods approach allowed us the triangulation of 'thick' contextual data and generalizable representative statistics to critically reflect on the sustainability transitions practices by business communities, the key benefits, risks and barriers of such transitions and required resources.

3 Findings

3.1 Russian Corporate Sustainability Practices (Retrospective and Prospective)

The study results have shown that every sixth company runs its own sustainability initiatives, programmes or standards, while another 23% of enterprises have implemented them at the level of informal initiatives (Fig. 1).

The most popular implemented sustainability initiatives (as in 2019) include: the use of water metering devices (52%), paper saving, duplex printing (49%): "*… all our printers are automatically set on duplex black and white printing. In order to print in a different way, you have to spend time on changing the settings*". Other priority practices include separate waste collection (46%), heat and lighting regulation (43%), collection of batteries, waste paper, light bulbs (38%), use of products made of recycled materials (37%), reduction/refusal of disposable plastic use (32%): "*we use reusable mugs in all our factories, disposable tableware use in the office has been prohibited*".

Less popular initiatives include sustainability awareness training (32%), use of eco-labelled products and materials (31%), use of environmentally friendly detergents (27%), and environmentally friendly office furniture (21%).

In the case of smaller and medium businesses, the most popular sustainability initiatives are the use of water metering devices, paper saving/duplex printing, heat and lighting regulation, and separate waste collection. According to managers, medium-sized businesses use eco-labelled products (47%) and environmentally

Initiative	%
Water metering devices	52
Paper saving, duplex printing	49
Separate waste collection	46
Heat and lighting regulation	43
Batteries, waste paper, light bulbs collection	38
Use of products made of recycled materials	37
Reduction/refusal of disposable plastic use	32
Sustainability awareness training	32
Use of eco-labeled products and materials	31
Use of environmentally friendly detersives	27
Environmentally friendly furniture	21
Use of alternative sources of energy	17
Green' logistics	16
Bike parking arrangement	12
CO_2 monitoring	7
Environmental management certification	7
Shuttle buses for the employers	6
Environmental certification of the office	6

Fig. 1 The structure of sustainability initiatives, implemented in Russian enterprises, in %

friendly office furniture (37%) more often than smaller and large-scale companies, furthermore, they also collect batteries, light bulbs, and waste paper (59%) more often than other types of enterprises.

Less popular initiatives, mainly implemented in mid-sized and large-scale companies, are alternative energy sources (17%), 'green' logistics (17%), bike parking arrangements (12%): *"there is a parking lot and an opportunity to take a shower after a cycle ride in every office"*. These practices also include carbon dioxide monitoring (7%), implementation of certified environmental management systems (7%), running shuttle buses for the employees (6%), and environmental certification of the office (LEED/BREEAM etc.) (6%). All these practices are mainly implemented in companies located in Moscow.

Many of these practices were mentioned in the *Zeleniy Ofis* ('Green Office') programme, which aims to create the office in conformity with best international 'green' standards. Remarkably, more than half of the managers of the surveyed companies (56%) are unaware of the programme. With only 9% of companies participating in the initiative in Moscow, and 2% taking part in it in Kazan, the programme mostly involves large businesses. On the other hand, managers of medium and smaller businesses tend to believe that their companies do not cause much harm to the environment, and for this reason they feel that they do not have to adopt any additional sustainability initiatives.

According to most managers (54%), there have been no changes in the implementation of sustainability practices over the past five years: *"I believe that the trends are not positive enough. Science is moving forward, but there is nothing to apply"*. However, it is worthwhile noting that 46% of managers pointed out positive trends. Primarily, it is in cases of large enterprises located in Moscow (68%). The implementation of such changes is mostly associated with *"the growth of environmental concerns among employees themselves"*, and *"the growth of the number and availability of eco-products in the market"*. Managers point out *"a noticeable economic effect of the use of sustainable technologies"*, and *"the company's respect and prestige growth, improvement of its trust reputation as a result of green technologies implementation"*. Many managers noted the geographical expansion of their project: *"30% of cities participated in our industries earlier, but now almost all of them joined the company"*.

According to managers, such initiatives as the use of environmentally friendly office furniture (30%) and detergents (30%), alternative energy sources (29%), bike parking arrangements (24%), eco-labelled products and materials (23%), disposable plastic reduction (23%) have highest possible levels of potential implementation over the next two years.

3.2 Key Rationales for the Implementation of Sustainability-Led Practices

Next, it was important to analyze key rationales for implementing these initiatives (Fig. 2). The key reason is environmental concern (72%), which is far ahead of the economic factors (56%) that was mostly mentioned by managers in Kazan (61%): "*mercury-filled lamps were replaced with LED ones, which resulted in the reduction of energy consumption. This initiative has already paid off*".

Human capital and labour protection (employees' health care) (51%) are also among the priorities of managers: "*We have spent a lot of money on air purification and an air conditioning system. As a manager, I consider it one of my priorities…*".

Another important factor is compliance with mandatory environmental standards, which is the case of companies trading internationally: "*In order to export to some countries, companies must have certain environmental certificates…*". In most cases, following mandatory standards protects companies from being fined, and this reason is more characteristic of enterprises in Moscow (36%): "*ecological equipment is bought, because they can impose rather large fines… A one-time fine can amount to 500.000 rubles followed by the subsequent closure of a company, until it addresses all the recommendations*".

It also leads to the increase of a company's attractiveness for employees and consumers (24%) and attracts new investors (5%).

The factors of environmental concern (94%), increased attractiveness for employees and consumers (42%), and attracting new investors (20%) are more characteristic of larger companies than smaller and medium ones. Factors that are of

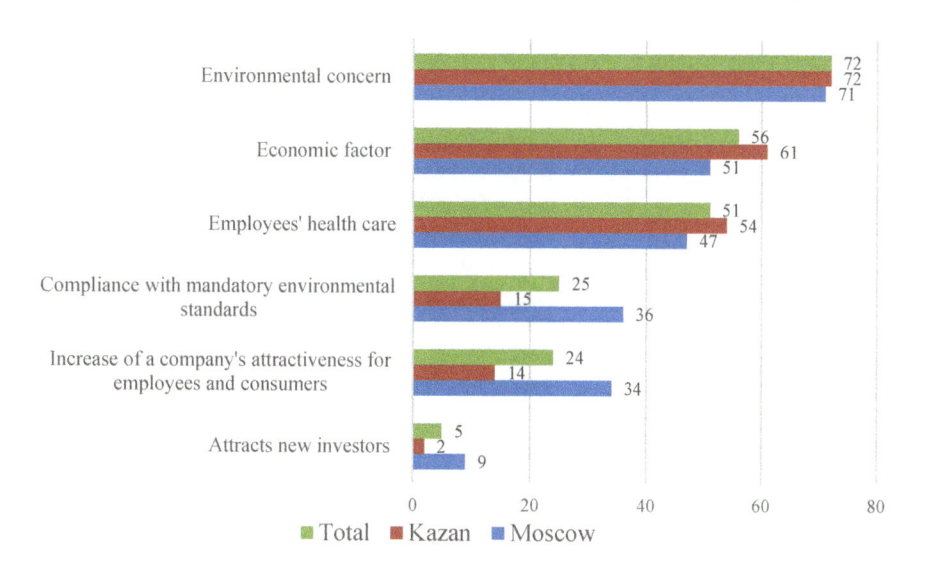

Fig. 2 Key rationales for the implementation of sustainability-led practices, in %

greater importance for smaller organizations include economic factors (62%), while medium companies are often motivated by mandatory standards compliance (37%).

3.3 Opportunities, Barriers and Drivers Associated with Adopting Sustainability-Led Practices

The reduction of environmental damage in the region is seen by Russian managers as the most important benefit of the implementation of sustainability initiatives (47.3%). This factor is followed by increasing attractiveness for consumers (30.2%), reduction of operating expenses (30.2%), improving corporate image in the region (28%), increasing energy and resource efficiency (25.8%), increasing attractiveness for employees (25.1%) and increasing profits (24.4%) (Fig. 3).

According to managers, some initiatives offer a double benefit. For example, *"monitoring the resources consumption itself is economically viable"*. Even those who have not implemented environmental programmes are aware of their economic benefits: *"when electricity got more expensive, we invested 50.000 in energy-saving technologies implementation. As per our calculation, it should pay off within six months, so the initiative seems reasonable..."*, *"we have calculated, that we would have lost about 1 million rubles, if we did not install environmentally efficient equipment"*.

The image factor is very important for companies located in Moscow: according to their managers, increasing attractiveness for consumers (40.7%) is of the greatest importance. This factor is followed by the aspect of environmental damage reduction (40%) and the improvement of the corporate image in the region (37.9%). The study of Kazan companies' motivation has shown different results: most managers (54.8%)

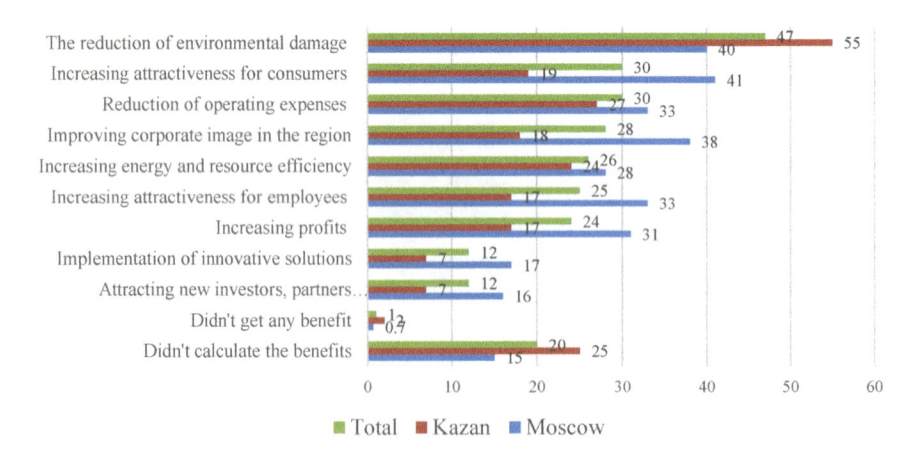

Fig. 3 Benefits associated with adopting sustainability-led practices, in %

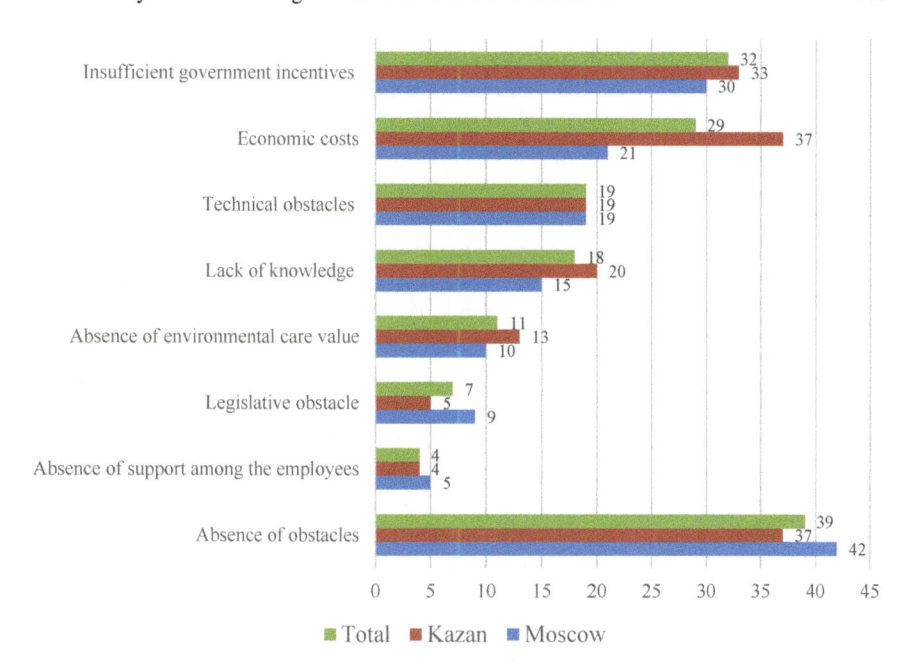

Fig. 4 Barriers associated with adopting sustainability-led practices, in %

Barrier	Total	Kazan	Moscow
Insufficient government incentives	32	33	30
Economic costs	29	37	21
Technical obstacles	19	19	19
Lack of knowledge	18	20	15
Absence of environmental care value	11	13	10
Legislative obstacle	7	5	9
Absence of support among the employees	4	4	5
Absence of obstacles	39	37	42

selected environmental damage reduction as the most important benefit of the implementation of sustainability initiatives. This option is followed by the reduction of operating expenses (27.4%).

The largest barriers to the implementation of sustainability initiatives are seen to be insufficient government incentives (31.6%) and economic costs associated with the implementation of initiatives (29%), as well as technical obstacles (19.1%): "...*everything in Russia is usually implemented on the basis of fines and punishments for those who do not abide*", "*we have also faced technical obstacles connected with separate waste collection: the people are not ready yet...*" (Fig. 4).

At the same time, Kazan managers have considered economic costs a barrier to the implementation of sustainability initiatives much more often than business representatives in Moscow (37.2% versus 20.7%, respectively). Next, the lack of knowledge and insufficient sustainability awareness has been mentioned in Kazan somewhat more often than in Moscow (20.4% against 14.8%). Legislative obstacles have been mentioned in Moscow a bit more often than in Kazan (8.9% versus 5.1%, respectively). Much attention has been paid to the way of ensuring that "*everyone in the supply chain changes their business model*".

Almost 40% (39.3%) of managers do not see any obstacles for the realization of sustainability initiatives. The absence of obstacles is noted by managers of smaller and medium companies more often than those of bigger ones (45.5% versus 28.6% respectively). In 4.4% of companies these practices have not been supported by employees.

Managers have explained that certain initiatives cannot be carried out because of the lack of infrastructure: "*Separate waste collection. Here you need to think of where to dispose of it. There is a company that picks up our wastepaper. We could collect plastic, but who would take care of its disposal?*" Big companies' initiatives are mostly hindered by insufficient government incentives (44.4%) and associated costs (34.9%): "*We are ready and have set ourselves a goal: by 2025 we want to use 35% of recyclables in all our packages. But they are 30% more expensive, and there is no subsidization or support in Russia*". Medium enterprises mention the absence of environmental care value more often than large and small businesses (16.3% versus 9.5% and 9.3%, respectively).

Every seventh enterprise in our sample does not implement any environmental initiatives. This is primarily due to the lack of financial resources and incentives (41.2%), which is more characteristic of Kazan than of Moscow (46.2% versus 36%, respectively): "*There is no fair competition, which discourages us... By introducing any initiative, you lose your advantage, while your competitors are not going to follow the same rules...*". Such factors as the economic crisis, the lack of environmental protection measures among corporate priorities and the lack of knowledge (15.7% each) are also important. Many companies in Moscow do not implement any practices, since, according to their managers, their firm's activities do not have any negative environmental impact, while managers in Kazan mention the absence of time and opportunity: "*you have to think about the economic result of your activity, not the environmental one*". A half of the managers of large companies have mentioned that environmental protection is not among their corporate priorities (50%), another 40% have pointed out the lack of government support or financial resources/incentives. A half of the representatives of medium businesses explained their inaction by the economic crisis and the lack of knowledge (50% each). In cases of the smaller businesses, the lack of financial resources and incentives are the main reason for environmental passivity (40%).

Tax exemptions could be the most important incentive for the implementation of sustainability initiatives (58.8%). Another 21.6% of business representatives have indicated new competitive advantages and opportunities. Remarkably, managers of Moscow companies qualify this factor as much more significant than managers in Kazan (32% versus 11.5%, respectively). Improving the company's image is considered a reason for the realization of environmental initiatives by 11.8% of companies, mostly located in Moscow (20% versus 3.8% for Kazan companies).

3.4 The Self-perceived Role of Business in Contributing to the Environment

Most company leaders have mentioned the introduction of separate waste collection and its recycling as being important (23%) in response to the open-ended question about what business can do today in Russia to protect the environment. The decrease in waste production and emissions, as well as refuse treatment, its neutralization and management (18%) come second. The importance of introducing sustainability initiatives in industrial sites and in the office, raising the employees' level of sustainability awareness is recognized by 16% of companies, especially large ones (26%). Every seventh manager has stated the necessity to invest in environmental protection, developments and education. Saving resources in industrial sites and in the office (water, electricity, wood and paper, etc.), as well as the use of degradable raw materials, safe technologies, and the refusal of plastic are equally important for 12% of organizations. The importance of environmental awareness, education and public relations is important for 9% of companies: "*I think the business has to take certain educational responsibility. It should honestly and openly tell consumers about its achievements, instead of engaging in greenwashing…*". A paternalistic position has been taken by 6% of organizations, whose managers have stated that "*business cannot and should not do anything, since this is the task of the state*".

3.5 Typology of Russian Enterprises: 'Green' Versus 'Brown'

For the purpose of subsequent analysis enterprises have been grouped based on the degree and nature of their sustainability practices. All practices that are typical of companies of different scales have been categorized as follows: (1) solution of the problem of waste reduction and its disposal (indicators 1, 2, 13), (2) resource efficiency (indicators 3, 4, 5), (3) environmentally friendly choice of materials (indicators 6, 7), (4) environmentally friendly office furniture (indicators 8, 9). The sustainability education of employees, environmental certification of the office and green logistics have been assigned separate indicators.

The survey results have shown that such practices as carbon dioxide monitoring, use of alternative energy sources, environmental management certification (ISO 14,001, FSC, etc.), bike parking arrangements, crew bus provision are only applicable in cases of medium and large companies. Therefore, they have not been included in the factor analysis and have been considered separately.

Categorical principal components analysis (CATPCA) has been implemented in order to reduce the data dimensionality. This choice is due to the nature of the nonmetric scale used in the survey: CATPCA makes it possible to digitize variables and to remove this restriction. The obtained values of factor variables were subsequently used for k-means clustering.

Enterprises have been grouped into two clusters with the following centres as a result of k-means clustering (Table 1).

The first cluster includes enterprises which have introduced various sustainability-led practices ('Green' enterprises, n = 228, 57%). This type is represented by large and medium companies. Accordingly, they declare the importance of environmental care. At the same time, representatives of medium and large enterprises have stated reputational benefits (46%) and adherence to mandatory standards (42%) as the

Table 1 End cluster centers

	Cluster	
	1	2
Waste reduction	,69	-,92
Resource saving	,61	-,81
Environmentally friendly materials	,71	-,94
Environmentally friendly furniture and detergents	,69	-,91
Sustainability awareness training	,60	-,80
Environmental management certification of the office	,37,678	-,49,945
Green logistics	,22,547	-,29,888

reason for the implementation of certain practices more often than managers of smaller companies.

The second group of companies either do not introduce any practices (about one third of such organizations in the cluster: 55), or they have introduced (or plan to do it) sporadic practices, which, in general, means that they cannot be characterized by thorough implementation of sustainability-led initiatives ('Brown' companies, n = 172, 43%). As a rule, this type is represented by microenterprises (57%).

Representatives of different clusters differently assess the potential of sustainability-led initiatives, not yet realized, but deserving wide implementation in Russian companies. Among the second type of representatives, only separate waste collection (65%) and reduction/refusal of the disposable plastic use (54%) were chosen by more than half of the respondents, while other practices find less support. Managers who do not implement and do not plan to implement any sustainability-led practice, were asked about the reason for their non-compliance and what could be an incentive for them to implement any such practices. In answer to this question they most often pointed out the lack of financial resources/incentives (40%) as the reason for the existing situation. It is worth mentioning that every fifth microenterprise and 13% of medium and large companies fall into this group. Other important factors for the low take up of sustainability practices are the lack of knowledge (15.4%), the economic crisis (15.4%), corporate priorities excluding environmental care (15.4%), the lack of state support (13.5%), technical obstacles (11.5%), and lack of stakeholders' support/pressure (9.6%). Accordingly, most managers noted tax exemptions (59%) as a viable incentive (Table 2).

Sustainability-led practices are differently distributed among microenterprises located in Moscow and in Kazan: all the practices (apart from separate waste collection) are more often implemented or planned in Kazan microenterprises. On the other hand, it is worth mentioning that the most significant discrepancy between the two cities has been revealed for the resource saving practice: paper saving (42% microenterprises in Kazan and only 22% of their counterparts in Moscow have introduced this practice), heat and lighting regulation (37% and 9%, respectively), water metering devices (48% and 28%, respectively).

Table 2 "Which sustainability initiatives deserve to be widely implemented in Russian companies, but are rarely used in reality?", %

	Enterprise clusters		
	'Green'	'Brown'	Total
1. Separate waste collection	71.9	65.1	69.0
2. Disposable plastic use reduction/refusal	66.7	53.5	61.0
3. Water metering devices	57.0	48.3	53.3
4. Heat and lighting regulation (motion detectors, recovery, etc.)	60.5	44.2	53.5
5. Paper saving, duplex printing	60.1	44.8	53.5
6. Use of eco-labelled products and materials	50.9	29.1	41.5
7. Use of products made of recycled materials	57.9	36.0	48.5
8. Environmentally friendly furniture	50.0	30.2	41.5
9. Cleaning the office by means of environmentally friendly detergents	55.3	33.1	45.8
10. Use of alternative sources of energy	53.5	34.3	45.3
11. Sustainability awareness training	56.1	34.9	47.0
12. Bike parking arrangements	40.4	22.7	32.8
13. Batteries, wastepaper, light bulb collection	59.2	43.0	52.3
14. Crew bus provision	37.3	22.1	30.8
15. 'Green' logistics (it takes into account environmental friendliness and fuel economy)	47.4	22.1	36.5
16. Environmental certification of the office (LEED/BREEAM etc.)	35.1	20.3	28.8
17. CO_2 monitoring	39.0	26.7	33.8
18. Environmental management certification (ISO 14,001, FSC etc.)	29.4	18.6	24.8

4 Conclusions and Discussions

The study provides holistic insights on the Russian business communities' corporate sustainability practices based on a representative survey (n = 400) and semi-structured interviews (n = 30) with Russian senior executives across various sectors, scales and cities. The given research added to the scarce academic debates on Russian corporate sustainability practices by providing managers' perspectives on these practices in retrospective and prospective terms. The study revealed managers' reflections on the key rationales and drivers for the implementation and enhancement of sustainable practices, opportunities and barriers against adopting sustainability-led initiatives and their self-perceived role towards achieving a better environment. Another unique contribution is the authors' typology of the Russian enterprises into 'Green' and 'Brown' groups based on companies' environmental performance.

The findings show that the most popular sustainability initiatives, mostly implemented in medium and large enterprises located in Moscow, include water metering devices, paper saving, duplex printing, separate waste collection, heat and lighting regulation, and the collection of batteries, light bulbs and waste paper. Approximately equal numbers of managers have noted either the absence of a positive trend in the realization of such practices or positive dynamics (Moscow companies mostly), which is due to the employees' environmental concern, positive economic results of the implementation of initiatives and availability of eco-products in the market. Environmentally friendly office furniture and detergent use as well as alternative energy sources and bike parking arrangements have been referred to as the initiatives that have the highest possible levels of potential implementation in the Russian business communities over the next two years. The typology of the enterprises based on their environmental performance distinguished two clusters, which are labelled by the authors as 'Green' and 'Brown' enterprises. 'Green' enterprises are environmentally friendly and run various types of green initiatives in the workplace and are mainly represented by middle- and large-scale firms. A 'Brown' type of enterprise has no or only few environmental activities and is presented by small and micro enterprises.

The key rationale behind the implementation of environmental initiatives is the environmental care factor. Economic factors, following the principal rationale, can be explained by either operating costs reduction or avoidance of fines and sanctions. It is worth noting that large companies are more motivated by environmental care, investment attractiveness and the pressure of stakeholders, while smaller companies are motivated by reduction of operating costs and resource savings to a greater extent. On the other hand, medium firms implement sustainability initiatives to avoid economic sanctions.

The main benefits provided by the implementation of sustainability-led practices include the decrease of the environmental impact in the region. According to the managers, many sustainability initiatives offer multiple benefits: they contribute to both the environmental improvement and create a positive corporate image in the eyes of customers. They also allow the reduction of operating costs and increase the energy and resource efficiency of the company.

The key barriers in the implementation of sustainability-led practices include the lack of governmental supportive measures, and economic and technical difficulties associated with running these practices. The lack of necessary infrastructure in place is also mentioned as an external barrier towards change. Therefore, tax incentives and other state administrative support could become the most important incentives for introducing environmental initiatives.

The role of businesses in contributing to improvement of the environment is seen through separate waste collection and its recycling implementation, reduction of waste production and emissions, refuse treatment and neutralization, waste disposal, introducing sustainability initiatives in the workplace and in the office, environmental culture awareness training, investing in environmental protection, development, and education.

Collectively, the findings reveal that the most common sustainability practices for the Russian executives are those that are convenient, economically and symbolically rewarding and financially and technically easy to integrate and maintain further. Corporate initiatives that are more economically, legislatively and technically challenging to implement such as the usage of alternative energy, certification of the enterprises with international standards, and CO2 monitoring are less attractive unless they are compulsory for the companies' certification and operation activities. Some of the corporate sustainability practices that usually are quite popular for international companies, such as shuttle buses that pick up employees near their homes at set times and bring them to and from work, organized carpooling systems with fellow colleagues, and bike parking arrangements, are not that attractive for Russian business communities due to the cold climate and sociocultural reasons.

Moreover, the findings show that sustainability initiatives and the rationales behind their implementation for small-scale, middle-scale and large-scale enterprises are different. The large-scale industrial enterprises who are the main CO_2 emitters are obliged by eco-regulations, market and stakeholders' pressure and have enough resources to be environmentally responsible while middle-scale and small-scale enterprises which are burdened in Russia with numerous bureaucratic procedures and inspections prioritize meeting such standards when they are economically beneficial for them. Thus, the main measures towards the enhancement of sustainability-led practices in Russian enterprises should be in receiving complex financial, legislative, technical and educational support in the form of private-governmental partnership.

The authors would like to state certain limitations associated with the current study that could be acknowledged and overcome in future research on the topic. First of all, the case study approach with only two regional cases in the sample does not allow a reliable generalization of the results for other broader Russian business community. Secondly, due to the environmental focus of the study, we did not cover other dimensions of corporate sustainability such as social and economic sustainability.

Acknowledgements This work was supported by the Russian Science Foundation under grant number: 17-78-20106 'Russian megacities in the context of new social and environmental challenges: building complex interdisciplinary model of an assessment of 'green' cities and strategies for their development in Russia'.

References

Andreassen N (2016) Arctic energy development in Russia—how "sustainability" can fit? Energy Res Soc Sci 16:78–88. https://doi.org/10.1016/j.erss.2016.03.015

Belyaeva ZS (2013) Transformation processes of the corporate development in Russia: social responsibility issues. Syst Prac Act Res 26(6):485–496. https://doi.org/10.1007/s11213-013-9298-4

Bobylev SN, Solov'eva SV (2017) Sustainable development goals for the future of Russia. Stud Rus Econ Dev 28(3):26–33

Crotty J, Hall SM (2014) Environmental awareness and sustainable development in the Russian Federation. Sustain Dev 22(5):311–320

Crotty J, Rodgers P (2012) Sustainable development in the Russia Federation: the limits of greening within industrial firms. Corp Soc Respon Environ Manag 19:178–190

Crotty J, Rodgers P (2012) The continuing reorganization of Russia's environmental bureaucracy. Probl Post-Communism 59:15–26

Ermolaeva P, Ermolaeva Y, Kuznetsova I, Basheva O, Korunova V (2020) Environmental issues in Russian cities: towards the understanding of regional and national mass media discourse. Rus J Commun 12:48–65

Fedosova MV (2020) Features of development of ecological strategy in Russian industrial companies. J Econ Bus 1:4–3

Finley JT, Verenikina AY, Verenikin AO (2019) Evaluation of environmental impact and responsibility of Russian largest companies. In: 8th international conference on industrial technology and management (ICITM), Cambridge, United Kingdom, pp 95–99. https://doi.org/10.1109/ICITM.2019.8710677

Furrer O, Egri CP, Ralston DA, Danis W, Reynaud E, Naoumova I, Molteni M, Starkus A, Darder FL, Dabic M, Furrer-Perrinjaquet A (2010) Attitudes toward corporate responsibilities in Western Europe and in central and East Europe. Manag Int Rev 50:379–398

Graves LM, Sarkis J, Gold N (2019) Employee proenvironmental behavior in Russia: the roles of top management commitment, managerial leadership, and employee motives. Res Conserv Recyc 54–64

Molchanova TK, Yashalova NN, Ruban DA (2020) Environmental concerns of Russian businesses: top company missions and climate change agenda. Climate 8:56

Orazalin N, Mahmood M (2018) Economic, environmental, and social performance indicators of sustainability reporting: evidence from the Russian oil and gas industry. Energ Pol 121:70–79. https://doi.org/10.1016/j.enpol.2018.06.015

Orazalin N, Mahmood M, Narbaev T (2019) The impact of sustainability performance indicators on financial stability: evidence from the Russian oil and gas industry. Environ Sci Pollut Res 26:8157–8168. https://doi.org/10.1007/s11356-019-04325-9

Ponomarenko TV, Wolnik R, Marinina OA (2016) Corporate social responsibility in coal industry (practices of Russian and European companies). Zap Gorn Inst 222:882–891

Salmi O (2008) Drivers for adopting environmental management systems in the post-Soviet mining industry. Int Environ Agreements 8(1):51–77

Schaltegger S, Burritt RL (2006) Corporate sustainability accounting. In: Sustainability accounting and reporting. Springer Publishing, Dordrecht, pp 37–59

Schaltegger S, Burritt RL (2010) Sustainability accounting for companies: catchphrase or decision support for business leaders? J Wor Bus 45:375–384

Tret'yak OA (2015) Corporate social responsibility of Russian companies. Russ J Manag. http://www.marketing.spb.ru/mr/business/Corporate_Governance.htm. Data of access 27.06.2020

Tynkkynen N (2005) Russia, a great ecological power? On Russian attitudes to environmental politics at home and abroad. In: Rosenholm A, Autio-Sarasmo S (eds) Understanding Russian nature: representations, values and concepts. Aleksanteri Papers 4:277–296

Visser O, Kurakin A, Nikulin A (2019) Corporate social responsibility, coexistence and contestation: large farms' changing responsibilities vis-à-vis rural households in Russia. Canad J Devel Stud 40(4):580–599

Polina Ermolaeva is an Associate Professor at the Department of Sociology of Kazan Federal University, Research Fellow at the Federal Center of Theoretical and Applied Sociology of the Russian Academy of Sciences and the Center of Advanced Economic Research. Her research interests center on sustainability transitions, climate research, environmental governance, sustainable cities and volunteering in emergency. Polina led over 50 global projects for such organizations as UBER, the World Bank Foundation, Open Society Foundation, British Petroleum, Fulbright, IREX, Russian Scientific Foundation and others

Ksenia Agapeeva is the project manager at Levada Center—Russian non-governmental research organization; she manages public opinion research projects including on topics related to attitudes and environmental practices of the population. Ksenia also takes part in international projects such as GEM (Global Entrepreneurship Monitor) and ISSP, specializes in development project methodology, data analysis and report preparing

Short-termism—The Causes and Consequences for the Sustainable Development of the Financial Markets

Małgorzata Janicka, Artur Sajnóg, and Tomasz Sosnowski

Abstract Sustainable finance is not the same as green finance understood as financing a low-carbon economy. It is a broader concept that focuses on the importance of long-term investments that include environmental, social, and governance factors. Owing to these long-term investments, the financial system develops in a harmonious way, serving sustainable economic growth. One of the obstacles that stand in the way of long-term investments is short-termism: the focus of managers and investors on the short time horizon, prioritizing a near-time shareholder interest over the long-term growth of the company. This characteristic feature of modern financial markets is seen as a suboptimal state that prevents companies from reaching their full potential and the goals of sustainable development. The aim of this comparative study across EU countries is to examine whether the abolition of the quarterly reporting requirement by public companies domiciled in selected EU member states resulted in companies withdrawing from the publication of quarterly reports.

Keywords Sustainable finance · Short-termism · Quarterly reporting

1 Introduction

In the context of sustainable economic development, the financial sector is perceived primarily from the perspective of financing pro-ecological, low-carbon investments, including ESG (Environmental, Social and Governance) factors (see Interim Report 2017, p. 8). These investments are also referred to as Socially Responsible Investment

---

M. Janicka (✉) · A. Sajnóg · T. Sosnowski
Faculty of Economics and Sociology, Department of International Finance and Investment, University of Lodz, Rewolucji 1905 Street 41, 90-255 Lodz, Poland
e-mail: malgorzata.janicka@uni.lodz.pl

A. Sajnóg
e-mail: artur.sajnog@uni.lodz.pl

T. Sosnowski
e-mail: tomasz.sosnowski@uni.lodz.pl

(SRI). SRI is an investment discipline that adds concerns about social or environmental issues to the normal ones of risk and return as determinants of equity portfolio construction or activity (Sparks 2008), or green finance, which is defined as any financial instrument or financial services activity that results in positive change for the environment and society over the long term (sustainability). The most basic "greenness" criterion of a company or project is that it contributes to reducing the emission of Greenhouse Gases (Wardle et al. 2019, p. 2). These concepts are often mistakenly equated to "sustainable finance."

This chapter focuses on the links between traditional and modern approaches to finance, targeting them towards sustainability. The shifts in the approach to the creation and implementation of business models, driven mainly by technological changes, make it necessary to adjust the existing paradigm towards innovation, and they point to the need to adopt a much wider perspective on sustainable development in the area of finance. This modern perspective on sustainable finance can help to provide some important insights into sustainable development innovations with regard to various EU priorities, institutional regulations, or the effects of their implementation.

Sustainable finance is a broader concept that also includes elements related to the functioning of financial markets as such and their role in sustainable development. Thus, determinants of the activities of investors (capital providers) and enterprises (borrowers) are part of sustainable finance on the financial market and, more specifically, the consequences of diverging investment horizons for the functioning of a sustainable economy.

Pro-ecological investments are long-term. As extending the investment horizon increases investment risk, long-term investors expect a sufficiently high rate of return on their investment to compensate for the increase in risk. In modern financial markets, investments over six years are perceived as long-term (ESMA 2019, p. 9), although some define them even as over three years (Atherton et al. 2007, p. 2) Shortening the period of investors' financial involvement means that companies need to shift their focus from achieving long-term goals and strategies to achieving short-term goals and results. As a result, the allocation of investment funds in the economy may be suboptimal from the point of view of sustainable development goals. The focus of managers and investors on the short time horizon, prioritizing a near-time shareholder interest over the long-term growth of the company, is called "short-termism." One of the factors that are conducive to short-termism on the financial markets was the EU's adoption of quarterly reporting by public companies in 2004, which was lifted in 2014.

The aim of this study is to examine whether the abolition of the quarterly reporting requirement by public companies domiciled in selected EU member states resulted in companies withdrawing from the publication of quarterly reports. To fulfill the aim of the paper we posit the research hypothesis: Despite the formal elimination of quarterly reporting requirements for public companies in most EU Member States (*de jure* aspect), companies still maintain them (*de facto* aspect).

Thus, this paper fits into the stream of current research on the concept of financial reporting and its importance for the development of sustainable finances. It is the eighth Sustainable Deveopment Goal that directly guides the need to integrate financial aspects, both on micro-economic and macro-economic scales, into a complex process of transformation aimed at balancing environmental, social and business issues. Therefore, this topic is directly within the scope indicated by the Sustainable Development Goals, especially in matters related to the promotion of sustained, inclusive, and sustainable economic growth.

This study is structured in four parts. In the first section, there is a brief review of existing studies of two concepts—sustainable finance and short-termism. The views expressed in the second part contain deliberations over the requirements regarding the obligation to publish financial data across the EU countries. The third part outlines the methodology deployed, which includes the specification of 13 European regulated stock markets and indices. The analysis of empirical results based on companies listed on these markets (from 2007 to 2019) is shown in the last part.

2 Sustainable Finance and Short-Termism—Characteristics of the Concepts

The idea of sustainable economic development was born at the turn of the 1980s and 1990s, and it is a development that enables the current needs of societies to be met without limiting this possibility for future generations (WCED 1987). However, the concept of sustainable finance still lacks one clear definition. Experts from the High-Level Expert Group on Sustainable Finance indicate that sustainable finance is about two imperatives (Final Report 2018, p. 6):

- improving the contribution of finance to sustainable and inclusive growth as well as the mitigation of climate change,
- strengthening financial stability by incorporating environmental, social, and governance (ESG) factors into investment decision-making.

And further-sustainable finance means a commitment to the longer term, as well as patience and trust in the value of investments that need time for their value to materialize (Final Report 2018, p. 9). According to the experts from the Expert Panel on Sustainable Finance, sustainable finance comprises capital flows, risk management activities, and financial processes that assimilate environmental and social factors as a means of promoting sustainable economic growth and the long-term stability of the financial system (Government of Canada 2019, p. 2). For the research conducted in this chapter, the key is to indicate the significance of the long-term perspective in the investment process. There are also definitions that more closely reflect the broad meaning of the concept and more closely link it to the functioning of enterprises and financial markets. Such an approach can be seen in the World Bank study, in which sustainable finance is understood as a strong commitment from owners and managers to make sustainability considerations a primary component of business strategy, incorporating sustainability strategies into the process to allocate resources—both the firms' own capital and intermediated resources—in support of creating new sustainable businesses lines, fostering the growth of existing ones, and moving away from activities not aligned with sustainability (World Bank 2017, p. 11).

To sum up, in a broad sense, the concept of "sustainable finance" mainly refers to strengthening financial stability in the economy by taking into account the ESG factors and preferring long-term investments in the decision-making process on the financial markets. In the narrow sense, it refers to the reallocation of funds for pro-ecological investments that positively affect the environment and society. Thus, the phenomenon of short-termism, which is examined in this article, is part of a broader concept.

Short-termism means focusing on short time horizons by both corporate managers and the financial markets, and prioritizing near-time shareholder interest over the long-term growth of the companies (Reilly et al. 2016). Although shareholders' decisions are considered to be the primary source of short-termism (the micro level), its consequences concern the functioning of the entire financial market and the economy (the macro level). According to EY, short-termism means decision-makers focusing excessively on short-term goals at the expense of long-term objectives. Short-termism results in insufficient attention being paid to the strategy, fundamentals, and the long-term value creation of a firm or an institution (Ernst & Young 2014, p. 7). It is a suboptimal state that prevents companies from fully using their potential in the long term and achieving sustainable development goals.

The management team myopia and primacy of actions in the near term that have detrimental outcomes for the long term have consequences on wide range of companies' actions (Marginson and Mcaulay 2008). Numerous empirical studies show the impact of short-termism pressures on making distorted investment decisions in listed companies (Asker et al. 2014; Edmans et al. 2013; Ladika and Sautner 2019). Hence, it reduces stock companies productivity, and consequently, industry and national productivity (Willey 2019). Moreover, short-termism in capital markets appears to come with a potentially significant influence on the assets valuation (Davies et al. 2014) and even limits the primary equity issuances (Taking Stock 2016). For Kay (2012, p. 9), short-termism was an underlying problem in UK equity markets, principally caused by a misalignment of incentives within the investment chain and the displacement of trust relationships by a culture based on trading.

3 Quarterly Reporting in the EU: A Study of the Current State

The obligation for public companies to publish quarterly reports in the EU was a consequence of the deepening integration of the European financial markets. Investors across the EU should be provided with the same decision-making conditions and level of protection. Increasing the transparency of operations was also designed to bring measurable benefits for companies, i.e., lower capital costs and increased stock liquidity (by decreasing the level of information asymmetries). Requirements regarding the obligation to publish financial data varied across the EU countries (see Table 1).

Table 2 presents data on the publication of quarterly reports by public companies in the EU-15 before the implementation of the obligation. They largely reflect the diversity of legal regulations in force in individual countries (see Table 1).

In 1999, the European Commission presented a plan to deepen the integration of financial markets: "Financial Services: Implementing the Framework for Financial Markets: Action Plan" (Communication of the European Commission 1999). One element was to increase transparency on the financial market. In 2004, a directive was adopted (implementation date: 2007), under which public companies were obliged to publish quarterly reports (Directive 2004/109/EC). Tables 3 and 4 contain information on the transposition of EU regulations into national systems.

The analysis of the above results showed that this requirement has not been fully implemented, for example, by the stock exchanges in Belgium, the Netherlands, France (belonging to Euronext), or Ireland. The legitimacy of the adopted solutions was questioned by the European Commission itself in 2011. In 2013, the directive was amended, and the obligation of short-term reporting was abolished (Directive 2013/50/EU, entered into force in 2014). Some countries retained reporting obligations for public companies listed on the main markets (Bulgaria, Croatia, and Poland; Austria abolished this requirement for the Prime Market only in 2019).

Following the implementation of the concept of the Capital Markets Union (European Commission 2015), in October 2016, the European Commission launched a High-Level Expert Group (HLEG) on Sustainable Finance. The HLEG pointed out that sustainability cannot develop in a context where investment is dominated by short-term considerations. This is because delivering sustainability in economic, social, and environmental dimensions requires large-scale investments in physical assets that are amortized not over a few months but over several years (Final Report 2018, p. 45). The concentration of companies on the implementation of short-term goals does not serve long-term investments that take into account ESG factors.

4 Methodology of the Empirical Study

The empirical research is aimed at finding answers to the research question regarding the frequency pattern for reporting financial information by companies whose shares are listed on European stock exchanges. As a result of the analysis, we want to find out whether, despite the abolition of the obligation to report financial results quarterly on most European stock markets, companies still decide to disclose information about their results at this frequency.

Our research sample is composed of companies listed on selected European regulated markets. Among the European exchanges, we analyze in detail the markets

Table 1 Frequency of periodic reporting in the EU (status for 2003)

Member state	Annual reporting	Semi-annual reporting	Quarterly reporting
Belgium	Yes	Yes	Yes—for companies listed in Next Prime and New Economy (Euronext)
Denmark	Yes	Yes	No
Germany	Yes	Yes	Requested by Deutsche Boerse for companies listed in the DAX segment and for the Neuer Markt
Greece	Yes	Yes	Yes
Spain	Yes	Yes	Yes
France	Yes	Yes	Turnover figures
Ireland	Yes	Yes	Preliminary statements
Italy	Yes	Yes	Yes
Luxembourg	Yes, plus foreign law provisions from the country where the company is registered	No	No
The Netherlands	Yes	Yes	Yes—for companies listed in Next Prime and New Economy (Euronext)
Austria	Yes	No	Yes—for companies listed in the Official Market
Portugal	Yes—management report, financial information, etc. and a report on corporate governance	Yes—management report, balance sheet, report by a registered auditor	Yes—firm's operations, profit/loss situation, financial situation
Finland	Yes	Yes	Yes
Sweden	Yes	Yes	Yes—requested by the Stockholm Stock Exchange
United Kingdom	Yes	Yes	Preliminary statements

Source Based on European Commission further to a question by MEP Chris Huhne, July 2003 in: Lannoo and Khachaturyan (2003), Disclosure Regulation in the EU The Emerging Framework, CEPS Task Force Report No. 48, Annex 2, p. 54

Table 2 Frequency of quarterly reporting in the EU (status for 2003)

Member state	Number of domestic listed companies (main market)	Number of listed companies providing quarterly reports or at least quarterly information	Share (%)
Belgium[a]	125	43	34.4
Denmark[b]	197	127	64.5
Germany	715	429	60.0
Greece	344	344	100.0
Spain	273	273	100.0
France	796	12[c]	1.5
Ireland	70	1	1.4
Italy	279	265	95.0
Luxembourg[d]	45	8	17.8
The Netherlands	179	75	41.9
Austria	110	90	81.8
Portugal	57	55	96.5
Finland	143	135	94.4
Sweden	350	340	97.1
United Kingdom[e]	2177	350	16.1
EU-15	5860	2547	43.5

[a]Not including Nasdaq Europe; [b]also includes foreign issuers; [c]figure referring to the CAC40, Euronext Paris. In addition, according to French law, all listed companies were requested to provide net turnover data on a quarterly basis; [d]also includes foreign issuers whose shares are admitted to trading on the regulated market; [e]32 issuers in the FTSE 100, 17 issuers under the United Kingdom Listing Rules requiring quarterly reports from companies with a trade record of fewer than three years

Source European Commission further to a question by MEP Chris Huhne, July 2003 in: Lannoo and Khachaturyan (2003), Disclosure Regulation in the EU The Emerging Framework, CEPS Task Force Report No. 48, p. 22 and own calculations

regulated by the European Union. It is worth noting that we have excluded the London Stock Exchange from the research sample, as the UK is no longer a member of the EU. However, due to the numerous similarities in terms of stock exchange regulations, corporate governance rules, as well as the accounting and reporting standards, in order to provide a broader view of the investigated issue, we decided to include the Swiss stock exchange (SIX Swiss Exchange) in our research sample. Therefore, our research sample ultimately consists of companies listed on 13 regulated European stock markets.

Due to the significant heterogeneity present on the European financial market, the scope of the analysis was limited to recognizing the reporting pattern among the largest and most liquid companies listed on particular stock exchanges. Thus, the companies included in the portfolio of the main and most important stock indices on 21 May 2020 were qualified to the research sample.

Table 3 Quarterly reporting environment in EU-15

Member state	Transparency Directive (TD) transposition	Quarterly financial reporting mandatory?		Quarterly reporting rules
		Before TD	After TD	
Belgium	1 September 2008	No	No—only IMS	No
Denmark	1 June 2007	No	No—only IMS	No
Germany	20 January 2007	No	No—only IMS	No—except companies listed in the Prime Standard
Greece	30 June 2007	Yes	Yes	Obligatory for all listed firms
Spain	20 December 2007	Yes	No—only IMS	No
France	19 December 2007	No	No—only IMS	No
Ireland	13 June 2007	No	No—only IMS	No
Italy	24 August 2009	Yes	No—only IMS	No
Luxembourg	19 January 2008	No	No—only IMS	No
The Netherlands	1 January 2009	No	No—only IMS	No
Austria	26 April 2007	No	No—only IMS	No—except companies listed in the Prime Market
Portugal	1 November 2007	Yes	Yes	Obligatory for all listed firms
Finland	15 February 2007	Yes	Yes	Obligatory for all listed firms
Sweden	1 July 2007	Yes	Yes	Obligatory for all listed firms on two regulated markets
United Kingdom	20 January 2007	No	No—only IMS	No

Source Link (2012), pp. 201–204

The use of this filter allows us to maintain relative comparability between individual markets, as the companies included in the portfolio of individual indices of the largest listed companies usually try to adapt to all the reporting procedures and standards implemented in these markets. The largest stock companies are subject to many formal and legal requirements, as well as information policy regulations. The importance of presenting financial reports seems to be the most significant among them, as they are obliged by law to disclose certain information within a certain period. As we have pointed above, this obligation is regulated both on the basis of European Union regulations and national legislative solutions. Moreover, the common feature of these companies is the fact that they are in the scope of interest of investors with

Table 4 Quarterly reporting environment in the new EU member states (which joined in 2004 and later)

Member state	Date of joining the EU	Quarterly financial reporting mandatory?	Quarterly reporting rules
Bulgaria	1 January 2007	Yes	Yes
Croatia	1 July 2013	Yes	Yes
Republic of Cyprus	1 May 2004	No	No—except companies listed on the Main Market
Czech Republic	1 May 2004	No	No—except companies listed on the Prime Market
Estonia	1 May 2004	Yes	Obligatory for firms listed on the regulated market
Hungary	1 May 2004	Yes/no	It depends on the capitalization or the number of shareholders
Latvia	1 May 2004	No	No
Lithuania	1 May 2004	No	No
Malta	1 May 2004	No	No
Poland	1 May 2004	Yes	Yes
Romania	1 January 2007	No	No
Slovakia	1 May 2004	No	Interim statement
Slovenia	1 May 2004	No	Interim statements or key operations data

Source Own study, based on materials from national exchanges and different studies

a similar investment profile, especially institutional investors who operate globally on financial markets.

It is worth mentioning that the number of the largest and most liquid companies is not always in line with the name of the index. For example, the PSI 20 index of Portuguese companies is composed of 18 rather than 20 companies, while the BELEX15 index is composed of only 10 companies. Finally, the research sample consists of 427 companies, of which the most numerous subsamples (40 companies) concern the OMX NORDIC 40, CAC 40, and FTSE MIB indices, and the least numerous are BELEX15, PX and SOFIX. Table 5 shows more detailed cross-sectional information on the study sample.

Our data concerning the financial reporting behavior covers the years 2007–2019 and is hand-collected from the websites of individual companies, where issues related to investor relations are presented. The sample period starts from 2007 due to the availability of information presented by companies, as data for earlier years were very limited in most cases.

Table 5 Sample characteristics: specification of markets and indices

EU-regulated markets	Stock index	Number of companies
Deutsche Boerse—Prime and General Standard	DAX	30
Nasdaq Nordic—Main	OMX NORDIC 40	40
Euronext	AEX	25
	BEL20	20
	CAC 40	40
	ISEQ 20	20
	PSI 20	18
SIX Swiss Exchange	SLI	30
Borsa Italiana—Main	FTSE MIB	40
Oslo Bors and Oslo Axess	OBX	25
Wiener Boerse	ATX	20
Warsaw—Main	WIG30	30
BME (Spanish Exchange)—Main	IBEX35	35
Sofia	SOFIX	15
Bucharest	BET	17
Belgrade Stock Exchange	BELEX15	10
Prague	PX	12
Total number of companies		427

Source Authors' own study

5 Empirical Results

The managers of listed companies make various business decisions under conditions of pressure and expectations, often divergent, which are expressed by various groups of stakeholders. The conditions of the public stock market force them to achieve and report better and better financial results. Thus, the pursuit of meeting the growing information needs of stock market investors may prompt them to undertake activities aimed at achieving short-term financial effects at the expense of the long-term strategic goals (Flammer and Bansal 2017; Thanassoulis and Somekh 2016; Zhang and Gimeno 2016) and sustainable development of the company, and this short-termism is reflected, for example, in their practices of financial disclosure. In particular, the publication of quarterly reports is especially important in this context. Table 6 shows which activities in this field are undertaken by the largest and most liquid companies listed on EU-regulated markets.

A thorough analysis of the reporting patterns indicates that on the European markets, one can see a diverse approach to the disclosure of quarterly financial information. Among the analyzed companies, the dominant strategy of disclosing

Table 6 Publication of quarterly financial reports by the largest companies on the European stock markets

Stock index	Quarterly reports (in any year)		No quarterly reports (in any year)	
	Number of companies	Share (%)	Number of companies	Share (%)
DAX	30	100.00	0	0.00
OMX NORDIC 40	40	100.00	0	0.00
AEX	12	48.00	13	52.00
BEL20	7	35.00	13	65.00
CAC 40	8	20.00	32	80.00
ISEQ 20	1	5.00	19	95.00
PSI 20	18	100.00	0	0.00
SLI	12	40.00	18	60.00
FTSE MIB	36	90.00	4	10.00
OBX	25	100.00	0	0.00
ATX	20	100.00	0	0.00
WIG30	30	100.00	0	0.00
IBEX35	30	85.71	5	14.29
SOFIX	12	80.00	3	20.00
BET	17	100.00	0	0.00
BELEX15	6	60.00	4	40.00
PX	9	75.00	3	25.00
Total	313	73.30	114	26.70

Source Authors' own study

current financial information is to publish relevant reports after the end of the quarterly accounting period. Such an information policy is recorded for 73.3% of the research sample.

In seven markets, all companies included in the portfolio of the most important stock exchange index have quarterly financial reports in at least some of the 52 analyzed quarters over the entire 13-year period, namely in the main markets of Deutsche Boerse, Nasdaq Nordic, Euronext Lisbon, Oslo Bors and Oslo Axess, Wiener Boerse, the Warsaw Stock Exchange, and the Bucharest Stock Exchange. Such results are consistent with previous findings regarding quarterly reporting rules in individual countries (see Tables 3 and 4). For instance, in Germany and Austria, such a requirement is still maintained for companies from the main market, while in Poland, Sweden, or Finland, it applies to all listed companies. A very interesting case is that of Romanian companies, which, despite the lack of obligation to report quarterly, present these reports on their websites in all studied years.

Based on the research results, it can also be stated that in the case of 5 markets, i.e., the stock exchanges located in Italy, Spain, Bulgaria, Serbia, and the Czech Republic, companies that publish quarterly reports containing relevant financial information prevail. Importantly, only in Bulgaria and the Czech Republic does such a reporting requirement arise from the regulations of a given country (see Table 4), while companies in other countries have adopted this practice despite the fact that neither the European Union nor national regulations require the presentation of quarterly financial statements.

Interestingly, Euronext stands out. Portuguese companies tend to disclose quarterly financial reports consistently, which can be explained by the obligation under national law in force on the main market on this stock exchange (see Table 3). However, issuers representing the remaining four stock markets included in this multinational financial services corporation seem to oppose the information pressure exerted by stock investors. For example, the vast majority of French and Irish companies do not disclose any quarterly reports throughout the whole period under investigation in this study. A similar tendency can also be observed for Dutch and Belgian companies, where 52 and 65% of companies decided not to publish such information, respectively. A comparable observation can be made by analyzing information for companies listed on the SIX Swiss Exchange. Therefore, the lack of mandatory quarterly reporting in EU or national regulations creates a number of inconsistencies in the reporting policy of companies.

Another aspect of the adopted information policy regarding the disclosure of quarterly financial information is the consistency of the management team in applying the once adopted approach. Table 7 presents information relating to this issue, presenting it separately for all companies in a given market and for companies that have chosen to disclose such information to stock exchange investors in general.

In general, the analysis of the pattern of disclosing quarterly financial information indicates that the majority of companies included in the research sample decided to publish a financial statement for each of the 52 quarters covered by the research. As much as 55.27% of all analyzed companies did so, and in the case of the subsample consisting of companies reporting financial results on a quarterly basis, it is 75.4%, respectively. Taking into account all analyzed markets, the prevailing ones are those where companies consistently reported quarterly financial information in all periods. Changes in decisions on quarterly reporting may have been influenced by changing EU recommendations. As shown in the previous section, the EU first adopted the relevant directive in 2007, according to which public companies were obliged to publish quarterly reports, but then abolished it in 2013. Thus, the national legal provisions to which the public companies must apply throughout the period 2007–2019 should be considered more consistent.

6 Conclusions and Discussion

Sustainable finance is a very broad topic of many interests, but some of the aspects of this issue have not been extensively studied so far. In this regard, we pay special attention to the issue of short-termism as a current practice of many companies. We also investigate the impact of changing EU requirements regarding the disclosure of

Table 7 Continuity of reporting quarterly financial information by the largest companies on the European stock markets

Stock index	Quarterly reports over the entire period (in every quarter)		
	Number of companies	Share (in relation to the whole sample) (%)	Share (in relation to companies reporting on a quarterly basis) (%)
DAX	30	100.00	100.00
OMX NORDIC 40	36	90.00	90.00
AEX	7	28.00	58.33
BEL20	4	20.00	57.14
CAC 40	4	10.00	50.00
ISEQ 20	1	5.00	100.00
PSI 20	10	55.56	55.56
SLI	7	23.33	58.33
FTSE MIB	19	47.50	52.78
OBX	21	84.00	84.00
ATX	20	100.00	100.00
WIG30	23	76.67	76.67
IBEX35	21	60.00	70.00
SOFIX	11	73.33	91.67
BET	10	58.82	58.82
BELEX15	4	40.00	66.67
PX	8	66.67	88.89
Total	236	55.27	75.40

Source Authors' own study

quarterly financial reports on the adjustment of the information policy of the largest and most liquid stock companies listed on EU-regulated markets. This is an important issue, as short-sighted decisions and actions taken by boards of directors often lead to a disruption of the company's sustainability path and, consequently, to financial instability from the perspective of the whole economy. This problem may increase, especially in the case of large enterprises, affecting the functioning of the national market.

Our findings lead us to formulate a number of insights regarding the short-term reporting policy among the European stock markets. First, although the EU recognizes this problem in the activities of companies and tries to encourage them to change their reporting policy, market leaders still disclose quarterly financial information. Second, the publication of quarterly reports is sometimes more important due to national legal provisions than EU recommendations. Third, one can notice many changes in decisions on quarterly reporting from 2009 till 2017. These conclusions are in line with our hypothesis that contrasts the *de jure* with the *de facto* aspect.

In general, there is a consensus that achieving sustainable development goals requires some changes in the rules and habits of frequently disclosing new financial information. Therefore, more decisive actions from policymakers should be expected because, as our research shows, the current, persistent state of inconsistency is unlikely to be conducive to sustainable finance, at least in the group of leading companies in the European market. Our research may serve to increase concerns regarding the effective integration of sustainable development policies with financial priorities and objectives. The lack of transition from short-termism to a longer perspective of investment can hinder the advancement of sustainable development and its results.

The conclusions of our research do not provide an unequivocal opinion on the usefulness or uselessness of quarterly reports for investors, and these results are consistent with findings from previous studies. Gigler et al. (2012) found that when financial results are reported too frequently, capital market pricing becomes equivalent to a premature evaluation of managerial actions whose benefits arrive mostly in later periods. Consequently, actions that produce large, short-term benefits become more attractive, and actions that do not immediately produce such benefits, but that would ultimately create more value for the firm, become less attractive. Thus, frequent reporting could become dysfunctional even though such reporting provides more information to the capital market.

Nallareddy et al. (2017) found no evidence that the initiation of mandatory quarterly reporting in UK firms in 2008 had any effect on investment decisions by UK public companies. Nor did they find any significant change in the level of investment by UK firms shifting from quarterly to semi-annual reporting effective 2015. An interesting approach was presented by Rosen et al. (2018). In their opinion, in the US, where quarterly reporting has been obligatory on the equity markets since 1970, the best system for the US companies would be "bifurcated" reporting, whereby smaller companies would be allowed to report on a semi-annual basis. It will contribute to enhancing long-term shareholder value. In contrast to the fear of the effects of short-termism, Barzuza and Talley (2019) think that corporate managers often fall prey to long-term bias—excessive optimism about their own long-term projects. In consequence, shareholder activism, even if unambiguously myopic, can provide a symbiotic counter-ballast against managerial long-termism. The CFA Institute's position is that fully functioning capital markets rely on complete, timely, and accurate information. It means that all companies with any type of securities listed on regulated markets should have to publish financial information quarterly, as timely and accurate financial information is "the lifeblood of financial markets" (CFA Institute 2019, p. 5).

References

Asker J, Farre-Mensa J, Ljungqvist A (2014) Corporate investment and stock market listing: a puzzle? Rev Finan Stud 28(2):342–390. https://doi.org/10.1093/rfs/hhu077

Atherton A, Lewis J, Plant R (2007) Causes of short-termism in the finance sector. Institute for Sustainable Futures, University of Technology, Sydney

Barzuza M, Talley E (2019) Long-term Bias, European Corporate Governance Institute (ECGI) Law Working Paper No. 449/2019; Virginia Law and Economics Research Paper No. 2019-05. https://scholarship.law.columbia.edu/faculty_scholarship/2288

CFA Institute (2019) In: Schacht KN, Peters SJ (eds) Comment on Earnings Releases and Quarterly Reports. https://www.sec.gov/comments/s7-26-18/s72618-5406001-184491.pdf

Communication of the European Commission (1999) Financial services: implementing the framework for financial markets: action plan, COM(1999)232

Davies R, Haldane AG, Nielsen M, Pezzini S (2014) Measuring the costs of short-termism. J Financ Stab 12:16–25. https://doi.org/10.1016/j.jfs.2013.07.002

Directive 2004/109/EC of the European Parliament and of the Council of 15 December 2004 on the harmonisation of transparency requirements in relation to information about issuers whose securities are admitted to trading on a regulated market and amending Directive 2001/34/EC, OJ L 390, 31.12.2004, pp. 38–57

Directive 2013/50/EU of the European Parliament and of the Council of 22 October 2013 amending Directive 2004/109/EC of the European Parliament and of the Council on the harmonisation of transparency requirements in relation to information about issuers whose securities are admitted to trading on a regulated market, OJ L 294, 6.11.2013, pp. 13–27

Edmans A, Fang V, Lewellen K (2013) Equity vesting and managerial myopia. NBER Working Papers 19407, National Bureau of Economic Research, Inc. https://doi.org/10.3386/w19407

Ernst & Young (2014) Short-termism in business: causes, mechanisms, and consequences. EY Poland report. http://www.ey.com/Publication/vwLUAssets/EY_Poland_Report/$FILE/Short-termism_raport_EY.pdf

ESMA (2019) Undue short-term pressure on corporations. Report of European Securities and Markets Authority, ESMA30-22-762. https://www.esma.europa.eu/sites/default/files/library/esma30-22-762_report_on_undue_short-term_pressure_on_corporations_from_the_financial_sector.pdf

European Commission (2015) Action plan on building a capital markets union. Communication from the Commission to the European Parliament, the Council, the European Economic and Social Committee and the Committee of the Regions, Brussels, COM/2015/0468 final

European Commission (2018) Action plan: financing sustainable growth. Communication from the Commission to the European Parliament, the Council, the European Economic and Social Committee and the Committee of the Regions, Brussels, COM/2018/097 final

Final Report (2018) Financing a sustainable European economy, the high-level expert group on sustainable finance. The European Commission. https://ec.europa.eu/info/sites/info/files/180131-sustainable-finance-final-report_en.pdf

Flammer C, Bansal PT (2017) Does long-term orientation create value? Evidence from a regression discontinuity. Strateg Manag J 38(9):1827–1847

Gigler F, Kanodia Ch, Sapra H, Venugopalan R (2012) How frequent financial reporting causes managerial short-termism: an analysis of the costs and benefits of reporting frequency. J Acc Res 52(2):357–387. https://doi.org/10.1111/1475-679X.12043

Government of Canada (2019) Mobilizing finance for sustainable growth. Final Report of the Expert Panel on Sustainable Finance, Minister of Environment and Climate Change, Canada 2019. https://law-ccli-2019.sites.olt.ubc.ca/files/2019/08/final.report.Expert.Panel_.Sustainable.Finance.June_.2019.pdf

Interim Report (2017) Financing a sustainable European economy, the high-level expert group on sustainable finance. The European Commission. https://ec.europa.eu/info/sites/info/files/170713-sustainable-finance-report_en.pdf

Kay J (2012) The Kay review of UK equity markets and long-term decision making. Final Report, July 2012. https://assets.publishing.service.gov.uk/government/uploads/system/uploads/attachment_data/file/253454/bis-12-917-kay-review-of-equity-markets-final-report.pdf

Ladika T, Sautner Z (2019) Managerial short-termism and investment: evidence from accelerated option vesting. Rev Finan 24(2):305–344. https://doi.org/10.1093/rof/rfz012

Lannoo K, Khachaturyan A (2003) Disclosure regulation in the EU: the emerging framework. CEPS Task Force Report No. 48, Centre for European Policy Studies, Brussels

Link B (2012) The struggle for a common interim reporting frequency regime in Europe. Acc Europe 9(2):191–226. https://doi.org/10.1080/17449480.2012.720874

Marginson D, Mcaulay L (2008) Exploring the debate on short-termism: a theoretical and empirical analysis. Strateg Manag J 29(3):273–292. https://doi.org/10.1002/smj.657

Nallareddy S, Pozen R, Rajgopal S (2017) Consequences of mandatory quarterly reporting: the U.K. experience. Soc Sci Res Netw. https://papers.ssrn.com/sol3/papers.cfm?abstract_id=2817120

Reilly G, Souder D, Ranucci R (2016) Time horizon of investments in the resource allocation process: review and framework for next steps. J Manag 42(5):1169–1194

Rosen T, Shire E, Winston B (2018) Should U.S. companies adopt semi-annual reporting? An analysis of quarterly reporting requirements and the practice of earnings guidance. Brown University, Providence, Rhode Island

Sparks R (2008) Socially responsible investment volume. In: Handbook of finance. Wiley. https://doi.org/10.1002/9780470404324.hof002014

Taking Stock: Going Public in Uncertain Times, Merger Market Report 2016. https://www.reedsmith.com/en/perspectives/2016/09/taking-stock-going-public-in-volatile-times

Thanassoulis J, Somekh B (2016) Real economy effects of short-term equity ownership. J Int Bus Stud 47(2):233–254

Wardle M, Ford G, Lallemand B, Mainelli M, Mills S (2019) The global green finance index 3, Z/Yen, Long Finance Initiative, Finance Watch. MAVA Foundation

Willey KM (2019) What are the harms of short-termism? In: Stock market short-termism. Springer International Publishing, pp. 195–222. https://doi.org/10.1007/978-3-030-22903-0_7

World Bank Group Initiative (2017) Roadmap for a sustainable financial system. UN Environment. http://unepinquiry.org/wp-content/uploads/2017/11/Roadmap_for_a_Sustainable_Financial_System.pdf

World Commission on Environment and Development (WCED) (1987) Our common future, report Brundtland. Oxford University Press, Oxford and New York

Zhang Y, Gimeno J (2016) Earnings pressure and long-term corporate governance: can long-term-oriented investors and managers reduce the quarterly earnings obsession? Organ Sci 27(2):354–372

Małgorzata Janicka Ph.D., is Associate Professor in Economics at the University of Lodz (Poland). She is the Director of the Institute of International Economics (http://www.instytutgm.uni.lodz.pl), the Head of the Department of International Finance and Investment and the Head of the Centre for Capital Investment Research at the University of Lodz, Faculty of Economics and Sociology. Her research interest covers international economics and international finance (specially international capital flows, international monetary systems, and sustainable finance).

Artur Sajnóg Ph.D., is an Assistant Professor of University of Lodz, Department of International Finance and Investments, Faculty of Economics and Sociology. His main interest areas include corporate finance, capital markets, financial accounting and reporting. His currently research is focused on three projects: "Predictive power of comprehensive income", "Family Office—theory and practice on the international markets", "Short-termism—the causes and consequences for the sustainable development of the economy and financial markets."

Tomasz Sosnowski Ph.D., is an active academic teacher and Assistant Professor in the Department of International Finance and Investments, University of Lodz, Poland. He is one of the leaders of the Centre for Capital Investment Research at the University of Lodz. In his everyday work, he combines scientific knowledge and professional practice in the field of management accounting, corporate finance and building effective strategies for the companies' development. His scientific interests focus on the divestment strategies of private equity funds, the implementation of initial public offerings (IPOs) and earnings management.

CPSIA information can be obtained
at www.ICGtesting.com
Printed in the USA
LVHW080017211022
731175LV00006B/256

9 783030 788278